OPERATIONAL CALCULUS

OPERATIONAL CALCULUS

BASED ON THE TWO-SIDED LAPLACE INTEGRAL

BY

BALTH. VAN DER POL, D.Sc.

Director of the International Radio Consultative Committee (C.C.I.R.), Geneva
Formerly Professor of Theoretical Electrotechnics in the Technische Hogeschool, Delft, and Director of Fundamental Radio Research, N. V. Philips' Gloeilampenfabrieken, Eindhoven, Netherlands

AND

H. BREMMER, D.Sc.

Member of the Research Laboratory, N. V. Philips' Gloeilampenfabrieken, Eindhoven, Netherlands

'La langue de l'analyse, la plus parfaite de toutes les langues, étant par elle-même un puissant instrument de découvertes; ses notations, lorsqu' elles sont nécessaires et heureusement imaginées, sont les germes de nouveaux calculs.'

LAPLACE, *Théorie analytique des probabilités*, Paris, 1812, p. 7

CHELSEA PUBLISHING COMPANY
NEW YORK, N. Y.

THIRD EDITION

This Third edition is textually the same as the Second (1955), apart from some small improvements, including correction of errata. It is published at New York, N.Y. in 1987 and it is printed on 'long-life' acid-free paper.

First published 1950

Reprinted with the permission of Cambridge University Press

International Standard Book No. 0-8284-0327-9
Library of Congress Catalog No. 86-71694

Printed in the United States of America

PREFACE

This book was developed from a series of lectures [by van der Pol] given at the 'Technische Hogeschool' of Delft during 1938 and following years, and from a second series given during the first half of 1940 at the 'Philips Research Laboratories', Eindhoven.

The second author [Bremmer] made extensive lecture reports on the latter series; subsequently the material was jointly extended during the German occupation of the Netherlands. In this period the original manuscript, in Dutch, was practically completed, while several problems founded on it were published from 1940 on in *Wiskundige Opgaven* of the Netherlands 'Wiskundig Genootschap'. The English translation of the original manuscript was edited by Dr C. J. Bouwkamp, of this Laboratory.

Primarily this book is intended for application of the operational calculus in its modern form to mathematics, physics and technical problems. We have therefore given not only the basic principles, ideas and theorems as clearly as possible (and rather extensively), but also many worked-out problems from purely mathematical and physical as well as from technical fields. In order to limit the size of the book, proofs of some of the deeper theorems have been omitted, and for these the reader is referred to the mathematical literature.

None the less, it is believed that the purely mathematical treatment is more advanced than is usual in books devoted primarily to practical applications. The Abel and Tauber theorems, for example, are extensively considered, with many examples taken from pure mathematics as well as from technical problems.

It is therefore hoped that the book may be of value to those pure mathematicians who are interested in a rapid and simple derivation of complicated and unexpected relations between various mathematical functions, as well as to the engineer in search, for example, of a very simple treatment of transient phenomena in electrical networks, such as filters. In both cases the operational calculus appears to its best advantage.

Furthermore, several applications of the operational calculus to the theory of numbers are to be found in this book, a field of application which appears to be of the greatest heuristic value and which at present seems to be far from exhausted.

We have endeavoured to give the operational calculus a rigorous mathematical basis; on the other hand, we have tried to give the subject-matter such a form that it can be applied simply to practical problems.

This led us to treat the operational calculus *ab initio*, by means of the bilateral or two-sided Laplace integral, contrary to the usual practice based on the one-sided Laplace integral. This procedure was greatly stimulated by extensive discussions with Dr Ph. le Corbeiller, now at Harvard University, Cambridge, Mass.

The foundation of the theory on the two-sided Laplace transform caused us to introduce at an early stage:

(*a*) the Heaviside unit function, $U(t)$, defined by $U(t) = 0$ for $t < 0$, $\frac{1}{2}$ for $t = 0$, and 1 for $t > 0$,

(*b*) the Dirac δ-function, $\delta(t)$.

Further, the use of the two-sided Laplace transform requires, for each operational relation, the stipulation of the band of $\mathrm{Re}\, p$ within which this relation is valid. It is felt, however, that the latter complication is more than compensated for by the following advantages:

(i) the class of functions suited to an operational treatment becomes much larger,

(ii) the 'transformation rules' are considerably simplified,

(iii) the entire treatment becomes more rigorous than the usual presentation of the one-sided integral in technical books.

The rapid way in which solutions of complicated problems can be found with operational calculus is often astounding. This is mainly due to the fact that discontinuous functions $h(t)$ of a real variable t, which frequently occur in the treatment of electrical and mechanical transients as well as in the theory of numbers, have an operational 'image' $f(p)$ that is analytic in some band of finite or infinite width of the complex p-plane. Simple transformations of these smooth, analytic, functions $f(p)$ then correspond uniquely to operations on the discontinuous functions $h(t)$, and so the complicated handling of these discontinuous functions can be replaced by extremely simple transformations of the corresponding analytic functions.

The treatment in this book is not limited to the Laplace transform of functions of one single variable; an extensive chapter is also devoted to the multidimensional Laplace transforms. This 'simultaneous operational calculus' enabled us, for example, to treat Green's functions in potential and wave problems; this part was mainly developed by the second author. Thus familiar solutions of the Maxwellian equations are obtained by a few extremely simple algebraic transformations in the p-field.

In the Introduction the reader will find a summary of the subjects treated in the various chapters.

The authors wish to express their thanks to several friends who read parts or the whole of the manuscript. In the first place we wish to thank Dr C. J. Bouwkamp for many remarks, which, we believe, have improved the clarity of the exposition, and for the translation of the original manuscript. Further we wish to thank Prof. N. G. de Bruÿn, whose remarks have contributed materially to the rigour of the treatment.

In conclusion, we would add that we shall be most grateful for remarks and criticism from readers. We feel that this first treatment of the practical operational calculus on the basis of the two-sided Laplace transform, with so many applications to both pure and applied mathematics, and a very great part of which we believe to be essentially new, is bound to show marks of immaturity typical of young scientists as well as of young sciences.

B. v. D. P.
H. B.

PHILIPS RESEARCH LABORATORY
EINDHOVEN

November 1947

PREFACE TO THE SECOND EDITION

The necessity of a second edition, four years after the appearance of the first one, gave us the opportunity to insert a number of corrections and improvements, scattered throughout the book. The greater part of the corrections are due to suggestions by several correspondents. In connexion herewith we wish to thank particularly Prof. S. Colombo (Paris); Prof. A. Erdelyi (Pasadena); Dr J. H. Pearce (London); and Mr H. van der Weg (Eindhoven).

We have also inserted the following new paragraphs:

Rules for the treatment of *correlation functions*,
A note on the theory of *distributions*,
A note on the *Wiener-Hopf technique*,

as all three, modern, subjects lend themselves well to a concise treatment in the 'language' of operational calculus.

Finally we wish to thank the Cambridge University Press for the care again shown in the preparation of this second edition.

B. v. D. P.
H. B.

PALAIS WILSON, GENEVA
PHILIPS' RESEARCH LABORATORIES, EINDHOVEN

September 1954

CONTENTS

CHAPTER I

GENERAL INTRODUCTION

1. History of the operational calculus

The operational calculus, a modern treatment of which is aimed at in the present book, can be traced back as far as the work of Oliver Heaviside (1850–1925). Though many scientists (Leibniz, Lagrange, Cauchy, Laplace, Boole, Riemann, and others) preceded Heaviside in introducing operational methods into analysis†, a systematic use of it in physical and technical problems was stimulated only by Heaviside's work.

Heaviside‡ was a 'self-made man', deprived of regular study at the university or the engineering college. Nevertheless, his curious methods, created by himself as they often were, led him to results in technics and theoretical physics that are undoubtedly among the most important ever reached. In this connexion let us remember that Heaviside's work § already contains Maxwell's equations of the electromagnetic field in the modern, now current, vector notation. Also due to him is the conception of the 'Heaviside Layer', which is of the greatest importance in present-day radio communication. Moreover, independently of Lorentz, Heaviside enunciated the theory of the electronic motion in a magnetic field; he further introduced into Maxwell's theory that part of the total current which is due to convection. His concept of impedance, defined independently of Kennelly, is much more general than that of the conventional alternating-current technique. The notion of 'negative resistance', now common property in electrical engineering (e.g. arc lamp, radio valve), is often put forward in his papers, and for the first time in 1895.

But it may be stated that even to-day Heaviside's papers, difficult to read as perhaps they are, still contain a great many views and hidden things, of both mathematical and physical interest, which are not yet very well known and which, therefore, have not met with proper appreciation. Certainly this is largely due to the strange manner in which Heaviside often derives and announces his results. Moreover, the fact that Heaviside was not a university man raised a barrier, a certain antagonism, between him and his contemporaries. The latter reproached him, rightly, with his great

†Compare, for instance, H. T. Davis, *The Theory of Linear Operators,* Bloomington, Indiana, 1936.

‡For a survey of the life and work of Heaviside the reader is referred to E. T. Whittaker, *Bull. Calcutta Math. Soc.* XX, 216, 1928-9 (reprinted in O. Heaviside, *Electromagnetic Theory,* 3rd Ed., New York, 1971, vol. I, pp. XIII-XXXIV); Balth. van der Pol, *Ned. Tijdschr. Natuurkunde,* V, 269, 1938; *The Heaviside Centenary Volume,* London, 1950.

§O. Heaviside, *Electrical Papers,* vols. I and II, London, 1892, 2nd ed (corrected reprint), New York, 1971; *Electromagnetic theory,* vols. I, II, and III (1892-1912), 2nd ed., 3 vols. in 1, New York, 1950, 3rd ed., New York, 1971.

lack of mathematical rigour. Yet Heaviside did develop an abundance of mathematical and physical methods and results which afterwards, on critical elaboration by various scientists, proved to be substantially true and have been approved as such. Though perhaps reasonable, it is regrettable that such a barrier existed between Heaviside and his fellow-mathematicians. Equally regrettable, but certainly unreasonable, is the point of view occasionally taken by modern mathematicians with regard to Heaviside's work; in many respects it is far superior to the later contributions to this part of science, both for the methods as well as for the results arrived at†.

Fortunately, there are other records too. For instance, Whittaker (loc. cit.) wrote, after discussing the difference in views on mathematics between Heaviside and the pure mathematician:

'Looking back on the controversy after thirty years, we should now place the Operational Calculus with Poincaré's discovery of automorphic functions and Ricci's discovery of the Tensor Calculus as the three most important mathematical advances of the last quarter of the nineteenth century. Applications, extensions and justifications of it constitute a considerable part of the mathematical activity of to-day.'

It is this Operational Calculus to which the present book is devoted.

2. The operational calculus based on the Laplace transform

Heaviside's ideas concerning the operational calculus may perhaps best be interpreted as follows‡. Imagine a linear electrical network originally at rest. Let an electromotive force $E(t)$ be applied to it, where $E(t)$ is an arbitrary function of the time t. The response current, $i(t)$, is then determined by

$$i(t) = Y(D_t)\,E(t), \tag{1}$$

in which $D_t = d/dt$. The function $Y(D_t)$ is an *operator function* applied to the *operand* $E(t)$, to give the current $i(t)$. If $E(t)$ is constant with time, $i(t)$ will be constant too; under these circumstances $Y(D_t)$ degenerates into the reciprocal of an ohmic resistance.

The question arises at once of how we are to interpret the operator function when, for instance, it is of the following form:

$$Y(D_t) = \frac{1}{1+D_t}.$$

† Heaviside was more than 'ein englischer Elektroingenieur', in spite of his (and his successor's) methods being 'mathematisch sehr unzulänglich' and 'allerdings mathematisch unzureichend'. Quotations from G. Doetsch, *Theorie und Anwendung der Laplace-Transformation*, Berlin, 1937, and New York, 1943, pp. 337, 421.

‡ *Proc. Roy. Soc.* LII, 504, 1892–3; LIV, 105, 1893.

Should we take it in the sense of

$$\frac{1}{1+D_t} = 1-D_t+D_t^2-D_t^3+\ldots, \tag{2}$$

or rather in the sense of

$$\frac{1}{1+D_t} = \frac{1}{D_t}-\frac{1}{D_t^2}+\frac{1}{D_t^3}-\ldots? \tag{3}$$

In the first case it is reasonable to interpret D_t^n as d^n/dt^n. Similarly, when applying (3) we would take $1/D_t$ to mean $\int_a^t d\tau$; and in so doing a second question has presented itself: What value of the constant of integration, a, is required? Guided by practical experience, Heaviside came to the conclusion that, if possible, the form (3) should be chosen rather than (2). He further concluded that in discussing switch-on phenomena in electrical networks (when the electromotive force does not come into action before $t = 0$; that is, $E(t) = 0$ when $t < 0$), the lower limit of integration, a, has to be equated to zero. However, in Heaviside's work, we have not been able to find any rigorous statements concerning this question in general.

A modern treatment of the operational calculus requires, therefore, a much more rigorous base. This is furnished by the Laplace transform, as was already pointed out by Heaviside himself†, though he did not use it extensively. The same Laplace transform was the starting-point of later writers such as Carson‡, Bush§, Humbert||, Doetsch¶, Wagner††, Droste‡‡, McLachlan§§, and Widder||||. When $h(t)$ is supposed to be given, then the Laplace transform of $h(t)$ is the function $f(p)$, defined by the following integral:

$$f(p) = p\int_0^\infty e^{-pt}h(t)\,dt. \tag{4}$$

By so doing we let the function f of the variable p correspond to the function h of the variable t. Conversely, as particularly discussed by Carson, the formula (4) may be considered as an integral equation for the unknown function $h(t)$, when $f(p)$ is supposed to be given.

† *Electromagnetic Theory*, III, 236.

‡ John R. Carson, *Electric Circuit Theory and the Operational Calculus*, New York, 1926.

§ V. Bush, *Operational Circuit Analysis*, New York, 1929.

|| P. Humbert, *Le calcul symbolique*, Paris, 1934.

¶ G. Doetsch, *Theorie und Anwendung der Laplace-Transformation*, Berlin, 1937, and New York, 1943.

†† K. W. Wagner, *Operatoren Rechnung*, Leipzig, 1940.

‡‡ H. W. Droste, *Die Lösung angewandter Differentialgleichungen mittels Laplacescher Transformation*, Berlin, 1939.

§§ N. W. McLachlan, *Complex Variable and Operational Calculus*, Cambridge, 1939.

|||| D. V. Widder, *The Laplace Transform*, Princeton, 1941.

A somewhat different point of view is taken by Bromwich†, who started from the complex integral

$$h(t) = \frac{1}{2\pi i}\int_{c-i\infty}^{c+i\infty} e^{pt}\frac{f(p)}{p}dp; \tag{5}$$

this integral, it may be noted in passing, was known to Riemann‡ as early as 1859. A complete survey of Bromwich's work is to be found in Jeffreys' book §. Wagner|| also based his contribution upon the integral (5). Further impact to the calculus is owed to Lévy¶, who pointed out that the solution of (4), considered as an integral equation for $h(t)$, is given by (5), and vice versa. Thus by Lévy's work the two different points of view came together in one consistent theory.

Also based on the Laplace transform (4), with zero as lower limit of integration, are the former investigations of Van der Pol†† and of Van der Pol and Niessen‡‡.

Henceforth the transformation (4) will be called the unilateral or *one-sided* Laplace transform. Contrary to the earlier investigations, this book will be based on the *two-sided* Laplace transform

$$f(p) = p\int_{-\infty}^{\infty} e^{-pt}h(t)\,dt, \tag{6}$$

to obtain a wider basis for the operational calculus, as will be discussed in detail in the next chapter. The two-sided Laplace integral has its lower limit of integration equal to $-\infty$ instead of 0. This generalization proves very advantageous, and includes the earlier calculus as a special case. In the first place, the operational rules are considerably simplified by the generalization and, secondly, a much larger class of functions (and phenomena) becomes accessible.

It is worth while to remark that, whether we use (4), (5) or (6) as the basis of the operational calculus, the indefinite concepts of operator and operand wholly disappear. Instead of the vague formulation of the early operational calculus there comes the functional transform (6), by which there corresponds to any given function $h(t)$ a new function $f(p)$ of the complex variable p. In the Volterra sense, $f(p)$ is a 'fonction de ligne' or 'fonctionnelle', indicating that the form of the function $f(p)$ depends on the

† T. J. I'a. Bromwich, *Proc. Lond. Math. Soc.* xv, 401, 1916.

‡ The integral occurs in Riemann's classical paper of only eight pages: 'Ueber die Anzahl der Primzahlen unter einer gegebenen Grösse', *Monatsber. Berl. Akad.* Nov. 1859; see also *Gesammelte Werke*, Leipzig, 1876, p. 136.

§ H. Jeffreys, *Operational Methods in Mathematical Physics*, Cambridge, 1927.

|| K. W. Wagner, *Arch. Elektrotech.* IV, 159, 1916.

¶ P. Lévy, *Le calcul symbolique d'Heaviside*, Paris, 1926.

†† Balth. van der Pol, *Phil. Mag.* VII, 1153, 1929; VIII, 861, 1929; XXVI, 921, 1938; *Physica,'s-Grav.*, IV, 585, 1937.

‡‡ Balth. van der Pol and K. F. Niessen, *Phil. Mag.* XI, 368, 1931; XIII, 537, 1932.

whole set of values which $h(t)$ assumes on the complete real axis of t, $-\infty < t < \infty$.

It is to be emphasized that (6) is essentially a *linear* functional transform, since in the integrand of (6) the function $h(t)$ occurs linearly. As a consequence, the operational calculus is applicable only to linear problems such as switch-on phenomena in linear networks, problems of small vibrations, heat diffusion, potential theory, and electrical cables.

As far as the general outlines of the theory are concerned, this book is restricted to giving an extensive survey; the more complicated theorems underlying the theory are usually stated without proof. For proofs the reader is always referred to existing literature, cited in the text. Our main aim is to demonstrate the vigour of the operational calculus in its applications, by giving many examples. The discussion will not be confined to applications in physics and technics; many problems of pure mathematics will be included too.

If the reader has made himself familiar with the fundamental principles of the calculus presented here, he will certainly become aware of the strength of this mathematical tool of almost unrestricted heuristic-analytic value; he will be guided by many examples illustrating the general theory; he will often be able to construct new analytic relations by quite simple means.

3. Survey of the subject-matter

The starting-point of chapter II is the Fourier integral, on which the foundation of the operational calculus is built. We are then led back to the fundamental expressions (5) and (6). In chapter III some elementary 'operational relations' are derived which prove useful in the course of the subsequent investigations. In chapter IV we shall establish elementary 'operational rules', indicating how certain changes of the p-function correspond to others of the t-function. Chapter V is devoted to a detailed discussion of the unit function, $U(t)$, and the delta or impulse function, $\delta(t)$. The latter was introduced by Dirac in quantum mechanics; but Heaviside had already used it extensively before him. The impulse function is particularly important in relation to the Green function of differential equations. It is formulated in terms of the general concept of the Stieltjes integral. Chapter VI should be considered as a deepening and extension of chapter II. It contains a detailed investigation of questions of convergence, particularly in connexion with the summing of series and integrals by the well-known methods of Abel and Cesàro. In chapter VII, especially the asymptotic expressions for 'image' and 'original' are outlined, as well as related topics. Further, chapter VIII concerns the operational treatment of differential equations having constant coefficients. This matter is extended to a system of equations in chapter IX. These two chapters also include the theory of linear electrical networks, together with the corresponding transient phenomena. Differential equations with variable coefficients are treated in chapter X. Applications are made to Legendre polynomials, Bessel functions, etc. The matter of chapter XI must be considered as a generalization of that given in chapter IV; general rules of more complicated character are discussed. Chapter XII is devoted to the study of step functions, with applications, amongst others, to number-theoretic functions. In chapters XIII and XIV we consider the operational calculus applied to difference equations and integral equations respectively. Chapters XV and XVI concern applications of the theory to problems in several independent variables, particularly with respect to linear partial differential equations. Chapter XVI is thereby based on the simultaneous transposition of more than one variable, which leads to the *simultaneous operational calculus*.

It is clear from the survey given above that the subject-matter in any chapter is determined by some specific part of mathematics to which the operational calculus is successfully applicable in one way or another. It may thus happen that closely interrelated 'operational rules', on the one hand, and 'operational relations' concerning some definite type of function, on the other, are discussed at several places scattered through the book. This may hamper further applications, and the material presented must therefore be made more readily available. We have done this by listing the most important results in an appendix at the end of the book. We have thus an opportunity to supply the reader with some additional results which have not been given explicitly in the course of the work. The first list contains the 'operational rules'; it forms the 'grammar' of the operational calculus. The second list is the 'dictionary', helpful in translating the language of t into that of p and vice versa. The 'operational relations' are ordered so that those which concern related functions are grouped together.

The division indicated is such that some chapters and sections may be omitted by those who are interested in the applications of the operational calculus to technical problems only. Similarly, other chapters may be omitted by the pure mathematician.

The practical man will find in the following chapters and sections an almost complete course for his purpose:

II, §§ 5, 6, 7; III, §§ 1, 2, 3, 5, 6; IV, except § 7; V, except §§ 4, 5, 7; VI, § 3; VII, except §§ 5, 12; VIII; IX; X, §§ 1, 2, 4; XI, except § 5; XII, §§ 1, 2, 3; XIV, §§ 1, 2, 3, 5; XV; XVI.

On the other hand, the mathematician will find a survey of the subjects of interest to him by reading only the following chapters and sections:

II; III; IV; V; VI; VII; VIII, §§ 1, 2, 3, 4; X; XI; XII; XIII; XIV; XV, except § 4; XVI.

Both may well find it helpful to read the complete text, since the parts otherwise omitted will serve to throw light on the recommended selection.

CHAPTER II

THE FOURIER INTEGRAL AS BASIS OF THE OPERATIONAL CALCULUS

1. The Fourier integral

The fundamental formulae of the operational calculus as developed in this book are closely connected with the Fourier integral, the theory of which will be recalled in this chapter. In subsequent chapters the usefulness of the operational calculus is demonstrated by means of simple examples and rules, and then, in chapters VI and VII, the foundations of the operational calculus are studied once more and in greater detail. Since the fundamental principles will be discussed first, inasmuch as they are directly connected with the Fourier integral, and since the Fourier integral may be considered as the limiting case of the Fourier series, we shall start with a short account of the latter.

It is well known that every periodic function—subject to proper conditions, which are always satisfied in applications to physical problems—can be expanded into an infinite series of trigonometric functions, viz.

$$h(t) = \sum_{n=0}^{\infty} a_n \cos(n\omega t) + \sum_{n=1}^{\infty} b_n \sin(n\omega t),$$

in which $2\pi/\omega$ denotes the period of the function $h(t)$ and ω is the so-called *angular frequency*. Replacing the sine and cosine functions by their exponential equivalents we obtain just one simple Fourier series:

$$h(t) = \sum_{n=-\infty}^{\infty} c_n \, e^{in\omega t}, \tag{1}$$

where the coefficients c_n are to be determined from

$$c_n = \frac{\omega}{2\pi} \int_{-\pi/\omega}^{\pi/\omega} e^{-in\omega\tau} h(\tau)\, d\tau.$$

The Fourier expansion (1) indicates that a single periodic function of frequency ω is virtually composed of an infinite number of trigonometric functions of angular frequencies ω, 2ω, 3ω, ..., respectively. As the original period increases (thus $\omega \to 0$) these frequencies get nearer and nearer to one another, whilst the individual coefficients c_n tend to zero as well. Therefore, in the limit, the Fourier series (1) is transformed into the *Fourier integral*, which reads as follows:

$$h(t) = \frac{1}{2\pi} \int_{-\infty}^{\infty} d\omega \, e^{i\omega t} \int_{-\infty}^{\infty} d\tau \, e^{-i\omega\tau} h(\tau). \tag{2}$$

Provided that in (2) the integration with respect to the variable ω be taken in the sense of Cauchy's 'valeur principale' (principal value), i.e.

$$\int_{-\infty}^{\infty} d\omega = \operatorname*{Lim}_{\lambda\to\infty}\int_{-\lambda}^{\lambda} d\omega,$$

then a rigorous investigation† of the Fourier integral (2)—often called *Fourier identity*—shows that the following system of (sufficient) conditions guarantees the validity of the Fourier identity at $t = t_0$:

(1) $h(t)$ has limited total fluctuation in the neighbourhood of $t = t_0$;

(2) $h(t)$ is integrable in any finite interval; moreover, the integral $\int_{-\infty}^{\infty} | h(t) | \, dt$ exists;

(3) at points of discontinuity the value of the function $h(t)$ is equal to the corresponding *mean* value, thus (see fig. 1)

$$h(t_0) = \operatorname*{Lim}_{\epsilon\to 0}\frac{h(t_0+\epsilon)+h(t_0-\epsilon)}{2}. \qquad (3)$$

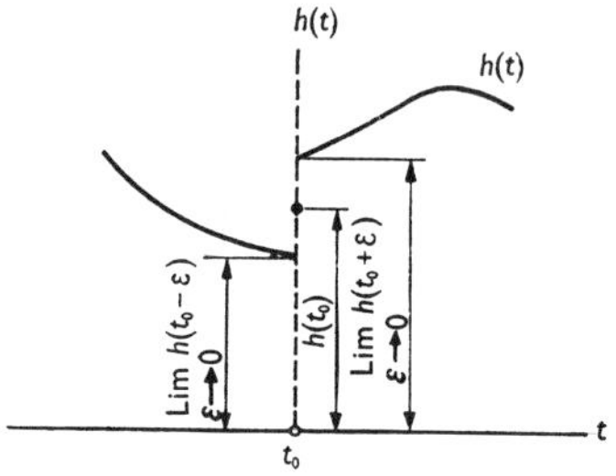

Fig. 1. Mean value of a discontinuous function.

The conditions quoted above are less stringent than the well-known earlier conditions of Dirichlet, and this is made possible by the introduction of the modern concept of 'limited total fluctuation'. As to the definition of this important concept, the function $h(t)$ is said to have limited total fluctuation in the interval $a < t < b$ if, for *every* set of subdivisions according to

$$a = t_0 < t_1 < t_2 < \ldots < t_{n-1} < t_n = b,$$

the sum $\sum_{m=1}^{n} | h(t_m) - h(t_{m-1}) |$ has an upper bound K, independent of n. The existence of the mean value as referred to in condition (3) follows from the fact that $h(t)$ has limited total fluctuation in an (arbitrarily) small interval around $t = t_0$. It should also be noted that, as regards the validity of the Fourier identity (2) at some definite point $t = t_0$, only the behaviour of $h(t)$ in the vicinity of $t = t_0$ is decisive. On the other hand, if $h(t)$ has limited total fluctuation in *any* finite interval, then its integrability follows at once. Moreover, any function with limited total fluctuation can be considered as the sum of two monotonic functions, one of which is non-decreasing, the other non-increasing. It is important to notice that, with respect to the validity of the Fourier identity, $h(t)$ need not be a *periodic* function, as was required in the analogous Fourier-series expansion. Moreover, we

† For details the reader is referred to E. C. Titchmarsh, *Introduction to the Theory of Fourier Integrals*, Oxford, 1937; S. Bochner, *Vorlesungen über Fouriersche Integrale*, Leipzig, 1932.

would emphasize that the conditions we have mentioned are only of a sufficient character; they are by no means necessary, and there do exist less restrictive conditions, especially if the Fourier integral be taken in the sense of the Cesàro or Cauchy limits (cf. VI, §9). Finally, we may remark that the complex expression (2) can also be written in the equivalent real form

$$h(t) = \frac{1}{\pi}\int_0^\infty d\omega \int_{-\infty}^\infty d\tau \cos\{\omega(t-\tau)\}\, h(\tau).$$

Example. Consider the function $h(t)$ defined by $(a>0)$

$$h(t) = \begin{cases} 0 & \text{if} \quad t<-a, \\ b & \text{if} \quad -a<t<a, \\ 0 & \text{if} \quad t>a. \end{cases}$$

For obvious reasons (see fig. 2) this function is called the *rectangle function.* From (2) its Fourier integral follows readily:

$$h(t) = \frac{1}{2\pi}\int_{-\infty}^\infty d\omega\, e^{i\omega t}\int_{-a}^a d\tau\, e^{-i\omega\cdot}\, b = \frac{b}{\pi}\int_{-\infty}^\infty e^{i\omega t}\frac{\sin(\omega a)}{\omega}\, d\omega. \tag{4}$$

This integral represents the rectangle function in the complete interval $-\infty<t<\infty$, including the points of discontinuity, if, and only if, the rectangle function is defined to assume the mean value $\frac{1}{2}b$ at $t = \pm a$. The real Fourier integral of the rectangle function is given by

$$h(t) = \frac{2b}{\pi}\int_0^\infty \sin(\omega a)\cos(\omega t)\frac{d\omega}{\omega}.$$

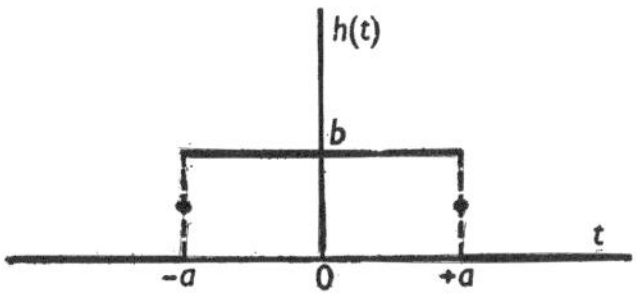

Fig. 2. The rectangle function.

2. A pair of integrals equivalent to the Fourier identity

In physical applications the variables t and ω often denote the time and the angular frequency, respectively. Then the Fourier integral gives the frequency spectrum (or spectral composition) of the original t-function; for, if the second integral in the Fourier identity is considered as a new function, viz.

$$g(\omega) = \frac{1}{\sqrt{(2\pi)}}\int_{-\infty}^\infty e^{-i\omega\tau}\, h(\tau)\, d\tau, \tag{5a}$$

then the Fourier integral itself can be written as follows:

$$h(t) = \frac{1}{\sqrt{(2\pi)}}\int_{-\infty}^\infty e^{i\omega t} g(\omega)\, d\omega. \tag{5b}$$

Formula (5b) indicates that $h(t)$ may be looked upon as a continuous sum of periodic components $e^{i\omega t}$, the strength of the individual components being proportional to $g(\omega)$. Actually $\frac{1}{\sqrt{(2\pi)}}g(\omega)\,d\omega$ represents the total strength of all those components which lie inside the infinitesimal frequency

band ω, $\omega+d\omega$. In general $g(\omega)$ is a complex function of the real variable ω; therefore one may write
$$g(\omega) = A(\omega)\,e^{i\phi(\omega)},$$
in which the modulus $A(\omega)$ and the argument $\phi(\omega)$ are both real functions of the frequency. Substituting this in (5b) we obtain
$$h(t) = \frac{1}{\sqrt{(2\pi)}}\int_{-\infty}^{\infty} A(\omega)\,e^{i\{\omega t+\phi(\omega)\}}\,d\omega,$$
from which it is evident that $\frac{1}{\sqrt{(2\pi)}}A(\omega)\,d\omega$ represents the total amplitude of all vibrations that are contained in $(\omega, \omega+d\omega)$ and have a common phase of amount $\phi(\omega)$ at the time $t = 0$. Summarizing, we may say that both amplitude and phase of the spectral components of $h(t)$ are completely determined by the corresponding frequency function $g(\omega)$.

Example. Returning to the rectangle function, we obtain from (4)
$$g(\omega) = b\sqrt{\left(\frac{2}{\pi}\right)}\frac{\sin(\omega a)}{\omega}. \tag{6}$$
In this case $\phi(\omega)$ vanishes identically because $g(\omega)$ is real; hence the spectral components are equally phased at $t = 0$.

The pair of integrals (5), which is completely equivalent to the Fourier identity (2), reveals a striking symmetry with respect to the variables ω and t on the one hand, and the functions $g(\omega)$ and $h(t)$ on the other. Indeed, formula (5b) can be obtained from that of (5a) by interchanging ω and t, and then substituting $h(x)$ for $g(x)$ and $g(-x)$ for $h(x)$, and vice versa. Therefore, there also exists a typical symmetry in physics between any time function and its spectral composition. This principle of duality may be illustrated by the rectangle function treated above. The counterpart of this time function is the frequency function $g(\omega)$ that has a non-vanishing constant amplitude inside, and a zero amplitude outside, some finite ω-interval. It should be noticed that this type of spectrum is well known in the theory of electrical filters (frequency response). Since the rectangle function has a spectrum given by $\frac{\sin(a\omega)}{\omega}$ it is obvious that the time function belonging to a frequency band of width $2a$ is proportional to $\frac{\sin(at)}{t}$.

Another remarkable feature of (5) is that the *kernel* $e^{\mp i\omega t}$ contains the variables ω and t only in the combination of their product, ωt. There exist, of course, other systems of integral transforms showing the same peculiarity; for instance,
$$g_1(\omega) = \int_0^{\infty} J_\nu\{2\sqrt{(\omega t)}\}\,h_1(t)\,dt \quad (\omega > 0), \tag{7a}$$
$$h_1(t) = \int_0^{\infty} J_\nu\{2\sqrt{(\omega t)}\}\,g_1(\omega)\,d\omega \quad (t > 0), \tag{7b}$$

in which $J_\nu(x)$ denotes the Bessel function of the first kind of order ν and argument x. This pair of functional transforms is based upon the *Hankel identity*

$$h(t) = \int_0^\infty d\omega\, J_\nu\{2\sqrt{(\omega t)}\}\int_0^\infty d\tau\, J_\nu\{2\sqrt{(\omega\tau)}\}\, h(\tau), \tag{8}$$

which clearly is the analogue of (2). There are many other identities of similar character which, though reducible to a pair of reciprocal functional transforms, usually contain the variables ω and t in a complicated manner. Serving as a simple example, the following system may be quoted:

$$g_2(\omega) = \int_0^1 \cot\{\pi(t-\omega)\}\, h_2(t)\, dt \quad (0<\omega<1),$$

$$h_2(t) = \int_0^1 \cot\{\pi(t-\omega)\}\, g_2(\omega)\, d\omega \quad (0<t<1).$$

Here the kernel depends on ω and t in the combination $(t-\omega)$. For the equations above to be valid it is required (amongst other conditions) that

$$\int_0^1 g_2(\omega)\, d\omega = \int_0^1 h_2(t)\, dt = 0,$$

whilst, on account of the singularity at the point $t = \omega$, both integrals should be taken in the sense of Cauchy's principal value.

So far the symmetry of the Fourier system (5) as referred to is not complete; there still remains some asymmetry with respect to the sign in the arguments of the kernel. However, if we confine ourselves to *even* functions $h(t) = E(t)$, such that $E(t) = E(-t)$, the symmetry is complete:

$$g(\omega) = \sqrt{\left(\frac{2}{\pi}\right)}\int_0^\infty \cos(\omega t)\, E(t)\, dt, \quad E(t) = \sqrt{\left(\frac{2}{\pi}\right)}\int_0^\infty \cos(\omega t)\, g(\omega)\, d\omega.$$

An analogous pair of equations, holding for *odd* functions, is obtained if 'cos' is replaced by 'sin'. Obviously the latter system is simply the specialization for $\nu = \frac{1}{2}$ of the Hankel system (7).

3. Parallel displacement of the path of integration in the integral (5 *b*)

In either of the integrals (5 *a*, *b*) the path of integration coincides with the corresponding real axis. In order to obtain a basic formula for the operational calculus we shall make some transformations. First, the path of integration in (5*b*) will be displaced parallel to itself over a distance c. To this end we put

$$\omega = \omega' + ic; \tag{9}$$

thus $\operatorname{Im}\omega' = -c$. Moreover, to obtain a simple notation we introduce new functions $h_1(t)$ and $g_1(\omega)$ by means of

$$h_1(t) = h(t)\, e^{ct}, \quad g_1(\omega) = g(\omega' + ic). \tag{10}$$

Making the necessary substitutions we obtain instead of (5) the following new pair of integrals:

$$g_1(\omega') = \frac{1}{\sqrt{(2\pi)}}\int_{-\infty}^{\infty} e^{-i\omega' t} h_1(t)\,dt \qquad (11a)$$
$$(\operatorname{Im}\omega' = -c),$$

$$h_1(t) = \frac{1}{\sqrt{(2\pi)}}\int_{-\infty-ic}^{\infty-ic} e^{i\omega' t} g_1(\omega')\,d\omega' \qquad (11b)$$
$$(\operatorname{Im} t = 0).$$

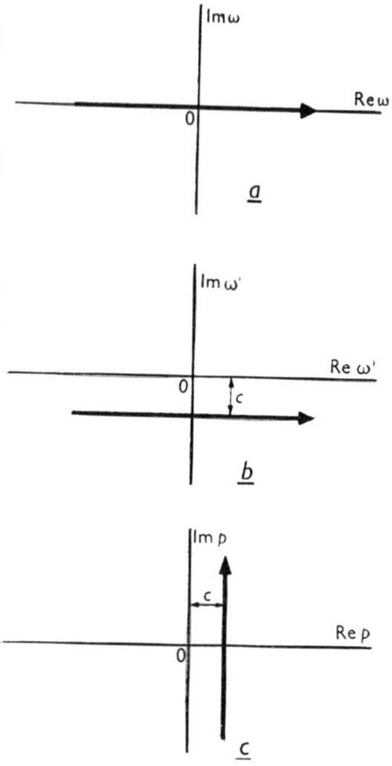

Fig. 3. Transformations of the path of integration in the Fourier integral.

It should be noticed that only the path of integration in the second integral has changed as a consequence of our substitution (see figs. 3*a* and 3*b*); that in the first integral is not altered. Since in physical applications ω denotes the frequency, the time function $h_1(t)$ is now no longer to be considered as composed of undamped vibrations; on the contrary, the components are either increasing $(c > 0)$ or decreasing $(c < 0)$ functions of the time. Because the function $h(t)$ in (2) is of quite general character, the function $h_1(t)$ is arbitrary as well (within certain limits, of course). There is only a formal difference between (5) and (11), $g(\omega)$ and $g_1(\omega)$ being defined in different domains of the respective variables.

4. Rotation of the path of integration in the integral (11 *b*)

After the introduction of the new complex variable ω' by means of the first transformation (9) it proves useful to make a second change of variable, thus arriving at the conventional notation of the operational calculus:

$$p = i\omega' = i\omega + c. \qquad (12)$$

Multiplication of ω' by a complex factor i means that the path of integration for ω' is rotated through an angle of $\frac{1}{2}\pi$. Finally, the path of integration in the complex plane of p is parallel to the imaginary axis, at a distance c on the right side of it. Summarizing, we see that the paths of integration have changed so far as indicated by

$$\operatorname{Im}\omega = 0, \quad \operatorname{Im}\omega' = -c, \quad \operatorname{Re} p = c$$

(compare fig. 3). The last substitution (12) is fully effective and leads to very simple formulae, if only the g-function is again replaced by a new one. Upon introducing the function f defined by

$$g_1(-ip) = \frac{1}{\sqrt{(2\pi)}} \frac{f(p)}{p},$$

and after dropping the suffix 1 in (11) we arrive at the final form of the set of functional transforms:

$$f(p) = p \int_{-\infty}^{\infty} e^{-pt} h(t)\, dt \quad (\operatorname{Re} p = c), \tag{13a}$$

$$h(t) = \frac{1}{2\pi i} \int_{c-i\infty}^{c+i\infty} e^{pt} \frac{f(p)}{p}\, dp \quad (\operatorname{Im} t = 0), \tag{13b}$$

where $f(p)$ and $h(t)$ are corresponding functions which can be given arbitrarily, apart from some very mild restrictions. It follows from the preceding that the constant c in (11a) does not differ from that in (11b). However, the set of equations (11a, b) may hold equally well for different values of c, as will be seen presently.

5. Strip of convergence of the Laplace integral

The integral shown in formula (13a) is an example of the so-called *Laplace integral*, generally defined by

$$\int_a^b e^{-pt} h(t)\, dt,$$

in which the limits of integration are quite arbitrary. Integrals of this type serve a useful purpose, for instance, in the solution of linear differential equations with variable coefficients. If, as in this book, the limits of integration are $-\infty$ and $+\infty$, we shall call them *two-sided* Laplace integrals; on the other hand, we shall speak of *one-sided* Laplace integrals if the limits of integration are 0 and ∞. It is tacitly assumed that the two-sided Laplace integral of formula (13a) is convergent for $\operatorname{Re} p = c$. Then the question arises whether or not the integral may be convergent for other values of p. The answer to this question is provided by an important theorem, proved† in the theory of the one-sided Laplace integral, viz.

THEOREM. *If the one-sided Laplace integral*

$$\int_0^{\infty} e^{-pt} h(t)\, dt$$

be convergent for $p = p_0$, *then it converges for* $\operatorname{Re} p > \operatorname{Re} p_0$.

† Cf. G. Doetsch, *Theorie und Anwendung der Laplace-Transformation*, Berlin, 1937, and New York, 1943, p. 15.

This theorem may be interpreted geometrically in the following manner. If the above integral is convergent in some point p_0 of the complex p-plane, then it converges everywhere in the region on the right-hand side of a straight line drawn through $p = p_0$ and parallel to the imaginary axis (see fig. 4). Consequently only the real part of p is decisive for the convergence of the one-sided Laplace integral; the imaginary part of p does not matter at all. Furthermore, if α be the greatest lower bound of values $\operatorname{Re} p$ for which the convergence holds, then the integral under consideration converges in the region that is bounded at the left by the straight line $\operatorname{Re} p = \alpha$ parallel to the imaginary axis, otherwise extending to infinity. This region of convergence is shaded in fig. 4. It is therefore adequate to call α the *abscissa of convergence* of the one-sided integral, irrespective of the yet undecided question whether or not there are points of convergence on the line $\operatorname{Re} p = \alpha$ itself (only in this case does the convergence of the integral depend on the imaginary part of p).

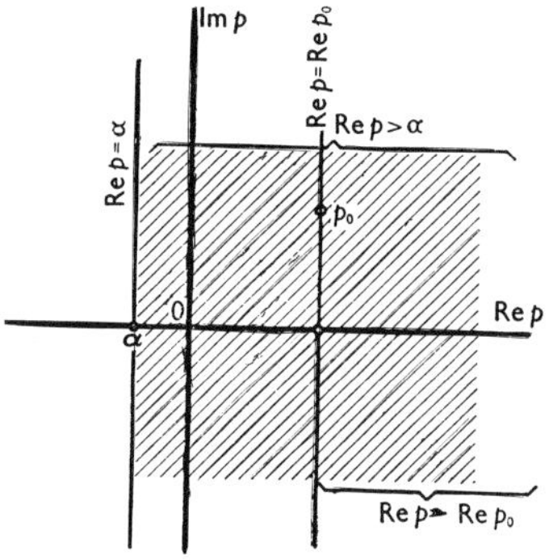

Fig. 4. Strip of convergence of a one-sided Laplace integral.

Guided by the preceding geometrical interpretation, we can now easily make a general survey of all possibilities with respect to questions of convergence of the *two-sided* Laplace integral. The two-sided integral may be considered as the sum of two one-sided integrals, viz.

$$\begin{aligned} f(p) &= p\int_{-\infty}^{0} e^{-pt}h(t)\,dt + p\int_{0}^{\infty} e^{-pt}h(t)\,dt \\ &= p\int_{0}^{\infty} e^{pt}h(-t)\,dt + p\int_{0}^{\infty} e^{-pt}h(t)\,dt. \end{aligned} \tag{14}$$

The second integral is a one-sided Laplace integral with abscissa of convergence α, say, and therefore converges for $\operatorname{Re} p > \alpha$. The first integral is a one-sided Laplace integral of the parameter $-p$; if its abscissa of convergence be denoted by $-\beta$, then it converges for $\operatorname{Re}(-p) > -\beta$, i.e. for $\operatorname{Re} p < \beta$, that is, in a region extending to infinity at the *left* of the vertical $\operatorname{Re} p = \beta$ (see fig. 5). For the two-sided Laplace integral (13 a) to have some domain of convergence, either of the one-sided integrals must converge somewhere, except when the two-sided integral is taken as a principal value with respect to $t = 0$ (cf. VI, §2). Moreover, if $h(t)$ is such that $\alpha < \beta$, the regions of convergence of the individual one-sided integrals in fact overlap, resulting in a common strip of the complex p-plane where (13a) thus converges as well. Therefore, provided that the two-sided Laplace integral converges somewhere, then it converges only in a region bounded by two verticals, viz.

$$\alpha < \mathrm{Re}\,p < \beta.$$

This region is called the *strip of convergence* of the two-sided Laplace integral under consideration (see fig. 5). If, in particular, $\alpha = \beta$, then the strip with boundaries reduces to a single vertical line (see VI, § 13). Finally, if $h(t)$ is such that $\alpha > \beta$, then the separate one-sided integrals will have no common region of convergence at all, and in this case $h(t)$ has no two-sided Laplace transform. It may incidentally be noticed that the imaginary axis of the p-plane may lie inside the strip of convergence, as well as outside it; in fig. 5 the imaginary axis is drawn at the left of the strip, though it may also be at the other side.

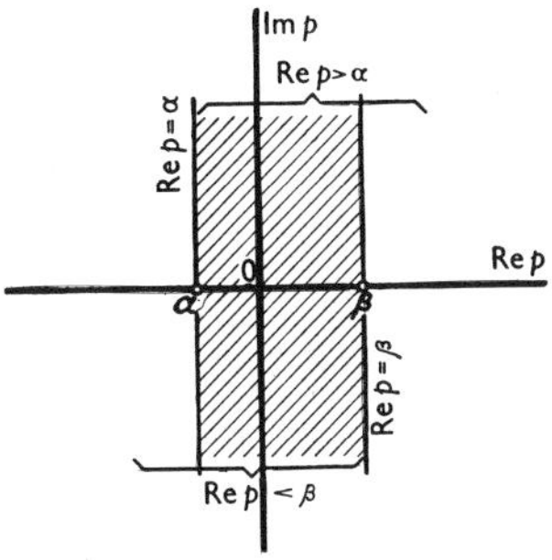

Fig. 5. Strip of convergence of a two-sided Laplace integral.

Example. As a simple example we will discuss the strip of convergence of a function $h(t)$ defined by

$$h(t) = \begin{cases} e^{\alpha t} & (t>0), \\ e^{\beta t} & (t<0). \end{cases}$$

This function is continuous at $t = 0$; its derivative is discontinuous (see fig. 6) unless $\alpha = \beta$. In this case the two-sided Laplace integral (13*a*) becomes

$$p\int_{-\infty}^{0} e^{(\beta-p)t}\,dt + p\int_{0}^{\infty} e^{(\alpha-p)t}\,dt,$$

the first term of which converges for $\mathrm{Re}\,p<\beta$, the second for $\mathrm{Re}\,p>\alpha$. If $\alpha<\beta$, the strip of convergence is determined by $\alpha<\mathrm{Re}\,p<\beta$. If, however, $\alpha>\beta$ (as is supposed in fig. 6), the integral (13*a*) converges nowhere.

Let us return to the expressions shown in (13). Obviously they state a relationship between $h(t)$ and $f(p)$ for values of t and p that satisfy the relations t real and $\mathrm{Re}\,p = c$. So far only a single definite value of c has been considered. To get rid of this restriction we may remember that the system (13) holds under conditions similar to those mentioned for the Fourier identity. The sufficient conditions of § 1 are easily transformed so as to be directly applicable to (13). As $h(t)$ of (13) is simply the original h-function of (2) multiplied by e^{ct}, we can readily state sufficient conditions for the legitimacy of (13*b*) for all real values of t: $h(t)$ must have

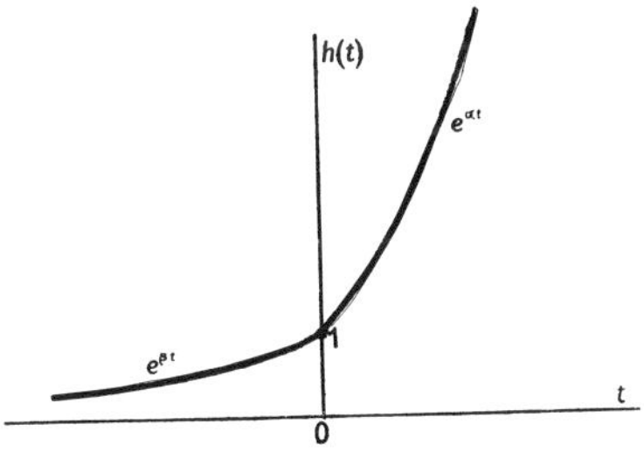

Fig. 6. A function $h(t)$ consisting of two exponential functions.

limited total fluctuation and be equal to its mean value everywhere, and $\int_{-\infty}^{\infty} |h(t)| e^{-ct} dt$ must be convergent. As to the influence of the number c, we observe that the above conditions require the absolute convergence of (13a). Since in general (13a) converges in some strip $\alpha < \mathrm{Re}\, p < \beta$, it will be clear that c may be equal to any of these values of $\mathrm{Re}\, p$, provided (13a) is *absolutely* convergent throughout the mentioned strip. We finally arrive at a pair of integrals that can be considered as an extension of the system (13), namely,

$$f(p) = p \int_{-\infty}^{\infty} e^{-pt} h(t)\, dt, \quad \alpha < \mathrm{Re}\, p < \beta, \tag{15a}$$

$$h(t) = \frac{1}{2\pi i} \int_{c-i\infty}^{c+i\infty} e^{pt} \frac{f(p)}{p}\, dp, \quad \alpha < c < \beta. \tag{15b}$$

We would like to emphasize that the above formulae are basic for the operational calculus; consequently, we shall return to them often during our further investigations.

As already indicated in the preceding paragraphs we can state at once sufficient conditions for (15b) to exist as a principal value, viz.

(1) the integral (15a) must be absolutely convergent in $\alpha < \mathrm{Re}\, p < \beta$,

(2) $h(t)$ must have limited total fluctuation in any finite interval, and

(3) $h(t)$ must be everywhere equal to its mean value.

On the other hand, it should be noted that the first condition of absolute convergence is far from necessary. Ordinary convergence is usually sufficient, provided that (15b) is taken as a first-order Cesàro limit (see VI, § 11).

Example. Let us introduce the discontinuous *unit function* $U(t)$. This function (see fig. 7), which is closely related to the rectangle function of fig. 2, is defined by

$$U(t) = \begin{cases} 1 & (t > 0), \\ \frac{1}{2} & (t = 0), \\ 0 & (t < 0). \end{cases} \tag{16}$$

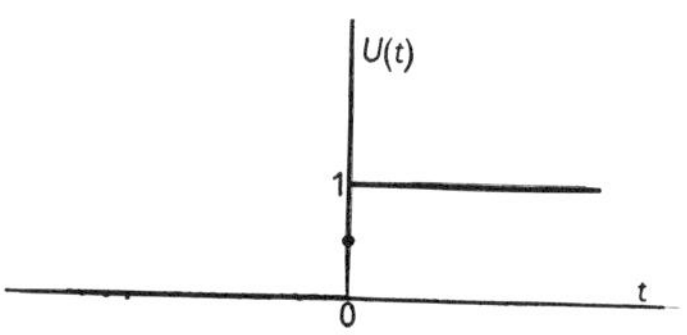

Fig. 7. The unit function.

From formula (15a) we then obtain

$$f(p) = p \int_{-\infty}^{\infty} e^{-pt} U(t)\, dt = p \int_{0}^{\infty} e^{-pt} dt = 1,$$

hence $f(p) = 1$. Moreover, since the integral converges for $\mathrm{Re}\, p > 0$, we have in this example $\alpha = 0$, $\beta = \infty$. According to (15b) we next arrive at

$$U(t) = \frac{1}{2\pi i} \int_{c-i\infty}^{c+i\infty} e^{pt} \frac{dp}{p} \quad (c > 0), \tag{17}$$

in which c may be any positive number. Apparently, the integral in (17) does not depend on the special value of c.

In chapter VI (§ 4) we shall derive formulae by means of which we easily determine the strip of convergence of any given function $h(t)$ in a straightforward manner. Once the strip of convergence is known, we can take for the path of integration in (15b) any vertical line inside the former strip, always getting the same value, i.e. $h(t)$. If c_1 and c_2 are two numbers inside the interval (α, β) (see fig. 8), but otherwise arbitrary, then the remarkable property mentioned above can be expressed as follows:

$$\int_{c_1-i\infty}^{c_1+i\infty} e^{pt}\frac{f(p)}{p}\,dp = \int_{c_2-i\infty}^{c_2+i\infty} e^{pt}\frac{f(p)}{p}\,dp.$$

This result is based on some theorems concerning the analytic behaviour of the function $f(p)$ in the strip of convergence. It can be proved† that $f(p)/p$ is analytic there and, moreover, it can be shown that the contributions of the horizontal segments N_1, N_2 (see fig. 8) tend to zero if these segments recede to infinity:

$$\int_{N_1} = \int_{N_2} = 0.$$

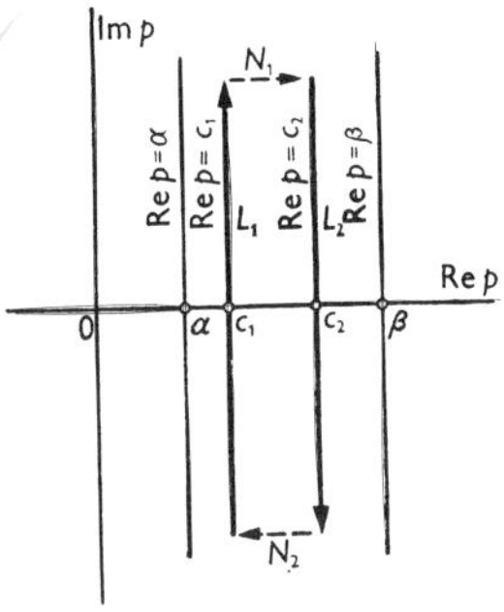

Fig. 8. Parallel shift of the path of integration in the integral (15b).

Using the theorem of Cauchy—stating that the integral along any closed path of integration, inside which the integrand is analytic throughout, is necessarily zero—it is evident that the integrals along the verticals L_1 and L_2 (see fig. 8) have opposite values in the limiting case $N_1, N_2 \to \infty$. In chapter VI the analytic properties of $f(p)/p$ inside the strip of convergence will be examined in greater detail, as well as the question whether the vertical boundaries of the strip may or may not, in whole or in part, belong to the region of convergence.

In concluding this section we may remark that the set of equations (15) is equivalent to the well-known inversion formulae of Mellin. To see this, let $t = -\log x$, and let functions ϕ and g be defined by means of

$$\phi(p) = \frac{f(p)}{p}, \quad g(x) = h(-\log x);$$

then (15) is transformed into

$$\phi(p) = \int_0^\infty x^{p-1} g(x)\,dx, \quad \alpha < \operatorname{Re} p < \beta, \tag{18a}$$

$$g(x) = \frac{1}{2\pi i}\int_{c-i\infty}^{c+i\infty} x^{-p}\,\phi(p)\,dp, \quad \alpha < c < \beta. \tag{18b}$$

These formulae are due to Mellin and Riemann, as has already been mentioned in chapter I. It will be obvious from the foregoing that the Mellin formulae are also closely connected with the Fourier identity (2).

† Cf. Doetsch, loc. cit. p. 104.

6. The language of the operational calculus

As already emphasized in the preceding section, we consider (15) as the fundamental equations of the operational calculus. Henceforth the relationship between $f(p)$ and $h(t)$ as expressed by ($15a, b$) will be written symbolically as follows:

$$f(p) \risingdotseq h(t), \quad \alpha < \mathrm{Re}\, p < \beta, \tag{19a}$$

or, alternatively,

$$h(t) \fallingdotseq f(p), \quad \alpha < \mathrm{Re}\, p < \beta. \tag{19b}$$

The symbols $\fallingdotseq$ and $\risingdotseq$ are introduced only in order to have a convenient and short abbreviation for the relations (15) which we write down once more for reference:

$$f(p) = p \int_{-\infty}^{\infty} \mathrm{e}^{-pt} h(t)\, dt, \quad \alpha < \mathrm{Re}\, p < \beta, \tag{20a}$$

$$h(t) = \frac{1}{2\pi i} \int_{c-i\infty}^{c+i\infty} \mathrm{e}^{pt} \frac{f(p)}{p}\, dp, \quad \alpha < c < \beta. \tag{20b}$$

The function $h(t)$ is called the *original*, and the corresponding function $f(p)$ its *image*. Almost immediately the question arises whether there exists a one-to-one correspondence between original and image. To this end, let us first consider the integral ($20a$) which is the actual foundation of our calculus and which we therefore call the *definition integral*. It is obvious that at most one image, if any, belongs to a given original. For a one-to-one correspondence, however, it is necessary that no other original will lead to the same image. Although it may actually occur that one image originates from different originals, this is certainly impossible within the same strip of convergence (apart from some academic refinements; e.g. when the originals differ from each other by a 'null function'; see VI, § 15. Null functions can always be added to an original without disturbing the functional relationship (20)). Therefore we may state that the definition integral ($20a$) actually establishes a one-to-one correspondence between the original and the image.

Let us now examine ($20b$). In general ($20b$) holds true if it is understood that $f(p)$ is determined from ($20a$) and $h(t)$ is subjected to some mild conditions (see VI, § 11). It is therefore wholly legitimate to call ($20b$) the *inversion integral* corresponding to the definition integral ($20a$). Still the definition integral is the more important one, since the legitimacy of the inversion by ($20b$) is not always ascertained, even if ($20a$) does exist as a convergent integral. On the other hand, it should be borne in mind that the inversion integral, too, is applicable in all practical cases, provided it be taken as a first-order Cesàro limit (cf. VI, § 11).

It is further to be noted that the upper dot in the symbols of (19) points to the original, whilst the lower dot points to the image; it is a matter of indifference whether we write the image at the right or at the left of the symbol.

We have already found that the unit function $h(t) = U(t)$ as original leads to the image $f(p) = 1$; moreover, the corresponding integral (20 *a*) converges for $\operatorname{Re} p > 0$. Therefore we have in our symbolic notation

$$1 \doteqdot U(t), \quad 0 < \operatorname{Re} p < \infty, \tag{21}$$

or, alternatively, $$U(t) \doteqdot 1, \quad 0 < \operatorname{Re} p < \infty.$$

These equations may be read as follows: 'the constant 1 is the image of the unit function', or, alternatively, 'the unit function is the original of the constant 1'. Because of this simple rule we decided to define the Laplace integral as $f(p)/p$, rather than $f(p)$ as many authors do.

We may once more emphasize that the strip of convergence should always be specified, since completely different originals may lead to one and the same image in different strips of convergence. The following simple example will serve as illustration. On the one hand we have

$$\frac{1}{p} \doteqdot \begin{cases} t & (t > 0) \\ 0 & (t < 0) \end{cases} \quad 0 < \operatorname{Re} p < \infty,$$

whereas on the other

$$\frac{1}{p} \doteqdot \begin{cases} 0 & (t > 0) \\ -t & (t < 0) \end{cases} \quad -\infty < \operatorname{Re} p < 0,$$

as is readily verified by application of (20 *a*).

In many physical applications the independent variable of the original is the time; therefore we have denoted it by t, though in pure mathematics it is usually written x. The independent variable of the image, however, is generally—following Heaviside—denoted by p.

It is evident from the definition integral that the original has to be given only for *real* values of t; the corresponding image (if existing at all) is then automatically determined in some strip of convergence in the complex p-plane. The operational transformation original→image generates a correspondence between a function defined on the real t-axis and a function of p in the complex p-plane. Even if the original is discontinuous, its image (if existing) is an analytic function of p in the strip of convergence. The last property is already indicated by the preceding example of the discontinuous unit function $U(t)$ leading to the image $1 (\operatorname{Re} p > 0)$, which is obviously analytic. Since analytic functions are more tractable than discontinuous functions, the operational calculus provides us with a method of studying properties of discontinuous functions with the aid of the milder image functions. In electrical-network theory discontinuous t-functions play an important part (e.g. a voltage suddenly applied at $t = 0$ represents such a discontinuous function). In this field of physics the operational calculus has proved its great value.

In actual applications of the operational calculus we first transpose the problem under consideration, supposed to be formulated in terms of the variable t, into a new problem in the p-language; i.e. the functional relations (such as difference, differential or integral equations), in which the problem is first stated, is transformed into its operational equivalent. This comes down to a multiplication of the given functional relation by the function $p\,e^{-pt}$, followed by an integration from $t = -\infty$ to $t = +\infty$. This procedure may conveniently be called the *transition from the t-language to the p-language*. In many problems, such as are encountered in the study of transient phenomena in electrical networks, the new formulation (in p) is considerably simpler than the old one (in t), and therefore usually leads to a solution in an easy way. Once the problem is solved in terms of p, it only remains, as the second stage of the method, to translate back the p-solution into the physical language of t.

For a rapid and efficient procedure it is necessary to have a 'dictionary' at hand; this dictionary is an assembly of the simplest originals together with their corresponding images. A 'grammar' of rules would also be helpful, indicating how functional relations between originals are transposed into those between the corresponding images, and vice versa. A short dictionary is constructed in chapter III, and a grammar in chapter IV. They will contain only the most important relations and rules, simply for a better understanding of the operational calculus. At the end of the book, however, the reader will find a more extensive list of rules and operational relations; a part of them has been treated in the complete text. It is to be emphasized that, in applying the dictionary and grammar, the integrals (20) need no longer be used explicitly, although they and only they, of course, are basic for the construction of both grammar and dictionary.

Finally, we may remark that—though it is still a matter of indifference whether (20) or some other system (see (7), for instance) is used as basis of the calculus—by the choice of (20) as fundamental relations, operations such as differentiation and integration can easily be translated; this would not be the case in any other kind of operational calculus.

7. Operational calculus based on one-sided Laplace integrals

Our definition integral contains $-\infty$ and $+\infty$ as limits of integration. In previous treatments by Heaviside, Carson, and others, the lower limit is 0 instead of $-\infty$; therefore the original $h^*(t)$ needed to be defined only for $t > 0$, since its values for $t < 0$ do not enter into the old definition integral. On introducing the unit function, this comes down to

$$h(t) = h^*(t)\,U(t), \tag{22}$$

since substitution of (22) in (20 a) automatically leads to the old one-sided integral. Functions $h(t)$ that vanish identically for $t<0$ in the sense of (22) are called *one-sided functions* (see fig. 9); those not vanishing for $t<0$ are called *two-sided functions.*

It will be observed that our treatment is to be preferred, since in the old theory two-sided originals could not be dealt with at all. On the other hand, this extension of the basis of the operational calculus necessarily requires a proper specification of the strip of convergence: $\alpha < \operatorname{Re} p < \beta$. According to the theorem mentioned in § 5, the one-sided Laplace integral converges in a region extending to infinity at the right of the vertical $\operatorname{Re} p = \alpha$. Therefore β is always infinitely large for one-sided originals, and the equations (20) become

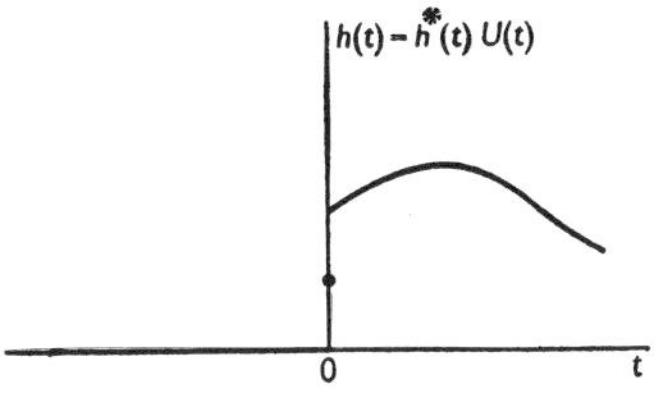

Fig. 9. A one-sided original.

$$f(p) = p\int_0^\infty e^{-pt} h^*(t)\,dt, \quad \alpha < \operatorname{Re} p < \infty, \tag{22a}$$

$$h^*(t)\,U(t) = \frac{1}{2\pi i}\int_{c-i\infty}^{c+i\infty} e^{pt}\frac{f(p)}{p}\,dp, \quad \alpha < c < \infty. \tag{22b}$$

It is significant to notice that in the new theory the factor $U(t)$ in one-sided originals should never be omitted, since the image of a one-sided original depends upon its continuation through negative values of t, unless it would vanish there identically.

Example. As to the function $e^{-\alpha t^2}$ $(\alpha > 0)$, we have

$$e^{-\alpha t^2} \doteq p\int_{-\infty}^{\infty} e^{-\alpha t^2} e^{-pt}\,dt.$$

Upon substituting in the image integral $s = t\sqrt{\alpha} + \dfrac{p}{2\sqrt{\alpha}}$, and using the well-known result of Poisson

$$\int_{-\infty}^{\infty} e^{-s^2}\,ds = \sqrt{\pi}, \tag{23}$$

we obtain

$$e^{-\alpha t^2} \doteq \sqrt{\left(\frac{\pi}{\alpha}\right)}\, p\, e^{p^2/4\alpha}, \quad -\infty < \operatorname{Re} p < \infty. \tag{24}$$

On the other hand, for the one-sided original,

$$e^{-\alpha t^2} U(t) \doteq p\int_0^\infty e^{-pt} e^{-\alpha t^2}\,dt = \frac{p}{\sqrt{\alpha}}\, e^{p^2/4\alpha}\int_{\frac{p}{2\sqrt{\alpha}}}^{\infty} e^{-s^2}\,ds, \quad -\infty < \operatorname{Re} p < \infty.$$

Introducing, finally, the conventional notation existing for the well-known probability function (error function), we arrive at

$$e^{-\alpha t^2} U(t) \doteq \frac{1}{2}\sqrt{\left(\frac{\pi}{\alpha}\right)}\, p\, e^{p^2/4\alpha} \operatorname{erfc}\left(\frac{p}{2\sqrt{\alpha}}\right), \quad -\infty < \operatorname{Re} p < \infty. \tag{25}$$

Having thus developed the fundamental ideas of the operational calculus by means of the Fourier integral, we shall deal in subsequent chapters with the simplest applications and general theorems of our calculus.

CHAPTER III

ELEMENTARY OPERATIONAL IMAGES

1. Introduction

In this chapter we shall discuss the images of some elementary functions which are frequently used in physical and other applications. Two simple rules have to be considered first, however. Since the definition integral (II, 20 *a*) involves the original $h(t)$ only in a linear way we have manifestly

(1) Multiplication of the original by any constant C corresponds in the p-language to multiplication of the image by C.

(2) The image of a sum of two originals is equal to the sum of the separate images.

Clearly, the strip of convergence does not alter after application of the first rule. The second rule, however, can (at first sight) only be applied if the strips of convergence overlap, and consequently the sum rule will hold only in the common domain of convergence. Thus, starting with the operational relations

$$f_1(p) \doteqdot h_1(t), \quad \alpha_1 < \operatorname{Re} p < \beta_1,$$

$$f_2(p) \doteqdot h_2(t), \quad \alpha_2 < \operatorname{Re} p < \beta_2,$$

we find for the sum (see also fig. 10)

$$f_1(p) + f_2(p) \doteqdot h_1(t) + h_2(t), \quad \max(\alpha_1, \alpha_2) < \operatorname{Re} p < \min(\beta_1, \beta_2), \tag{1}$$

provided, of course, there is a common strip of convergence.

It is still possible that (1) holds true in some strip of convergence larger than that indicated; for instance, if a common pole of the functions $\frac{1}{p} f_1(p)$, $\frac{1}{p} f_2(p)$ is cancelled out after the addition. This peculiarity is illustrated in the following example (cf. § 3), in which a denotes any positive number:

$$\frac{p}{p-a} \doteqdot e^{at}\, U(t), \quad \operatorname{Re} p > a,$$

$$\frac{a}{p-a} \doteqdot (e^{at} - 1)\, U(t), \quad \operatorname{Re} p > a.$$

By subtraction we obtain the operational relation $1 \doteqdot U(t)$, already mentioned, and holding in fact for $\operatorname{Re} p > 0$.

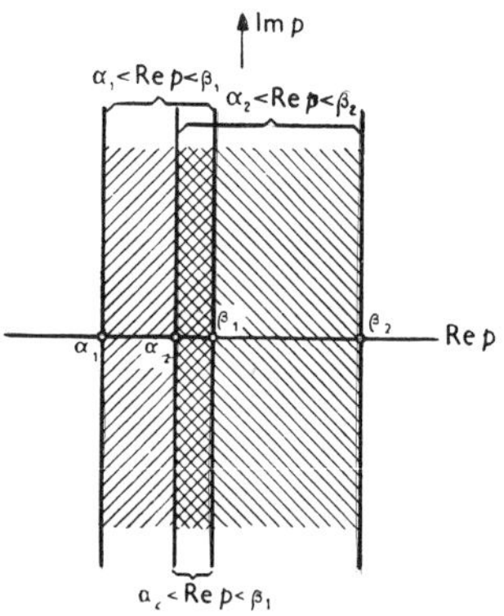

Fig. 10. Common strip of convergence of two operational relations.

By successive application of the sum rule we obtain the image of a finite, or even infinite, number of originals. Thus the image function $\sum_{n=0}^{\infty} f_n(p)$ corresponds to the original $\sum_{n=0}^{\infty} h_n(t)$. With regard to convergence questions, etc., the reader is referred to chapter VII, §§ 8–11.

2. Images of polynomials; the Γ-function

According to the sum rule of § 1 we can easily determine the image of any polynomial in t, once the image of t^n is known. The two-sided Laplace integral of t^n does not exist; so we have to confine ourselves to one-sided polynomials, i.e. polynomials multiplied by the unit function $U(t)$.

Let us thus start with a discussion of $t^\nu U(t)$; its image is, according to definition,

$$p\int_{-\infty}^{\infty} e^{-pt} t^\nu U(t)\,dt = p\int_0^{\infty} e^{-pt} t^\nu\,dt.$$

On substituting $s = pt$ and using Euler's integral for the Γ-function, we find, provided $p > 0$ and $\nu > -1$,

$$\frac{1}{p^\nu}\int_0^{\infty} e^{-s} s^\nu\,ds = \frac{\Pi(\nu)}{p^\nu} = \frac{\Gamma(\nu+1)}{p^\nu}. \tag{2}$$

The transformation given above holds also for complex values of p, provided $\operatorname{Re} p > 0$. The new upper limit of integration, obtained when the substitution $s = pt$ is performed, and which lies in the right half-plane of s, can be replaced by $+\infty$ without changing the value of the integral. Moreover, the condition $\operatorname{Re}\nu > -1$ is already sufficient, if it is understood that the many-valued function p^ν is defined as $e^{\nu \log p}$, where $\log p$ is real for $p > 0$. On the other hand, the integral diverges if either $\operatorname{Re} p < 0$ or $\operatorname{Re}\nu < -1$. Summarizing, we may say that the indicated extension of (2) leads to the following operational relation:

$$t^\nu U(t) \risingdotseq \frac{\Pi(\nu)}{p^\nu}, \quad 0 < \operatorname{Re} p < \infty \ (\operatorname{Re}\nu > -1). \tag{3}$$

It is to be noticed that in (3) the negative power of p is equal to the positive power of t; this simple result is a consequence of having defined the Laplace integral (II, 20a) equal to $\frac{1}{p} f(p)$ rather than to $f(p)$.

It is thus established that the one-sided original of the simple function t^ν, whether ν be integral or not, leads to an image involving the famous Γ-function $\Gamma(\nu+1) = \Pi(\nu)$ as a numerical coefficient. With regard to further investigations, it may be worth while to recall some of its principal properties. As is well known, $\Gamma(n+1) = \Pi(n) = n!$ for any positive integer n. A closer study reveals that the Π-function is throughout positive if $\nu > -1$.

By analytic continuation into the domain $\mathrm{Re}\,\nu < -1$, for which values of ν the function cannot be represented by Euler's integral, it can be shown that $\Pi(\nu) = \Gamma(\nu+1)$ has simple poles at $\nu = -1, -2, \ldots$. For real values of ν the function $\Pi(\nu)$ is plotted in fig. 11.

Integration by parts of the Euler integral leads to a very important relation; in terms of the Π-function:

$$\Pi(\nu) = \nu\Pi(\nu-1), \tag{4}$$

in accordance with the familiar $n! = n(n-1)!$.

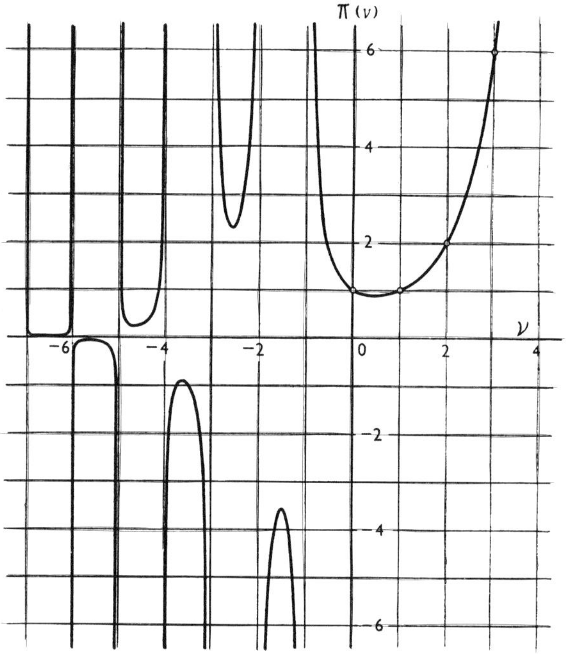

Fig. 11. The Π-function.

The functional equation (4) is very characteristic for the Π-function. As shown by Hölder, $\Pi(\nu)$ does not satisfy any differential equation with algebraic coefficients. According to Artin† $\Pi(\nu)$ is the only solution of the difference equation (4) that is logarithmically convex for $\nu > -1$, satisfying the normalizing condition $\Pi(0) = 1$.

The property of logarithmic convexity is essential. In this connexion we may consider the example

$$f(\nu) = \tfrac{1}{3}\{\cos(2\pi\nu)+2\}\,\Pi(\nu),$$

being a function satisfying the relations $f(0) = 1$ and $f(\nu) = \nu f(\nu-1)$. Therefore it coincides completely with the function $\Pi(\nu)$ for positive

† E. Artin, *Einführung in die Theorie der Gammafunktionen*, Leipzig, 1931.

integral values of ν; its logarithm $\log f(\nu)$, however, is not convex with respect to the ν-axis; this can be readily verified.

As is already indicated by the formula (2), two different notations, $\Pi(\nu)$ and $\Gamma(\nu)$, are in use; they are related by

$$\Pi(\nu) = \Gamma(\nu+1).$$

In this book both notations are used alternatively, depending on which of the two leads to the simpler result in a given case. We may further remark that even for ν equal to an integral multiple of $\frac{1}{2}$ the numeric $\Pi(\nu)$ can be given explicitly. To see this, we remember the equivalence of Euler's integral for $\Pi(-\frac{1}{2})$ with Poisson's integral (II, 23)

$$\Pi(-\tfrac{1}{2}) = \Gamma(\tfrac{1}{2}) = \sqrt{\pi}. \tag{5}$$

For other half-integral values of the argument, according to the recurrence relation (4),

$$\Pi(\tfrac{1}{2}) = \tfrac{1}{2}\Pi(-\tfrac{1}{2}) = \tfrac{1}{2}\sqrt{\pi}, \quad \Pi(\tfrac{3}{2}) = \tfrac{3}{2}\Pi(\tfrac{1}{2}) = \tfrac{3}{4}\sqrt{\pi}, \text{ etc.}$$

So, for instance, if ν is taken equal to $\frac{5}{2}$, (3) yields

$$t^{\frac{5}{2}}\, U(t) \doteqdot \frac{15}{8}\frac{\sqrt{\pi}}{p^{\frac{5}{2}}}, \quad 0 < \operatorname{Re} p < \infty.$$

In the same manner any finite sum of one-sided powers of t (exceeding -1), whether they are integral or rational, can be transposed. As just another example we may quote

$$(3t^2 + 7t^{\frac{5}{2}})\, U(t) \doteqdot \frac{6}{p^2} + \frac{105}{8}\frac{\sqrt{\pi}}{p^{\frac{5}{2}}}, \quad 0 < \operatorname{Re} p < \infty,$$

which can be written down almost at once.

Having briefly indicated the importance of the Γ-function in elementary operational calculus we may state that the calculus, in its turn, can also give useful information about the Γ-function. A simple example will illustrate this statement. Since t^ν is positive on the path of integration, the definition integral of (3) converges absolutely; moreover, the function $t^\nu U(t)$ has limited total fluctuation in any finite interval, and is equal to its mean value. Consequently (the sufficient conditions of II, § 5 being fulfilled) the existence and the validity of the inversion integral are certain. Thus we have

$$\frac{t^\nu}{\Pi(\nu)} U(t) = \frac{1}{2\pi i}\int_{c-i\infty}^{c+i\infty} \frac{e^{pt}}{p^{\nu+1}}\,dp, \quad 0 < c < \infty \ (\operatorname{Re}\nu > -1).$$

It should be noted that the above integral must vanish automatically for negative values of t, because the left-hand side does so.

Let us now consider the special case $t = 1$; then

$$\frac{1}{\Pi(\nu)} = \frac{1}{2\pi i}\int_{c-i\infty}^{c+i\infty} \frac{e^{p}}{p^{\nu+1}}\,dp \quad (\operatorname{Re}\nu > -1;\ c > 0). \tag{6}$$

This integral representation of the inverse Π-function is closely related to a well-known formula of Hankel, viz.

$$\frac{1}{\Pi(\nu)} = \frac{1}{2\pi i}\int_L \frac{e^p}{p^{\nu+1}}\,dp, \tag{7}$$

which defines the Π-function in the whole plane of ν. The path of integration L starts at $p = -\infty$; then it turns around the origin, and finally goes back to $p = -\infty$, in such a way that the negative axis of p is never crossed (see fig. 12). Obviously the integrals (6) and (7) differ only with respect to their paths of integration. They are both valid for $\operatorname{Re}\nu > -1$, indicating that for these values of ν the path L in (7) can be transformed into a straight vertical line, parallel to and at the right of the imaginary axis, without changing the value of the integral. For other values of ν such a transformation is not allowed, since then the factor $p^{-\nu-1}$ does not decrease sufficiently for $|p| \to \infty$. It will be evident from this example that even the most elementary functions can lead to interesting analytical results, by a proper application of the operational calculus.

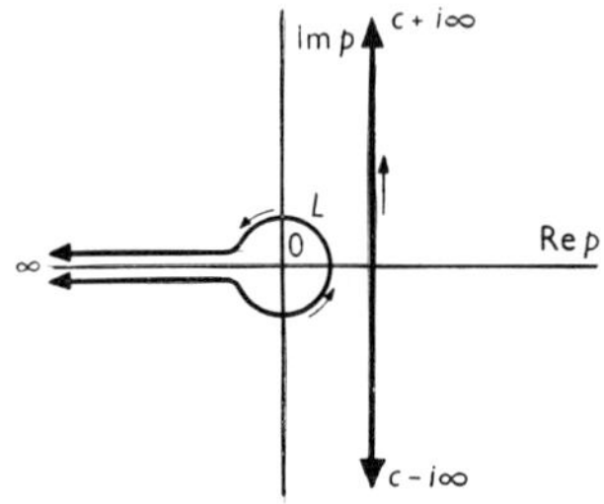

Fig. 12. Paths of integration of Hankel's integral and of the inversion integral.

3. Images of simple exponential functions

As will be seen later on, the images of the simple exponential functions are of great importance in physical applications. We shall begin our study with the simplest exponential original, viz. the one-sided $e^{-\alpha t}\,U(t)$. Applying the definition integral, we obtain

$$p\int_{-\infty}^{\infty} e^{-pt}\,e^{-\alpha t}\,U(t)\,dt = p\int_0^{\infty} e^{-(p+\alpha)t}\,dt = \frac{p}{p+\alpha}$$

for $\operatorname{Re}(p+\alpha) \geqslant 0$. Consequently the following operational relation holds for arbitrary values of α:

$$e^{-\alpha t}\,U(t) \doteq \frac{p}{p+\alpha}, \qquad -\operatorname{Re}\alpha < \operatorname{Re}p < \infty. \tag{8}$$

After writing down the relation (8) with α replaced by $-\alpha$, and performing a simple addition or subtraction, we are led at once to the following pair of operational relations:

$$\cosh(\alpha t)\,U(t) \doteq \frac{p^2}{p^2-\alpha^2}, \qquad |\operatorname{Re}\alpha| < \operatorname{Re}p < \infty, \tag{9}$$

$$\sinh(\alpha t)\,U(t) \doteq \frac{\alpha p}{p^2-\alpha^2}, \qquad |\operatorname{Re}\alpha| < \operatorname{Re}p < \infty, \tag{10}$$

in which the proper strips of convergence can readily be accounted for.

It is to be noticed that the analogous two-sided t-functions (thus ignoring the unit function $U(t)$) have no images at all, unless a degenerated strip of convergence should be allowed (see VI, § 13). On the other hand, the image remains equally simple when t is replaced in the exponent by its absolute value, and $U(t)$ is again omitted; namely,

$$\tfrac{1}{2}e^{-\alpha|t|} \doteqdot \frac{\alpha p}{\alpha^2-p^2}, \qquad -\operatorname{Re}\alpha < \operatorname{Re}p < \operatorname{Re}\alpha, \tag{11}$$

valid only for $\operatorname{Re}\alpha > 0$, of course.

In comparing (11) with (10) we observe once more that two different originals may belong to one and the same image, though in different strips of convergence only. We may finally remark that the old theory, based upon the one-sided Laplace integral, did not give an image for $\frac{1}{2}e^{-\alpha|t|}$ at all.

4. Images of complicated exponential functions

As to originals involving exponential functions in a complicated way, their images may sometimes be found by an adequate change of variables in the initial Laplace integral. The following examples will illustrate this in some detail.

Example 1. The pair of related functions

$$\frac{1}{e^t+1} \quad \text{and} \quad \frac{U(t)}{e^t+1}$$

again shows how the character of the image is influenced by properly accounting for the unit function. In the first place we have

$$\frac{1}{e^t+1} \doteqdot p\int_{-\infty}^{\infty} \frac{e^{-pt}}{e^t+1}\,dt = K(p), \quad \text{say.}$$

The substitutions $t = \tau \pm i\pi$ lead to two new integrals with identical integrands, but different paths of integrations. The contour integral representing their difference is easily evaluated from the residue at $\tau = 0$ and results in the equation

$$2i\sin(\pi p)\frac{K(p)}{p} = -2\pi i,$$

which is equivalent to the operational relation:

$$\frac{1}{e^t+1} \doteqdot -\frac{\pi p}{\sin(\pi p)}, \qquad -1 < \operatorname{Re}p < 0. \tag{12}$$

Secondly, for the other original, after the substitution $e^{-t} = \sqrt{u}$ in the definition integral, we have

$$\frac{p}{2}\int_0^1 \frac{u^{\frac{1}{2}(p-1)}}{\sqrt{u}+1}\,du. \tag{13}$$

This integral can be expressed in terms of the logarithmic derivative of the Π-function, as follows from Γ-function theory. For

$$C+\psi(z) = C+\frac{d}{dz}\{\log\Pi(z)\} = \int_0^1 \frac{1-t^z}{1-t}\,dt \quad (\operatorname{Re}z > -1), \tag{14}$$

in which C denotes Euler's constant

$$C = \operatorname*{Lim}_{n\to\infty}\left(1+\frac{1}{2}+\frac{1}{3}+\ldots+\frac{1}{n}-\log n\right) = 0{\cdot}5772\ldots.$$

Since (13) may be written as the difference of two integrals of the type (14), we at once arrive at the required image; thus the second operational relation reads

$$\frac{U(t)}{\mathrm{e}^t+1} \doteqdot \frac{p}{2}\left\{\psi\left(\frac{p}{2}\right)-\psi\left(\frac{p-1}{2}\right)\right\}, \quad -1<\mathrm{Re}\,p<\infty. \tag{15}$$

Finally, it is interesting to compare the respective inversion integrals with regard to the influence of the unit-function factor in the originals. They are, of course,

$$\frac{1}{\mathrm{e}^t+1} = -\frac{1}{2i}\int_{c-i\infty}^{c+i\infty}\frac{\mathrm{e}^{pt}}{\sin(\pi p)}\,dp \quad (-1<c<0),$$

$$\frac{U(t)}{\mathrm{e}^t+1} = \frac{1}{4\pi i}\int_{c-i\infty}^{c+i\infty}\mathrm{e}^{pt}\left\{\psi\left(\frac{p}{2}\right)-\psi\left(\frac{p-1}{2}\right)\right\}dp \quad (c>-1).$$

It is evident that these integrals are equal in value for positive values of t, whilst the lower integral, unlike the upper one, vanishes identically for all negative values of t.

Example 2. We will now discuss another example showing that complicated functions may often lead to quite simple images. To this end, let us consider the original $h(t) = \mathrm{e}^{-\mathrm{e}^{-t}}$ whose definition integral is transformed into

$$p\int_0^\infty \mathrm{e}^{-s}s^{p-1}\,ds$$

after the introduction of the new variable of integration $s = \mathrm{e}^{-t}$. Next, on account of Euler's integral for the Γ-function, this is equal to $p\Pi(p-1) = \Pi(p)$ if $\mathrm{Re}\,p>0$. Consequently we then obtain the interesting operational relation

$$\mathrm{e}^{-\mathrm{e}^{-t}} \doteqdot \Pi(p), \quad 0<\mathrm{Re}\,p<\infty. \tag{16}$$

Another remarkable formula is found on application of the corresponding inversion integral; first, we have

$$\mathrm{e}^{-\mathrm{e}^{-t}} = \frac{1}{2\pi i}\int_{c-i\infty}^{c+i\infty}\mathrm{e}^{pt}\,\Pi(p-1)\,dp \quad (c>0),$$

and secondly, for the special values $c = 1$, $t = 0$ after changing the variable of integration by $p = 1+i\omega$,

$$\frac{1}{2}\int_{-\infty}^{\infty}\Pi(i\omega)\,d\omega = \frac{\pi}{e},$$

thus involving the ratio of the transcendental numbers π and e.

Example 3. We will consider once more the function of the second example, now multiplied by the unit function. Again substituting $s = \mathrm{e}^{-t}$ in the definition integral, we obtain in this case

$$\mathrm{e}^{-\mathrm{e}^{-t}}U(t) \doteqdot p\int_0^1 \mathrm{e}^{-s}s^{p-1}\,ds \quad (\mathrm{Re}\,p>0). \tag{17}$$

If the upper limit in (17) were ∞, then the integral would be equal to $\Pi(p)$. Since the contribution of the part $1<s<\infty$ is missing now, the resulting function is customarily called an *incomplete* Γ-function. Generally, if the path of integration $0<s<\infty$ is split up into two parts at $s = \rho$, then the following incomplete Γ-functions come into existence:

$$P(\nu,\rho) \equiv \int_0^\rho e^{-s} s^{\nu-1}\,ds, \tag{18a}$$

$$Q(\nu,\rho) \equiv \int_\rho^\infty e^{-s} s^{\nu-1}\,ds, \tag{18b}$$

which functions are usually labelled as *Prym functions*†. Obviously, their sum is equal to the Γ-function

$$P(\nu,\rho)+Q(\nu,\rho) = \Gamma(\nu).$$

In terms of incomplete Γ-functions we may then write (17) as follows:

$$e^{-e^{-t}}\,U(t) \doteqdot pP(p,1), \quad 0<\mathrm{Re}\,p<\infty. \tag{19a}$$

The other Prym function is involved in an analogous operational relation, obtained when the original of the second example is multiplied by $U(-t)$ instead of $U(t)$, namely,

$$e^{-e^{-t}}\,U(-t) \doteqdot pQ(p,1), \quad -\infty<\mathrm{Re}\,p<\infty. \tag{19b}$$

An interesting application of the sum rule of § 1 is established in the following manner. By expanding the left-hand side of (19*a*) into a power series in e^{-t} we obtain the equivalent relation

$$\sum_{n=0}^{\infty} \frac{(-1)^n}{n!} e^{-nt}\,U(t) \doteqdot pP(p,1), \quad 0<\mathrm{Re}\,p<\infty.$$

Now, according to (8), the individual terms at the left have images $\dfrac{(-1)^n}{n!}\dfrac{p}{p+n}$ for $-n<\mathrm{Re}\,p<\infty$. After adding the infinite number of separate images we arrive at an operational relation valid in the common strip of convergence, that is, $\mathrm{Re}\,p>0$. Identification of the different expressions for the image finally leads to

$$\sum_{n=0}^{\infty} \frac{(-1)^n}{n!}\frac{1}{p+n} = P(p,1). \tag{20}$$

A proof of the validity of the sum rule in this case will be delayed for the moment. Yet we may mention that, though we have made (20) plausible only for $\mathrm{Re}\,p>0$, the expansion (20) holds for general values of p. It will also be evident that $P(p,1)$ and $\Gamma(p)$ have a common set of simple poles, $p = 0, -1, -2, \ldots$, with corresponding equal residues. Therefore, the difference of $\Gamma(p)$ and $P(p,1)$ is necessarily an integral function of p. This function is $Q(p,1)$; in accordance with $Q(p,1)$ being an integral function of p, we see that (19*b*) holds for all finite values of p. Thus the separation of the Euler integral into the Prym functions is, from the point of view of analysis, a separation into a meromorphic and an integral function; the specialization $\rho = 1$ is not essential in this connexion.

5. Images of trigonometric functions

The images of trigonometric functions are of primary importance with regard to physical applications. Therefore a short account of them may be given here.

With the aid of (8) we can write down almost at once

$$\cos t\,U(t) \doteqdot \frac{p^2}{p^2+1}, \quad 0<\mathrm{Re}\,p<\infty, \tag{21}$$

$$\sin t\,U(t) \doteqdot \frac{p}{p^2+1}, \quad 0<\mathrm{Re}\,p<\infty, \tag{22}$$

† The notations are those of N. E. Nörlund, *Differenzenrechnung*, Berlin, 1924, p. 392.

if use is made of the familiar formulae

$$\cos t = \tfrac{1}{2}e^{it} + \tfrac{1}{2}e^{-it}, \quad \sin t = \frac{1}{2i}e^{it} - \frac{1}{2i}e^{-it},$$

and (8) is applied for the special values $\alpha = i$, $\alpha = -i$, followed by an addition or subtraction.

It is to be remarked that other simple trigonometric functions, such as $\tan t U(t)$ and $\cot t U(t)$, have no operational images, because the corresponding Laplace integrals diverge for all p. This is because the poles of $h(t)$ lie on the path of integration. Though it might be possible to assign a definite meaning to integrals of this type by properly taking them in the sense of a principal value, we will not go into further details here. On the other hand, there exist many operational relations involving the elementary trigonometric functions in a more complicated way; the latter can usually be derived by adequate superposition of simple exponential and trigonometric relations or by applying the rules of the next chapter. As an example we quote the image of the function $\sin^{2n} t\, U(t)$ which can be verified, amongst others, by induction.

$$\sin^{2n} t\, U(t) \doteq \frac{(2n)!}{(p^2+2^2)(p^2+4^2)\ldots(p^2+4n^2)}, \quad 0 < \operatorname{Re} p < \infty. \tag{23}$$

6. Images of logarithmic functions

In discussing this class of functions we have an opportunity to develop a procedure which often allows of deriving essentially new operational relations from others already investigated. To this end, let us start with some operational relation involving a parameter ν, so that for $\nu_1 \leqslant \nu \leqslant \nu_2$:

$$h(t, \nu) \doteq f(p, \nu), \quad \alpha < \operatorname{Re} p < \beta.$$

In most cases it will be permitted to differentiate either side of the relation above with respect to the parameter ν, leading to the new operational relation

$$\frac{\partial h(t, \nu)}{\partial \nu} \doteq \frac{\partial f(p, \nu)}{\partial \nu}, \quad \alpha < \operatorname{Re} p < \beta \ (\nu_1 \leqslant \nu \leqslant \nu_2). \tag{24}$$

As to the legitimacy of this procedure, it is sufficient to require that, in the ν-interval mentioned, (i) the definition integral belonging to (24) is uniformly convergent at both end-points, and (ii) $\dfrac{\partial}{\partial \nu} h(t, \nu)$ is a continuous function of t and ν.

Let us return to the operational relation (3) involving the parameter ν. Introducing the same notation of the logarithmic derivative of the Π-function as before in (14), viz.

$$\psi(\nu) = \frac{\Pi'(\nu)}{\Pi(\nu)},$$

we have by application of the procedure (24)

$$\{\log t - \psi(\nu)\}\frac{t^\nu}{\Pi(\nu)}U(t) \doteq -\frac{\log p}{p^\nu}, \quad 0 < \operatorname{Re} p < \infty \ (\operatorname{Re}\nu > -1). \tag{25}$$

A simple specialization of this relation is found when ν is made zero; since $\psi(0) = -C$ (Euler's constant, see also (14) for $z = 0$), we obtain

$$(\log t + C)\,U(t) \doteq -\log p, \quad 0 < \operatorname{Re} p < \infty, \tag{26}$$

or, alternatively,

$$-\log t\,U(t) \doteq \log p + C, \quad 0 < \operatorname{Re} p < \infty. \tag{27}$$

7. Well-known integrals of the Laplace type

Very often one is led to integrals of the Laplace type—if necessary, after some trivial modification. Of course, these integrals may be interpreted in terms of the operational calculus. We shall therefore discuss some simple examples from this point of view.

Example 1. In the conventional notation of Bessel-function theory†, we have, for the Hankel function of imaginary argument (cf. x, § 4),

$$K_\nu(a) = \frac{\pi i}{2} e^{\frac{1}{2}\nu\pi i} H_\nu^{(1)}(ia), \tag{28}$$

the following integral representation

$$K_\nu(a) = \frac{1}{2}\int_{-\infty}^{\infty} e^{-a\cosh t - \nu t}\,dt \quad (\operatorname{Re} a > 0). \tag{29}$$

Let us now replace the parameter ν by p; then we arrive at the following two-sided Laplace integral:

$$2pK_p(a) = p\int_{-\infty}^{\infty} e^{-pt} e^{-a\cosh t}\,dt \quad (\operatorname{Re} a > 0).$$

We have thus established the operational relation

$$e^{-a\cosh t} \doteq 2pK_p(a), \quad -\infty < \operatorname{Re} p < \infty \ (\operatorname{Re} a > 0), \tag{30}$$

into which the parameter p of the operational calculus enters as the order of the Hankel function of imaginary argument. Still other relations can be derived from the integral (29) involving, however, the parameter p in the argument of the Hankel function rather than in its order. For instance, take $a = 2\sqrt{p}$ in (29) and substitute $s = e^t/\sqrt{p}$. The modified integral is then easily recognized as the definition integral corresponding to the operational relation

$$2p^{\frac{1}{2}(\nu+1)}K_{\nu-1}(2\sqrt{p}) \doteq \frac{e^{-1/t}}{t^\nu}U(t), \quad 0 < \operatorname{Re} p < \infty \ (-\infty < \operatorname{Re}\nu < \infty). \tag{31}$$

In addition, the specialization $\nu = \frac{1}{2}$ leads to the quite simple operational relation

$$\sqrt{p}\, e^{-2\sqrt{p}} \doteq \frac{e^{-1/t}}{\sqrt{(\pi t)}}U(t), \quad 0 < \operatorname{Re} p < \infty. \tag{32}$$

† G. N. Watson, *Theory of Bessel Functions*, Cambridge, 1922, p. 182.

Example 2. Another example showing that an integral defining some transcendental function may be transformed into one of the Laplace type is provided by the well-known transcendent Ei, which is conventionally called the exponential integral (it can also be written li so as to denote the logarithmic integral, of different argument of course), viz.

$$\operatorname{Ei}(-p) = \operatorname{li}(e^{-p}) = \int_{\infty}^{p} \frac{e^{-s}}{s}\, ds. \tag{33}$$

This integral is easily changed into the appropriate form by the substitution $s = pt$. For $\operatorname{Re} p > 0$ the result becomes

$$-p \operatorname{Ei}(-p) = p \int_{1}^{\infty} \frac{e^{-pt}}{t}\, dt = p \int_{-\infty}^{\infty} e^{-pt} \frac{U(t-1)}{t}\, dt.$$

According to definition we therefore have the following operational relation:

$$-p \operatorname{Ei}(-p) = -p \operatorname{li}(e^{-p}) \doteqdot \frac{U(t-1)}{t}, \quad 0 < \operatorname{Re} p < \infty. \tag{34}$$

Example 3. Still another integral representation admitting almost immediate interpretation as a definition integral is encountered in ζ-function theory. The generalized ζ-function, of arguments ν and a, is customarily defined by means of an infinite series as follows:

$$\zeta(\nu, a) = \sum_{n=0}^{\infty} \frac{1}{(n+a)^{\nu}}, \tag{35}$$

which series converges for $\operatorname{Re} \nu > 1$. The ζ-function of Riemann is obtained by especially choosing $a = 1$, namely,

$$\zeta(\nu, 1) = \zeta(\nu) = \sum_{n=1}^{\infty} \frac{1}{n^{\nu}} \quad (\operatorname{Re} \nu > 1). \tag{36}$$

As to the generalized ζ-function, we have for $\operatorname{Re} \nu > 1$ and $\operatorname{Re} a > 0$ the integral representation

$$\zeta(\nu, a) = \frac{1}{\Gamma(\nu)} \int_{0}^{\infty} \frac{x^{\nu-1} e^{-ax}}{1 - e^{-x}}\, dx. \tag{37}$$

Upon replacing a by p, and multiplying through by $p\Gamma(\nu)$, one will easily find the equivalent relation

$$\frac{t^{\nu-1}}{1-e^{-t}} U(t) \doteqdot \Gamma(\nu)\, p\zeta(\nu, p), \quad 0 < \operatorname{Re} p < \infty \ (\operatorname{Re} \nu > 1). \tag{38}$$

It will be obvious that (38) does not lead to an operational relation (in the proper sense) for the restricted Riemann ζ-function, since for that the operational variable p should be taken equal to 1. In order to obtain an operational relation involving the Riemann ζ-function we substitute $a = 1$ in (37); thus

$$\Gamma(\nu)\, \zeta(\nu) = \int_{0}^{\infty} \frac{x^{\nu-1}}{e^{x} - 1}\, dx \quad (\operatorname{Re} \nu > 1), \tag{39}$$

and this integral can be moulded into a proper definition integral by substituting $x = e^{-t}$ and taking $\nu = p$, leading finally to the operational relation

$$\Pi(p)\, \zeta(p) \doteqdot \frac{1}{e^{e^{-t}} - 1}, \quad 1 < \operatorname{Re} p < \infty. \tag{40}$$

CHAPTER IV

ELEMENTARY RULES

1. Introduction

This chapter will be devoted to establishing simple rules by which the operational calculus enables us to derive, in a simple manner, known relations as well as new ones between various mathematical functions. As will be seen in due course, a simple relationship between two images often leads to corresponding, though perhaps rather intricate, relations between the originals. In this as well as in subsequent chapters we shall often have an opportunity to demonstrate this important property of the operational calculus by means of various examples.

We might again place emphasis on the requirement that any rule is complete only if the corresponding strip of convergence is properly indicated. It should further be noticed in advance that the rules, if formulated with respect to two-sided Laplace transforms (II, 20 *a*), are usually much more simple than they would be if the theory were based on the one-sided integral (II, 22 *a*). The old theory was treated some time ago by one of the present authors†; the rules that follow may be considered as an extension and simplification of the earlier results.

In simple applications we shall often use results of later chapters; for instance, the equality of two images in some strip of convergence implies that of their respective originals. This property holds if the originals obtained are sectionally continuous, the indeterminateness of the original with respect to the question of null functions being of no importance (see VI, § 15).

2. The similarity rule

In the preceding chapter (§ 1) we have already formulated two very simple rules. The similarity rule which will be discussed now is of equal simplicity. It indicates how the image alters if in the original the variable t is multiplied by a *real* constant λ:

If $$h(t) \risingdotseq f(p), \quad \alpha < \operatorname{Re} p < \beta,$$

then we have

1st for $\lambda > 0$: $$h(\lambda t) \risingdotseq f\left(\frac{p}{\lambda}\right), \quad \lambda\alpha < \operatorname{Re} p < \lambda\beta; \tag{1 a}$$

2nd for $\lambda < 0$: $$h(\lambda t) \risingdotseq -f\left(\frac{p}{\lambda}\right), \quad \lambda\beta < \operatorname{Re} p < \lambda\alpha. \tag{1 b}$$

The proof of this rule is easy; it follows from the definition integral for the new original $h(\lambda t)$ after substituting $s = \lambda t$.

† Balth. van der Pol, 'On the operational solution of linear differential equations and an investigation of the properties of these solutions', *Phil. Mag.* VIII, 861, 1929.

Example 1. Let us start with the operational relation

$$U(t) \doteq 1, \quad 0 < \operatorname{Re} p < \infty.$$

On applying the above rule for $\lambda = -1$, we obtain

$$-U(-t) \doteq 1, \quad -\infty < \operatorname{Re} p < 0. \tag{2}$$

The left-hand function of t is equal to -1 for negative values of t, and zero if $t > 0$ (see fig. 13). Thus, again, we have an example showing that one image ($f(p) = 1$) may have different originals in different strips of convergence.

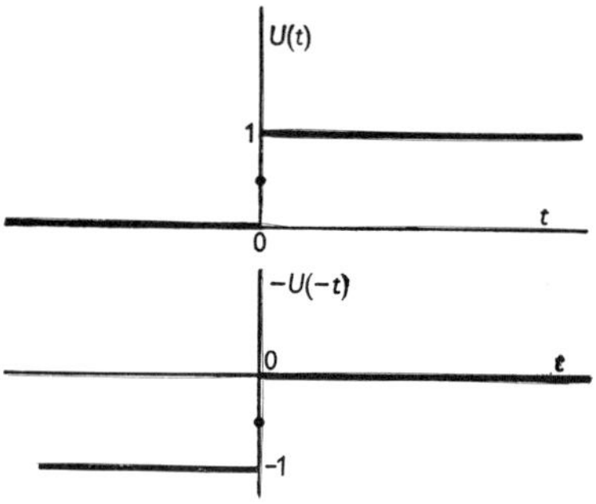

Fig. 13. The unit functions of positive and negative arguments.

Example 2. From the example (III, 23), in which the image has poles on the imaginary axis at the *even* integral multiples of i, we deduce with $\lambda = \frac{1}{2}$ a new operational relation involving an image with poles at *all* integral multiples of i, viz.

$$(2 \sin \tfrac{1}{2}t)^{2n}\, U(t) \doteq \frac{(2n)!}{(p^2+1^2)(p^2+2^2)\dots(p^2+n^2)}, \quad 0 < \operatorname{Re} p < \infty. \tag{3}$$

3. The shift rule

This rule shows the effect of adding a constant λ to the argument t of the original:

Let
$$h(t) \doteq f(p), \quad \alpha < \operatorname{Re} p < \beta,$$

then, for real values of λ,

$$h(t+\lambda) \doteq e^{\lambda p} f(p), \quad \alpha < \operatorname{Re} p < \beta. \tag{4}$$

It is to be noticed that here the strip of convergence does not change.

This rule too is easily derived with the aid of the definition integral; for

$$h(t+\lambda) \doteq p \int_{-\infty}^{\infty} e^{-pt} h(t+\lambda)\, dt = e^{\lambda p}\, p \int_{-\infty}^{\infty} e^{-ps} h(s)\, ds,$$

the right-hand member being $e^{\lambda p}$ times the image $f(p)$ of the original $h(t)$.

In physics the shift rule is frequently applied to the treatment of retardation phenomena (Lorentz retardation); it is also particularly useful in the solution of difference equations, as will be seen in chapter XIII.

It is just the shift rule that was less tractable in the old theory of the one-sided Laplace integral, owing to the change of the lower limit of integration in the definition integral after the substitution $s = t+\lambda$. In the old theory we had the following rule:

Let $$h(t)\,U(t) \doteqdot f(p), \quad \alpha < \operatorname{Re} p < \infty,$$

then, in the same strip of convergence,

1st for $\lambda > 0$: $$h(t+\lambda)\,U(t) \doteqdot e^{\lambda p} f(p),$$

provided $h(t) = 0$ for $0 < t < \lambda$,

2nd for $\lambda < 0$: $$h(t+\lambda)\,U(t+\lambda) \doteqdot e^{\lambda p} f(p).$$

Incidentally, we may remark that a different formulation of the shift rule in the case of one-sided originals is possible. To this end let $[[\phi(p)]]$ denote that part of the Laurent expansion of $\phi(p)$ which consists of the non-positive integral powers of p. Then it follows that

$$(t+a)^n\,U(t) \doteqdot n! \sum_{r=0}^{n} \frac{a^r}{r!\,p^{n-r}} = n! \left[\left[\frac{e^{ap}}{p^n}\right]\right] \quad (\operatorname{Re} p > 0),$$

which operational relation can be obtained after expansion of the original according to the binomial formula. Provided the original is expressible as a Maclaurin series for all values of t, we easily derive

$$h(t+a)\,U(t) = U(t) \sum_{n=0}^{\infty} \frac{(t+a)^n}{n!} h^{(n)}(0) \doteqdot \sum_{n=0}^{\infty} h^{(n)}(0) \left[\left[\frac{e^{ap}}{p^n}\right]\right] = \left[\left[e^{ap} \sum_{n=0}^{\infty} \frac{h^{(n)}(0)}{p^n}\right]\right].$$

Now the last series is simply the image $f(p)$ of the original $h(t)\,U(t)$; consequently, for the shift rule in the case of one-sided originals:

$$h(t+a)\,U(t) \doteqdot [[e^{ap} f(p)]], \quad 0 < \operatorname{Re} p < \infty. \tag{4a}$$

As to the application of the general shift rule, the following few examples may prove useful.

Example 1. With the aid of the unit function we are able to arrive at the image of the rectangle function of fig. 2. Since the latter function may be looked upon as the difference of two suitable unit functions, namely (see fig. 14),

$$h(t) = bU(t+a) - bU(t-a),$$

we have from $U(t) \doteqdot 1$ $(\operatorname{Re} p > 0)$, applying the shift rule,

$$U(t+a) \doteqdot e^{ap} \quad (\operatorname{Re} p > 0),$$

and $$U(t-a) \doteqdot e^{-ap} \quad (\operatorname{Re} p > 0).$$

Therefore the image of the rectangle function becomes

$$b(e^{ap} - e^{-ap}) = 2b \sinh(ap) \quad (\operatorname{Re} p > 0), \tag{5}$$

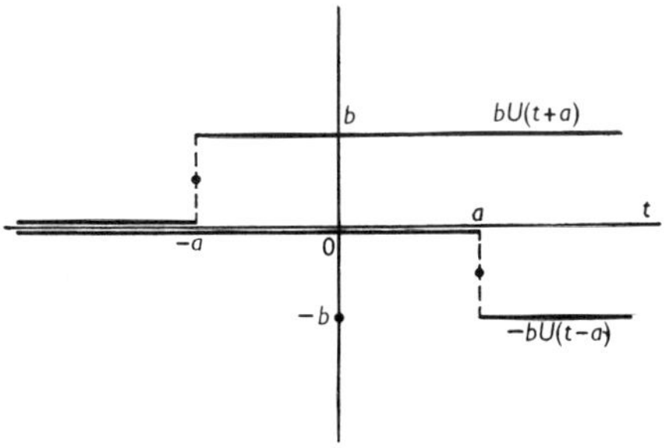

Fig. 14. Splitting of the rectangle function into two unit functions.

which is also found by direct application of the definition integral, though in the last way the strip of convergence appears to be wider, namely, $-\infty < \operatorname{Re} p < \infty$. This example shows that by addition of two images a common pole of $f(p)/p$, which in this case lies at $p = 0$, may cancel out, so as to lead to a strip of convergence wider than that expected from the sum rule.

Example 2. The shift rule provides us with a method of transposing one-sided periodic functions which originate by periodic continuation of a given function defined in some finite interval. For instance, let $h(t)$ be given in the range $0<t<1$ (see fig. 15a), and suppose its image $f(p)$ to be known. It is understood that $h(t)$ is replaced by zero outside the mentioned interval; in other words, we assume the relation

$$h(t)\{U(t)-U(t-1)\}\fallingdotseq f(p),$$

which will be valid for all p since its Laplace integral can be taken over the finite range of integration $0\leqslant t\leqslant 1$. Replacing t by $t-1$ we obtain the shifted function of fig. 15b. According to the shift rule the new function satisfies the relation

$$h(t-1)\{U(t-1)-U(t-2)\}\fallingdotseq e^{-p}f(p).$$

Proceeding in the same manner, by repeated applications of the shift rule, we displace $h(t)$ to subsequent intervals. After addition of the resulting countable set of functions we obtain the 'periodic' function shown in fig. 15c. Obviously, this function can be written as $h(t-[t])\,U(t)$, if $[t]$ denotes the greatest integer smaller than t (for integer values of t it is suitable to define h as the mean value). The image of this function becomes

$$h(t-[t])\,U(t)\fallingdotseq f(p) + e^{-p}f(p)+e^{-2p}f(p)+\dots,$$

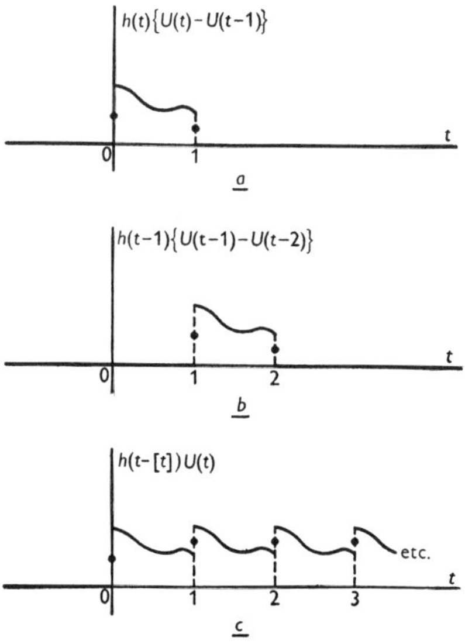

Fig. 15. Periodic continuation of a function given for $0<t<1$.

provided the series above converges; actual convergence obtains for $\operatorname{Re}p>0$, thus

$$h(t-[t])\,U(t)\fallingdotseq\frac{f(p)}{1-e^{-p}},\quad 0<\operatorname{Re}p<\infty. \tag{6}$$

With the aid of the similarity rule one can equally well deal with periodic functions of period different from unity. Of equal simplicity is the transposition of *step functions*; these functions are sectionally constant and can therefore be built up by suitably displaced unit functions. Step functions are discussed in some detail in chapter XII; we shall here give only a simple example.

The one-sided function $$\frac{\sin t}{|\sin t|}\,U(t)$$

is equal to $+1$ in the intervals $0<t<\pi$, $2\pi<t<3\pi$, etc., and equal to -1 in the remainder $\pi<t<2\pi$, $3\pi<t<4\pi$, etc. At the points of discontinuity it assumes the mean values of $\frac{1}{2}$ at $t=0$ and zero at the other jumps (see fig. 16). For positive values of t this discontinuous function has many features in common with the trigonometric $\sin t$; it is therefore usually called the square-sine $\mathfrak{Sin}\,t$. The corresponding image can be obtained either directly from the definition integral or by application of (6), if in the latter case it is remembered that, for $t>2\pi$, $\mathfrak{Sin}\,t$ is the periodic continuation of the function

$$U(t)-2U(t-\pi)+U(t-2\pi)\fallingdotseq 1-2e^{-\pi p}+e^{-2\pi p}=(1-e^{-\pi p})^2.$$

The result is easily shown to be

$$\mathfrak{Sin}\, t\, U(t) \doteqdot \tanh\left(\frac{\pi p}{2}\right), \quad 0 < \operatorname{Re} p < \infty. \tag{7}$$

This may serve as an example of an *analytic* image function obtained from an original having an infinite number of discontinuities.

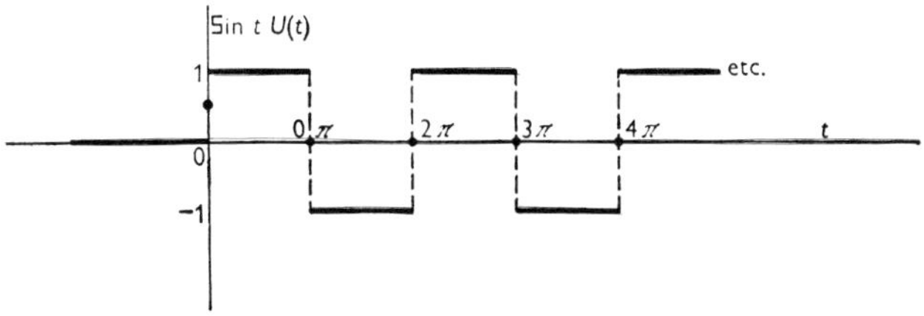

Fig. 16. The square-sine function.

Example 3. Another application of the shift rule will be studied now. It leads to a well-known transformation formula for series, originally due to Euler, by means of brief operational reasoning.

Let $h(t) \doteqdot f(p)$ for $\operatorname{Re} p < 0$, then

$$h(t) - h(t+1) + h(t+2) - h(t+3) + \ldots \doteqdot f(p)\,(1 - e^p + e^{2p} - e^{3p} + \ldots) = \frac{f(p)}{e^p + 1}.$$

Expanding the image as follows $\displaystyle f(p) \sum_{n=0}^{\infty} \frac{(1-e^p)^n}{2^{n+1}}$,

and then transposing this result term by term, we deduce the identity

$$h(t) - h(t+1) + h(t+2) - h(t+3) + \ldots$$
$$= \frac{h(t)}{2} + \frac{h(t) - h(t+1)}{2^2} + \frac{h(t) - 2h(t+1) + h(t+2)}{2^3} + \ldots.$$

Finally, if t is taken equal to zero, and $h(n)$ replaced by a_n, we arrive at Euler's formula:

$$a_0 - a_1 + a_2 - a_3 + \ldots = \frac{a_0}{2} + \frac{a_0 - a_1}{2^2} + \frac{a_0 - 2a_1 + a_2}{2^3} + \ldots = \sum_{n=0}^{\infty} \frac{(-\Delta)^n a_0}{2^{n+1}}.$$

In this formula the asymmetrical difference operator Δ is defined by

$$\Delta a_k = a_{k+1} - a_k,$$

which will be of frequent use in later investigations.

Example 4 (*Möbius inversion*). A slightly more complicated application of the shift rule leads to the determination of $h(x)$ from the expression

$$H(x) = \sum_{n=1}^{\infty} h\left(\frac{x}{n}\right) \quad (x > 0).$$

At first sight it seems reasonable to use here the similarity rule with respect to $x = t$. But it is usually much more simple first to reduce the division of x by n to a shift in the variable t defined by $x = e^t$, and next apply the shift rule; in other words, let us start from

$$H(e^t) = \sum_{n=1}^{\infty} h(e^{t - \log n}).$$

Furthermore, let $$H(e^t) \risingdotseq F(p), \quad h(e^t) \risingdotseq f(p), \tag{8}$$

and assume the existence of a common strip of convergence. Then the shift rule yields

$$F(p) = f(p) \sum_{n=1}^{\infty} \frac{1}{n^p}.$$

The series above (being convergent for $\operatorname{Re} p > 1$) is equal to Riemann's ζ-function $\zeta(p)$; therefore a simple division,

$$f(p) = \frac{F(p)}{\zeta(p)}, \tag{9}$$

gives the image of the still unknown original $h(t)$. An explicit expression could be obtained by means of the inversion integral (if convergent). On the other hand, however, we have still a different method. First we have

$$\frac{1}{\zeta(p)} = \sum_{n=1}^{\infty} \frac{\mu(n)}{n^p} = 1 - \frac{1}{2^p} - \frac{1}{3^p} - \frac{1}{5^p} + \frac{1}{6^p} - \frac{1}{7^p} + \frac{1}{10^p} - \cdots \quad (\operatorname{Re} p > 1), \tag{10}$$

in which $\mu(n)$ denotes the *Möbius function* of number theory ($\mu(n) = 0$ for any n containing a square factor; $\mu(n) = +1$ or -1 according as the number of different prime factors of n is even or odd; $\mu(1) = 1$). Then (9) can be written as

$$f(p) = \sum_{n=1}^{\infty} \mu(n) \frac{F(p)}{n^p} = \sum_{n=1}^{\infty} \mu(n) F(p) e^{-p \log n}.$$

Secondly, from (8), applying the shift rule once more,

$$h(e^t) = \sum_{n=1}^{\infty} \mu(n) H\left(\frac{e^t}{n}\right).$$

Replacing e^t by the old variable x, we then obtain finally the following set of corresponding equations:

$$H(x) = \sum_{n=1}^{\infty} h\left(\frac{x}{n}\right), \quad h(x) = \sum_{n=1}^{\infty} \mu(n) H\left(\frac{x}{n}\right). \tag{11}$$

An example such as that given above indicates clearly the heuristic importance of the operational calculus in solving rapidly quite an intricate problem. As to our derivation, it is sufficient to assume that $h(e^t)$ and $H(e^t)$ have a common strip of convergence inside the domain $\operatorname{Re} p > 1$. Once the solution of a problem is found by an heuristic method, it may afterwards be verified in a different and (if necessary) more rigorous manner. Thus (11) can also be proved if one starts from the basic identity

$$\sum_{d/k} \mu(d) = 0 \qquad (k > 1).$$

A rigorous proof along these lines requires the legitimacy of changing the order of summation in a double series.

4. The attenuation rule

Whereas the shift rule indicates the change of the image function when a constant λ is added to the argument t of the *original*, we shall now discuss the change caused by the addition of a constant λ to the argument p of the *image*. The corresponding new rule reads:

Given $$h(t) \risingdotseq f(p), \quad \alpha < \operatorname{Re} p < \beta,$$

then it follows

$$e^{-\lambda t} h(t) \doteqdot \frac{p}{p+\lambda} f(p+\lambda), \quad \alpha - \operatorname{Re}\lambda < \operatorname{Re} p < \beta - \operatorname{Re}\lambda. \tag{12}$$

In contradistinction to the case of § 3 the strip of convergence has now changed; moreover, the constant λ will in general be complex since $f(p)$ is essentially a function of the complex variable p. Since $f(p)$ transforms into $f(p+\lambda)$ and therefore $p+\lambda$ takes the place of p, it is obvious that the new strip of convergence is as indicated.

The proof of (12) is easy; it follows from the definition integral:

$$e^{-\lambda t} h(t) \doteqdot p \int_{-\infty}^{\infty} e^{-(p+\lambda)t} h(t)\, dt = \frac{p}{p+\lambda} (p+\lambda) \int_{-\infty}^{\infty} e^{-(p+\lambda)t} h(t)\, dt.$$

The right-hand member is actually that of (12) in view of the definition integral of $f(p+\lambda)$, just in the strip of convergence mentioned.

This rule is called the attenuation rule because it is often used in physical applications to describe attenuated vibrations if t denotes time and λ is positive; if $\lambda < 0$ the factor $e^{+\lambda t}$ is characteristic for a vibration continually increasing in amplitude, the attenuation then being negative.

Example. The operational relation (III, 8) for the exponential function is easily obtained by means of the attenuation rule; if one starts with

$$U(t) \doteqdot 1 \quad (\operatorname{Re} p > 0),$$

then at once

$$e^{-\alpha t} U(t) \doteqdot \frac{p}{p+\alpha}, \quad -\operatorname{Re}\alpha < \operatorname{Re} p < \infty.$$

5. The composition product

We shall next discuss one of the most important rules of the operational calculus, which will be useful in the determination of the original corresponding to the product of two images whose separate originals are known. Its formulation in case of two-sided Laplace integrals reads:

Given the two operational relations

$$f_1(p) \risingdotseq h_1(t), \quad \alpha_1 < \operatorname{Re} p < \beta_1,$$

$$f_2(p) \risingdotseq h_2(t), \quad \alpha_2 < \operatorname{Re} p < \beta_2,$$

then it follows that

$$\frac{1}{p} f_1(p) f_2(p) \risingdotseq \int_{-\infty}^{\infty} h_1(\tau)\, h_2(t-\tau)\, d\tau \tag{13}$$

is valid in the common strip of convergence, provided that the latter exists and the corresponding definition integral is convergent. Furthermore, the common strip of convergence is then as specified below:

$$\max(\alpha_1, \alpha_2) < \operatorname{Re} p < \min(\beta_1, \beta_2).$$

The right-hand side of (13) is called the *composition product* or *convolution* (in German: *Faltung*) of the functions $h_1(t)$ and $h_2(t)$. It has already been applied by Duhamel, Hopkinson, Rayleigh, Borel, Lévy, Vito Volterra, and others. We would remark that, in our opinion, Heaviside himself was not acquainted with the composition product, since there seems no reference to it either in his papers or his unpublished manuscripts†.

The relation (13) can be deduced by first replacing one of the functions $f(p)$, say f_1, in the left of (13) by its definition integral; thus

$$\frac{1}{p}f_1(p)f_2(p) = f_2(p)\int_{-\infty}^{\infty} e^{-p\tau}h_1(\tau)\,d\tau = \int_{-\infty}^{\infty} e^{-p\tau}f_2(p).h_1(\tau)\,d\tau, \qquad \alpha_1 < \operatorname{Re} p < \beta_1, \quad (14)$$

and then determining the original corresponding to the right-hand side of (14). The last can be accomplished readily, since the integrand, involving the non-operational parameter τ, can be transposed with the aid of the shift rule

$$e^{-p\tau}f_2(p).h_1(\tau)\,d\tau \risingdotseq h_2(t-\tau).h_1(\tau)\,d\tau \quad (\alpha_2 < \operatorname{Re} p < \beta_2).$$

The validity of the composition-product rule is proved if it is legitimate to integrate the above relation with respect to τ from $-\infty$ to $+\infty$. In general, the legitimacy of integrating an operational relation like

$$f(p,\tau) \risingdotseq h(t,\tau)$$

is in effect that of changing the order of integration in corresponding repeated integrals:

$$\int_{-\infty}^{\infty} d\tau \int_{-\infty}^{\infty} dt\, e^{-pt}h(t,\tau) = \int_{-\infty}^{\infty} dt\, e^{-pt}\int_{-\infty}^{\infty} d\tau\, h(t,\tau).$$

It can be stated that in almost all practical applications of the operational calculus these transformations are actually admissible. For, in the first place, (13) holds‡ under the conditions:

> $h_1(t)$, $h_2(t)$ and the composition product of h_1 and h_2 have convergent Laplace integrals in a common strip of convergence,

and secondly, it holds § under the conditions:

> $h_1(t)$ and $h_2(t)$ have absolutely convergent Laplace integrals in a common strip of convergence.

In the latter case the composition product of h_1 and h_2 exists, and is equal to the original of $\frac{1}{p}f_1(p)f_2(p)$. In chapter VI it is indicated that (13) holds everywhere inside the common strip of convergence of h_1 and h_2 provided the integrals are taken as Cesàro limits.

† In the library of the Institution of Electrical Engineers, London.

‡ For a proof concerning one-sided originals, see G. Doetsch, *Theorie und Anwendung der Laplace-Transformation*, Berlin, 1939, and New York, 1943, p. 163.

§ Doetsch, loc. cit. p. 162; for one-sided originals even the absolute convergence of one of the two integrals is already sufficient; cf. Doetsch, loc. cit. p. 165.

Returning now to the form of the rule (13), one observes that it is symmetric with respect to the functions $h_1(t)$ and $h_2(t)$, though not explicitly for the right-hand side of expression (13). Actually, however, neither of the functions h_1, h_2 plays a special role, since by a change of integration variable $(\tau = t-\tau')$ the right-hand side of (13) is transformed into its analogue

$$\int_{-\infty}^{\infty} h_1(t-\tau')\, h_2(\tau')\, d\tau'.$$

It is still of some importance to discuss the composition product from the point of view of one-sided integrals. To this end, let

$$f_1(p) \doteqdot h_1(t)\, U(t) \quad (\operatorname{Re} p > \alpha_1),$$

$$f_2(p) \doteqdot h_2(t)\, U(t) \quad (\operatorname{Re} p > \alpha_2),$$

and, therefore, on applying the rule (13),

$$\frac{1}{p} f_1(p) f_2(p) \doteqdot \int_{-\infty}^{\infty} h_1(\tau)\, U(\tau)\, h_2(t-\tau)\, U(t-\tau)\, d\tau.$$

On account of the factor $U(\tau)$ the lower limit of integration can be replaced by zero; the further effect of $U(t-\tau)$ is to make the right-hand side equal to zero for negative values of t, whilst for positive values of t the upper limit of integration may be changed into t. Consequently, the rule of the composition product now reads:

Given

$$f_1(p) \doteqdot h_1(t)\, U(t), \quad \alpha_1 < \operatorname{Re} p < \infty,$$

$$f_2(p) \doteqdot h_2(t)\, U(t), \quad \alpha_2 < \operatorname{Re} p < \infty,$$

then it follows that

$$\frac{1}{p} f_1(p) f_2(p) \doteqdot U(t) \int_0^t h_1(\tau)\, h_2(t-\tau)\, d\tau, \quad \max(\alpha_1, \alpha_2) < \operatorname{Re} p < \infty. \tag{15}$$

Example 1. *Euler's integral of the first kind.* As an exercise, some properties of Γ-functions will now be derived.

Let us first consider (III, 3) for two different values μ and ν of the parameter,

$$\frac{1}{p^\mu} \doteqdot \frac{t^\mu}{\Pi(\mu)}\, U(t), \quad \operatorname{Re} p > 0 \ (\operatorname{Re} \mu > -1),$$

$$\frac{1}{p^\nu} \doteqdot \frac{t^\nu}{\Pi(\nu)}\, U(t), \quad \operatorname{Re} p > 0 \ (\operatorname{Re} \nu > -1).$$

The composition-product rule (15) for one-sided functions then yields

$$\frac{1}{p}\frac{1}{p^\mu}\frac{1}{p^\nu} = \frac{1}{p^{\mu+\nu+1}} \doteqdot U(t) \int_0 \frac{\tau^\mu}{\Pi(\mu)} \frac{(t-\tau)^\nu}{\Pi(\nu)}\, d\tau.$$

Now the original of the left-hand member can easily be written down, leading to the equality

$$\frac{t^{\mu+\nu+1}}{\Pi(\mu+\nu+1)}U(t) = U(t)\int_0^t \frac{\tau^\mu}{\Pi(\mu)}\frac{(t-\tau)^\nu}{\Pi(\nu)}d\tau.$$

Suppressing the common factor $U(t)$ for $t>0$, and substituting $\tau = ts$, we obtain the following identity:

$$\int_0^1 s^\mu(1-s)^\nu ds = \frac{\Pi(\mu)\,\Pi(\nu)}{\Pi(\mu+\nu+1)} \quad (\operatorname{Re}\mu > -1;\ \operatorname{Re}\nu > -1). \tag{16}$$

The integral in (16) is the well-known first integral of Euler, which is usually denoted by $B(\mu+1, \nu+1)$, the beta integral.

A second property of the Γ-function concerns the product $\Pi(p)\,\Pi(-p)$. In order to apply the rule of the composition product one would probably try to start from (III, 16) together with the similar relation for $\Pi(-p)$, namely,

$$\Pi(p) \risingdotseq e^{-e^{-t}} \quad (\operatorname{Re}p>0),$$

$$\Pi(-p) \risingdotseq -e^{-e^{t}} \quad (\operatorname{Re}p<0).$$

A composition product cannot be formed directly, since the relations do not refer to a common strip of convergence. This, however, can easily be effected by first applying the attenuation rule to the first relation

$$\frac{p}{p+1}\Pi(p+1) = p\Pi(p) \risingdotseq e^{-t}e^{-e^{-t}} \quad (\operatorname{Re}p> -1).$$

If then the corresponding relation (13) is written down, and in the integral the substitution $s = e^{-\tau}$ is performed, we arrive at the following result:

$$\Pi(p)\,\Pi(-p) \risingdotseq -\frac{1}{e^t+1} \quad (-1<\operatorname{Re}p<0).$$

The image of the right-hand side is already known from (III, 12). After identification of the different forms of the image, we finally get

$$\Pi(p)\,\Pi(-p) = \frac{\pi p}{\sin(\pi p)}. \tag{17}$$

Though actually proved for $-1<\operatorname{Re}p<0$, it is valid for all values of p, on the principle of analytic continuation.

As a third property of the Γ-function, the *duplication formula* will now be discussed. This formula states a relationship between $\Pi(p)$, $\Pi(p-\frac{1}{2})$ and $\Pi(2p)$, and can be deduced with the aid of the composition-product rule applied to the originals of $\Pi(p)$ and $\{p/(p+\frac{1}{2})\}\,\Pi(p+\frac{1}{2}) = p\Pi(p-\frac{1}{2})$, the original of the latter image being found from that of the former by means of the attenuation rule. Performing the substitution $s = e^{\frac{1}{2}\tau-\frac{1}{4}t}$ in the resulting composition-product integral we obtain

$$\Pi(p)\,\Pi(p-\tfrac{1}{2}) \risingdotseq 2e^{-\frac{1}{4}t}\int_0^\infty \exp\left\{-e^{-\frac{1}{2}t}\left(s^2+\frac{1}{s^2}\right)\right\}ds \quad (\operatorname{Re}p>0).$$

Further, we can take half the sum of this expression and that obtained from it by replacing s by s^{-1}. A second substitution $u = e^{-\frac{1}{4}t}\left(s-\dfrac{1}{s}\right)$ then leads to

$$\Pi(p)\,\Pi(p-\tfrac{1}{2}) \risingdotseq \sqrt{\pi}\exp\{-2e^{-\frac{1}{4}t}\} \quad (\operatorname{Re}p>0),$$

in which the right-hand member, after the shift rule and the similarity rule have been applied to (III, 16), turns out to be the original of

$$\frac{\sqrt{\pi}}{4^p}\Pi(2p).$$

Consequently, for $\operatorname{Re} p > 0$, we have the required duplication formula

$$\Pi(p)\,\Pi(p-\tfrac{1}{2}) = \frac{\sqrt{\pi}}{4^p}\Pi(2p), \tag{18}$$

which also holds for general complex values of p, on account of the analytic character of both sides.

Example 2. For another application of the composition-product rule, let us return to the cylinder function K defined in (III, 29) and take the composition product of (III, 30) and its analogue obtained when a is replaced by b. It is not difficult to transform the resulting composition product into a new integral of exactly the type (III, 29); thus we are led to the operational relation

$$2pK_p(a)\,K_p(b) \doteq K_0\{\sqrt{(a^2+b^2+2ab\cosh t)}\}, \quad -\infty < \operatorname{Re} p < \infty \quad (\operatorname{Re} a > 0;\ \operatorname{Re} b > 0). \tag{19}$$

6. Repeated composition product

The preceding composition-product rule (13) contains the product of only two images. It is equally possible to form a composition product involving three images, say f_1, f_2, f_3, if one starts with the relations

$$f_1(p) \doteq h_1(t), \quad \alpha_1 < \operatorname{Re} p < \beta_1,$$
$$f_2(p) \doteq h_2(t), \quad \alpha_2 < \operatorname{Re} p < \beta_2,$$
$$f_3(p) \doteq h_3(t), \quad \alpha_3 < \operatorname{Re} p < \beta_3.$$

To this end we first write down the composition product of h_1 and h_2 and its corresponding image, τ_2 denoting the variable of integration:

$$\frac{f_1(p)f_2(p)}{p} \doteq \int_{-\infty}^{\infty} h_1(t-\tau_2)\,h_2(\tau_2)\,d\tau_2, \quad \max(\alpha_1,\alpha_2) < \operatorname{Re} p < \min(\beta_1,\beta_2). \tag{20}$$

Introducing the abbreviation $\phi(t)$ for the integral in (20), we have for the composition product of $\phi(t)$ and $h_3(t)$:

$$\frac{1}{p}\frac{f_1(p)f_2(p)}{p}f_3(p) \doteq \int_{-\infty}^{\infty} h_3(\tau_3)\,\phi(t-\tau_3)\,d\tau_3 = \int_{-\infty}^{\infty} h_3(t-\tau_3)\,\phi(\tau_3)\,d\tau_3,$$
$$\max(\alpha_1,\alpha_2,\alpha_3) < \operatorname{Re} p < \min(\beta_1,\beta_2,\beta_3).$$

The integral that contains $\phi(t-\tau_3)$ is the more useful; if $\phi(t-\tau_3)$ is replaced by its value according to (20), then the original can be written as a double integral, viz.

$$\frac{f_1(p)f_2(p)f_3(p)}{p^2} \doteq \int_{-\infty}^{\infty} d\tau_2 \int_{-\infty}^{\infty} d\tau_3\, h_1(t-\tau_2-\tau_3)\,h_2(\tau_2)\,h_3(\tau_3),$$
$$\max(\alpha_1,\alpha_2,\alpha_3) < \operatorname{Re} p < \min(\beta_1,\beta_2,\beta_3).$$

The above procedure can obviously be repeated indefinitely, thus leading to a repeated-composition-product rule involving a $(n-1)$-uple integral as original of a product of n images:

$$\frac{f_1(p)f_2(p)\dots f_n(p)}{p^{n-1}}$$
$$\doteqdot \int_{-\infty}^{\infty} d\tau_2 \int_{-\infty}^{\infty} d\tau_3 \dots \int_{-\infty}^{\infty} d\tau_n\, h_1(t-\tau_2-\tau_3-\dots-\tau_n)\, h_2(\tau_2)\dots h_n(\tau_n),$$
$$\max(\alpha_1, \alpha_2, \dots, \alpha_n) < \operatorname{Re} p < \min(\beta_1, \beta_2, \dots, \beta_n). \quad (21)$$

As clearly indicated by the specification of admissible values of $\operatorname{Re} p$, the repeated-composition product is significant only if the separate originals involved have a common strip of convergence; in this case (21) itself also holds.

Another form of the rule (21) is very interesting, both with respect to mathematical elegance and geometrical interpretation of the repeated composition product. In order to get as many factors p in the denominator as there are images in the numerator of the left-hand side of (21), we first take $n+1$ functions instead of n, and afterwards specialize $f_1(p)$, so as to make it equal to 1; thus $h_1(t) = U(t)$, $\alpha_1 = 0$, $\beta_1 = \infty$. Lowering the suffixes of all the remaining functions by one, we finally obtain the symmetrical form

$$\frac{f_1(p)f_2(p)\dots f_n(p)}{p^{n}}$$
$$\doteqdot \int_{-\infty}^{\infty} d\tau_1 \int_{-\infty}^{\infty} d\tau_2 \dots \int_{-\infty}^{\infty} d\tau_n\, U(t-\tau_1-\tau_2-\dots-\tau_n)\, h_1(\tau_1)\dots h_n(\tau_n),$$
$$\max(0, \alpha_1, \alpha_2, \dots, \alpha_n) < \operatorname{Re} p < \min(\beta_1, \beta_2, \dots, \beta_n). \quad (22)$$

This can still be shortened somewhat; the unit-function factor in the integrand indicates that actually the integration extends over that part of the n-dimensional space ($\tau_1, \tau_2, \dots, \tau_n$ refer to an orthogonal cartesian co-ordinate system) which is subject to

$$\tau_1 + \tau_2 + \dots + \tau_n < t.$$

This means that the integration has to be carried out over the part of space that lies on one side of the hyperplane

$$\tau_1 + \tau_2 + \dots + \tau_n = t. \quad (22a)$$

Summarizing, we thus have the rule

$$\frac{f_1(p)f_2(p)\dots f_n(p)}{p^{n}} \doteqdot \underset{\tau_1+\tau_2+\dots+\tau_n<t}{\int d\tau_1 \int d\tau_2 \dots \int} d\tau_n\, h_1(\tau_1)\, h_2(\tau_2)\dots h_n(\tau_n),$$
$$\max(\alpha_1, \alpha_2, \dots, \alpha_n, 0) < \operatorname{Re} p < \min(\beta_1, \beta_2, \dots, \beta_n), \quad (23)$$

in which, if required, the limits of integration can also be written as indicated below:

$$\int_{-\infty}^{t} d\tau_1 \int_{-\infty}^{t-\tau_1} d\tau_2 \int_{-\infty}^{t-\tau_1-\tau_2} d\tau_3 \ldots \int_{-\infty}^{t-\tau_1-\tau_2-\ldots-\tau_{n-1}} d\tau_n\, h_1(\tau_1)\, h_2(\tau_2) \ldots h_n(\tau_n).$$

In general the domain of integration in (23) extends to infinity in all directions on one side of the hyperplane (22 *a*). This domain becomes automatically finite if all the originals involved are one-sided functions, making the integrand non-vanishing for positive values of all variables τ. These are the reasons why the repeated composition product can serve a useful purpose in the computation of n-dimensional volumes, an example of which will be given first.

Example 1. The calculation of the volume of an n-dimensional sphere. Let $x_1, \ldots, x_n$ refer to an orthogonal cartesian co-ordinate system in n-dimensional space. The volume $V_n(R)$ of the corresponding hypersphere (radius R) is determined by

$$\int dx_1 \int dx_2 \ldots \int dx_n,$$

where the integration is restricted to the domain

$$x_1^2 + x_2^2 + \ldots + x_n^2 < R^2.$$

On account of the symmetry with respect to the centre of the sphere we may also write

$$V_n(R) = 2^n \overbrace{\int_0 dx_1 \int_0 dx_2 \ldots \int_0 dx_n}^{x_1^2+x_2^2+\ldots+x_n^2<R^2},$$

or, after a change of variables, $x_k = \sqrt{\tau_k}$:

$$V_n(R) = \overbrace{\int_0 d\tau_1 \int_0 d\tau_2 \ldots \int_0 d\tau_n}^{\tau_1+\tau_2+\ldots+\tau_n<R^2} \frac{1}{\sqrt{(\tau_1 \tau_2 \ldots \tau_n)}}.$$

This is nothing but the repeated composition product (23) of n ~~mutually~~ identical functions:

$$h_1(t) = h_2(t) = \ldots = h_n(t) = \frac{U(t)}{\sqrt{t}},$$

provided R^2 be identified with t. Since any of these functions has the image

$$\Pi(-\tfrac{1}{2})\sqrt{p} = \sqrt{(\pi p)} \quad (\operatorname{Re} p > 0),$$

we infer from (23) that $$V_n(\sqrt{t}) \risingdotseq \frac{\{\sqrt{(\pi p)}\}^n}{p^n} = \left(\frac{\pi}{p}\right)^{\frac{1}{2}n}.$$

But, from the known original of this p-function, we have further

$$V_n(\sqrt{t}) = \frac{(\pi t)^{\frac{1}{2}n}}{\Pi(\frac{1}{2}n)}\, U(t),$$

and, since $t = R^2 > 0$, we obtain finally

$$V_n(R) = \frac{\pi^{\frac{1}{2}n}}{\Pi(\frac{1}{2}n)} R^n. \tag{24}$$

In particular, for $n = 1, 2, 3, 4$, we have the well-known formulae

$V_1 = 2R$ (length of a line segment extending from $-R$ to R),

$V_2 = \pi R^2$ (area of a circle),

$V_3 = \frac{4}{3}\pi R^3$ (volume of a sphere),

$V_4 = \frac{1}{2}\pi^2 R^4$ (volume of a four-dimensional sphere).

Example 2. We shall next discuss an example in which the repeated composition product involves an infinite number of integrations; it is linked with Gauss's infinite product for the Γ-function.

Consider the image of $(1-e^{-t})^\nu U(t)$. The corresponding definition integral can be transformed into Euler's integral of the first kind (16) by means of the substitution $s = e^{-t}$:

$$(1-e^{-t})^\nu U(t) \doteqdot \frac{\Pi(\nu)\,\Pi(p)}{\Pi(p+\nu)}, \quad 0 < \operatorname{Re} p < \infty \ (\operatorname{Re}\nu > -1). \tag{25}$$

In particular, if ν is any positive integer n, (25) becomes

$$(1-e^{-t})^n U(t) \doteqdot \frac{n!}{(p+1)(p+2)\dots(p+n)}, \quad 0 < \operatorname{Re} p < \infty. \tag{26}$$

Replacing t by $t + \log n$, and applying the shift rule we obtain

$$\left(1-\frac{e^{-t}}{n}\right)^n U(t+\log n) \doteqdot \frac{n!\,n^p}{(p+1)(p+2)\dots(p+n)} \quad (\operatorname{Re} p > 0). \tag{27}$$

Let n tend to infinity. Then the original in (27) reduces to the function $e^{-e^{-t}}$ (for all values of t) of which the image for $\operatorname{Re} p > 0$ is $\Pi(p)$. We therefore obtain the limit

$$\Pi(p) = \operatorname*{Lim}_{n\to\infty} \frac{n!\,n^p}{(p+1)(p+2)\dots(p+n)}. \tag{28}$$

This is the well-known infinite product of Gauss, which is derived here in a simple manner, though not at all rigorously.

Let us next return to (27) for finite values of n. The right-hand side can be written as

$$n!\,e^{p\log n}\frac{1}{p^n}\left(\frac{p}{p+1}\right)\left(\frac{p}{p+2}\right)\dots\left(\frac{p}{p+n}\right).$$

It follows from (23) that the left-hand side of (27) is identical with a repeated composition product, viz.

$$\left(1-\frac{e^{-t}}{n}\right)^n U(t+\log n) = n!\int_0 d\tau_1\int_0 d\tau_2\dots\int_0^{\tau_1+\tau_2+\dots+\tau_n<t+\log n} d\tau_n \exp\{-(\tau_1+2\tau_2+\dots+n\tau_n)\}.$$

This formula holds for any positive integral n; if n tends to infinity we arrive at a repeated composition product involving an infinite number of integrals, viz.

$$e^{-e^{-t}} = \operatorname*{Lim}_{n\to\infty} n!\int_0 d\tau_1\int_0 d\tau_2\dots\int_0^{\tau_1+\tau_2+\dots+\tau_n<t+\log n} d\tau_n \exp\{-(\tau_1+2\tau_2+\dots+n\tau_n)\},$$

or, by the substitutions $e^{-t} = x$, $e^{\tau_k} = u_k$ $(k = 1, 2, \dots, n)$

$$e^{-x} = \operatorname*{Lim}_{n\to\infty} n!\int_1 du_1\int_1 du_2\dots\int_1^{u_1u_2\dots u_n<n/x} du_n \frac{1}{u_1^2u_2^3\dots u_n^{n+1}}.$$

7. The image of the product of two originals

As we have already stressed, there exists a definite relationship between the shift rule and the attenuation rule, inasmuch as they deal with the effect produced by the addition of a constant λ to the argument of, respectively, the original and the image. Similarly, to the composition-product rule, which determines the original of the product of two *images*, there corresponds another rule which will give the image of the product of two *originals*. This new rule reads as follows:

Let

$$h_1(t) \risingdotseq f_1(p), \quad \alpha_1 < \operatorname{Re} p < \beta_1,$$

$$h_2(t) \risingdotseq f_2(p), \quad \alpha_2 < \operatorname{Re} p < \beta_2,$$

then the following relation holds

$$h_1(t)\, h_2(t) \risingdotseq \frac{p}{2\pi i} \int_{c-i\infty}^{c+i\infty} \frac{f_1(s)}{s} \frac{f_2(p-s)}{p-s}\, ds, \quad \alpha_2 + c < \operatorname{Re} p < \beta_2 + c \; (\alpha_1 < c < \beta_1). \tag{29}$$

The derivation of the above formula is quite analogous to that of the composition product (13); the first factor in $h_1(t)\,h_2(t)$ is replaced by its inversion integral; then the variable of integration p is changed into s, leading to

$$h_1(t)\, h_2(t) = \frac{h_2(t)}{2\pi i} \int_{c-i\infty}^{c+i\infty} e^{st} \frac{f_1(s)}{s}\, ds \quad (\alpha_1 < c < \beta_1).$$

The image of the integrand can be given at once if the attenuation rule is applied (s is a parameter), namely,

$$\frac{1}{2\pi i} \frac{f_1(s)}{s} e^{st} h_2(t) \risingdotseq \frac{1}{2\pi i} \frac{f_1(s)}{s} \frac{p}{p-s} f_2(p-s), \quad \alpha_2 + \operatorname{Re} s < \operatorname{Re} p < \beta_2 + \operatorname{Re} s.$$

Finally, the rule (29) is obtained after an integration with respect to s along the line $\operatorname{Re} s = c$. Further details concerning a rigorous proof will be omitted here, since the rule in question will hardly be used in what follows. Yet it may be worth while to discuss one simple example, which leads to a relation for the Γ-function.

Example. Let us apply (29) to the operational relations

$$e^{-at} e^{-e^{-t}} \risingdotseq \frac{p}{p+a} \Pi(p+a) = p\Gamma(p+a) \quad (\operatorname{Re} p > -\operatorname{Re} a),$$

$$e^{-bt} e^{-e^{-t}} \risingdotseq p\Gamma(p+b) \quad (\operatorname{Re} p > -\operatorname{Re} b).$$

With the new rule the result is

$$e^{-(a+b)t} e^{-2e^{-t}} \risingdotseq \frac{p}{2\pi i} \int_{c-i\infty}^{c+i\infty} \Gamma(s+a)\, \Gamma(p-s+b)\, ds \quad (\operatorname{Re} p > c - \operatorname{Re} b;\; c > -\operatorname{Re} a).$$

There is a different way, however, of expressing the image of the left-hand side; it can be found directly with the aid of the shift rule and the attenuation rule. A mere identification of the different expressions for the image then leads to

$$\frac{\Gamma(p+a+b)}{2^{p+a+b}} = \frac{1}{2\pi i}\int_{c-i\infty}^{c+i\infty} \Gamma(s+a)\,\Gamma(p-s+b)\,ds, \quad -\operatorname{Re} a < c < \operatorname{Re}(p+b),$$

or, when taking $p = 0$,

$$\frac{\Gamma(a+b)}{2^{a+b-1}} = \frac{1}{\pi i}\int_{c-i\infty}^{c+i\infty} \Gamma(a+s)\,\Gamma(b-s)\,ds, \quad -\operatorname{Re} a < c < \operatorname{Re} b. \tag{30}$$

As to the strip of convergence in (30), it is obvious that the path of integration lies somewhere between the verticals through the poles, $s = -a$ and $s = b$, of the integrand.

In particular, for $a = b = 1$ we find

$$\frac{\pi i}{2} = \int_{c-i\infty}^{c+i\infty} \Pi(s)\,\Pi(-s)\,ds, \quad -1 < c < 1,$$

which, for $c = 0$, and using (17), can be transformed into the simple integral

$$\frac{1}{2} = \int_{-\infty}^{\infty} \frac{t}{\sinh(\pi t)}\,dt.$$

8. The differentiation rule

We are now going to discuss a set of rules that involve differentiation or integration, to be carried out in both original and image. The simplest and most extensively used is the *differentiation rule*, which we state in the following way:

Given
$$h(t) \doteqdot f(p), \quad \alpha < \operatorname{Re} p < \beta,$$

then it follows, inside the strip of convergence of $h'(t)$ (if this strip exists),

$$h'(t) \doteqdot p f(p). \tag{31}$$

Consequently, a differentiation of the original with respect to t corresponds to a multiplication of the image by a factor p; it should be stated that this simple and important rule is characteristic of the Laplace transform here chosen as basis of the operational calculus.

A proof of the differentiation rule may be based upon the shift rule, for we immediately find

$$\frac{h(t)-h(t-\epsilon)}{\epsilon} \doteqdot \frac{(1-\mathrm{e}^{-\epsilon p})}{\epsilon} f(p), \quad \alpha < \operatorname{Re} p < \beta.$$

Assuming that this operational relation remains valid in the limit for ϵ tending to zero, we at once obtain the differentiation rule. In general, however, it is not at all certain whether operational relations containing a parameter

$$h(t,\epsilon) \doteqdot f(p,\epsilon)$$

actually hold in the limit. This would require the justification of changing the order of the limit and integral signs:

$$\operatorname*{Lim}_{\epsilon\to 0} \int_{-\infty}^{\infty} \mathrm{e}^{-pt} h(t,\epsilon)\,dt = \int_{-\infty}^{\infty} \mathrm{e}^{-pt} \operatorname*{Lim}_{\epsilon\to 0} h(t,\epsilon)\,dt.$$

On the other hand, it is definitely possible that the strip of convergence of (31) is wholly different from that of $h(t) \doteqdot f(p)$. We shall give an example of all of the following cases:

(a) the strip of convergence does not change,

(b) the strip of convergence widens,

(c) the strip of convergence narrows.

(a) The relation (II, 24) is valid for any complex p. The same holds true for the operational relation obtained by differentiation, viz.

$$2t\,e^{-t^2} \doteqdot -\sqrt{\pi}\,p^2\,e^{\frac{1}{4}p^2}, \quad -\infty < \operatorname{Re} p < \infty.$$

Here the strip of convergence is not influenced.

(b) If the attenuation rule is applied to the relation (III, 16)

$$e^{-e^{-t}} \doteqdot \Pi(p), \quad 0 < \operatorname{Re} p < \infty,$$

then it follows that

$$e^{-t}e^{-e^{-t}} \doteqdot \frac{p}{p+1}\Pi(p+1) = p\Pi(p), \quad -1 < \operatorname{Re} p < \infty. \tag{32}$$

The new original could equally well be obtained by means of the differentiation rule. Therefore, this is an example in which the strip of convergence is wider than before; the new relation also holds in $-1 < \operatorname{Re} p < 0$.

(c) In addition to the operational relation

$$\sin(e^{-t}) \doteqdot \sin\left(\frac{\pi p}{2}\right)\Pi(p), \quad -1 < \operatorname{Re} p < 1,$$

which will be treated in chapter VI, we have the following:

$$e^{-t}\cos(e^{-t}) \doteqdot -p\sin\left(\frac{\pi p}{2}\right)\Pi(p), \quad -1 < \operatorname{Re} p < 0.$$

The latter can be obtained by differentiation of the former; notice the narrowing of the strip of convergence.

The above examples teach us that we should be very careful in handling the differentiation rule†.

The differentiation rule may also be investigated with the aid of a partial integration of the definition integral of $h'(t)$. We have

$$p\int_{-\infty}^{\infty} e^{-pt}h'(t)\,dt = p\int_{-\infty}^{\infty} e^{-pt}\,dh(t) = p\,e^{-pt}h(t)\Big|_{-\infty}^{\infty} + p\left\{p\int_{-\infty}^{\infty} e^{-pt}h(t)\,dt\right\}.$$

† In this connexion it may be remarked that theorem 2 of Doetsch, loc. cit. p. 154 is only valid under the proviso that s_2 is negative, a condition not stated by Doetsch.

Obviously the validity of the differentiation rule is connected with that of the limit

$$\lim_{t\to+\infty}\{e^{-pt}h(t)\}-\lim_{t\to-\infty}\{e^{-pt}h(t)\}=0.$$

It will be obvious that the differentiation rule can be applied repeatedly. Once again differentiating (31) leads to

$$\frac{d^2}{dt^2}h(t)\doteqdot p^2f(p),$$

and in general (n denoting any positive integer),

$$\frac{d^n}{dt^n}h(t)\doteqdot p^nf(p). \tag{33}$$

It is impossible, however, to indicate the corresponding strips of convergence.

The general rule (33) is closely connected with the Taylor-series development. For, if the right-hand member of the shift rule, $h(t+\lambda)\doteqdot e^{\lambda p}f(p)$, is expanded into powers of λ,

$$f(p)+\lambda pf(p)+\frac{\lambda^2}{2!}p^2f(p)+\frac{\lambda^3}{3!}p^3f(p)+\ldots,$$

and afterwards the original of each separate term is taken according to (33), then we get

$$h(t+\lambda)=h(t)+\lambda h'(t)+\frac{\lambda^2}{2!}h''(t)+\frac{\lambda^3}{3!}h'''(t)+\ldots.$$

Incidentally, it will be clear that the shift rule is more general than the Taylor series in the sense that, even when the Taylor-series expansion is impossible (e.g. in cases of discontinuous functions), yet the original $h(t+\lambda)$ of the image function $e^{\lambda p}f(p)$ may exist. As to the one-sided functions in the case of $f(p)\doteqdot h(t)\,U(t)$, provided both $h(t)$ and $h^{(n)}(t)$ are expansible into a Maclaurin series, we can also deduce the following relation by transposing the series term by term:

$$h^{(n)}(t)\,U(t)\doteqdot[[p^nf(p)]], \tag{33a}$$

where [[ ]] is the symbol already introduced in (4a).

Example. As a simple example showing how one can deduce new rules by combining rules already known we may mention

$$\frac{d^n}{d(-e^{-t})^n}h(t)=\left(e^t\frac{d}{dt}\right)^n h(t)\doteqdot p(p-1)(p-2)\ldots(p-n+1)f(p-n), \tag{33b}$$

which originates from the repeated process of using the differentiation and attenuation rules alternately.

ns for the original

$(\tau)(t-\tau)^{n-1}\,d\tau$

$(\operatorname{Re} p > 0)$. (35)

ay of extending
of n; we merely
uestion raised by
be attributed to

at an extension of
guments. To this
y. We then find

). (36)

al for the lower
t) it reads

n the definition

the Γ-function,
$^{p-1}$ and a sum
y to make the
is not new; it
t manners.

be discussed.

a; $\operatorname{Re}\nu > 1$).

(37)

ζ-function
the latter
umber t—
III, § 4).

f the original leads to multiplica-
ration rule amounts to a division
t of integration is properly taken
according as $\operatorname{Re} p > 0$ or $\operatorname{Re} p < 0$.
f two parts:

$$< \operatorname{Re} p < \beta,$$

$$\max(\alpha, 0) < \operatorname{Re} p < \beta, \quad (34a)$$

$$\alpha < \operatorname{Re} p < \min(\beta, 0). \quad (34b)$$

se formulae, we first take $f_1(p) = f(p)$
roduct rule (13). We must further take
the strip of convergence, such that we

$$U(t-\tau)\,d\tau = \int_{-\infty}^{t} h(\tau)\,d\tau.$$

hen according to (2) $h_2(t) = -U(-t)$, and

$$-t)\,d\tau = -\int_{t}^{\infty} h(\tau)\,d\tau = \int_{+\infty}^{t} h(\tau)\,d\tau.$$

specialization of the composition-product
t it is valid without restrictions, unlike the
stence was not always assured. In the above
le holds generally†.
s split up into two different parts it will be
l strip of convergence contains the imaginary
ain is divided into a pair of separate strips,
e and the same p-function represents the image
s.
plied repeatedly, leading to multiple integrals
ng to $\frac{1}{p^n} f(p)$. Instead of integrating n times we
osition-product rule to $\frac{1}{p}\frac{1}{p^{n-1}} f(p)$ in order to

idder, *The Laplace Transform*, Princeton, 1941, p. 239,
is taken in the sense of a Stieltjes integral.

derive the original more directly. The resulting expressio
in question are for $\operatorname{Re} p > 0$:

$$\frac{1}{p^n} f(p) \doteqdot \int_{-\infty}^{t} d\tau_1 \int_{-\infty}^{\tau_1} d\tau_2 \ldots \int_{-\infty}^{\tau_{n-1}} d\tau_n h(\tau_n) = \frac{1}{(n-1)!} \int_{-\infty}^{t}$$

Incidentally, the identity above indicates a possible
the definition of n-uple integration to non-integer values
replace $(n-1)!$ by $\Gamma(n)$ in (35). This is closely related to a q
Heaviside, amongst others, as to what significance may
derivatives and integrals of fractional order.

Example 1. The integration rule may be used in order to arrive
Euler's integral (III, 2) for the Γ-function in the case of negative a
end let us apply the rule to (32) for $\operatorname{Re} p > 0$ and $\operatorname{Re} p < 0$ separatel

$$\Pi(p) \doteqdot \begin{cases} \displaystyle\int_{-\infty}^{t} e^{-\tau} e^{-e^{-\tau}} d\tau = e^{-e^{-t}}, & 0 < \operatorname{Re} p < \infty, \\ \displaystyle\int_{+\infty}^{t} e^{-\tau} e^{-e^{-\tau}} d\tau = e^{-e^{-t}} - 1, & -1 < \operatorname{Re} p < \end{cases}$$

The upper expression is equal to (III, 16), but the definition integ
one leads to a new relation. By a proper change of variables $(s = e^{-}$

$$\Gamma(p) = \int_0^\infty s^{p-1}(e^{-s} - 1)\, ds, \quad -1 < \operatorname{Re} p < 0.$$

Multiplying (36) by e^{-t} once more, integrating, and then writing dow
integral, we obtain further

$$\Gamma(p) = \int_0^\infty s^{p-1}(e^{-s} - 1 + s)\, ds, \quad -2 < \operatorname{Re} p < -1.$$

This procedure can be continued; the result is always an integral for
valid in a p-strip of unit width, whilst the integrand is the product of
containing just as many terms of the power series of e^{-s} as are necessa
integral convergent in the corresponding strip. Of course, the result
has already been obtained by Saalschütz and Hermite in quite differen

Example 2. Another application, involving the ζ-function, will now
Performing the shift rule on (III, 38), we find

$$\frac{e^{-at} t^{\nu-1}}{1 - e^{-t}} U(t) \doteqdot \Gamma(\nu)\, p\zeta(\nu, p + a), \quad \operatorname{Re} p > -\operatorname{Re} a \ (\operatorname{Re} \nu > 1),$$

which after integration becomes

$$U(t) \int_0^t \frac{e^{-ax} x^{\nu-1}}{1 - e^{-x}} dx \doteqdot \Gamma(\nu)\, \zeta(\nu, p + a) = \int_0^\infty \frac{e^{-(p+a)x} x^{\nu-1}}{1 - e^{-x}} dx \quad (\operatorname{Re} p > -\operatorname{Re}$$

We have thus established an operational relation between a generalize
(as image) and its corresponding 'incomplete' ζ-function (as original)
being defined by replacing the upper limit of integration $+\infty$ by a finite
as in the case of the Γ-function and its corresponding Prym functions (cf.

10. Rules for multiplication by t^n

Whereas (33) shows that multiplication of the image by a positive integral power of p corresponds to differentiating the original a number of times, we have, conversely, that the multiplication of the original by t^n (n positive integral) corresponds to the differentiation of the image a number of times. The new rule in question reads as follows:

Given
$$h(t) \risingdotseq f(p), \quad \alpha < \operatorname{Re} p < \beta,$$

then we have
$$t^n h(t) \risingdotseq p\left(-\frac{d}{dp}\right)^n \left\{\frac{f(p)}{p}\right\}, \quad \alpha < \operatorname{Re} p < \beta. \tag{38}$$

The above rule is readily proved if the definition integral, written in the form

$$\frac{f(p)}{p} = \int_{-\infty}^{\infty} \mathrm{e}^{-pt} h(t)\, dt,$$

is differentiated n times successively (under the sign of integration) with respect to $-p$; after multiplication by p it is merely the definition integral for $t^n h(t)$. The differentiation is always allowed on account of $f(p)/p$ being analytic in the strip of convergence (cf. VI, § 2).

A combination of the rule (38) and the differentiation rule (provided the latter may be applied) leads to

$$\left(t\frac{d}{dt}\right)^n h(t) \risingdotseq \left(-p\frac{d}{dp}\right)^n f(p). \tag{39}$$

This rule contains image and original in a symmetrical way except for the factor $(-1)^n$; it can also be written in the equivalent form

$$\left(\frac{d}{d\log t}\right)^n h(t) \risingdotseq \left(-\frac{d}{d\log p}\right)^n f(p).$$

11. Rules for division by t, and related integrals

The last section of this chapter will be devoted to rules concerning the image of $h(t)/t$ and integrals of this function. We shall discuss in particular the simplest and most useful case, namely, that of one-sided originals. Let us assume that
$$f(p) \risingdotseq h(t)\, U(t), \quad \alpha < \operatorname{Re} p < \infty,$$

then in the same strip of convergence it follows that, provided the corresponding originals exist,

(1) for α negative

$$\int_0^p \frac{f(s)}{s}\, ds \risingdotseq U(t) \int_t^{\infty} \frac{h(\tau)}{\tau}\, d\tau, \quad \alpha < \operatorname{Re} p < \infty, \tag{40}$$

(2) for any real value of α,

if $\operatorname{Re} p > 0$: $$\int_p^\infty \frac{f(s)}{s}\,ds \doteq U(t)\int_0^t \frac{h(\tau)}{\tau}\,d\tau, \tag{41 a}$$

if $\operatorname{Re} p < 0$: $$\int_p^\infty \frac{f(s)}{s}\,ds \doteq \int_\infty^{\max(0,t)} \frac{h(\tau)}{\tau}\,d\tau, \tag{41 b}$$

and finally for any p: $$p\int_p^\infty \frac{f(s)}{s}\,ds \doteq \frac{h(t)}{t}\,U(t). \tag{42}$$

The above rules can be proved easily if the definition integral is applied in the form

$$\frac{f(s)}{s} = \int_0^\infty \mathrm{e}^{-su}\,h(u)\,du.$$

For instance, (40) is obtained by integrating the above identity with respect to s between 0 and p, and then changing the order of integration:

$$\int_0^p \frac{f(s)}{s}\,ds = \int_0^\infty du\,h(u)\int_0^p ds\,\mathrm{e}^{-su} = \int_0^\infty du\,\frac{h(u)}{u}\,(1-\mathrm{e}^{-pu}).$$

If one determines further the original of the integrand one is led to (40). We shall not attempt to construct a rigorous proof; the rule in question may be considered heuristic, producing new operational relations which eventually can be verified more rigorously. It should at the same time be borne in mind, however, that for some images the integral representing the original may diverge; an example of this peculiarity is provided by

$$t\sin t\,U(t) \doteq \frac{2p^2}{(p^2+1)^2} \quad (\operatorname{Re} p > 0),$$

for which (40) leads to the non-convergent integral $\int_t^\infty \sin\tau\,d\tau$.

The other rules, (41) and (42), can be proved in the same manner. It is to be noted here that the analogous rules formulated for two-sided functions are more complicated; this is partly due to the circumstance that the integral representing the original often converges only in the sense of a principal value referred to $t = 0$. It will suffice to give one simple example:

$$\int_0^p \frac{f(s)}{s}\,ds \doteq \int_t^{\infty\, t/|t|} \frac{h(\tau)}{\tau}\,d\tau, \quad \alpha < \operatorname{Re} p < \beta. \tag{43}$$

Example. With the aid of the rules discussed above we readily come to the images of the following one-sided transcendents:

$$\operatorname{Ci}(t) = \int_\infty^t \frac{\cos\tau}{\tau}\,d\tau \quad \text{(cosine integral)},$$

$$\operatorname{Si}(t) = \int_0^t \frac{\sin\tau}{\tau}\,d\tau \quad \text{(sine integral)},$$

$$\operatorname{Ei}(-t) = \operatorname{li}(\mathrm{e}^{-t}) = \int_\infty^t \frac{\mathrm{e}^{-\tau}}{\tau}\,d\tau \quad \text{(exponential or logarithmic integral)},$$

of which the last function has already been met in (III, 33). Applying (40) to (III, 21), we find

$$\mathrm{Ci}\,(t)\,U(t) \risingdotseq \log\frac{1}{\sqrt{(p^2+1)}}, \quad 0<\mathrm{Re}\,p<\infty. \tag{44}$$

Furthermore, by (III, 22) and (41 a),

$$\mathrm{Si}\,(t)\,U(t) \risingdotseq \operatorname{arc\,cot} p, \quad 0<\mathrm{Re}\,p<\infty \ (0<\mathrm{Re}\operatorname{arc\,cot} p<\tfrac{1}{2}\pi), \tag{45}$$

and finally, if (40) is applied to $h(t) = \mathrm{e}^{-t}U(t)$,

$$-\mathrm{Ei}\,(-t)\,U(t) \risingdotseq \log(p+1), \quad -1<\mathrm{Re}\,p<\infty, \tag{46}$$

which is the counterpart of (III, 34), where the exponential integral occurs in the image rather than in the original.

12. Rules for the treatment of correlation functions

As an example we consider the following function

$$C(t) = \int_{-\infty}^{\infty} h(s)\,h(t+s)\,ds = \int_{-\infty}^{\infty} h(-s)\,h(t-s)\,ds,$$

which, apart from a constant, defines an autocorrelation function. It constitutes the composition product of $h(-t)$ and $h(t)$. For functions cut off at both sides, the images of $h(t) \risingdotseq f(p)$ and $h(-t) \risingdotseq -f(-p)$ converge everywhere; hence, in view of (13),

$$C(t) \risingdotseq -\frac{1}{p}f(p)f(-p).$$

This relation expresses a fundamental property. Since $f(i\omega)/i\omega$ represents the Fourier transform of $h(t)$, the corresponding transform of $C(t)$ appears to be given by

$$\frac{f(i\omega)}{i\omega}\,\frac{f(-i\omega)}{-i\omega},$$

which equals the square of the modulus of the transform of $h(t)$ itself, provided $h(t)$ is real.

The simplest operational rules have been discussed now. We shall deal with some identities connected with them in VII § 6, and with more complicated rules in chapter XI.

CHAPTER V

THE DELTA OR IMPULSE FUNCTION

1. Introduction

This chapter is devoted to properties of the *δ-function*. This function $\delta(t)$ has some features in common with a discontinuous function, since '$\delta(t)$ is equal to zero for $t \neq 0$, and infinitely large at $t = 0$, making the integral $\int_{-\infty}^{\infty} \delta(t)\,dt$ equal to 1'. The δ-function, often called the *impulse function*, plays an important part in the operational calculus as may be inferred from the simplicity of its image, which is p. Whereas the physicist usually never hesitates to treat the δ-function as an ordinary function, the mathematician often denies its existence. One of the main problems, therefore, is to indicate how the physical δ-function can be interpreted in rigorous mathematical terms. As in other chapters, the general theory will be accompanied by specific examples and applications. Before starting the discussion of the δ-function itself, however, we shall deal briefly with properties of the unit function $U(t)$, already defined, in order to facilitate the introduction of the improper δ-function in question.

2. The unit function

This real function has already been often used in the foregoing chapters. In (II, 16) it was defined as follows:

$$U(t) = \begin{cases} 1 & (t > 0), \\ \frac{1}{2} & (t = 0), \\ 0 & (t < 0), \end{cases}$$

and its image was given by the relation (II, 21)

$$U(t) \doteqdot 1, \quad 0 < \operatorname{Re} p < \infty.$$

As emphasized before, this simple discontinuous function (see fig. 7) is of great importance in both mathematics and physics. For instance, the time dependence of an electromotive force of unit strength, switched on instantaneously at $t = 0$, is described by $U(t)$. The unit function was in use by Heaviside, though in a different notation; but even Heaviside was not the first, since Cauchy† knew the unit function in the definition

$$U(t) = \frac{1}{2}\left(1 + \frac{t}{\sqrt{t^2}}\right),$$

which was called by him '*coefficient limitateur*' or '*restricteur*'.

† *Encyklopädie Math. Wiss.*, Leipzig, 1904–1916, vol. II, 1, 2, p. 1324.

The discontinuous unit function may be obtained in many different ways as the limiting case of some suitably chosen continuous function. This is analogous to the fact that physical phenomena, though ultimately of a continuous character, can often be described approximately by discontinuous functions which may make the analytical treatment more simple.

We shall now give some examples of continuous functions that approximate the behaviour of the unit function when the parameter involved grows indefinitely. In any of the four following examples the unit function is obtained in the limit for $\lambda \to \infty$:

(1) $$U(t) = \lim_{\lambda \to \infty} \left\{ \frac{1}{2} + \frac{1}{\pi} \arctan(\lambda t) \right\}, \tag{1}$$

in which the many-valued function arc tan is given its principal value according to which $-\frac{1}{2}\pi < \arctan < \frac{1}{2}\pi$. This function is drawn in fig. 17 for some values of λ; the larger λ, the closer the approximation to the unit function.

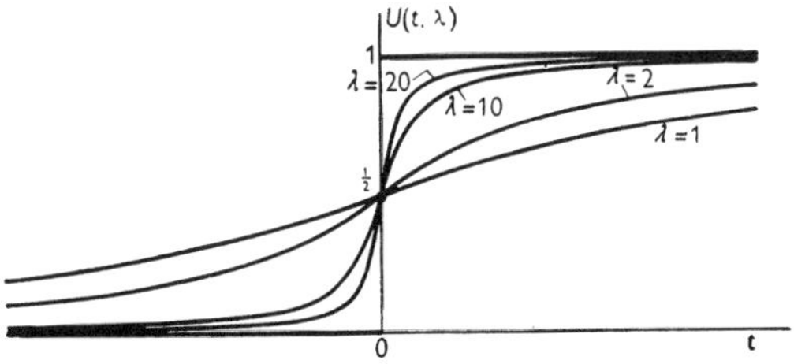

Fig. 17. Approximations to the unit function.

(2) $$U(t) = \lim_{\lambda \to \infty} \tfrac{1}{2} \operatorname{erfc}(-\lambda t) = \lim_{\lambda \to \infty} \frac{1}{\sqrt{\pi}} \int_{-\lambda t}^{\infty} e^{-\tau^2} d\tau. \tag{2}$$

The validity of this limiting relation is evident; the lower limit of integration tends to $-\infty$ or $+\infty$, depending on whether t is negative or positive.

(3) $$U(t) = \lim_{\lambda \to \infty} \frac{1}{\pi} \int_{-\infty}^{\lambda t} \frac{\sin \tau}{\tau} d\tau. \tag{3}$$

This also is evident since, according to the Dirichlet integral, we have

$$\int_{-\infty}^{\infty} \frac{\sin \tau}{\tau} d\tau = \pi. \tag{4}$$

(4) As a fourth example we mention

$$U(t) = \lim_{\lambda \to \infty} 2^{-e^{-\lambda t}}. \tag{5}$$

It should be noticed that all of the above limits actually lead to $U(0) = \frac{1}{2}$. There are, however, other approximations which, though unity for $t > 0$ and zero for $t < 0$, do not yield the mean value $\frac{1}{2}$ at $t = 0$; the function

$$\lim_{\lambda \to \infty} e^{-e^{-\lambda t}},$$

which for $t = 0$ is equal to $1/e$, may serve as an example.

Also from the point of view of the operational calculus the functions mentioned are close approximations to the unit function, since the image of any of them tends to 1, the image of the unit function, if λ tends to infinity and $\operatorname{Re} p > 0$. For instance, with the aid of (II, 24) and the integration rule, we have

$$\tfrac{1}{2}\operatorname{erfc}(-\lambda t) \doteq e^{p^2/4\lambda^2} \quad (\operatorname{Re} p > 0), \tag{6}$$

whilst from (III, 16) and the shift and similarity rules it follows that

$$2^{-e^{-\lambda t}} \doteq \frac{\Pi(p/\lambda)}{(\log 2)^{p/\lambda}} \quad (\operatorname{Re} p > 0).$$

In both cases the image in the right member approaches unity for $\lambda \to \infty$ and $\operatorname{Re} p > 0$.

Many properties of the unit function become almost evident if it is remembered that any expression like $U(\phi(t))$ is equal to 1 for all values of t that make $\phi(t)$ positive, and is equal to zero for any t with $\phi(t) < 0$; as simple examples we quote the following:

$$U(e^t) = 1,$$
$$U(t - \log \alpha) = U(e^t - \alpha),$$
$$U\{(t-a)(t-b)\} = U\{t - \max(a,b)\} + U\{\min(a,b) - t\}.$$

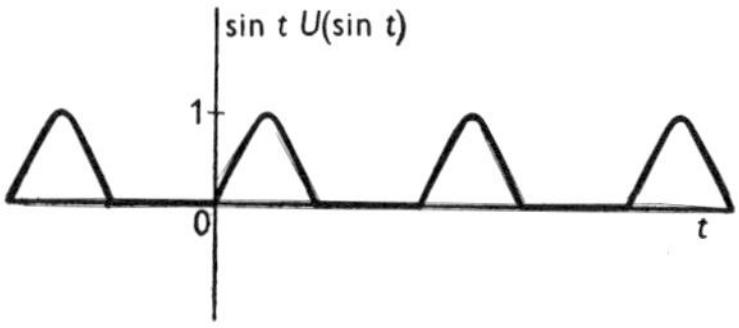

Fig. 18. A rectified alternating current.

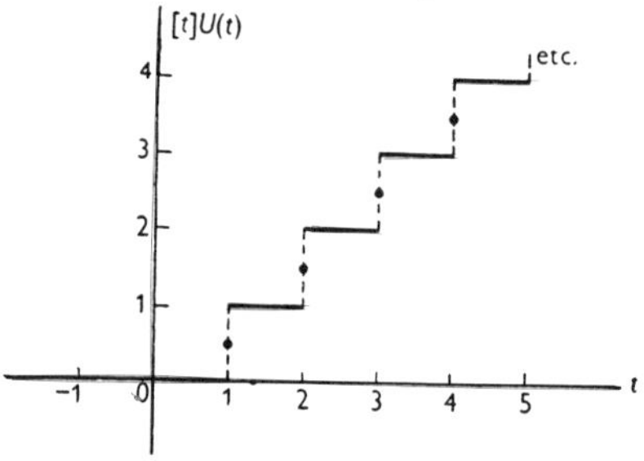

Fig. 19. The one-sided step function $[t]\,U(t)$.

A typical application of the unit function in physics is further provided by the function $\sin t\, U(\sin t)$, which, when drawn in a diagram as in fig. 18, represents a rectified alternating-current signal. Furthermore, as already stressed (IV, § 3, example 2), any step function can be built up by means of unit functions, as will be demonstrated explicitly in the case of the function $[x]$. This function is defined as equal to the greatest integral number below x, whilst at integral values of x ($x = n$) the mean value ($n - \frac{1}{2}$) is assumed. The corresponding one-sided function (see fig. 19) can be written as an infinite series, namely

$$[t]\,U(t) = U(t-1) + U(t-2) + U(t-3) + \ldots, \tag{7}$$

the image of which, for $\operatorname{Re} p > 0$, is found with the help of the shift rule to be

$$e^{-p} + e^{-2p} + e^{-3p} + \ldots.$$

A mere summation then leads to the operational relation

$$[t]\,U(t) \fallingdotseq \frac{1}{e^p-1}, \quad 0<\operatorname{Re} p<\infty. \tag{8}$$

Related to the unit function is the function *signum x*, defined by

$$\operatorname{sgn} x = \begin{cases} 1 & (x>0), \\ 0 & (x=0), \\ -1 & (x<0), \end{cases}$$

which can also be expressed in terms of the well-known Dirichlet integral as follows:

$$\operatorname{sgn} x = \frac{1}{\pi}\int_{-\infty}^{\infty} \frac{\sin(xu)}{u}\,du. \tag{9}$$

Its connexion with the unit function is simply

$$\operatorname{sgn} x = 2U(x)-1.$$

3. The δ-function as derivative of the unit function

We shall next introduce the δ-function as the first-order derivative of the unit function and discuss some of the main properties of $\delta(t)$ accordingly, a more rigorous investigation being deferred until § 5.

With the exception of the point $t = 0$, the unit function $U(t)$ admits of differentiation anywhere, and obviously the derivative vanishes everywhere. Whereas we cannot differentiate the unit function at its point of discontinuity $t = 0$, we can do so in the case of any of the four approximations above. Let us consider, for instance, the derivative of the function in (1):

$$U(t,\lambda) = \frac{1}{2}+\frac{1}{\pi}\arctan(\lambda t) \quad (|\arctan| \leqslant \tfrac{1}{2}\pi).$$

The derivative is simply

$$\delta(t,\lambda) = \frac{d}{dt}U(t,\lambda) = \frac{\lambda}{\pi(\lambda^2t^2+1)}. \tag{10}$$

Just as we may look upon $U(t)$ as the limit for $\lambda\to\infty$ of the function $U(t,\lambda)$, so we introduce the 'derivative' $\delta(t)$ of $U(t)$ as the limit of

$$\delta(t,\lambda) = \frac{d}{dt}U(t,\lambda)$$

for $\lambda\to\infty$; that means

$$\delta(t) = \frac{d}{dt}U(t) = \operatorname*{Lim}_{\lambda\to\infty} \frac{\lambda}{\pi(\lambda^2t^2+1)}. \tag{10a}$$

This 'function', henceforth called the δ-function or impulse function, is apparently zero for $t\neq 0$; moreover, it is $+\infty$ for $t = 0$, since $\delta(0,\lambda) = \lambda/\pi$.

Fig. 20 illustrates how the derivative $\delta(t, \lambda)$ gradually tends to zero for non-vanishing values of t when λ increases indefinitely. At the same time it reveals that the value λ/π assumed in $t = 0$ increases more and more, whilst the peak becomes narrower, but without diminishing the area below the curve; for, according to (10), we have, independently of λ,

$$\int_{-\infty}^{\infty} \delta(t, \lambda)\, dt = 1.$$

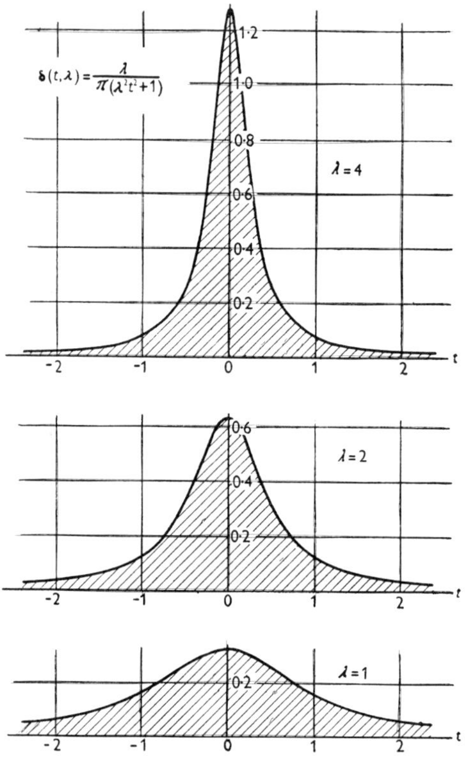

Fig. 20. Approximations to the δ-function.

Since we are now considering $\delta(t)$ as the limit of $\delta(t, \lambda)$ for $\lambda \to \infty$, we will attribute the same property to the δ-function *itself*; thus

$$\int_{-\infty}^{\infty} \delta(t)\, dt = 1. \tag{11}$$

The δ-function is consequently a 'function' that vanishes for $t \neq 0$ and is equal to $+\infty$ for $t = 0$, whilst its integral from $-\infty$ to $+\infty$ is equal to unity. It should be borne in mind, however, that our considerations are significant only if these properties are, to a certain extent, independent of the special choice of the function $U(t, \lambda)$ which is representative for the unit function.

In this respect the limit function derived from the differential quotient of (2), that is,

$$\delta(t) = \operatorname*{Lim}_{\lambda \to \infty} \frac{\lambda}{\sqrt{\pi}} e^{-\lambda^2 t^2}, \tag{12}$$

leads precisely to the same result.

The usefulness of the impulse function in practice depends on the integration property:

$$\int_{-\infty}^{\infty} h(t-\tau)\,\delta(\tau)\,d\tau = \int_{-\infty}^{\infty} h(\tau)\,\delta(t-\tau)\,d\tau = h(t); \tag{13}$$

this formula, from the point of view of the mathematician, may be taken only in the sense of the rigorous

$$\operatorname*{Lim}_{\lambda \to \infty} \int_{-\infty}^{\infty} h(t-\tau)\,\delta(\tau,\lambda)\,d\tau = h(t). \tag{14}$$

The physicist, however, would always change the order of limit and integration in (14), and thus prefer the symbolic form (13).

The property indicated by (14) or (13) holds for a large class of functions $h(t)$ subject to some mild conditions only. Symbolically, the δ-function acts as if it were a sieve; after multiplying an arbitrary function $h(t-\tau)$ by $\delta(\tau)$, and then integrating over the real t-axis, we just pick out the value $h(t)$ at $\tau = 0$. We will refer to this property as the *sifting* property of the δ-function. It results from the fact that the integrand $h(t-\tau)\,\delta(\tau)$ can be replaced at $\tau = 0$ by $h(t)\,\delta(\tau)$; this also holds for $\tau \neq 0$, since then both $h(t-\tau)\,\delta(\tau)$ and $h(t)\;\delta(\tau)$ are equal to zero. Consequently

$$\int_{-\infty}^{\infty} h(t-\tau)\,\delta(\tau)\,d\tau = \int_{-\infty}^{\infty} h(t)\,\delta(\tau)\,d\tau = h(t)\int_{\infty}^{\infty} \delta(\tau)\,d\tau = h(t).$$

The second integral relation in (13) is obtained after substituting $t-\tau = \tau'$ in the first one.

The sifting property of the δ-function according to the second integral relation (13) is illustrated by fig. 21. It shows that $h(\tau)$ is approximately constant (equal to $h(t)$) in the region where the impulse function is large; the integral is accordingly equal to $h(t)$ times the area $(= 1)$ below the $\delta(t-\tau)$-function curve. It must also be noticed that in 'sifting integrals' like (13) the limits of integration $-\infty$ and $+\infty$ may be replaced by any pair of finite numbers a and b, provided the critical point (here $\tau = t$) lies in the interior of the interval (a, b). Virtually the sifting property also occurs in functions different from the δ-function; it holds, for instance, for the limit $\lambda \to \infty$ of the function

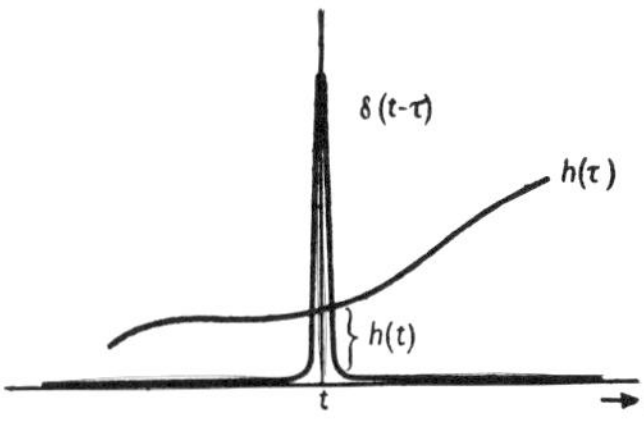

Fig. 21. The sifting property of the δ-function.

$$\frac{\sin(\lambda t)}{\pi t}, \tag{15}$$

which is found by differentiation of (3). This function is particularly important in Dirichlet's theory of the Fourier integral; we therefore call it the Dirichlet function. The limit of this Dirichlet function, however, is not identical with the δ-function, since for non-vanishing t it has no limit at all, let alone zero.

Example. The following physical problem leads quite naturally to the introduction of the δ-function. Let us determine the electric field between two parallel planes at $x = \pm c$, due to a charged plate at $x = 0$ (see fig. 22). Since all quantities depend only on the x-co-ordinate, the three-dimensional potential equation reduces to $d^2V/dx^2 = 0$, in the charge-free parts of space. The general solution is $V = ax+b$; in order to account properly for the surface charges on the plate $x = 0$, the constants of integration a, b are different on different sides of the plate. In the simplest case the potential is $V = |x|$. Let us now calculate the second-order derivative of V. First we have

$$\frac{dV}{dx} = \operatorname{sgn} x = 2U(x)-1, \tag{16}$$

and secondly
$$\frac{d^2V}{dx^2} = 2\delta(x).$$

This is nothing but the Poisson equation

$$d^2V/dx^2 = -4\pi\rho,$$

in which $\rho(x)$ denotes the volume-charge density. Therefore the solution $V = |x|$ corresponds to a density of amount $\rho(x) = -\frac{1}{2\pi}\delta(x)$. Indeed, as is indicated by the δ-function, the charge on the plate is wholly concentrated in the infinitely thin layer at $x = 0$. In other words, the surface-charge density on the plate is

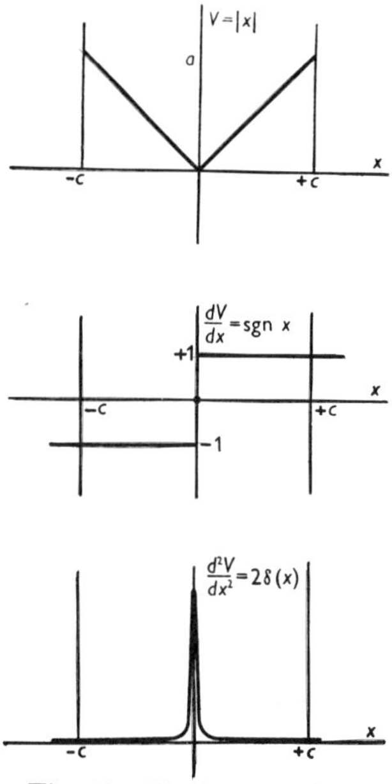

Fig. 22. The function $|x|$ and its derivatives.

$$\int_{-\infty}^{\infty} \rho(x)\,dx = -\frac{1}{2\pi}\int_{-\infty}^{\infty} \delta(x)\,dx = -\frac{1}{2\pi}.$$

4. History of the impulse function

The impulse function proves a powerful tool in present-day mathematical physics, particularly in quantum mechanics—in which it was introduced by Dirac. Its occurrence in pure mathematics, however, is of much earlier times, as will become apparent from the following quotations, which we do not claim to be exhaustive.

At the end of the last century Hermite† wrote:

'En même temps nous voyons que l'intégrale

$$\int_{\alpha}^{\beta} \frac{2i\lambda}{(t-\theta)^2+\lambda^2}\,dt \tag{17}$$

† *Cours de M. Hermite* (Faculté des Sciences de Paris), IVe édition, Paris, 1891, p. 155.

n'est pas toujours nulle avec $\lambda \to 0$. En supposant, en effet, θ compris entre α et β, elle est égale à $2\pi i$ pour une valeur infiniment petite de cette quantité. C'est ce qu'on appelle une intégrale singulière. Les éléments d'une pareille intégrale sont nuls, sauf l'élément unique et infini qui correspond à $t = \theta$. Les intégrales singulières ont été souvent employées par Cauchy et Poisson, mais elles n'ont plus un rôle aussi étendu dans les travaux analytiques de notre époque.'

If in (17) λ is replaced by $1/\lambda$ and θ is taken equal to zero, Hermite's statement in our notation reads

$$\lim_{\lambda\to\infty}\int_\alpha^\beta \frac{\lambda}{\lambda^2t^2+1}\,dt \neq 0,$$

provided α and β lie at different sides of the point $t = 0$. This is in complete agreement with the previous formula (11) if the approximation (10) is substituted for the function $\delta(t)$.

Also Cauchy's derivation of the Fourier-integral theorem, which Cauchy† obtained independently of Fourier, was based on the same type of impulse function. In this book the function (10) is therefore referred to as Cauchy's function. The proof of the Fourier identity by Cauchy's method will be reproduced in the first example of § 8.

Poisson‡ also discovered the Fourier-integral theorem independently; he followed exactly the same path as Cauchy, in a prize essay concerning this matter. In this respect Poisson wrote—if we take the liberty of changing the old notation into modernized form—

'Quelle que soit la fonction $f(x)$, continue ou discontinue, pourvue qu'elle ne devienne infinie pour aucune valeur réelle de x, on aura, pour toutes les valeurs réelles de cette variable

$$f(x) = \frac{1}{\pi}\int_{-\infty}^{\infty} d\alpha \int_0^\infty da\, e^{-ak} f(\alpha)\cos\{a(x-\alpha)\}$$

...et k, une quantité positive qu'on devra supposer infiniment petite ou nulle après l'intégration. En effet, ...on a

$$\int_0^\infty e^{-ak}\cos\{a(x-\alpha)\}\,da = \frac{k}{k^2+(x-\alpha)^2},$$

d'où il suit

$$\int_{-\infty}^{\infty} d\alpha \int_0^\infty da\, e^{-ak} f(\alpha)\cos\{a(x-\alpha)\} = \int_{-\infty}^{\infty}\frac{kf(\alpha)\,d\alpha}{k^2+(x-\alpha)^2}.$$

Or, $f(\alpha)$ ne devenant jamais infinie, il est évident que cette intégrale simple sera infiniment petite en même temps que k, excepté dans l'étendue des

† A. L. Cauchy, *Théorie de la propagation des ondes* (Prix d'analyse mathématique), Concours de 1815 et de 1816, pp. 140–2; see also pp. 281–2.

‡ S. D. Poisson, *Mémoire sur la théorie des ondes* (lu le 2 octobre et le 18 décembre 1815), pp. 85, 86.

valeurs de α qui diffèrent infiniment peu de x; il suffira donc d'intégrer depuis $\alpha = x-u$ jusqu'à $\alpha = x+u$, u étant une quantité positive et infiniment petite: entre ces limites $f(\alpha)$ sera censée constante et égale à $f(x)$; par conséquent on aura

$$\int_{-\infty}^{\infty} \frac{kf(\alpha)\,d\alpha}{k^2+(x-\alpha)^2} = f(x)\int_{-\infty}^{\infty} \frac{k\,d\alpha}{k^2+(x-\alpha)^2}$$

$$= f(x)\arctan\left(\frac{\alpha-x}{k}\right)\Bigg|_{-\infty}^{\infty} = \ldots = \pi f(x),$$

lorsqu'on y fait $k = 0$.'

The last sentence involves a derivation of the sifting property whereby the function

$$\frac{k}{\pi(k^2+x^2)}$$

is used as an approximation to the impulse function. The above function reduces to that of Cauchy if k is replaced by $1/\lambda$.

In addition to Cauchy, Poisson and Hermite, Kirchhoff† too was acquainted with the valuable impulse function, in view of his formulation of Huygens's principle in the wave theory of light. As an example of the impulse function he mentioned

$$f(t) = \frac{\lambda}{\sqrt{\pi}}\,e^{-\lambda^2t^2},$$

in which $\lambda\to\infty$; this is our function (12).

Later, Von Helmholtz in his *Vorlesungen über die elektromagnetische Theorie des Lichtes* employed the same Cauchy function in his proof of Kirchhoff's theorem.

Again, the 'heat source' of Lord Kelvin

$$f(x,t) = \frac{1}{\sqrt{(4\pi kt)}}\,e^{-x^2/4kt},$$

in which t tends to zero through positive values, is equivalent to the approximating function (12).

Later on the impulse function is put forward in Heaviside's work‡, especially in his symbolic calculus. As a physical example this author mentions, amongst others, the time function representing the current through a condenser under the influence of the electromotive force $U(t)$. Another example of a function possessing the sifting property is given by Heaviside§ in the form of an infinite series, namely,

$$u = \frac{2}{l}\sum_{n=1}^{\infty}\sin\left(\frac{n\pi x}{l}\right)\sin\left(\frac{n\pi y}{l}\right). \tag{18}$$

† G. R. Kirchhoff, *S.B. Kön. Akad. Wiss. Berlin vom* 22 *Juni* 1882, p. 641; see also *Wied. Ann.* XVIII, 663, 1883.

‡ See, for instance, O. Heaviside, *Electromagnetic Theory*, vol. II, pp. 54–5.

§ Heaviside, loc. cit. p. 92.

In our notation this series, if taken in the sense of a first-order Cesàro sum (see VI, § 8), is equal to

$$\sum_{n=-\infty}^{\infty} \{\delta(x-y-2nl)-\delta(x+y-2nl)\},$$

whence it follows that it is simply $\delta(x-y)$ in a sufficiently small interval around $x = y$. Concerning the sifting property of the function u, Heaviside writes:

'The function u... spots a single value of the arbitrary function in virtue of its impulsiveness.'

As already mentioned, the recent introduction of the impulse function in quantum mechanics is due to Dirac†; since then the function has been known by the name of delta function, and is customarily denoted by $\delta(t)$. Dirac himself writes as follows:

'This $\delta(p-q)$, we can say, is an improper function of the variable p, having the value zero for all values of p except q and the value infinity for $p = q$, the infinity being such that its integral is unity.'

And further:

'The introduction of the δ-function into our analysis will not be in itself a source of lack of rigour in the theory, since any equation involving the δ-function can be transcribed into an equivalent but usually more cumbersome form in which the δ-function does not appear. The δ-function is thus merely a convenient notation.'

On the other hand, Weyl remarked that such a function does not exist, though it is still possible to approximate it to any degree of accuracy‡. As Weyl remarks, the limit may be taken only after, and not before, all operations are performed. As is often the case, the difference in views between the mathematician and the physicist seems more serious at first sight than after close examination. Indeed, the Dirichlet function (15) occurring in the theory of the Fourier integrals, i.e.

$$\operatorname*{Lim}_{\lambda \to \infty} \frac{\sin(\lambda t)}{\pi t},$$

and the analogous function in the theory of the Fourier series, namely,

$$\operatorname*{Lim}_{n \to \infty} \frac{\sin\{(2n+1)\tfrac{1}{2}t\}}{2\pi \sin \tfrac{1}{2}t},$$

are closely related to the δ-function with which they have the sifting property in common. This is particularly emphasized by Lebesgue§ in the

† Cf. P. A. M. Dirac, *The Principles of Quantum Mechanics*, Oxford, 1930, pp. 63–6.
‡ See also T. Lewis, *Phil. Mag.* XXIV, 329, 1937.
§ H. Lebesgue, *Leçons sur les séries trigonométriques*, Paris, 1906, p. 74.

discussion of the general theory concerning Fourier series. At the same time this author mentions the existence of a large class of functions that, in the limit, have the δ-function properties. Many of these functions will be discussed in § 7.

In what follows, the impulse function will be employed extensively, since it lends itself pre-eminently to application in operational calculus, and leads to the simplification of many problems.

In the next section the sifting property of the impulse function is given a rigorous mathematical foundation in terms of the concept of the Stieltjes integral, whilst in § 7 this will be done by means of the limit of an ordinary Riemann integral, as already briefly indicated (see (14)).

5. The sifting integral as a Stieltjes integral

We shall first recall the concept of the Stieltjes integral of which the ordinary Riemann integral is a mere specialization†. In order to define the Stieltjes integral

$$\int_a^b f(x)\,dg(x),$$

the interval of integration $a < x < b$ is dissected by the points $a_1, a_2, \ldots, a_{n-1}$ into n subintervals ($a = a_0 < a_1 < a_2 \ldots < a_{n-1} < a_n = b$). The integral in question is then defined by

$$\operatorname*{Lim}_{\max|a_{i+1}-a_i|\to 0} \sum_{i=0}^{n-1} f(a_i)\{g(a_{i+1}) - g(a_i)\}.$$

In this limit the number of dissecting points grows indefinitely in such a manner that the length of the largest subinterval, denoted by $\max|a_{i+1} - a_i|$, tends to zero. Stieltjes himself showed the existence of the limit if $f(x)$ is continuous and $g(x)$ monotonic non-decreasing throughout the interval of integration. Further, since any function with limited total fluctuation can be written as the difference of two suitably chosen monotonic non-decreasing functions, the Stieltjes integral also exists for $f(x)$ continuous and $g(x)$ having limited total fluctuation. In particular, for any function that is differentiable in the whole interval of integration, the Stieltjes integral reduces to an ordinary Riemann integral, viz.

$$\int_a^b f(x)\,dg(x) = \int_a^b f(x)\,g'(x)\,dx,$$

whilst for $g(x) = x$ it is just the Riemann integral

$$\int_a^b f(x)\,dx.$$

† For a general survey the reader is referred to: T. J. Stieltjes, *Oeuvres complètes*, Groningen, 1918, vol. II, p. 469 a.f.; O. Perron, *Die Lehre von den Kettenbrüchen*, Leipzig, 1929, pp. 362–408; D. V. Widder, *The Laplace Transform*, Princeton, 1946, chap. I.

Of great importance in the discussion of the sifting integral is the case where $g(x)$ coincides with the unit function. Then, for any function with limited total fluctuation in the vicinity of the point t, the following identities hold:

$$\int_0^{\infty} h(t-\tau)\,dU(\tau) = \tfrac{1}{2}h(t-0), \quad \int_{-\infty}^{0} h(t-\tau)\,dU(\tau) = \tfrac{1}{2}h(t+0), \tag{19}$$

in which the right members actually exist, on account of

$$h(t-0) = \lim_{\epsilon\to 0} h(t-\epsilon), \quad h(t+0) = \lim_{\epsilon\to 0} h(t+\epsilon).$$

The proof of formulae (19) is easy; as follows at once from the general definition of the Stieltjes integral, in this special case only the neighbourhood of $t = 0$ contributes to the integrals, the factor $\tfrac{1}{2}$ being a consequence of the definition $U(0) = \tfrac{1}{2}$. After addition of formulae (19) the following identity is obtained:

$$\int_{-\infty}^{\infty} h(t-\tau)\,dU(\tau) = -\int_{-\infty}^{\infty} h(\tau)\,dU(t-\tau) = \frac{h(t+0)+h(t-0)}{2}, \tag{20}$$

that is, a Stieltjes integral for the mean value of any function with limited total fluctuation. We now observe the important fact that, by writing

$$dU(\tau) = \delta(\tau)\,d\tau, \tag{21}$$

formula (20) reduces to the sifting integral (13) with respect to the mean value of the function $h(t)$. In other words, the sifting integral (13) is the *formal* equivalent of the rigorous Stieltjes integral (20); just as in (21) we can take

$$\delta(t) = \frac{d}{dt}\,U(t)$$

in agreement with § 3, where the δ-function was looked upon as the derivative of the unit function.

Besides, on account of (19), the sifting integral under consideration can always be written as the sum of two separate integrals as follows:

$$\left.\begin{aligned} \int_0^{\infty} h(t-\tau)\,\delta(\tau)\,d\tau &= \int_{-\infty}^{t} h(\tau)\,\delta(t-\tau)\,d\tau = \tfrac{1}{2}h(t-0), \\ \int_{-\infty}^{0} h(t-\tau)\,\delta(\tau)\,d\tau &= \int_{t}^{\infty} h(\tau)\,\delta(t-\tau)\,d\tau = \tfrac{1}{2}h(t+0). \end{aligned}\right\} \tag{22}$$

In particular for the integral (11),

$$\int_{-\infty}^{0} \delta(\tau)\,d\tau = \int_0^{\infty} \delta(\tau)\,d\tau = \tfrac{1}{2}. \tag{23}$$

Furthermore, we readily find that

$$\int_a^b \delta(\tau)\,d\tau = \int_a^b dU(\tau) = U(b) - U(a),$$

which indicates that the integral of the δ-function over a finite interval is equal to unity or zero, depending on whether the origin $t = 0$ lies inside or outside the interval of integration.

Another important application of the concept of the Stieltjes integral is provided by the special choice $g(x) = [x]\,U(x)$. On account of (7) we then have

$$\int_0^\infty h(x)\,d[x] = \sum_{n=1}^{\infty} \int_0^\infty h(x)\,dU(x-n);$$

therefore, provided that $h(x)$ is continuous and everywhere equal to its mean value,

$$\int_0^\infty h(x)\,d[x] = \sum_{n=1}^{\infty} h(n). \tag{24}$$

Hence any infinite series can be transformed into a corresponding Stieltjes integral.

6. The image of the δ-function

The image of the impulse function follows immediately from the sifting integral

$$p\int_{-\infty}^{\infty} e^{-pt}\,\delta(t)\,dt = p.$$

This equation holds for any value of p, whence we deduce the operational relation

$$\delta(t) \risingdotseq p, \qquad -\infty < \operatorname{Re} p < \infty. \tag{25}$$

This result, too, is mathematically sound if the underlying definition integral is taken in Stieltjes's sense rather than Riemann's:

$$f(p) = p\int_{-\infty}^{\infty} e^{-pt}\,d\left\{\int_{-\infty}^{t} h(s)\,ds\right\}. \tag{26}$$

The operational relation (25) is then simply a shorthand notation for the following Stieltjes integral:

$$p = p\int_{-\infty}^{\infty} e^{-pt}\,dU(t).$$

This relation is further in agreement with the meaning of the δ-function as the 'derivative' of the unit function, since the differentiation rule applied to $U(t) \risingdotseq 1$ ($\operatorname{Re} p > 0$) yields

$$\delta(t) = \frac{d}{dt}\,U(t) \risingdotseq p\,.\,1 = p,$$

whilst, when applied to

$$-U(-t) \risingdotseq 1 \quad (\operatorname{Re} p < 0),$$

it again leads to

$$-\frac{d}{dt}U(-t) = \delta(-t) = \delta(t) \doteqdot p \quad (\operatorname{Re} p < 0).$$

In this use is made of the evenness of $\delta(t)$; this property can be found by differentiation of the identity $U(t) + U(-t) = 1$.

It should be noticed that the sifting integral (13) may be considered as the composition product of the arbitrary function $h(t)$ and the δ-function $\delta(t)$ (whose images are $f(p)$ and p respectively), since

$$f(p) = \frac{1}{p} f(p)\, p \doteqdot \int_{-\infty}^{\infty} h(t-\tau)\,\delta(\tau)\,d\tau = h(t).$$

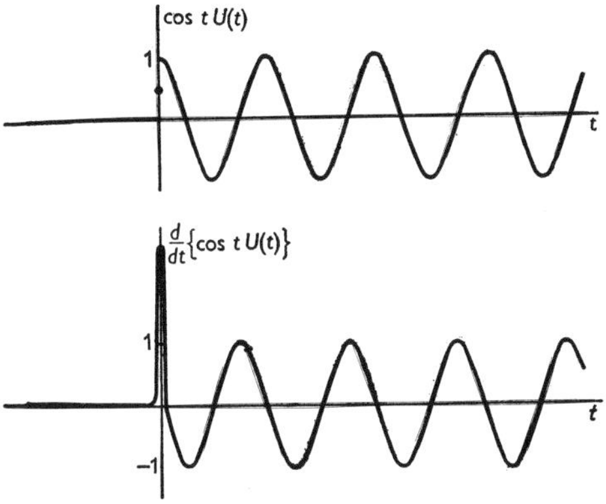

Fig. 23. The one-sided cosine function and its derivative.

The formalism is also consistent in the following manner. If in the definition integral (II, 20a) the integrand $p\,\mathrm{e}^{-ps} h(s)$ is transposed into the t-language with the help of the shift rule, then

$$p\,\mathrm{e}^{-ps} h(s) \doteqdot \delta(t-s)\,h(s),$$

whence it follows after integration that

$$f(p) = \int_{-\infty}^{\infty} p\,\mathrm{e}^{-ps} h(s)\,ds \doteqdot \int_{-\infty}^{\infty} \delta(t-s)\,h(s)\,ds = h(t).$$

Further, it is easy to verify with the aid of (25) and the similarity rule the important property

$$\delta(\alpha t) = \frac{\delta(t)}{|\alpha|},$$

whilst the evident equation

$$h(t)\,\delta(t) = h(0)\,\delta(t) \quad (h(0) \neq 0) \tag{27}$$

is often useful in the differentiation of operational relations involving unit functions. For instance, let us apply the differentiation rule to (III, 21), that is, to

$$\cos t\, U(t) \doteqdot \frac{p^2}{p^2+1} \quad (\operatorname{Re} p > 0).$$

Then we obtain

$$-\sin t\, U(t)+\cos t\,\delta(t) = -\sin t\, U(t)+\delta(t) \doteqdot \frac{p^3}{p^2+1} \quad (\operatorname{Re} p>0),$$

in which the original contains an impulse function of unit strength (see fig. 23). The above relation is consistent with that found by first writing the image as

$$\frac{p^3}{p^2+1} = -\frac{p}{p^2+1}+p,$$

and then transposing it term by term according to (III, 22) and (25).

7. Functions approximating the δ-function

Whereas in § 5 we have learned to consider the sifting integral as a Stieltjes integral, we shall now treat it as the formal equivalent of the previously mentioned limit of a definite Riemann integral (cf. (14)):

$$\operatorname*{Lim}_{\lambda\to\infty}\int_{-\infty}^{\infty} h(t-\tau)\,\delta(\tau,\lambda)\,d\tau = \frac{h(t+0)+h(t-0)}{2}. \tag{28}$$

Sufficient for the validity of (28) will be the existence of the following set of identities (cf. (22)):

$$\operatorname*{Lim}_{\lambda\to\infty}\int_{0}^{\infty} h(t-\tau)\,\delta(\tau,\lambda)\,d\tau = \operatorname*{Lim}_{\lambda\to\infty}\int_{-\infty}^{t} h(\tau)\,\delta(t-\tau,\lambda)\,d\tau = \frac{h(t-0)}{2}, \tag{29a}$$

$$\operatorname*{Lim}_{\lambda\to\infty}\int_{-\infty}^{0} h(t-\tau)\,\delta(\tau,\lambda)\,d\tau = \operatorname*{Lim}_{\lambda\to\infty}\int_{t}^{\infty} h(\tau)\,\delta(t-\tau,\lambda)\,d\tau = \frac{h(t+0)}{2}. \tag{29b}$$

Fortunately, there are numerous functions $\delta(t,\lambda)$ that make (29 a, b) true for any function $h(x)$ that has limited total fluctuation in some finite interval around $x = t$. The existence of a great number of such functions has been recognized by Lebesgue†. For instance, any function like

$$\delta(t,\lambda) = \frac{\lambda K(\lambda t)}{2\int_0^{\infty} K(s)\,ds} \quad (t>0) \tag{30}$$

can be used in (29 a), if only the integral in the denominator of (30) exists and is absolutely convergent with respect to its upper limit‡. It should be noticed that these functions $\delta(t,\lambda)$ not only lead to the sifting integral (29) for $\lambda\to\infty$, but, provided $K(0)\neq 0$, for $\lambda\to\infty$ they are identical with the δ-function itself; in this respect they behave as the preceding approxima-

† *Lecons sur les séries trigonométriques*, Paris, 1906, p. 74.
‡ Cf. S. Bochner, *Vorlesungen über Fouriersche Integrale*, Leipzig, 1932, p. 25.

tions (10a) and (12). For any function (30) shows, as $\lambda \to \infty$, the three characteristic features of the δ-function:

(1) $$\lim_{\lambda \to \infty} \delta(t, \lambda) = 0 \quad \text{for} \quad t \neq 0;$$

this is because the integral in (30) is absolutely convergent—$K(x)$ must be of the order $x^{-1-\epsilon}$ for $x \to \infty$ ($\epsilon > 0$).

(2) $$\int_0^\infty \delta(t)\, dt = \tfrac{1}{2},$$

provided that this property is considered as the limit as $\lambda \to \infty$ of

$$\int_0^\infty \delta(t, \lambda)\, dt = \tfrac{1}{2},$$

which is valid for any function (30).

(3) $$\lim_{\lambda \to \infty} \delta(0, \lambda) = \infty.$$

This is evident because $K(0) \neq 0$.

It will be obvious that, in addition to (29a), (29b) also holds, if for $t < 0$ we also have

$$\delta(t, \lambda) = \frac{\lambda K(\lambda t)}{2\int_{-\infty}^{0} K(s)\, ds} \quad (t < 0), \tag{31}$$

in which the integral in the denominator should now be absolutely convergent with respect to the lower limit. In the special case of $K(x)$ even, (30) implies (31).

Consequently, it is easy to construct many different functions approximating the δ-function, by merely choosing suitable functions $K(x)$. The following examples may be discussed particularly:

(1) The function $$K(x) = \frac{1}{x^2 + 1}$$

leads to the Cauchy function (10a)

$$\delta(t) = \lim_{\lambda \to \infty} \frac{\lambda}{\pi(\lambda^2 t^2 + 1)} = \lim_{\epsilon \to +0} \frac{\epsilon}{\pi(t^2 + \epsilon^2)}. \tag{32}$$

(2) $K(x) = U(x+1) - U(x-1)$ leads to

$$\delta(t) = \lim_{\lambda \to \infty} \frac{\lambda}{2}\{U(\lambda t + 1) - U(\lambda t - 1)\}.$$

(3) $K(x) = e^{-|x|^n}$; this function leads to

$$\delta(t) = \lim_{\lambda \to \infty} \frac{\lambda}{2\Pi(1/n)} e^{-|\lambda t|^n} \quad (n > 0). \tag{33}$$

The particular cases $n = 1, 2$ are very simple; in these cases the corresponding image can be written down explicitly, viz.

$$\delta_1(t,\lambda) = \frac{\lambda}{2}\mathrm{e}^{-|\lambda t|} \doteqdot \frac{p}{1-p^2/\lambda^2}, \quad -\lambda < \mathrm{Re}\, p < \lambda, \tag{33a}$$

$$\delta_2(t,\lambda) = \frac{\lambda}{\sqrt{\pi}}\mathrm{e}^{-\lambda^2 t^2} \doteqdot p\,\mathrm{e}^{p^2/4\lambda^2}, \quad -\infty < \mathrm{Re}\, p < \infty. \tag{33b}$$

The last function is that used by Poisson, Kirchhoff and Lord Kelvin, as already mentioned in §4. In (33a), as well as in (33b), the image tends to that of the δ-function for $\lambda \to \infty$, as was to be expected.

(4) The function $\quad K(x) = \int_0^1 \cos(xs)\,(1-s)^n\, ds$

leads to
$$\delta(t) = \operatorname*{Lim}_{\lambda\to\infty} \frac{1}{\pi}\int_0^\lambda \cos(ts)\left(1-\frac{s}{\lambda}\right)^n ds \quad (n>0), \tag{34}$$

the factor π of which arises from the integral $\int_0^\infty K(s)\,ds$ which can be evaluated as the Fourier integral at $t = 0$ of the function

$$(1-t)^n\, U(t)\, U(1-t).$$

This type of δ-function approximation is extremely important with regard to the Cesàro limits (see VI, §9). For integral values of n they reduce to simple functions involving exponentials, namely,

$$\delta_n(t,\lambda) = \frac{\lambda}{\pi}\mathrm{Re}\left\{\mathrm{e}^{-\omega}\frac{d^n}{d\omega^n}\left(\frac{\mathrm{e}^{\omega}-1}{\omega}\right)\right\}_{\omega=-i\lambda t}.$$

It is found, for instance, that

$$\delta_1(t,\lambda) = \frac{1}{\pi}\frac{1-\cos(\lambda t)}{\lambda t^2} = \frac{2}{\pi}\frac{\sin^2(\frac{1}{2}\lambda t)}{\lambda t^2},$$

$$\delta_2(t,\lambda) = \frac{2}{\pi}\frac{\lambda t - \sin(\lambda t)}{\lambda^2 t^3},$$

$$\delta_3(t,\lambda) = \frac{3}{\pi}\frac{\lambda^2 t^2 + 2\cos(\lambda t) - 2}{\lambda^3 t^4}.$$

These functions are even in t, whether n is integral or not; moreover, they are positive for $n \geqslant 1$, as may be seen for $n > 1$ from (34) after two integrations by parts:

$$\delta_n(t,\lambda) = \frac{2n(n-1)}{\pi\lambda^2 t^2}\int_0^\lambda \sin^2\left(\frac{ts}{2}\right)\left(1-\frac{s}{\lambda}\right)^{n-2} ds \quad (n>1).$$

Furthermore, the limiting cases $n = 0$, $n = \infty$ are just the functions of Dirichlet (15) and Cauchy (10), respectively. Fig. 24 shows the functions for $n = 0, 1, 2$, drawn to have the same maximum height.

(5) As an example showing that the requirement $K(0) \neq 0$ is essential, we may mention the function

$$K(x) = \frac{\mathrm{e}^{-1/x}}{x^{\nu+2}} \quad (\nu > -1).$$

The corresponding

$$\delta(t, \lambda) = \frac{e^{-(\lambda|t|)^{-1}}}{2\Pi(\nu)\,\lambda^{\nu+1}\,|t|^{\nu+2}} \quad (\nu > -1) \tag{35}$$

does have the sifting property; yet it is no approximation for the δ-function; though still vanishing for $\lambda \to \infty$ at any point $t \neq 0$, it does not increase indefinitely at $t = 0$ for $\lambda \to \infty$, since $K(0) = 0$.

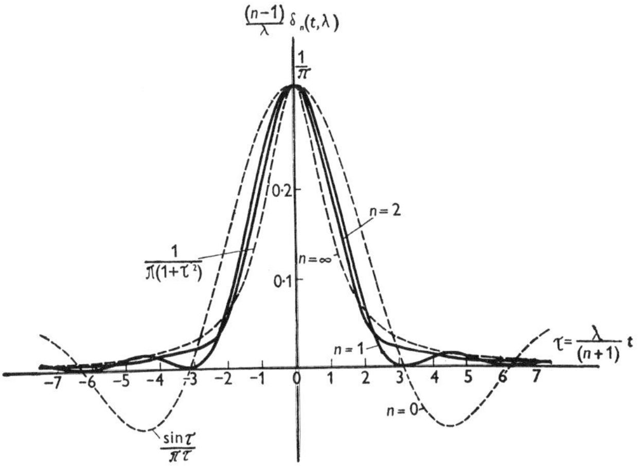

Fig. 24. The approximating functions

$$\delta_n(t, \lambda) = \frac{1}{\pi}\int_0^\lambda \cos(ts)\left(1-\frac{s}{\lambda}\right)^n ds.$$

Any of the examples given above is a special case of (30). There are still other functions, not belonging to the type (30), which have the sifting property (28) and which, moreover, are approximations of the δ-function. Two examples may illustrate this statement.

(1) $$\delta(t) = \lim_{\epsilon \to 0} \frac{\sinh \epsilon}{2\pi(\cosh \epsilon - \cos t)}\{U(t+\pi) - U(t-\pi)\},$$

or, after the substitution $r = e^{-\epsilon}$,

$$\delta(t) = \lim_{r \to 1-0} \frac{1}{2\pi}\frac{1-r^2}{1-2r\cos t + r^2}\{U(t+\pi) - U(t-\pi)\}. \tag{36}$$

This function plays an important role in potential theory. Let r, ϑ denote polar co-ordinates. The solution of the two-dimensional potential equation inside the unit circle with boundary values $V(\vartheta)$ at $r = 1$ is given by

$$V(r, \vartheta) = \frac{1}{2\pi}\int_{-\pi}^{\pi} V(\vartheta - u)\frac{1-r^2}{1-2r\cos u + r^2}du.$$

The sifting property of the function (36) is equivalent to the statement that the above integral representation of V is also valid on the unit circle itself.

(2)
$$\delta(t) = \operatorname*{Lim}_{\lambda\to\infty} \sqrt{\left(\frac{\lambda}{2\pi}\right)} e^{-\lambda t}(t+1)^{\lambda}\, U(t+1). \tag{37}$$

This approximating function is not even in t, unlike all the foregoing examples; it is very important for the inversion formula of Widder (VII, § 12).

Finally, we would like to draw attention to a class of functions of the form (30) which have in fact the property (28) but which cannot be considered as equivalent with the δ-function, since now the integral in the denominator of (30) will no longer be supposed to converge absolutely with respect to its upper limit. Without the requirement of absolute convergence these functions do not vanish for $t \neq 0$ in the limit $\lambda \to \infty$. For instance, a general class of such functions is generated by (30) if $K(x)$ is of the type

$$K(x) = a\frac{\cos x}{x^p} + b\frac{\sin x}{x^q} + H(x),$$

in which p and q are both positive and $H(x)$ absolutely integrable for $x \to \infty$. It can be proved† that the corresponding functions $\delta(t, \lambda)$ still have the sifting property. Also Dirichlet's function (15) belongs to the type under consideration, viz.

$$\frac{\sin(\lambda t)}{\pi t},$$

which does not have a definite value as $\lambda \to \infty$ $(t \neq 0)$ owing to its oscillatory character (see fig. 25). Another example of such a sifting function is the Bessel function

$$\frac{\lambda}{2} J_{\nu}(\lambda\,|\,t\,|) \quad (\nu > -1).$$

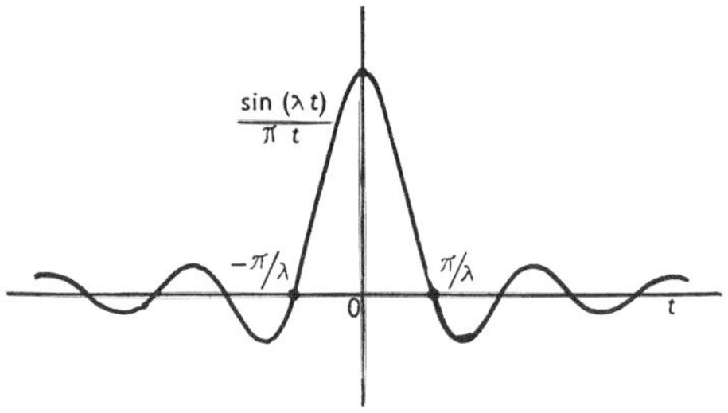

Fig. 25. The Dirichlet function.

We may finally mention an example of a function which, though zero rather than infinite at $t = 0$, still has the sifting property, namely,

$$\operatorname*{Lim}_{\lambda\to\infty} \frac{\lambda}{2} e^{-\lambda t}(t+1)^{\lambda-1}\,|\,t\,|\, U(t+1). \tag{38}$$

8. Applications of the δ-function

In this section we show with simple examples how, by suitable use of the δ-function properties, many problems can be solved, or at least surveyed most naturally.

† See Bochner, loc. cit. p. 25.

Example 1. *The Fourier integral.* Let us consider the following limit:

$$I = \operatorname*{Lim}_{\epsilon \to +0} \frac{1}{2\pi} \int_{-\infty}^{\infty} d\omega \, e^{-\epsilon|\omega|} \int_{-B}^{A} d\tau \, h(\tau) \, e^{i\omega(t-\tau)}, \tag{39}$$

which obviously differs from the Fourier integral in that the limits of integration in the second integral are finite instead of infinite, and that the factor $e^{-\epsilon|\omega|}$ is involved; this factor, increasing the chance of convergence with regard to the integral under consideration, will be called a *convergence factor*†. Granted for the moment that the order of integration and proceeding to a limit may be changed, we at once see that I represents a sifting integral for the mean value of $h(t)$ if $-B < t < A$; indeed, the integration with respect to ω leads to Cauchy's function

$$\frac{1}{2\pi} \int_{-\infty}^{\infty} d\omega \, e^{-\epsilon|\omega| + i\omega(t-\tau)} = \frac{\epsilon}{\pi\{(t-\tau)^2 + \epsilon^2\}}.$$

Example 2. *The Green functions.* In the case of inhomogeneous linear differential equations of mathematical physics the right-hand member often represents the exterior force acting on a system. The influence of a force concentrated in one single point is described by a Green function. For instance, consider the second-order differential equation of the Sturm-Liouville type:

$$\frac{d}{dx}\left\{f(x) \frac{d}{dx} y(x)\right\} - g(x) \, y(x) = -\phi(x),$$

which has to be satisfied in some interval $a < x < b$, whilst at the end-points $x = a, b$ certain homogeneous boundary conditions must be fulfilled. In this case the Green function $K(x, \xi)$ is uniquely determined by the following requirements:

(1) Outside $x = \xi$ it is a continuous function of x, satisfying both the homogeneous equation and the boundary conditions.

(2) Both $\partial K/\partial x$ and $\partial^2 K/\partial x^2$ are continuous functions outside $x = \xi$; $\partial K/\partial x$ shows a jump at $x = \xi$ of amount $-1/f(\xi)$ as indicated by

$$\frac{\partial}{\partial x} K(x, \xi) \Big|_{x=\xi-0}^{x=\xi+0} = -\frac{1}{f(\xi)}.$$

It turns out that these two conditions can be gathered into a single one, namely, $K(x, \xi)$ should be a solution of the inhomogeneous differential equation

$$\frac{d}{dx}\left\{f(x) \frac{\partial K}{\partial x}\right\} - g(x) \, K = -\delta(x-\xi), \tag{40}$$

which contains a δ-function in its right-hand member. This statement can be verified readily by integrating (40) with respect to x from $\xi - \epsilon$ to $\xi + \epsilon$, the integral on the right being considered as a Stieltjes integral.

Once the above Green function is known, it is easy to write down the required solution for an arbitrary function $-\phi$ in the right-hand member. Since $\phi(x)$ can be expressed as a sifting integral, namely,

$$\phi(x) = \int_a^b \delta(x-\xi) \, \phi(\xi) \, d\xi,$$

we have, on account of the linearity of the differential equation and the principle of superposition, the solution (fulfilling the boundary conditions)

$$y(x) = \int_a^b K(x, \xi) \, \phi(\xi) \, d\xi.$$

† For a detailed treatment of this limit, see A. Sommerfeld, *Die willkürlichen Funktionen in der mathematischen Physik*, Königsberg, 1891.

Example 3. *Continuous systems of orthogonal functions.* In numerous cases the system of orthogonal functions $\phi_n(x)$ used in practical problems is discrete; all of these functions are accounted for by letting n pass through the sequence of positive integers. Both the properties of orthogonality and normalization are usually expressed in the form

$$\int_a^b \bar{\phi}_m(\xi)\,\phi_n(\xi)\,d\xi = \begin{cases} 0 & (m \neq n), \\ 1 & (m = n) \end{cases} \tag{41}$$

($\bar{\phi}$ being the complex conjugate of ϕ). As a consequence, the coefficients in the expansion

$$h(x) = \sum_{n=-\infty}^{\infty} c_n \phi_n(x) \tag{42}$$

can be determined from

$$c_n = \int_a^b h(\xi)\,\bar{\phi}_n(\xi)\,d\xi. \tag{43}$$

As is well known†, continuous systems of orthogonal functions $\phi_n(x)$ also exist, in which n runs through the set of real numbers ν. In order to normalize these functions in the simplest possible way, the multiplication constants are so chosen that, in the analogue of (41), a δ-function appears in the right-hand member, viz.

$$\int_a^b \bar{\phi}_\mu(\xi)\,\phi_\nu(\xi)\,d\xi = \delta(\mu - \nu). \tag{44}$$

Again, the above relation expresses both the orthogonality and the normalization; after integration with respect to ν the customary equivalent of (44) is obtained, namely,

$$\int_a^b d\xi\,\bar{\phi}_\mu(\xi) \int_{\nu_1}^{\nu_2} d\nu\,\phi_\nu(\xi) = \begin{cases} 0 & (\mu < \nu_1), \\ 1 & (\nu_1 < \mu < \nu_2), \\ 0 & (\mu > \nu_2). \end{cases}$$

Instead of (42) we now have for any function $h(x)$

$$h(x) = \int_{-\infty}^{\infty} c_\nu \phi_\nu(x)\,d\nu, \tag{45}$$

in which, analogously to (43), we find the coefficients from

$$c_\nu = \int_a^b h(x)\,\bar{\phi}_\nu(x)\,dx.$$

The validity of the expansion (45) can be verified by eliminating $h(x)$ from the two relations above and using (44).

The most important continuous system of normalized orthogonal functions is in fact that of

$$\phi_\nu(x) = \frac{1}{\sqrt{(2\pi)}}\,e^{i\nu x}$$

in the interval $-\infty < x < \infty$. In this case (45) is nothing but the Fourier integral; moreover, if $\nu - \mu$ is put equal to x, then (44) becomes

$$\frac{1}{2\pi}\int_{-\infty}^{\infty} e^{ix\xi}\,d\xi = \delta(x).$$

This expression, convergent in the sense of a Cesàro limit, is the inversion integral corresponding to (25).

† For a clear discussion of continuous systems of orthogonal functions, see A. Sommerfeld, *Atombau und Spektrallinien*, Braunschweig, 1939, vol. II, p. 752.

Another simple continuous system of orthonormal functions is generated by the δ-function itself, namely,

$$\phi_\nu(x) = \delta(x-\nu),$$

again in the interval $-\infty < x < \infty$; here (44) reduces to

$$\int_{-\infty}^{\infty} \delta(\xi-\mu)\,\delta(\xi-\nu)\,d\xi = \delta(\mu-\nu),$$

which has been given by Dirac†. Furthermore, (45) is the sifting integral (13) for an arbitrary function; it is thus simply the expansion of $h(x)$ in terms of a special set of continuous orthonormal functions.

Finally, the functions above can be considered from the point of view of eigenvalue theory. In this respect it is remembered that the eigenfunctions $\psi_\nu(x)$ of a functional operator L are such as to make

$$L\{\psi_\nu(x)\} = \lambda_\nu \psi_\nu(x),$$

where the constant λ_ν is the corresponding eigenvalue. In this way the functions $\frac{1}{\sqrt{(2\pi)}}e^{i\nu x}$ of the Fourier integral are eigenfunctions of the operator $L = d/dx$ with eigenvalues $i\nu$, whilst $\delta(x-\nu)$ are those of the simple operator $L = x$, that is, multiplication by x, as follows from (27).

Example 4. *Stieltjes's integral of moments.* A direct application of an approximate impulse function is encountered in the problem of determining $h(x)$ from the integral equation

$$F(z) = \int_{-\infty}^{\infty} \frac{h(s)}{s-z}\,ds, \tag{46}$$

in which $F(z)$ is given and $\operatorname{Im} z > 0$. Actually, the integral (46) diverges for any real value x of z, unless $h(x) = 0$. On the other hand, it has a definite limiting value if the complex number z approaches the point x from above, namely,

$$F(x+i0) = \operatorname*{Lim}_{\epsilon\to+0} \int_{-\infty}^{\infty} \frac{h(s)}{s-x-i\epsilon}\,ds.$$

The imaginary part, provided that $h(x)$ is real, becomes

$$\operatorname{Im} F(x+i0) = \operatorname*{Lim}_{\epsilon\to+0} \int_{-\infty}^{\infty} h(s)\frac{\epsilon}{(s-x)^2+\epsilon^2}\,ds,$$

and this is equivalent to

$$h(x) = \frac{1}{\pi}\operatorname{Im} F(x+i0), \tag{47}$$

on account of the sifting property of Cauchy's function (32). Accordingly, we may say that the unknown function $h(x)$ is now solved from the integral equation (46) in terms of the given function F; we will return to this matter in example 1 of chapter XIV, § 4.

The integral (46) was first considered by Stieltjes, particularly in connexion with the problem of moments, where he discussed to what extent the function $h(x)$ is determined by its infinite number of moments‡

$$\int_{-\infty}^{\infty} h(x)\,x^n\,dx \quad (n = 0, 1, 2, \ldots);$$

these moments occur in the expansion of $F(z)$ into powers of z^{-1} (see also (VII, 41)).

† *Proc. Roy. Soc.* A, CXIII, 626, 1926.

‡ See, for further details: T. J. Stieltjes, *Oeuvres complètes*, Groningen, 1918, vol. II, pp. 398–566; O. Perron, *Die Lehre von den Kettenbrüchen*, Leipzig-Berlin, 1929, pp. 372, 408; this author accomplishes the inversion of (46) when the integral is taken in the sense of a Stieltjes integral, i.e. $h(s)\,ds$ replaced by $d\psi(s)$.

Examining the derivation of (47) again we notice that, if (46) is given for $\operatorname{Im} z < 0$, we find the analogous solution

$$h(x) = -\frac{1}{\pi} \operatorname{Im} F(x - i0). \tag{48}$$

As will be evident from the preceding, the real axis will be a cut for any complex function $F(z)$ defined by an integral like (46) in so far as $h(x)$ is different from zero. Further, since the real part of $F(z)$ does not jump, that is, $\operatorname{Re} F(x+i0) = \operatorname{Re} F(x-i0)$, we find for the difference of the F-values at both sides of the cut

$$F(x+i0) - F(x-i0) = 2\pi i\, h(x). \tag{49}$$

Two simple examples may illustrate equation (49):

(*a*) By means of the substitution $s = u + p = u - z$ the exponential integral (III, 33) is changed into

$$-e^{-z} \operatorname{Ei}(z) = \int_0^\infty \frac{e^{-u}}{u-z}\, du, \tag{50}$$

which is clearly of the form (46). Consequently, $\operatorname{Ei}(z)$ has a cut along the positive part of the real axis; moreover,

$$\operatorname{Ei}(x+i0) - \operatorname{Ei}(x-i0) = -2\pi i \quad (x > 0). \tag{50a}$$

(*b*) The well-known relation between the Legendre functions (see X, § 6) P_n, Q_n

$$Q_n(z) = \frac{1}{2} \int_{-1}^{1} \frac{P_n(s)}{z-s}\, ds \tag{51}$$

immediately shows that $Q_n(z)$ has a cut along the real segment $-1 \leqslant x \leqslant 1$; it follows from (49) that

$$Q_n(x+i0) - Q_n(x-i0) = -\pi i\, P_n(x) \quad (-1 < x < 1).$$

9. Series of impulse functions

Until now we have only discussed impulse functions that are singular at one single point; as a consequence only the value of $h(t-\tau)$ at the point $\tau = 0$ was spotted by the sifting integral

$$\int_{-\infty}^{\infty} h(t-\tau)\, \delta(\tau)\, d\tau.$$

We shall next consider impulsive functions picking out more than one value. A simple example of a series of impulse functions can be obtained by 'differentiation' of (7) (see fig. 26):

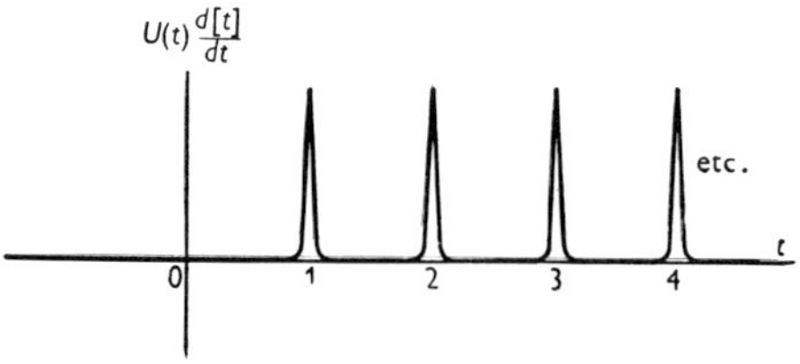

Fig. 26. The 'derivative' of $[t]\, U(t)$.

$$U(t) \frac{d[t]}{dt} = \delta(t-1) + \delta(t-2) + \delta(t-3) + \ldots.$$

This infinite series of equidistant impulse functions has a very simple operational image; as can be found by application of the shift rule, we have

$$\delta(t-1) + \delta(t-2) + \delta(t-3) + \ldots \doteqdot \frac{p}{e^p - 1}, \quad 0 < \operatorname{Re} p < \infty. \tag{52}$$

It is often possible to express such a sum of impulse functions by means of a single impulse function, if only the argument $\phi(t)$ of the latter is suitably chosen. Let us assume, for instance, that $\phi(t)$ is real throughout and that it changes sign at the points $t_1, t_2, \ldots$. Then the function $U\{\phi(t)\}$ (see fig. 27) can be written as a sum of elementary unit functions, viz.

$$U\{\phi(t)\} = U(t-t_1) - U(t-t_2) + U(t-t_3) - U(t-t_4) + \ldots.$$

A mere differentiation together with (27) then easily leads to

$$\delta\{\phi(t)\} = \Sigma \frac{\delta(t-t_n)}{|\phi'(t_n)|}, \qquad (53)$$

in which the summation should be extended over all points t_n where $\phi(t)$ changes sign. Some simple special cases of the above relation are

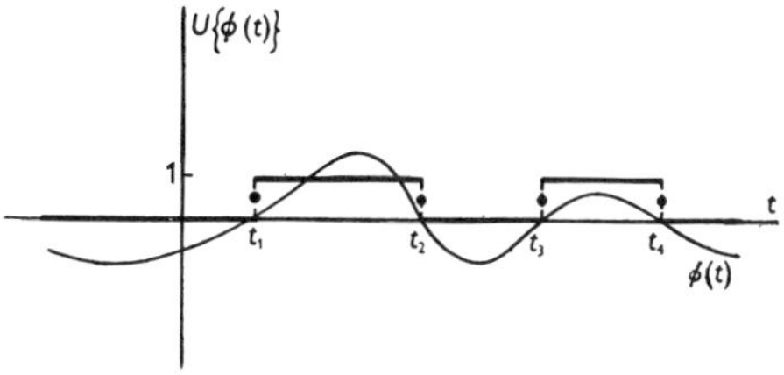

Fig. 27. Unit function having an arbitrary function as argument.

$$\delta\{(t-a)(t-b)\} = \frac{\delta(t-a)+\delta(t-b)}{|a-b|} \quad (a \neq b), \qquad (54)$$

$$\pi\delta\{\sin(\pi t)\} = \sum_{n=-\infty}^{\infty} \delta(t-n).$$

It is not difficult to see that approximate impulse functions may also lead to multiple impulses. In this connexion it is to be remarked that the previous function (36), related to Poisson's integral of potential theory, was intentionally multiplied by the rectangle function, in order to produce only one impulse. If, however, that factor is dropped, the result is an infinite series of impulse functions, namely,

$$\lim_{r\to 1-0} \frac{1}{2\pi} \frac{1-r^2}{1-2r\cos t + r^2} = \sum_{n=-\infty}^{\infty} \delta(t-2\pi n). \qquad (55)$$

This function is discussed by Heaviside† in the form of a Fourier power series:

$$v(t,r) = \frac{1}{2\pi}\left\{1 + 2\sum_{n=1}^{\infty} r^n \cos(nt)\right\}.$$

As to its sifting property, this author remarks that (r being finite) the value of the integral

$$\int_{-\pi}^{\pi} v(x-t,r)\,F(t)\,dt$$

gradually tends to the value $F(x)$ when r tends to unity from below. Concerning the limit $r \to 1$, by which $v(x-t,r)$ changes into the impulse function $\delta(x-t)$, he writes: 'I know no reason why a failure should occur just as perfection is reached.'

† *Electromagnetic Theory*, vol. II, 119.

10. Derivatives of the δ-function

By analogy with the reasoning in § 3, where by introducing the δ-function as the limit as $\lambda \to \infty$ of the derivative of the approximate unit function of (1), we were led to Cauchy's function (10a), the derivative of the δ-function itself can be defined as the limit of the derivative of Cauchy's function. One then obtains

$$\delta'(t) = -\lim_{\lambda\to\infty} \frac{2}{\pi} \frac{\lambda^3 t}{(\lambda^2 t^2+1)^2} = -\lim_{\epsilon\to+0} \frac{2}{\pi} \frac{\epsilon t}{(t^2+\epsilon^2)^2}.$$

The corresponding approximations ($\epsilon \neq 0$) for $\delta'(t)$ (see fig. 28), which, like those of fig. 20 for $\delta(t)$ itself, do tend to zero ($t \neq 0$) if $\lambda \to \infty$ or $\epsilon \to 0$, behave quite differently (in comparison with $\delta(t)$) in the vicinity of $t = 0$. Though crossing the t-axis at $t = 0$, they show to the left and the right of it at $t = -\epsilon/\sqrt{3}$ and $t = \epsilon/\sqrt{3}$ respectively, a sharp maximum and minimum, of absolute amount $\frac{9}{8\pi\sqrt{3}}\frac{1}{\epsilon^2}$. For ϵ tending to zero, both peaks increase in height and approach the origin. Therefore, the limit function $\delta'(t)$ is a 'function' that is $+\infty$ at $t = 0^-$ and $-\infty$ at $t = 0^+$; in the neighbourhood of $t = 0$ its behaviour is comparable to that of the function $-1/t$.

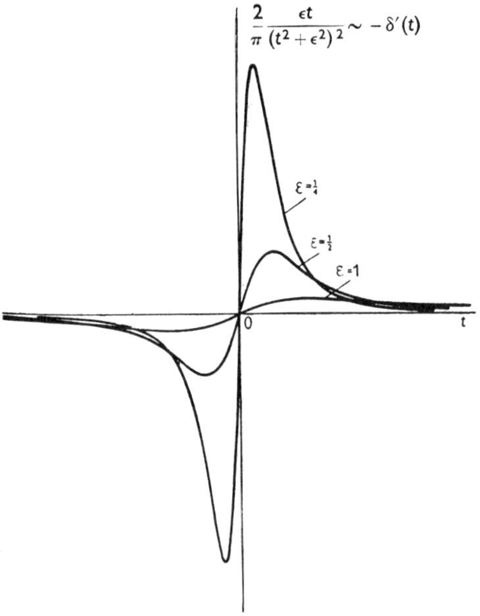

Fig. 28. Continuous approximations of the 'derivative' of the δ-function.

Example. As an exercise we would discuss a physical example involving the derivative of the δ-function. In the example of § 3 we have seen that an infinitely thin metal plate at $x = 0$ with surface-charge density $-1/2\pi$ led to a potential distribution $V(x) = |x|$ and a volume-charge density $\rho(x) = -\frac{1}{2\pi}\delta(x)$. Let us next consider the field of two neighbouring plates (at $x = -\frac{1}{2}\epsilon$ and $x = \frac{1}{2}\epsilon$) with specific surface charges $-\frac{1}{2\pi\epsilon}$ and $\frac{1}{2\pi\epsilon}$, respectively (see fig. 29). In this case the total volume-charge density is obviously

$$\rho(x) = -\frac{1}{2\pi\epsilon}\delta(x+\tfrac{1}{2}\epsilon) + \frac{1}{2\pi\epsilon}\delta(x-\tfrac{1}{2}\epsilon).$$

If now the plates approach each other, then in the limit a single plate is obtained; the two sides of this plate maintain equal but opposite charges in such a way that the product of surface density and mutual distance ϵ, that is, the moment, remains constant ($=1/2\pi$). Therefore, to the plane $x = 0$ charged with dipoles of moment $1/2\pi$ per unit area there belongs the following volume-charge density:

$$\rho(x) = -\frac{1}{2\pi}\lim_{\epsilon\to+0}\frac{\delta(x+\frac{1}{2}\epsilon)-\delta(x-\frac{1}{2}\epsilon)}{\epsilon} = -\frac{1}{2\pi}\delta'(x),$$

which is a function proportional to the 'derivative' of the δ-function.

In a similar way we can define derivatives of the δ-function of any integral order. Again using Cauchy's function as starting-point, we find

$$\delta^{(n)}(t) = \lim_{\epsilon\to+0}\frac{(-1)^n n!}{\pi}\,\frac{\sin\left\{(n+1)\arctan\dfrac{\epsilon}{t}\right\}}{(t^2+\epsilon^2)^{\frac{1}{2}(n+1)}}. \tag{56}$$

These limits have only a symbolic meaning; they become more significant when explicit reference to the corresponding sifting integrals is given. Let us first write the earlier sifting integral (28) in the following form with finite, instead of infinite, limits of integrations:

$$\lim_{\lambda\to\infty}\int_{-B}^{A} h(\tau)\,\delta(t-\tau,\lambda)\,d\tau = \frac{h(t+0)+h(t-0)}{2}.$$

This identity then holds for $-B<t<A$ if $h(x)$ has limited total fluctuation in the same interval, as follows from (28) when $h(t)$ is replaced by zero for $t<-B$ and $t>A$.

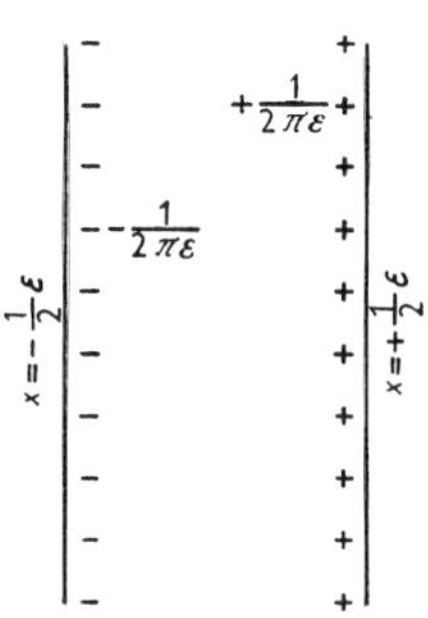

Fig. 29. Two metal plates with opposite charges.

Let us now apply the formula to the nth-order derivative $h^{(n)}(t)$; then

$$\lim_{\lambda\to\infty}\int_{-B}^{A} h^{(n)}(\tau)\,\delta(t-\tau,\lambda)\,d\tau = \frac{h^{(n)}(t+0)+h^{(n)}(t-0)}{2}. \tag{57}$$

As a consequence of having introduced finite limits of integration we can easily perform n successive integrations by parts. For the integrated terms vanish in the limit as $\lambda\to\infty$, provided that

$$\lim_{\lambda\to\infty}\delta^{(k)}(t,\lambda) = 0 \quad (t \neq 0)\ (k = 0,1,2,\dots,n-1), \tag{58}$$

in which (k) denotes the order of the differential quotient with respect to t, λ being kept constant. The conditions (58) are all satisfied in the case of the approximating functions (30) of § 7, owing to the absolute convergence of the integral in the denominator, which implies that $|K(t)|$ tends to zero at least as $t^{-1-\alpha}$ ($\alpha > 0$). Thus we arrive at the following sifting integral:

$$\lim_{\lambda\to\infty}\int_{-B}^{A} h(\tau)\,\delta^{(n)}(t-\tau,\lambda)\,d\tau = \frac{h^{(n)}(t+0)+h^{(n)}(t-0)}{2}, \tag{59}$$

valid in the interval $-B<t<A$ if $h^{(n)}(t)$ has limited total fluctuation and $\delta(t,\lambda)$ is any of the functions (30).

Further, if the original infinite limits of integration are introduced, we obtain the identity

$$\lim_{\lambda\to\infty}\int_{-\infty}^{\infty} h(\tau)\,\delta^{(n)}(t-\tau,\lambda)\,d\tau = \frac{h^{(n)}(t+0)+h^{(n)}(t-0)}{2}, \tag{60}$$

which, again, is valid for each of the approximations (30) and any function $h(t)$ whose nth-order derivative has limited total fluctuation in every finite interval. Similarly to the case of the δ-function itself, we write

$$\int_{-\infty}^{\infty} h(\tau)\,\delta^{(n)}(t-\tau)\,d\tau = \int_{-\infty}^{\infty} h(t-\tau)\,\delta^{(n)}(\tau)\,d\tau = \frac{h^{(n)}(t+0)+h^{(n)}(t-0)}{2} \tag{61}$$

as the formal equivalent of (60). It should be remarked that this extension of the sifting integral to higher-order derivatives of the impulse function was given by Dirac†.

The transposition of (61) into operational language is quite simple and illustrative, particularly if it is applied to the definition integral for $\delta^{(n)}(t)$; we easily find

$$p\int_{-\infty}^{\infty} \mathrm{e}^{-p\tau}\,\delta^{(n)}(\tau)\,d\tau = p\left(\frac{d^n}{dt^n}\mathrm{e}^{pt}\right)_{t=0} = p^{n+1},$$

and consequently the operational relation

$$\delta^{(n)}(t) \doteqdot p^{n+1}, \qquad -\infty < \operatorname{Re} p < \infty \tag{62}$$

which fully agrees with the result obtained when the differentiation rule is applied to (25). More generally, (61) can be interpreted as the composition product of the arbitrary function $h(t) \doteqdot f(p)$ and $\delta^{(n)}(t) \doteqdot p^{n+1}$. The image of the left-hand side of (61) then is $\frac{1}{p}f(p)\,p^{n+1}$, and this is indeed equal to $p^n f(p)$, the image of the right-hand side of (61).

If necessary, the extended sifting integral can be taken in the sense of a Stieltjes integral; further details, however, will be omitted here.

The behaviour at $t = 0$ of the 'derivatives' of the δ-function is the more complicated the higher the order, as may be seen from the corresponding approximations. To this end let us start with the image function

$$\left\{\frac{\sinh(\epsilon p)}{\epsilon}\right\}^{n+1}, \tag{63}$$

which tends to p^{n+1} for $\epsilon \to 0$; hence it may be expected that the original of (63) is some approximation for $\delta^{(n)}(t)$. The original in question is a step function which can most easily be surveyed if it is remembered that multiplication of any image $f(p)$ by the factor $2\sinh(\epsilon p)$ corresponds, according to the shift rule, to the operation

$$\underset{2\epsilon}{\Delta}\, h(t) = h(t+\epsilon) - h(t-\epsilon) \tag{64}$$

† Loc. cit. p. 626.

being performed on the original $h(t)$. Since (63) contains $n+1$ such factors, whilst the remaining factor $\frac{1}{(2\epsilon)^{n+1}}$ has the original $\frac{U(t)}{(2\epsilon)^{n+1}}$ ($\mathrm{Re}\, p > 0$), we at once obtain the following approximation:

$$\delta^{(n)}(t) = \operatorname*{Lim}_{\epsilon \to 0} \frac{1}{(2\epsilon)^{n+1}} \Delta^{n+1}_{2\epsilon}\, U(t),$$

in which the right-hand member can, of course, be written out with the aid of the binomial formula. So, for instance, if $n = 1$,

$$\delta'(t) = \operatorname*{Lim}_{\epsilon \to 0} \frac{U(t+2\epsilon) - 2U(t) + U(t-2\epsilon)}{4\epsilon^2},$$

which is plotted in fig. 30. The analogue for the third-order derivative $\delta'''(t)$ is also given, in fig. 31; the complicated behaviour of the higher-order derivatives of the δ-function is apparent.

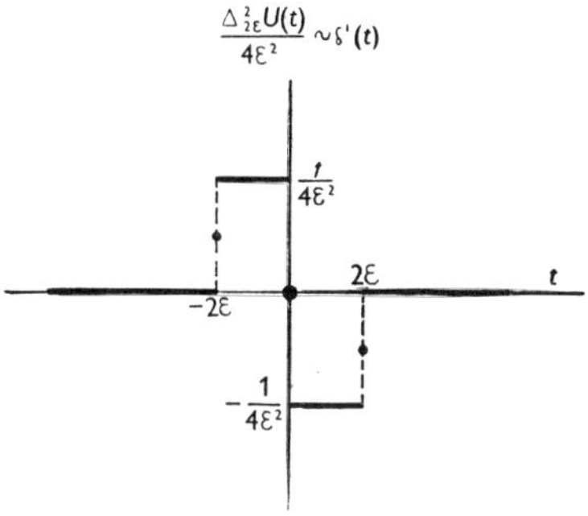

Fig. 30. A discontinuous approximation to the 'derivative' of the δ-function.

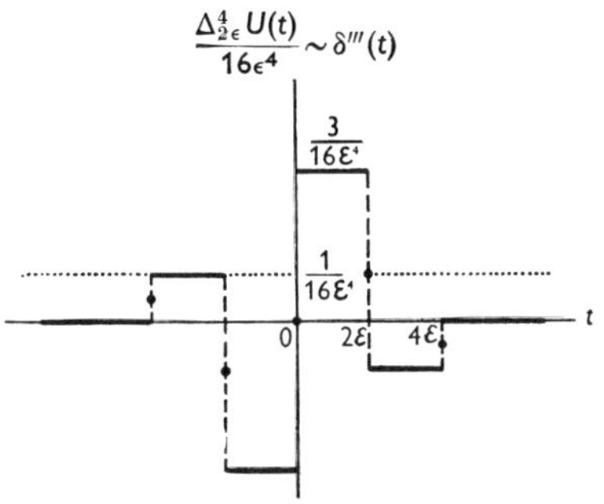

Fig. 31. A discontinuous approximation to the second 'derivative' of the δ-function.

As another approximation for the higher-order δ-functions we may mention that obtained by differentiating (33 *b*) n times. If He_n denotes the Hermite polynomial

$$\mathrm{He}_n(x) = (-1)^n\, \mathrm{e}^{x^2} \frac{d^n}{dx^n} (\mathrm{e}^{-x^2}),$$

the resulting operational relation is as follows:

$$\frac{\lambda^{n+1}}{\sqrt{\pi}}\, \mathrm{e}^{-\lambda^2 t^2}\, \mathrm{He}_n(-\lambda t) \doteqdot p^{n+1}\, \mathrm{e}^{p^2/4\lambda^2}, \quad -\infty < \mathrm{Re}\, p < \infty \tag{65}$$

which reduces to (62) if $\lambda \to \infty$.

We would like to emphasize that in the application of the operational calculus the preceding δ-function derivatives can be treated as if they were ordinary functions; they never lead to wrong results or discrepancies. In this way numerous relations between various symbolic functions can then be deduced. For instance, if (IV, 38) is applied to (62), we obtain

$$t^n \delta^{(n)}(t) \doteqdot (-1)^n n!\, p \quad (-\infty < \operatorname{Re} p < \infty),$$

and hence, since the original of the right-hand member also equals $(-1)^n n!\, \delta(t)$, it follows that

$$\delta^{(n)}(t) = (-1)^n n! \frac{\delta(t)}{t^n}. \tag{66}$$

In particular, for $n = 1, 2$ one gets

$$t\delta'(t) = -\delta(t), \quad t^2\delta''(t) = 2\delta(t).$$

On the other hand, from the operational relation

$$t^{n+1}\delta^{(n)}(t) \doteqdot (-1)^{n+1} p \frac{d^{n+1}}{dp^{n+1}}(p^n) = 0,$$

it follows that $\quad t\delta(t) = 0, \quad t^2\delta'(t) = 0, \quad t^3\delta''(t) = 0, \quad$ etc. (67)

We shall return to these relations in VI, § 15, where we deal with null functions.

Finally, we remark that the original of a polynomial in p is always a sum of impulse functions and their derivatives, for instance:

$$p(p-1)(p-2)\dots(p-n) \doteqdot \delta^{(n)}(1-e^{-t}), \quad -\infty < \operatorname{Re} p < \infty. \tag{68}$$

This relation is obtained from the derivative of $U(1-e^{-\frac{1}{2}}) \doteqdot 1$ by applying the rule (IV, 33b):

11. Note on the theory of distributions

All the properties discussed above of the delta function have obtained a rigorous mathematical basis in the 'theory of distributions' developed by Schwartz†. The distributions T are defined by fixing for every so-called testing function $\phi(x)$ (which has continuous derivatives of any order and which vanishes outside some finite interval) a corresponding number $T(\phi)$. The properties of those distributions T that can be defined with the aid of an integral

$$T(\phi) = \int_{-\infty}^{\infty} t(x)\,\phi(x)\,dx$$

are identical with those of an ordinary function $t(x)$. The delta function is not of this type, but corresponds to the distribution defined by $T\phi(x) = \phi(0)$. The extension of ordinary derivatives to derivatives T' of distributions is given by $T'(\phi) = -T(\phi')$. All the sifting integrals mentioned above, including those depending on derivatives of the delta function, can thus be interpreted in terms of distributions.

† L. Schwartz, 'Théorie des Distributions', *Actualités Scientifiques et Industrielles*, nos. 1091 and 1122, Paris, 1950–1.

CHAPTER VI

QUESTIONS CONCERNING THE CONVERGENCE OF THE DEFINITION INTEGRAL

1. Introduction

This chapter deals mainly with general properties of images and originals so far as they are connected with the intrinsic structure of the definition and inversion integrals. Accordingly, much of its contents may be omitted on a first reading. After briefly indicating in §§ 2 and 3 many special cases that can be included in an operational treatment, owing partly to a suitable extension of the definition integral, we shall discuss in detail questions about convergence of the definition integral in §§ 4–7. In §§ 8 and 9 methods of summing divergent series and integrals are recapitulated, so that we can investigate in §§ 10–14 the behaviour at the boundary of the convergence region as well as the validity of the inversion integral. Finally, the question of uniqueness of operational transformations is examined in § 15.

2. Extensions of the definition integral

The definition integral (II, 20 a) was initially introduced as a Riemann integral. On the other hand, in the theory of the δ-function, it was taken in the sense of a Stieltjes integral (see (V, 26)), whilst the definition integral for the derivatives of the δ-function (see (V, 62)) was again considered as a formal notation for the limit of a certain Riemann integral. Apparently still other extensions may be taken into consideration, to enable us to treat a much larger class of functions operationally. In this connexion we may remark that Widder† has taken the Stieltjes integral as the general basis of his calculus. This author starts with the integral

$$\int_{-\infty}^{\infty} e^{-pt}\, d\{\alpha(t)\}, \tag{1}$$

which is defined as the sum of two limits of Stieltjes integrals according to

$$f(p) = \lim_{\lambda \to \infty} \int_{0}^{\lambda} e^{-pt}\, d\{\alpha(t)\} + \lim_{\lambda \to \infty} \int_{-\lambda}^{0} e^{-pt}\, d\{\alpha(t)\}.$$

Moreover, it is supposed that $\alpha(t)$ has limited total fluctuation in any finite interval. It can be verified that in our notation (1) is equivalent to the operational relation $f(p) \doteqdot \alpha(t)$ for all functions $\alpha(t)$ that satisfy

$$e^{-pt}\, \alpha(t) \Big|_{-\infty}^{\infty} = 0.$$

† D. V. Widder, *The Laplace Transform*, Princeton, 1941.

General properties of the operational transforms, such as the existence of a strip of convergence, remain valid when the Stieltjes integral (1) is taken as the basis instead of the Riemann integral. In both cases the function $f(p)/p$ is analytic† within the strip of convergence $\alpha < \mathrm{Re}\, p < \beta$. Moreover, it can be proved that the derivatives of the image may be found by differentiation with respect to p under the sign of integration of the definition integral; thus, for instance, in the case of the Riemann definition integral

$$\frac{d^k}{dp^k}\left\{\frac{f(p)}{p}\right\} = \int_{-\infty}^{\infty} \mathrm{e}^{-pt}(-t)^k h(t)\, dt, \quad \alpha < \mathrm{Re}\, p < \beta.$$

As to applications of the operational calculus, it is of little importance whether or not the definition integral is extended in the way indicated (as a matter of fact we could even base the operational calculus upon the concept of the Lebesgue integral). Far more important is to define the definition integral in the sense of Cauchy's principal value in cases of divergence due to a finite or infinite number of singularities of $h(t)$. Thus, when $h(t)$ becomes infinitely large only at one point $t = t_0$,

$$p \,\overline{\int}_{-\infty}^{\infty} \mathrm{e}^{-pt} h(t)\, dt = \lim_{\epsilon \to +0} \left\{ p \int_{-\infty}^{t_0-\epsilon} \mathrm{e}^{-pt} h(t)\, dt + p \int_{t_0+\epsilon}^{\infty} \mathrm{e}^{-pt} h(t)\, dt \right\},$$

where a horizontal bar in the sign of integration may denote the principal value. In the following examples the definition integrals under consideration only exist as principal values.

Example 1. If for the original $\dfrac{\mathrm{e}^{-\nu|t|}}{\mathrm{e}^t - 1}$ the integration in the definition integral $\mathrm{Lim}\left(\int_{-\infty}^{-\epsilon} + \int_{\epsilon}^{\infty}\right)$ is reduced to one single integration along the positive t-axis, the image is found to be

$$p \lim_{\epsilon \to 0} \int_{\epsilon}^{\infty} \frac{\mathrm{e}^{-u(p+\nu)} - \mathrm{e}^{u(p-\nu+1)}}{\mathrm{e}^u - 1}\, du,$$

in which the limit sign may now be omitted, if at the same time the lower limit of integration is replaced by zero. After the substitution $s = \mathrm{e}^{-u}$ and using (III, 14), we obtain

$$p\{\psi(\nu - p - 1) - \psi(\nu + p)\} \doteqdot \frac{\mathrm{e}^{-\nu|t|}}{\mathrm{e}^t - 1}, \quad -\mathrm{Re}(\nu+1) < \mathrm{Re}\, p < \mathrm{Re}\, \nu. \tag{2}$$

In cases like this the results are always two-sided transforms that cannot be considered as the sum of two one-sided relations (whose individual strips of convergence would otherwise extend at the right to $+\infty$ and at the left to $-\infty$, respectively; see fig. 10).

Example 2. According to (V, 50 a) the exponential integral Ei (t) has a cut along the positive axis of t. On the cut itself the function may be adequately defined by a principal value as follows:

$$\overline{\mathrm{Ei}}(t) = \overline{\int}_{-\infty}^{t} \frac{\mathrm{e}^s}{s}\, ds = \overline{\int}_{-\infty}^{1} \frac{\mathrm{e}^{tu}}{u}\, du \quad (t > 0). \tag{3}$$

† Cf. the proof in Widder, loc. cit. pp. 57, 240.

For the corresponding one-sided function the image also exists as a principal value, viz.

$$\overline{\mathrm{Ei}}(t)\, U(t) \fallingdotseq -\log(p-1), \quad 1<\mathrm{Re}\, p<\infty, \tag{4}$$

which can be derived by integrating the relation

$$\frac{1}{u}\mathrm{e}^{ut}U(t) \fallingdotseq \frac{p}{u(p-u)}, \quad \mathrm{Re}\, p>u,$$

over the interval $-\infty<u<1$ of the parameter u (see page 40). This result should be compared with the analogue (IV, 46).

Incidentally we may mention an interesting application of (4). First we obtain with the aid of the shift rule

$$\mathrm{e}^{-t}\overline{\mathrm{Ei}}(t)\, U(t) \fallingdotseq -\frac{p}{p+1}\log p \quad (\mathrm{Re}\, p>0),$$

of which the right-hand member can be expanded as

$$-\log p \sum_{n=0}^{\infty} \frac{(-1)^n}{p^n} = \sum_{n=0}^{\infty} (-1)^n \frac{d}{dn}\left(\frac{1}{p^n}\right).$$

Secondly, we arrive easily at the corresponding original by applying formula (III, 3). Identification of the two different expressions found for the original then leads to the remarkable equality

$$\mathrm{e}^{-x}\,\overline{\mathrm{Ei}}(x) = \sum_{n=0}^{\infty} (-1)^n \frac{d}{dn}\left\{\frac{x^n}{\Pi(n)}\right\}. \tag{4a}$$

This expression for $\overline{\mathrm{Ei}}(x)$ is very simple indeed; when the differentiations are carried out logarithmic terms appear†.

3. The operational treatment of some special series

In this section we shall discuss several important infinite series from the point of view of the operational calculus; it is thus made possible to develop the theory of these series by means of the more general analysis of the Laplace transform.

(*a*) *Dirichlet series*

Considering them as image functions we write

$$f(p) = \sum_{n=1}^{\infty} a_n \mathrm{e}^{-p\lambda_n}, \tag{5}$$

in which the real exponents λ_n increase indefinitely and monotonically; moreover, $\lambda_1 \geqslant 0$. Since Dirichlet series like (5) are known to converge within (and also perhaps on the boundary of) some range $\mathrm{Re}\, p>\alpha$—in other words, have a strip of convergence similar to that of operational relations—it is quite natural to take them as images rather than originals. After transposing (5) term by term (as a matter of fact we shall only deal with $\mathrm{Re}\, p>0$), the following operational relation is found:

$$\sum_{n=1}^{\infty} a_n \mathrm{e}^{-p\lambda_n} \fallingdotseq \sum_{n=1}^{\infty} a_n U(t-\lambda_n) = \sum_{n=1}^{\lambda_n<t} a_n, \tag{6}$$

whilst the strip of convergence is specified by

$$\max(\alpha, 0) < \mathrm{Re}\, p < \infty. \tag{7}$$

† The analogous expansion for Ei(x) was derived operationally by Balth. van der Pol, *Wisk. Opgaven, Amst.*, XVIII, 95, 1943.

Since the numbers λ_n increase monotonically with n, the original at the right includes more and more terms as t increases. Apparently this original is a step function (with discontinuities at the points $t = \lambda_1, \lambda_2, \ldots$) which can be compared with an indefinite integral of which the upper limit t also increases monotonically. It should further be noticed that for $t < \lambda_1$ the original must be taken as zero, which could have been indicated by means of the factor $U(t-\lambda_1)$. In this book we prefer, however, to define any 'empty sum' equal to zero (if $t < \lambda_1$ no suffix n satisfies $\lambda_n < t$).

The legitimacy of the term-by-term transposition performed above will now be verified. We therefore start from the original in (6). The corresponding definition integral may be written in the form

$$\lim_{\mu \to \infty} p \int_{-\infty}^{\mu} e^{-pt} \sum_{n=1}^{\infty} a_n U(t-\lambda_n)\, dt.$$

For fixed values of μ the integral reduces to a finite sum which is easily written out. We thus obtain the operational relation

$$\sum_{n=1}^{\lambda_n < t} a_n \doteqdot \lim_{\mu \to \infty} \left\{ \sum_{j=1}^{N} a_j e^{-p\lambda_j} - e^{-p\mu} \sum_{j=1}^{N} a_j \right\}, \tag{8}$$

provided the integer N be chosen so as to have $\lambda_N < \mu < \lambda_{N+1}$. Since λ_n increases monotonically, N tends to infinity if μ does so. Therefore, the left-hand series in the right member of (8) actually yields the former Dirichlet series for any p within its range of convergence, $\operatorname{Re} p > \alpha$. The second term at the right of (8) in general only vanishes for $\mu \to \infty$ if $\operatorname{Re} p > 0$. This can be proved by using the following estimation,

$$\left| \sum_{j=1}^{N} a_j \right| \leqslant C e^{\alpha_1 \lambda_N} \quad (\alpha > 0),$$

which may be deduced by the method of partial summation; α_1 is a real number that exceeds α by an arbitrarily small amount, and C is independent of N. We thus arrive again at the operational relation (6) with domain of convergence (7), the term-by-term transposition being legitimate for $\operatorname{Re} p > \max(\alpha, 0)$. Nevertheless, it may happen that the right-hand member of (8) does exist even at the left of the abscissa of convergence $\operatorname{Re} p = \alpha$. In this case the operational relation (8) leads to an analytic continuation of the Dirichlet series into a region where the series itself diverges (see the example of the series (13) discussed below). It is not possible, however, that the same peculiarity occurs for the operational relation found by differentiation of (6), i.e.

$$p \sum_{n=1}^{\infty} a_n e^{-p\lambda_n} \doteqdot \sum_{n=1}^{\infty} a_n \delta(t-\lambda_n), \tag{9}$$

since here the definition integral for the original, as sifting integral, becomes identical with the Dirichlet series multiplied by p.

(*b*) *Restricted Dirichlet series*

These are obtained by the specialization $\lambda_n = \log n$. In this case the operational relation (6) becomes

$$\sum_{n=1}^{\infty} \frac{a_n}{n^p} \doteqdot \sum_{n=1}^{\infty} a_n\, U(t-\log n) = \sum_{n=1}^{\infty} a_n\, U(e^t - n) = \sum_{n=1}^{[e^t]} a_n, \quad \max(\alpha, 0) < \operatorname{Re} p < \infty. \tag{10}$$

The original is a step function jumping at $t = \log 1, \log 2$, etc.; if $a_n > 0$ the general behaviour may be surveyed from fig. 32. The classical example of such a restricted Dirichlet series is furnished by the ζ-function of Riemann (III, 36) for which $a_n = 1$, and thus from (10),

$$\zeta(p) \doteqdot \sum_{n=1}^{[e^t]} 1 = [e^t], \quad 1 < \operatorname{Re} p < \infty. \tag{11}$$

This important relation is closely connected† with the prime-number theorem to be discussed later (XII, § 5), together with other number-theoretical problems.

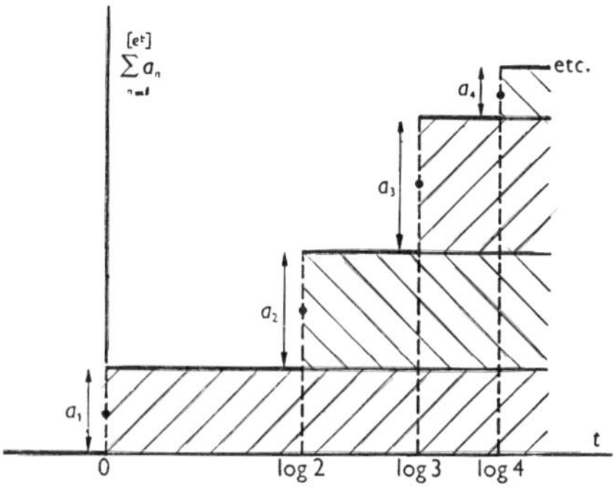

Fig. 32. The original of a restricted Dirichlet series.

As to the analytic continuation across the abscissa of convergence referred to above, let us return to (10). According to (8) the image of the right-hand function can be written as

$$\lim_{\mu\to\infty} \sum_{n=1}^{[e^\mu]} a_n \left(\frac{1}{n^p} - e^{-p\mu}\right). \tag{12}$$

Let us now specialize (10) and (12) by taking $a_1 = \frac{1}{2}$, $a_n = (-1)^{n+1}$ $(n \geqslant 2)$. In the first place (10) leads to

$$\frac{1}{2}\frac{1}{1^p} - \frac{1}{2^p} + \frac{1}{3^p} - \frac{1}{4^p} + \ldots \doteqdot -\tfrac{1}{2}(-1)^{[e^t]}\, U(t) \quad (\operatorname{Re} p > 0). \tag{13}$$

However, if μ runs through the numbers $\log N + \epsilon$ (ϵ arbitrarily small, positive), then the limit (12) leads to another series for the image of the right-hand member of (13), namely,

$$\frac{1}{2}\left(\frac{1}{1^p} - \frac{1}{2^p}\right) - \frac{1}{2}\left(\frac{1}{2^p} - \frac{1}{3^p}\right) + \frac{1}{2}\left(\frac{1}{3^p} - \frac{1}{4^p}\right) - \frac{1}{2}\left(\frac{1}{4^p} - \frac{1}{5^p}\right) + \ldots,$$

which converges even for $\operatorname{Re} p > -1$. Only if $\operatorname{Re} p > 0$ is it allowed to change the order of the terms; in that case the Dirichlet series of (13) is again obtained. Furthermore, if $\operatorname{Re} p > 1$, the series (13) can easily be evaluated

† See Balth. van der Pol, *Phil. Mag.* XXVI, 921, 1938.

in terms of the ζ-function by summing the positive and negative terms separately. The resulting p-function represents the image as an analytic function even in $-1 < \operatorname{Re} p < 1$; accordingly, we finally obtain the operational relation

$$(1-2^{1-p})\zeta(p)-\tfrac{1}{2} \doteq [e^t]-2[\tfrac{1}{2}e^t]-\tfrac{1}{2}U(t) = -\tfrac{1}{2}(-1)^{[e^t]}U(t), \quad -1 < \operatorname{Re} p < \infty.$$

(c) *Power series*

This kind of series admits of an operational attack in various ways. First they can be transformed into a Dirichlet series by the substitution $z = e^{-p}$:

$$\sum_{n=0}^{\infty} a_n z^n = \sum_{n=0}^{\infty} a_n e^{-pn},$$

and consequently, if (6) is applied with $\lambda_n = n$,

$$\sum_{n=0}^{\infty} a_n e^{-pn} \doteq \sum_{n=0}^{[t]} a_n. \tag{14}$$

Again, the operational relation (14) may have a region of convergence larger than that of the series itself. Certainly the image function cannot be defined by the series as it stands, outside its region of convergence. In the following argument, however, we shall always tacitly assume that in those cases the series is replaced by its analytic continuation. The domain of convergence of the series on the left of (14), that is, $\operatorname{Re} p > \alpha$, is transformed by the change of variables $p = -\log z$ into $|z| < e^{-\alpha}$; in other words, $\rho = e^{-\alpha}$ is the radius of convergence of the original power series. Conversely, the circular domain $|z| < \rho$, inside which the power series converges, corresponds (via the transformation $z = e^{-p}$) to the infinite strip of convergence $\alpha < \operatorname{Re} p < \infty$ of the Dirichlet series.

Secondly, the power series can be treated operationally by introducing the parameter p according to $z = 1/p$. Then the following operational relation is obtained:

$$\sum_{n=0}^{\infty} \frac{a_n}{p^n} \doteq U(t) \sum_{n=0}^{\infty} a_n \frac{t^n}{n!}, \quad 1/\rho < \operatorname{Re} p < \infty. \tag{15}$$

The translation into the t-language may be performed term by term. This follows from the property that the sum of any convergent series is in any way equal to its *Borel sum*†:

$$\sum_{n=0}^{\infty} c_n = \int_0^{\infty} e^{-t} \sum_{n=0}^{\infty} \frac{c_n}{n!} t^n \, dt, \tag{16}$$

and from the fact that the Borel sum for $c_n = a_n/p^n$ coincides with the definition integral for (15) for those real values of p that satisfy $p > \rho^{-1}$. Consequently, the definition integral of (15) certainly converges for $p > \rho^{-1}$,

† See, for instance, E. T. Whittaker and G. N. Watson, *A Course of Modern Analysis*, Cambridge, 1940, p. 154.

and thus for $\operatorname{Re} p > \rho^{-1}$; this is in agreement with the specification of the strip of convergence as indicated in (15). If necessary, the series $\Sigma(a_n/p^n)$ should be replaced by its analytic continuation, that is, its definition integral. Moreover, if the strip of convergence of (15) happens to be wider than $\operatorname{Re} p > \rho^{-1}$, then the definition integral can provide us with the analytic continuation of the power series which itself only converges for $|p| > \rho^{-1}$.

As an example let us consider the binomial series

$$\left(1-\frac{1}{p}\right)^{\nu} = \sum_{r=0}^{\infty} \binom{\nu}{r} \frac{(-1)^r}{p^r},$$

which converges for $|p| > 1$; the corresponding original

$$U(t) \sum_{r=0}^{\infty} \binom{\nu}{r} \frac{(-t)^r}{r!}$$

has a definition integral convergent for $\operatorname{Re} p > 0$; thus the strip of convergence is wider than the domain of convergence of the initial series. Especially for ν positive integral, when the series terminates, the right member is equal to the Laguerre polynomial $L_n(t)$ except for the factor $n!$; we thus have the operational relation

$$L_n(t)\, U(t) \risingdotseq n! \left(1-\frac{1}{p}\right)^{n}, \quad 0 < \operatorname{Re} p < \infty. \tag{17}$$

Other operational relations in connexion with power series of more complicated character will be derived in XI, § 6.

4. The strip of convergence of operational relations

Whereas in II, § 5 we derived the existence of a certain strip of convergence $\alpha < \operatorname{Re} p < \beta$ for the two-sided definition integral from the fact that the one-sided integral

$$\int_0^{\infty} e^{-pt} h(t)\, dt$$

in general has a domain of convergence $\alpha < \operatorname{Re} p < \infty$, we shall now develop explicit formulae for the strip in the case of any given original $h(t)$. For the sake of convenience we may first recall some properties of limits. A limit or limit-point L of a point set is any point with the property that any interval around L, no matter how small, contains an unlimited number of points belonging to the set in question. The meaning of the 'greatest of the limits' and of the 'least of the limits' is then obvious; they are denoted by $\overline{\text{Lim}}$ and $\underline{\text{Lim}}$, respectively. Thus, for instance, if $\overline{\text{Lim}}_{n\to\infty}\, a_n = A$, then it follows from the definition that, for any given positive number ϵ, a number $N(\epsilon)$

can be found such that $a_n < A + \epsilon$ for $n > N$; moreover, $a_n > A - \epsilon$ must hold for an infinite number of n's. For example, if x is real,

$$\overline{\operatorname{Lim}}_{x \to \infty} \sin x = 1, \quad \underline{\operatorname{Lim}}_{x \to \infty} \sin x = -1.$$

We now assume a definition integral converging for some value p_0 in the right-hand half of the p-plane ($\operatorname{Re} p_0 > 0$). Then it can be shown† that the integral $\int_{-\infty}^{t} h(s)\, ds$ exists for all finite values of t. With the help of this integral the definition integral can be written as

$$p \int_{-\infty}^{\infty} \mathrm{e}^{-pt}\, d\left\{\int_{-\infty}^{t} h(s)\, ds\right\},$$

from which it further follows, if α (supposed to be positive) denotes the left boundary of the domain of convergence,

$$\alpha = \overline{\operatorname{Lim}}_{x \to \infty} \left\{\frac{1}{x} \log \left| \int_{-\infty}^{x} h(s)\, ds \right| \right\}. \tag{18}$$

Similarly, starting from some p_0 in the left half of the p-plane, it can be shown that the analogous integral $\int_{+\infty}^{t} h(s)\, ds$ exists for any finite t. In this case α, which now is necessarily negative, is to be determined from

$$\alpha = \overline{\operatorname{Lim}}_{x \to \infty} \left\{\frac{1}{x} \log \left| \int_{+\infty}^{x} h(s)\, ds \right| \right\}. \tag{19}$$

The above formulae supply a complete specification of the strip of convergence $\alpha < \operatorname{Re} p < \infty$ of any right-sided integral $p \int_{0}^{\infty} \mathrm{e}^{-pt} h(t)\, dt$. Moreover, by transforming the left-sided integral $p \int_{-\infty}^{0} \mathrm{e}^{-pt} h(t)\, dt$ into a right-sided one (compare II, § 5), we obtain an analogous formula for the right-hand boundary β, simply by replacing in (18) and (19) $\overline{\operatorname{Lim}}_{x \to +\infty}$ by $\underline{\operatorname{Lim}}_{x \to -\infty}$.

The above results concerning α and β can be summarized in the following expression for the strip of convergence,

$$\overline{\operatorname{Lim}}_{x \to \infty} \left\{\frac{1}{x} \log \left| \int_{\mp\infty}^{x} h(s)\, ds \right| \right\} < \operatorname{Re} p < \underline{\operatorname{Lim}}_{x \to -\infty} \left\{\frac{1}{x} \log \left| \int_{\mp\infty}^{x} h(s)\, ds \right| \right\}, \tag{20}$$

if it is understood that in either side the lower limit of integration is chosen as $-\infty$, provided this will lead to a positive limit, and as $+\infty$, provided this will lead to a negative limit. At first sight there remains an ambiguity in that, for instance, the left-hand member might yield simultaneously a positive limit for $-\infty$ and a negative limit for $+\infty$. This, however, is impossible; it would imply the indeterminateness of the abscissa of con-

† For proofs concerning the following statements, when the theory is based upon the Stieltjes integral, see Widder, loc. cit. pp. 43–4.

vergence of the one-sided integral $\int_0^\infty e^{-pt}h(t)\,dt$. Only if $\alpha = 0$ would we find both limits in the left member to be equal to zero.

It is to be remarked that (20) can be used in two different directions. First, it provides us with the strip of convergence if $h(t)$ is given. Secondly, it may prove useful in determining properties of $h(t)$ when the strip of convergence is known. For instance, the existence of the limit (α, say) of the left-hand member implies that ($\alpha > 0$)

$$\left|\int_{-\infty}^{x} h(s)\,ds\right| \leqslant e^{(\alpha+\epsilon)x}, \tag{21}$$

which is valid for sufficiently large values of x, that is, for $x > x_0$, where x_0 depends on ϵ of course.

Furthermore, formula (20) includes the case that $h(t)$ does not have a two-sided Laplace integral at all. This will occur when the right-hand member turns out to be less than the left-hand member, in other words, when there is no space available in the p-plane for a proper strip of convergence. On the other hand, the formula (20) does not answer questions concerning convergence on the boundaries $\operatorname{Re} p = \alpha$, $\operatorname{Re} p = \beta$ themselves (for these questions, see § 10). Deferring more general questions concerning the strip of convergence to the next section, we shall demonstrate here the usefulness of (20), in a simple example.

Example. In order to determine the strip of convergence with respect to the original $h(t) = \sin(e^{-t})$ we have to know the behaviour for $x \to \pm\infty$ of the functions

$$\int_{\mp\infty}^{x} \sin(e^{-s})\,ds = -\int_{\substack{\infty\\0}}^{e^{-x}} \frac{\sin u}{u}\,du.$$

We obtain without much difficulty

$$\int_{-\infty}^{x} h(s)\,ds \sim \begin{cases} \tfrac{1}{2}\pi & (x \to +\infty),\\ e^{x}\cos(e^{-x}) & (x \to -\infty),\end{cases}$$

$$\int_{\infty}^{x} h(s)\,ds \sim \begin{cases} -e^{-x} & (x \to +\infty),\\ -\tfrac{1}{2}\pi & (x \to -\infty).\end{cases}$$

The first and the last of these relations follow from the Dirichlet integral; the second is obtained by means of a partial integration with respect to u from the integrated term, whilst the third relation results from the approximation $\sin u \sim u$ for $u \to 0$. We then find the four limits of (20) to be

$$\overline{\operatorname*{Lim}_{x\to\infty}} \left\{\frac{1}{x}\log\left|\int_{-\infty}^{x} h(s)\,ds\right|\right\} = 0,$$

$$\underline{\operatorname*{Lim}_{x\to-\infty}} \left\{\frac{1}{x}\log\left|\int_{-\infty}^{x} h(s)\,ds\right|\right\} = 1,$$

$$\overline{\operatorname*{Lim}_{x\to\infty}} \left\{\frac{1}{x}\log\left|\int_{\infty}^{x} h(s)\,ds\right|\right\} = -1,$$

$$\underline{\operatorname*{Lim}_{x\to-\infty}} \left\{\frac{1}{x}\log\left|\int_{\infty}^{x} h(s)\,ds\right|\right\} = 0.$$

Thus from (20) the strip of convergence is $-1 < \operatorname{Re} p < 1$.

We shall also derive the corresponding image, since that will be required later on. First the images of the exponential functions $e^{\pm ie^{-t}}$ are determined separately. Using the same substitution $u = e^{-t}$ as before, we find

$$p\int_0^\infty e^{\pm iu} u^{p-1} du = p\, e^{\pm \frac{1}{2} i\pi p} \int_0^{\mp i\infty} e^{-v} v^{p-1} dv.$$

The integral at the right can be transformed into Euler's Γ-function integral by putting

$$\int_0^{\mp i\infty} = \operatorname*{Lim}_{\lambda\to\infty} \left\{ \int_0^\lambda + \int_\lambda^{\mp i\lambda} \right\}.$$

The first integral tends to $\Gamma(p)$ if $\operatorname{Re} p > 0$ whilst the second is easily shown to vanish in the limit $\lambda \to \infty$, provided $\operatorname{Re} p < 1$. We have thus established the two operational relations

$$\exp(\pm i\, e^{-t}) \doteqdot e^{\pm \frac{1}{2} i\pi p}\, \Pi(p), \quad 0 < \operatorname{Re} p < 1. \tag{22}$$

Further, by addition or subtraction,

$$\cos(e^{-t}) \doteqdot \cos\left(\frac{\pi}{2} p\right) \Pi(p), \quad 0 < \operatorname{Re} p < 1, \tag{23}$$

$$\sin(e^{-t}) \doteqdot \sin\left(\frac{\pi}{2} p\right) \Pi(p), \quad -1 < \operatorname{Re} p < 1. \tag{24}$$

Although the image (24) is here derived only for $0 < \operatorname{Re} p < 1$ it also holds for $-1 < \operatorname{Re} p < 0$, since it is analytic and the strip of convergence is known to be $-1 < \operatorname{Re} p < 1$.

5. Particular cases of the strip of convergence

In this section we shall discuss some special cases of the general formula (20).

(1) *Originals integrable at $t = 0$, and one-sided originals*

We first notice that the integrals $\int_{-\infty}^x h(s)\, ds$ and $\int_{+\infty}^x h(s)\, ds$ in the left- and right-hand members of formula (20) may be replaced by $\int_0^x h(s)\, ds$ if the latter integral exists. The introduction of the former integrals is important only if the integrals have to be taken as principal values with respect to $t = 0$, as was done in the original of (2). In particular, one-sided originals $h(t)\, U(t)$ must be integrable at $t = 0$. For the latter the strip of convergence $\operatorname{Re} p > \alpha$ is determined by

$$(a)\ \text{if } \alpha > 0: \qquad \overline{\operatorname*{Lim}_{x\to\infty}} \left\{ \frac{1}{x} \log \left| \int_0^x h(s)\, ds \right| \right\} < \operatorname{Re} p < \infty, \tag{25a}$$

$$(b)\ \text{if } \alpha < 0: \qquad \overline{\operatorname*{Lim}_{x\to\infty}} \left\{ \frac{1}{x} \log \left| \int_x^\infty h(s)\, ds \right| \right\} < \operatorname{Re} p < \infty. \tag{25b}$$

(2) *The strip of convergence of $h'(t)$*

The strip of convergence of the derivative of an original can be given much more easily than that of the original itself, since the corresponding integrals $\int_{\mp\infty}^{x} h(s)\,ds$ can be solved explicitly. Thus the strip of convergence of $h'(t)$ is determinable from

$$\overline{\lim_{x\to\infty}}\left\{\frac{1}{x}\log|h(x)-h(\mp\infty)|\right\} < \operatorname{Re} p < \underline{\lim_{x\to-\infty}}\left\{\frac{1}{x}\log|h(x)-h(\mp\infty)|\right\}, \tag{26}$$

in which, again, the upper signs should be taken if this leads to a positive limit, and the lower signs if this leads to a negative limit. Of course, (26) determines the strip of $h(t)$ itself if $h(t)$ and $h'(t)$ are known to have the same domain of convergence.

(3) *The Dirichlet series*

Since we have seen that the strip of convergence of the Dirichlet series

$$\sum_{n=1}^{\infty} a_n e^{-p\lambda_n} \quad (\operatorname{Re} p > \alpha)$$

is equal to that of (9), we easily obtain for the abscissa of convergence α, by first putting in (25)

$$h(t) = \sum_{n=1}^{\infty} a_n \delta(t-\lambda_n)$$

the following formulae:

(*a*) if $\alpha > 0$:
$$\alpha = \overline{\lim_{x\to\infty}}\left\{\frac{1}{x}\log\left|\sum_{n=1}^{\lambda_n<x} a_n\right|\right\}, \tag{27a}$$

(*b*) if $\alpha < 0$:
$$\alpha = \overline{\lim_{x\to\infty}}\left\{\frac{1}{x}\log\left|\sum_{\lambda_n>x}^{\infty} a_n\right|\right\}. \tag{27b}$$

These formulae are well known, certainly, but the interesting point is that here they result from a specialization of the more general theory of the Laplace transform which again illustrates the importance of the latter theory. In particular, $\lambda_n = n$ leads to formulae for the radius of convergence of power series. For instance, Cauchy's formula is easily obtained in the way indicated; to this end, however, the series $\Sigma a_n e^{-np}$ should first be multiplied by $1-e^{-p}$ leading to a new series with the same domain of convergence.

(4) *Cut-off functions*

These functions are zero for $t > b$ as well as for $t < a$ $(a < b)$; their general form is therefore

$$h(t)\{U(t-a)-U(t-b)\} \quad (a < b).$$

The corresponding domain of convergence covers the whole complex plane of p, since the limits of integration in the definition integral are both finite. The same result is of course also obtained by application of formula (20). Less trivial are originals that have a convergent Laplace integral for any finite p though not belonging to the type of functions under consideration.

As an example (see also (II, 24) and (VI, 49)) we may mention the functions that behave like $e^{-|t|^{1+\alpha}}$ $(\alpha > 0)$ for $t \to \pm\infty$. It should further be remarked that the δ-function and its derivatives are limiting cases of cut-off functions; indeed, their domain of convergence is the whole p-plane.

6. Absolute convergence

Whereas in §4 we studied the definition integral with respect to ordinary convergence, we shall now deal with absolute convergence. We thus seek values of p for which the definition integral is absolutely convergent; that is, for which the following integral exists:

$$p\int_{-\infty}^{\infty} e^{-\mathrm{Re}\,p\,t}\,|\,h(t)\,|\,dt.$$

Since in this Laplace integral too only $\mathrm{Re}\,p$ is decisive for its convergence, we observe that absolute convergence for $h(t)$ coincides with ordinary convergence of the new original $|\,h(t)\,|$. Therefore the domain of absolute convergence can be found at once from formula (20) by simply replacing $h(t)$ by $|\,h(t)\,|$:

$$\overline{\mathrm{Lim}}_{x\to\infty}\left\{\frac{1}{x}\log\left(\pm\int_{\mp\infty}^{x}|\,h(s)\,|\,ds\right)\right\} < \mathrm{Re}\,p < \underline{\mathrm{Lim}}_{x\to-\infty}\left\{\frac{1}{x}\log\left(\pm\int_{\mp\infty}^{x}|\,h(s)\,|\,ds\right)\right\}, \tag{28}$$

in which the meaning of $\pm$ is as before. Henceforth this new domain of absolute convergence will be denoted by

$$\alpha_a < \mathrm{Re}\,p < \beta_a.$$

Since absolute convergence implies ordinary convergence, the strip of absolute convergence lies wholly inside that of ordinary convergence or, at best, coincides with it (see fig. 33); in other words

$$\alpha \leqslant \alpha_a \leqslant \beta_a \leqslant \beta.$$

It is often more simple to investigate the absolute convergence directly, instead of with the help of (28).

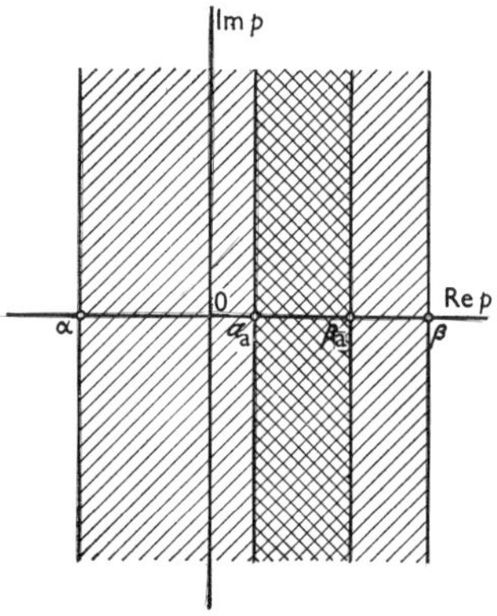

Fig. 33. The strips of ordinary and of absolute convergence.

Both strips of convergence, that of absolute and that of ordinary convergence, are bounded by vertical lines parallel to the imaginary axis of p. As to their mutual situation, the following may occur:

(1) The domain of absolute and that of ordinary convergence are *equal*. This obviously holds for all originals not changing sign. A non-trivial

example is furnished by the function $\sin t\, U(t)$; by dividing the interval of integration into equal parts of length π the definition integral for $|\sin t|\, U(t)$ is reduced to a series that can easily be summed for $\operatorname{Re} p > 0$. We then obtain

$$|\sin t|\, U(t) \doteqdot \frac{p}{p^2+1} \coth\left(\frac{\pi p}{2}\right), \quad 0 < \operatorname{Re} p < \infty, \tag{29}$$

in which the domain of convergence is just that of $\sin t\, U(t)$ itself. Furthermore, any operational relation that can be derived by means of the integration rule (IV, 34) has equal domains of absolute and ordinary convergence†.

(2) The domain of absolute convergence is only a *part* of that of ordinary convergence. As an example the previous function $\sin(e^{-t})$ may be mentioned. Here the definition integral is absolutely convergent if (substituting $u = e^{-t}$) the convergence of

$$p \int_0^\infty u^{p-1} |\sin u|\, du$$

is ascertained. The latter is true for $-1 < \operatorname{Re} p < 0$. Therefore the convergence is absolute only in the left half of the strip of convergence $-1 < \operatorname{Re} p < 1$.

(3) No domain of absolute convergence, of which the operational relation (23) is a simple example.

We may further remember that the Dirichlet series also has an abscissa of absolute convergence α_a; that is, (5) converges absolutely for $\operatorname{Re} p > \alpha_a$. An explicit expression for α_a is obtained from (27) by changing a_n into $|a_n|$.

7. Uniform convergence

In § 2 we stated that the image is an analytic function of p inside the strip of ordinary convergence. It can further be proved‡ that the one-sided integral $\int_0^\infty e^{-pt} h(t)\, dt$, as far as the upper limit of integration is concerned, converges uniformly in p inside any domain

$$-\tfrac{1}{2}\pi + \Delta \leqslant \arg(p - p_1) \leqslant \tfrac{1}{2}\pi - \Delta,$$

where p_1 is inside the strip of convergence and Δ arbitrarily small and positive. This means that for any given $\epsilon > 0$ a number $\lambda_0(\epsilon)$, independent of p, can be found, such that for any p subjected to the conditions above the following holds if $\lambda > \lambda_0$:

$$\left| \int_\lambda^\infty e^{-pt} h(t)\, dt \right| < \epsilon.$$

† For a proof in the case of one-sided originals, see G. Doetsch, *Theorie und Anwendung der Laplace-Transformation*, Berlin, 1937, and New York, 1943, p. 149.

‡ Doetsch, loc. cit. p. 41.

By considering the two-sided definition integral as the sum of two one-sided integrals, as we did before, we easily obtain the domain of uniform convergence (with respect to both upper and lower limits of integration) of the two-sided integral, namely,

$$-\tfrac{1}{2}\pi+\Delta_1<\arg(p-p_1)<\tfrac{1}{2}\pi-\Delta_1,$$
$$-\tfrac{1}{2}\pi+\Delta_2<\arg(p_2-p)<\tfrac{1}{2}\pi-\Delta_2,$$

in which p_1 and p_2 are now two points in the interior of the strip of ordinary convergence satisfying $\operatorname{Re}p_1<\operatorname{Re}p_2$. Such a domain is illustrated in fig. 34 by the region inside two sectors which are bounded by the dashed lines. Obviously any finite part G of the p-plane inside the strip of convergence $\alpha<\operatorname{Re}p<\beta$ (see the shaded part of fig. 34) is part of a suitably chosen sectorial domain. Therefore the definition integral divided by p supplies us with a uniformly convergent representation of $f(p)/p$ in any G.

One naturally asks whether such a domain of uniform convergence is also likely to include the infinite parts of the strip of convergence. If in the latter we put $p=c+i\mu$, the question amounts to whether the integral

$$\int_{-\infty}^{\infty} e^{-ct-i\mu t}\,h(t)\,dt$$

is uniformly convergent (with respect to the upper and lower limits of integration) for $\alpha<c<\beta$ and $-\infty<\mu<\infty$. It appears† that this condition is equivalent to the requirement that the definition integral for the new original, that is, $h(t)\,e^{-i\mu t}$, must ordinarily be convergent for *all* real values of μ *simultaneously*. We then find with the aid of (20) for the strip of uniform convergence

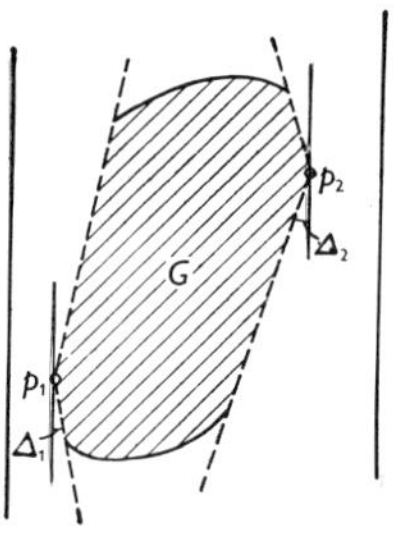

Fig. 34. A domain of uniform convergence of the Laplace integral.

$$\overline{\operatorname*{Lim}_{x\to\infty}}\left\{\frac{1}{x}\log\max\left|\int_{\mp\infty}^{x} h(s)\,e^{i\mu s}\,ds\right|\right\}<\operatorname{Re}p$$
$$<\underline{\operatorname*{Lim}_{x\to-\infty}}\left\{\frac{1}{x}\log\min\left|\int_{\mp\infty}^{x} h(s)\,e^{i\mu s}\,ds\right|\right\},\qquad(30)$$

or abbreviated, $\alpha_u<\operatorname{Re}p<\beta_u$, in which max and min refer to μ. This is again a domain bounded by lines parallel to the imaginary axis. Whereas the strips of ordinary convergence and of absolute convergence include any finite p of ordinary and absolute convergence, respectively, the strip of uniform convergence includes only those parts of the p-plane in which the uniform convergence holds, *not excepting infinity*; and further, the convergence is uniform in all other *finite* parts inside the strip of ordinary convergence. Moreover, it appears that the function $f(p)/p$ tends to zero uniformly in p if $|p|\to\infty$ with p inside the strip of uniform convergence‡.

† See Widder, loc. cit. p. 53.

‡ See Doetsch, loc. cit. p. 51.

Since uniform convergence implies ordinary convergence the domain of the former is necessarily inside that of the latter (the respective boundaries may coincide of course). In its turn the domain of absolute convergence lies inside that of uniform convergence. Consequently, there are in general three different strips of convergence with six vertical boundaries in total, such that $\alpha \leqslant \alpha_u \leqslant \alpha_a \leqslant \beta_a \leqslant \beta_u \leqslant \beta$. In fig. 35 all six boundaries are different, though various other combinations, in which the boundaries partly coincide, are equally possible.

Example. Let us return to the operational relation (24):

$$\sin(\mathrm{e}^{-t}) \doteq \sin\left(\frac{\pi p}{2}\right)\Pi(p),$$

whose domains of ordinary and absolute convergence were shown to be $-1<\operatorname{Re} p<1$ and $-1<\operatorname{Re} p<0$, respectively. Instead of using (30) we can determine its domain of uniform convergence more directly, whilst we may further confine ourselves to $0<\operatorname{Re} p<1$. The integral to be investigated is

$$\int_0^\infty u^{p-1}\sin u\,du.$$

The lower limit of integration needs no special attention; we have only to examine for what values of p the integral

$$\int_\lambda^\infty u^{p-1}\sin u\,du$$

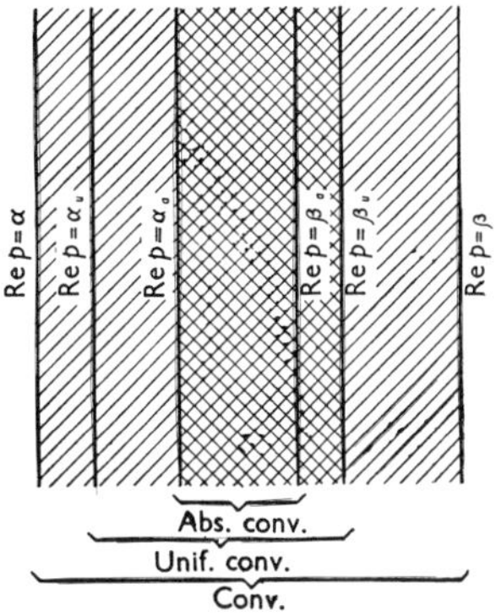

Fig. 35. The strips of ordinary, of absolute, and of uniform convergence.

tends to zero uniformly in p for $\lambda\to\infty$. This appears to be the case for $\operatorname{Re} p<\frac{1}{2}$, the abscissa $\operatorname{Re} p=\frac{1}{2}$ being the separation between the domains in which $f(p)/p$ tends to zero and to infinity, respectively, for $|\operatorname{Im} p|\to\infty$; the value $\operatorname{Re} p = \frac{1}{2}$ results from the asymptotic relation

$$|\Gamma(x+iy)| \sim \sqrt{2\pi}\,|y|^{x-\frac{1}{2}}\mathrm{e}^{-\pi|y|/2} \quad (|y|\to\infty).$$

Summarizing, we thus have for the definition integral of $\sin(\mathrm{e}^{-t})$:

ordinary convergence	if	$-1<\operatorname{Re} p<1$,
uniform convergence	if	$-1<\operatorname{Re} p<\frac{1}{2}$,
absolute convergence	if	$-1<\operatorname{Re} p<0$.

In this example the left-hand boundaries of all three domains of convergence coincide, whilst the right-hand boundaries are all different. An example having six different boundaries is now readily obtained by application of some elementary rules to the example above, viz.

$$\mathrm{e}^{\vartheta t}\sin(\mathrm{e}^{-t})+\mathrm{e}^{-\vartheta t}\sin(\mathrm{e}^{t}) \doteq p[\sin\{\tfrac{1}{2}\pi(p-\vartheta)\}\,\Gamma(p-\vartheta)-\sin\{\tfrac{1}{2}\pi(p+\vartheta)\}\,\Gamma(-p-\vartheta)],$$

in which ϑ is arbitrary but for $0<\vartheta<\frac{1}{4}$; in this relation there is

ordinary convergence	if	$-1+\vartheta<\operatorname{Re} p<1-\vartheta$,
uniform convergence	if	$-\frac{1}{2}-\vartheta<\operatorname{Re} p<\frac{1}{2}+\vartheta$,
absolute convergence	if	$-\vartheta<\operatorname{Re} p<\vartheta$.

8. Summable series

With a view to later investigations concerning the inversion integral and the boundaries of the strips of convergence we shall first recall, in this and the next section, something about the existing methods of 'summing' series and integrals. In 'summing' series we investigate what meaning, if any, can be attached to the 'sum' of a divergent series; of course, the new definition must always be such that for ordinarily convergent series the new sum is equal to the old (condition of permanence or consistency).

Let us first consider a series $\sum_0^\infty a_n$ whose partial sum $S_N = \sum_{n=0}^{N-1} a_n$ oscillates for large values of N without leading to a definite limit. It can generally be said that the new methods of summing amount to replacing the ordinary sum

$$S^{(0)} = \operatorname*{Lim}_{N\to\infty} S_N = \operatorname*{Lim}_{N\to\infty} \sum_{n=0}^{N-1} a_n$$

by the following:
$$S = \operatorname*{Lim}_{N\to\infty} \sum_{n=0}^{N-1} a_n \mu(n,N),$$

in which $\mu(n,N)$ is some suitably chosen 'weight function'.

Some special sum definitions all satisfying the condition of consistency are:

(1) The limit of the mean value of the partial sum S_N. In this case S becomes

$$S^{(1)} = \operatorname*{Lim}_{N\to\infty} \frac{1}{N} \sum_{n=0}^{N-1} S_n = \operatorname*{Lim}_{N\to\infty} \frac{1}{N} \sum_{n=1}^{N-1} \sum_{m=0}^{n-1} a_m$$

$$= \operatorname*{Lim}_{N\to\infty} \frac{1}{N} \sum_{n=0}^{N-2} a_n(N-n-1) = \operatorname*{Lim}_{N\to\infty} \frac{1}{N} \sum_{n=0}^{N-1} a_n(N-n). \quad (31)$$

This is equivalent to the introduction of the weight function

$$\mu_1(n,N) = 1-\frac{n}{N}.$$

The transformation of the double sum performed in (31) is based on reversing the summations over n and m.

(2) The limits of higher-order mean values (method of Hölder). That of the order r is obtained recurrently from that of the order $r-1$ in the same way as $S^{(1)}$ is obtained from $S^{(0)}$; in other words, $S^{(r)}$ is the limit of the mean values of the partial sums belonging to $S^{(r-1)}$. Therefore the rth-order Hölder sum is defined explicitly by

$$S^{(r)} = \operatorname*{Lim}_{n_0\to\infty} \frac{1}{n_0} \sum_{n_1=0}^{n_0-1} \frac{1}{n_1} \sum_{n_2=0}^{n_1-1} \frac{1}{n_2} \cdots \frac{1}{n_{r-1}} \sum_{n_r=0}^{n_{r-1}-1} \sum_{n_{r+1}=0}^{n_r-1} a_{n_{r+1}}.$$

(3) The rth-order Cesàro limits, which may best be defined as follows:

$$(C,r) = \operatorname*{Lim}_{N\to\infty} \frac{\sum\limits_{n_1=0}^{N-1}\sum\limits_{n_2=0}^{n_1}\sum\limits_{n_3=0}^{n_2}\cdots\sum\limits_{n_{r+1}=0}^{n_r} a_{n_{r+1}}}{\sum\limits_{n_1=0}^{N-1}\sum\limits_{n_2=0}^{n_1}\sum\limits_{n_3=0}^{n_2}\cdots\sum\limits_{n_r=0}^{n_{r-1}} 1}. \tag{32}$$

After a reversal of the order of summation this definition turns out to be equivalent to taking the weight function

$$\mu^{(r)}(n,N) = \left(1-\frac{n}{N}\right)\left(1-\frac{n}{N+1}\right)\cdots\left(1-\frac{1}{N+r-1}\right).$$

The Hölder and Cesàro limits of the first order are obviously equal. The same holds, however, for the higher-order limits, since it has been proved that the existence of the rth-order Hölder sum implies the existence (and equality) of the rth-order Cesàro sum, and vice versa†. Consequently, the two methods of summing are of equal value. Whereas the Hölder definition has the advantage that the sum can be simply formulated in terms of mean values, the Cesàro definition involves a much simpler weight function. It is further known that the existence of the rth-order sum implies the existence of those of any order exceeding r. The probability that a series can be summed increases, as it were, with increasing value of r. This is because for increasing r the weight function so changes that the higher terms of the series are more and more suppressed. This is schematically indicated in fig. 36, where the Cesàro weight functions of the orders 1 and 2 are shown together with the horizontal line $\mu = 1$ which represents the constant weight function in case of the ordinary definition of the sum ($r = 0$). It will further be clear that 'two-sided' series can be treated as well; for instance, the first-order sum is then defined by

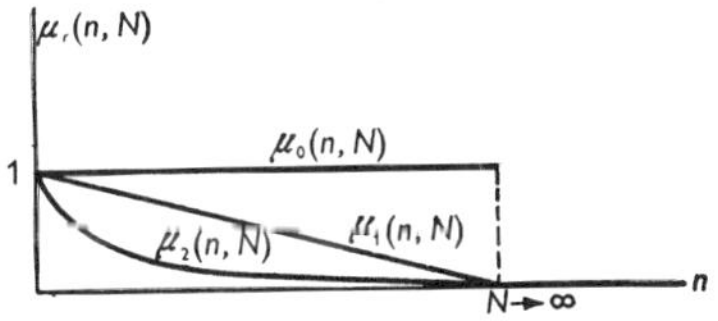

Fig. 36. The weight functions related to Cesàro's summations.

$$(C,1)\sum_{n=-\infty}^{\infty} a_n \equiv \operatorname*{Lim}_{N\to\infty}\sum_{n=-N+1}^{N-1} a_n\left(1-\frac{|n|}{N}\right).$$

Example. Let us consider the series $\sum\limits_{-\infty}^{\infty} e^{2\pi inx}$ which is not convergent in the elementary sense. Its partial sum is easily found to be

$$\sum_{n=-N+1}^{N-1} e^{2\pi inx} = \frac{\sin\{2\pi(N-\frac{1}{2})x\}}{\sin(\pi x)},$$

† See K. Knopp, Inaugural Diss. Berlin, 1907, p. 19; W. Schnee, *Math. Ann.* LXVII, 110, 1909.

which obviously has no limit for $N \to \infty$. Let us now take the first-order Cesàro limit, that is, the limit for $N \to \infty$ of

$$\sum_{n=-N+1}^{N-1} e^{2\pi inx}\left(1-\frac{|n|}{N}\right) = \frac{\sin^2(\pi Nx)}{N\sin^2(\pi x)}.$$

In the limit $N \to \infty$ the right-hand member becomes an infinite sum of equidistant δ-functions, thus

$$(C,1)\sum_{n=-\infty}^{\infty} e^{2\pi inx} = \sum_{n=-\infty}^{\infty} \delta(x-n). \tag{33}$$

The sifting property of this function is simply the equivalent of the following well-known formula of Poisson:

$$\sum_{n=-\infty}^{\infty}\int_{-\infty}^{\infty} h(x)\, e^{2\pi inx}\, dx = \sum_{n=-\infty}^{\infty} h(n), \tag{34}$$

for the validity of which (in the sense of ordinary sums: $r = 0$) it is sufficient that the series $\sum_{-\infty}^{\infty} h(n+x)$ is uniformly convergent in the interval $0 \leqslant x \leqslant 1$ and expansible into a Fourier series. Further, by r-times differentiation of the series (33) a new series is obtained which is summable $(C, r+1)$:

$$(C, r+1)\sum_{n=-\infty}^{\infty} (2\pi in)^r e^{2\pi inx} = \sum_{n=-\infty}^{\infty} \delta^{(r)}(x-n).$$

Compared with the Hölder-Cesàro method Abel's method is still more effective. The Abel sum of Σa_n is defined as the limit of the corresponding power series:

$$\operatorname*{Lim}_{x\to 1-0}\sum_{n=0}^{\infty} a_n x^n.$$

The existence of the Abel sum is ascertained when the series in question is known to be summable (C, r) for some value of r. Moreover, the condition of consistency is always satisfied; the same holds true with respect to the Borel sum mentioned in (16), which is still more effective than the Abel sum.

9. Summable integrals

We are now going to discuss analogous methods of summing divergent integrals. If the integral $\int_0^\infty h(x)\,dx$ diverges with respect to the conventional definition

$$\operatorname*{Lim}_{\lambda\to\infty}\int_0^\lambda h(x)\,dx,$$

it may yet happen that a convergent expression is obtained after introducing a suitable weight function $\mu(x, \lambda)$; we then conveniently define the integral in question as equal to the following limit:

$$\operatorname*{Lim}_{\lambda\to\infty}\int_0^\lambda h(x)\,\mu(x,\lambda)\,dx.$$

Again the condition of consistency has to be fulfilled. Corresponding to the first-order Cesàro-Hölder sum we now have for $\mu(x, \lambda) = 1-\dfrac{x}{\lambda}$

$$\operatorname*{Lim}_{\lambda\to\infty}\frac{1}{\lambda}\int_0^\lambda d\mu\int_0^\mu dx\, h(x).$$

Further, the analogue of (32) is the Cesàro limit of order r:

$$(C,r)\int_0^\infty h(x)\,dx \equiv \operatorname*{Lim}_{\lambda\to\infty} \frac{\int_0^\lambda d\mu_1 \int_0^{\mu_1} d\mu_2 \int_0^{\mu_2} d\mu_3 \dots \int_0^{\mu_r} d\mu_{r+1}\, h(\mu_{r+1})}{\int_0^\lambda d\mu_1 \int_0^{\mu_1} d\mu_2 \int_0^{\mu_2} d\mu_3 \dots \int_0^{\mu_{r-1}} d\mu_r}, \tag{35}$$

which according to (IV, 35) is equivalent to

$$(C,r)\int_0^\infty h(x)\,dx \equiv \operatorname*{Lim}_{\lambda\to\infty} \int_0^\lambda h(x)\left(1-\frac{x}{\lambda}\right)^r dx. \tag{36}$$

The advantage of the last expression is that it also furnishes Cesàro limits of non-integral orders. These integrals have properties quite analogous to the corresponding (C,r) sums. In the first place the principle of consistency is not violated, whilst, furthermore, the existence of the rth-order limit implies that of any order exceeding r (whether integral or not), and, moreover, the higher-order limits yield the same result†. Therefore also in the case of integrals the probability of a possible summing grows with increasing r. In addition, it can be proved that the Cesàro limit of any order r, if existing, is equal to the following expression:

$$\operatorname*{Lim}_{\epsilon\to+0} \int_0^\infty h(x)\, e^{-\epsilon x}\, dx,$$

which we would call Cauchy's limit, since the Fourier identity was derived by Cauchy as such a limit (see example 1 of V, § 8). The Cauchy limit is more effective than any Cesàro limit, as the former is the limiting case $r\to\infty$ of the latter. This follows from (36) by putting $\lambda = r/\epsilon$ and letting r tend to infinity. Moreover, if $e^{-\epsilon}$ is replaced by x, then the Cauchy limit is recognized as the analogue of the Abel sum in the theory of series.

So far we have treated the methods of summing divergent integrals that are of the greatest importance for the operational calculus. We may add that the corresponding treatment of two-sided integrals is also easy. For instance, the Cesàro limit of order r then becomes

$$(C,r)\int_{-\infty}^\infty h(x)\,dx \equiv \operatorname*{Lim}_{\lambda\to\infty} \int_{-\lambda}^\lambda h(x)\left(1-\frac{|x|}{\lambda}\right)^r dx.$$

In this respect we may consider the principal value, already dealt with in chapter II in the discussion of the Fourier identity, as a Cesàro limit of order zero.

Example. Summable in the Cesàro sense are the otherwise divergent integrals

$$\int_0^\infty \cos(\omega t)\, d\omega \quad \text{and} \quad \int_0^\infty \sin(\omega t)\, d\omega.$$

† See E. C. Titchmarsh, *Introduction to the Theory of Fourier Integrals*, Oxford, 1937, p. 27.

With definition (36) the first integral is found from (v, 34) which represents the Cesàro limit of arbitrary order $r>0$:

$$(C,r)\frac{1}{\pi}\int_0^\infty \cos(\omega t)\,d\omega = \delta(t) \quad (r>0). \tag{37}$$

This formula occurs in Heaviside's work†, though less rigorously. The second integral can be reduced to the first by an integration by parts, and turns out to be summable (C,r) for any positive order r:

$$(C,r)\int_0^\infty \sin(\omega t)\,d\omega = \begin{cases} 1/t & (t\neq 0) \\ 0 & (t=0) \end{cases} \quad (r>0). \tag{38}$$

Whereas (37), as a function of t, has the character of a resonance curve (describing, for instance, the amplitude of a forced oscillation versus frequency in the vicinity of resonance), the function (38) shows some features of a dispersion curve (cf. the refractive index versus frequency in the region of anomalous dispersion).

10. The behaviour of the definition integral on the boundaries of the strip of convergence

It is always certain that the definition integral converges throughout the *interior* of some strip of convergence $\alpha < \operatorname{Re} p < \beta$, and, moreover, that it diverges outside the said domain. But a general and definite statement concerning the behaviour on the boundaries $\operatorname{Re} p = \alpha$, $\operatorname{Re} p = \beta$ is not possible, except to state that at those points where the definition integral does converge the latter is the analytic continuation of the image function inside the strip of convergence (see VII, § 3). Several different cases concerning the behaviour on the boundaries are possible, just as for power series with respect to their circle of convergence. It is well known that the latter series have at least one singularity on the circle of convergence. The analogue of this property does hold for the Laplace transform, that is, the function $f(p)/p$ does necessarily have a singularity on the boundary of the strip of convergence. It should be borne in mind, however, that this singularity need not lie on the finite part of the boundary. On the other hand, it has been proved‡ that the real point $p = \alpha$ of the left-hand boundary of the strip of convergence is always a singularity of $f(p)/p$, provided the real and one-sided original $h(t)$ does not change sign for sufficiently large values of t. This theorem is an extension of the well-known property that the intersection of the positive real axis with the circle of convergence of any power series of positive coefficients is always singular.

It will thus be obvious that with respect to the nature of the singularities and that of the other points, the boundaries admit of various possibilities; among the singularities there may be found poles, branch points, logarithmic singularities, etc.; moreover, at non-singular points the defini-

† *Electromagnetic Theory*, vol. II, p. 100.
‡ See Doetsch, loc. cit. p. 59.

tion integral may exist no longer, or only in the sense of a Cauchy or Cesàro limit. In the last case the least order r required may still be different for different points, etc. This great variety of possibilities may now be illustrated by some examples.

(1) $$e^{\lambda t} U(t) \doteqdot \frac{p}{p-\lambda}, \quad \operatorname{Re}\lambda < \operatorname{Re}p < \infty.$$

In this case $f(p)/p$ has only one singularity on the boundary $\operatorname{Re}p = \operatorname{Re}\lambda$, namely, the pole at $p = \lambda$. On the remaining part of the boundary the definition integral is summable (C, r) for any > 0, as can be proved with the help of (37) and (38).

(2) $$\sum_{n=0}^{\infty} \delta(t-n) \doteqdot \frac{p}{1-e^{-p}}, \quad 0 < \operatorname{Re}p < \infty.$$

In this example the function $f(p)/p$ has an infinite number of equidistant poles on the boundary $\operatorname{Re}p = 0$, namely, at $p = 2\pi i n$, where n is integral. With the exception of these poles the definition integral is summable $(C, 1)$, since the first-order Cesàro limit then leads to

$$\operatorname*{Lim}_{\lambda\to\infty} i\omega \int_{-\lambda}^{\lambda} e^{-i\omega t} \sum_{n=0}^{\infty} \delta(t-n) \left(1 - \frac{|t|}{\lambda}\right) dt = \operatorname*{Lim}_{\lambda\to\infty} i\omega \sum_{n=0}^{[\lambda]} e^{-i\omega n} \left(1 - \frac{n}{\lambda}\right) = \frac{i\omega}{1-e^{-i\omega}}.$$

(3) $$t^{\nu} U(t) \doteqdot \frac{\Pi(\nu)}{p^{\nu}}, \quad 0 < \operatorname{Re}p < \infty \ (\operatorname{Re}\nu > -1).$$

The only singularity of $f(p)/p$ at $\operatorname{Re}p = 0$ is (unless ν is integral) the branch point $p = 0$. On the rest of the boundary the definition integral is summable $(C, \operatorname{Re}\nu)$, as was proved by Hardy†.

(4) $$-\operatorname{Ei}(-t)\, U(t) \doteqdot \log(p+1), \quad -1 < \operatorname{Re}p < \infty.$$

Again, one singularity on the boundary $\operatorname{Re}p = -1$, namely, the logarithmic singularity at $p = -1$. Apart from this point the definition integral is ordinarily convergent.

(5) $$\sin(e^{-t}) \doteqdot \sin\left(\frac{\pi p}{2}\right) \Pi(p), \quad -1 < \operatorname{Re}p < 1.$$

On the finite part of the right-hand boundary $\operatorname{Re}p = 1$ no singularity at all; $f(p)/p$ is singular for $p \to 1 + i\infty$, however, since its modulus tends to infinity.

(6) $$\left[\sqrt{\frac{t}{\pi}}\right] U(t) \doteqdot \sum_{n=1}^{\infty} e^{-\pi n^2 p} = \tfrac{1}{2}\{\theta_3(0,p) - 1\}, \quad 0 < \operatorname{Re}p < \infty. \qquad (39)$$

This operational relation is easily obtained by transposing the series for the θ-function term by term. From known properties of the θ-function it follows that $f(p)/p$ is singular at any point of the boundary $\operatorname{Re}p = 0$.

† *Proc. Lond. Math. Soc.* (2), IX, 126, 1910.

It follows that there is in general no ordinary convergence on the boundaries of the strip of convergence, and at best the definition integral exists only after application of some summing method or other. At first sight one could think it possible to attach a definite meaning to the divergent integral even *outside* the domain of proper convergence. This, however, can never be accomplished by means of the Cesàro method, since even the still more effective Cauchy method always fails. In fact, on account of the weight factor $e^{-\epsilon|t|}$ of the latter the only difference produced before taking the limit $\epsilon \to 0$ is the addition of $\pm\epsilon$ to p, by which the original strip of convergence is shifted over a distance ϵ; if afterwards ϵ is made equal to zero, the strip of convergence reduces to the initial one, so that finally nothing has changed.

11. The inversion integral

In the foregoing chapters we have always considered the definition integral (II, 20*a*) as the actual basis of the operational calculus. At the same time the inversion integral (II, 20*b*) was put forward as a means of reproducing, under suitable restrictions, the original for some given image. In II, § 5 we remarked that, provided that the definition integral is absolutely convergent for $\operatorname{Re} p = c$ and that the inversion integral is taken in the sense of a principal value, that is,

$$\operatorname*{Lim}_{\lambda\to\infty} \frac{1}{2\pi i}\int_{c-i\lambda}^{c+i\lambda} e^{pt}\frac{f(p)}{p}\,dp \quad (\alpha_a < c < \beta_a),$$

the latter integral yields the mean value $\frac{1}{2}h(t+0)+\frac{1}{2}h(t-0)$ of the original in those points t in the vicinity of which $h(t)$ has limited total fluctuation. In addition, we shall now give a survey of other conditions for the inversion integral which, after its introduction as a Cesàro limit, appears to be almost without restriction applicable.

We first discuss the case where the inversion integral holds as principal value, thus without the introduction of the Cesàro limit. Since we are continually required to speak of originals as having limited total fluctuation and being equal to their mean values, it will be convenient to call them 'normalized' functions of limited total fluctuation. The above statement can then be generally formulated as follows, no matter whether the original becomes infinitely large somewhere or not:

I. The inversion integral holds as principal value at those points where the original is a normalized function of limited total fluctuation, provided that $\alpha_a < c < \beta_a$ and that the original is absolutely integrable in any finite interval.

As an example for possible application of this theorem† we mention the original of

$$\sum_{n=1}^{\infty} \frac{U(t-n)}{\sqrt{(t-n)}} \doteqdot \frac{\sqrt{(\pi p)}}{e^{p}-1}, \quad 0<\operatorname{Re} p<\infty,$$

which, though infinitely large at the infinite number of points $t = 1, 2, 3, \ldots$, is absolutely integrable in any finite interval.

The most stringent condition in theorem I is the requirement $\alpha_a < c < \beta_a$, that is, the absolute convergence of the definition integral. This, however, is not serious in the case of operational relations that emerge from others by means of the integration rule, since then (see § 6) the strips of ordinary and absolute convergence are identical. For instance, the relation (6) for the Dirichlet series in the strip (7) follows by an integration of (9). In all cases like this the inversion integral applies without any further restriction. Thus at once from (11)

$$[e^{t}] = \operatorname*{Lim}_{\lambda \to \infty} \frac{1}{2\pi i} \int_{c-i\lambda}^{c+i\lambda} e^{pt} \frac{\zeta(p)}{p}\, dp \quad (c>1).$$

It has proved possible, however, to drop the condition of absolute convergence in almost all practical cases simply by introducing the first-order Cesàro limit. With this in mind the theorem above can be completed as follows‡:

II. The inversion integral holds as a first-order Cesàro limit at those points where the original is a normalized function of limited total fluctuation provided that $\alpha < c < \beta$ and that the original is absolutely integrable in any finite interval.

In order to apply the second theorem, that is,

$$\frac{h(t+0)+h(t-0)}{2} = \operatorname*{Lim}_{\lambda \to \infty} \frac{1}{2\pi i} \int_{c-i\lambda}^{c+i\lambda} e^{pt} \frac{f(p)}{p} \left(1-\frac{|p|}{\lambda}\right) dp,$$

only the strip of ordinary convergence has to be known; any line $\operatorname{Re} p = c$ parallel to the imaginary axis of p that lies inside the strip of ordinary convergence can be chosen as a path of integration. Even for the operational relation (v, 25) involving the δ-function, the inversion is legitimate, leading to

$$\delta(t) = \operatorname*{Lim}_{\lambda \to \infty} \frac{1}{2\pi i} \int_{c-i\lambda}^{c+i\lambda} e^{pt} \left(1-\frac{|p|}{\lambda}\right) dp \quad (-\infty<c<\infty).$$

This may be verified by transforming the integral in question into that of (v, 34), by putting $p = c+is$ and $n = 1$. It further appears that in this special case the inversion integral even holds as a Cesàro limit of arbitrary positive order. As to the operational relations (v, 62) involving the derivatives of the δ-function, the theorem above does not apply. This may

† For a proof, see Widder, loc. cit. p. 241.
‡ See Widder, loc. cit. p. 244.

easily be understood, since the odd function $\delta'(t)$, for example, is certainly not absolutely integrable. Yet the inversion integral of $\delta^{(n)}(t) \doteqdot p^{n+1}$ is valid when taken as a Cesàro limit of the order $n+1$.

In comparing theorem II with I we observe that the restriction of absolute convergence of the definition integral could be dropped in virtue of the introduction of the first-order Cesàro limit. This is quite analogous to the case of the Fourier identity for which it was first found that the existence of $\int_{-\infty}^{\infty} |h(t)|\, dt$ is sufficient for normalized functions of limited total fluctuation in order to guarantee the validity of the identity. Afterwards the supplementary condition was shown to be unnecessary provided the first-order Cesàro limit was introduced: The Fourier identity holds as a first-order Cesàro limit for any function that is integrable in the Lebesgue sense in any finite interval†, whilst the analogue for Fourier series is also valid in view of a well-known theorem of Fejér. It may also be remarked that Hardy‡ was able to extend the second theorem above after replacing the condition of absolute Riemann integrability by the less stringent Lebesgue integrability over any finite interval. In practice, however, Hardy's extension is of little importance.

Of greater importance is Amerio's theorem§ which was derived in a profound study of the inversion integral, viz.

III. In any t-interval where the functions

$$|h(t)|, \quad |h'(t)|, \quad \ldots, \quad |h^{(n-1)}(t)|$$

are continuous, the inversion integral is valid as a $(C, n+1)$ limit with respect to the operational relation $h^{(n)}(t) \doteqdot p^n f(p)$ which is obtained by differentiation n times of the operational relation under consideration $h(t) \doteqdot f(p)$.

In addition to the discussions above we shall mention another important property of the inversion integral, which is required later on in the treatment of Heaviside's *expansion theorem* (see VII, § 10). In many cases it is possible to alter (in the inversion integral) the path of integration $\text{Re}\, p = c$, such that for $t > 0$ (see fig. 37) or $t < 0$ the new contour encloses the negative or positive parts of the real p-axis, respectively. This is a consequence of the exponential decrease of the factor e^{pt} in the integrand for $|p| \to \infty$ in the left half of the p-plane ($t > 0$) or

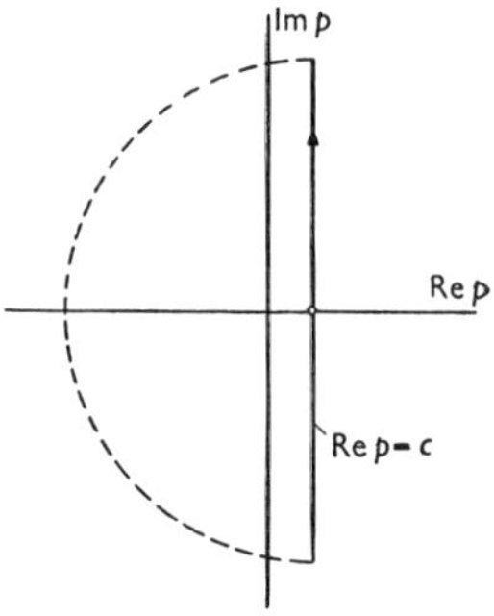

Fig. 37. Transformation of the inversion integral into a contour integral.

† See E. C. Titchmarsh, *Introduction to the Theory of Fourier Integrals*, Oxford, 1937, p. 13.

‡ *Mess. Math.* L, 165, 1921.

§ *Rendiconti dell' Acc. delle Scienze Fisiche e Matematiche, Napoli*, (4), x, 1940.

the right half ($t<0$). On account of a theorem due to Jordan† the deformations indicated are certainly permissible if $f(p)/p$ tends to zero, uniformly with respect to arg p, for $|p| \to \infty$ and p in the left (right) half of the complex plane. As to the operational relation (III, 3) for the Γ-function, this amounts to transforming the inversion integral (III, 6) into Hankel's integral (III, 7).

In concluding this section we may remark that, whereas thus far we have considered the inversion integral only in connexion with a given definition integral, the question arises whether, conversely, the original when given in form of a convergent inversion integral in some strip of convergence ($\alpha < c < \beta$, say) leads to the existence of the definition integral in $\alpha < \operatorname{Re} p < \beta$. Sufficient for this is‡: (i) $f(p)/p$ analytic and tending to zero uniformly for $|\operatorname{Im} p| \to \infty$ and (ii) the inversion integral given converges absolutely at $t = 0$. The inversion integral is assumed here as an ordinary integral, but the resulting definition integral may be valid only in the sense of a principal value

$$\operatorname*{Lim}_{\lambda \to \infty} p \int_{-\lambda}^{\lambda} e^{-pt} h(t)\, dt.$$

12. Adjacent strips of convergence for the same image

Whereas in the case of a given original the definition integral yields a uniquely defined image for some definite strip of convergence, *different* originals may correspond to *one* image, in different strips of convergence of course. It is then natural to ask what relationship, if any, exists between those originals. Quite a simple rule can be established with the aid of the inversion integral, if the singularities on the boundary of the strip of convergence are assumed to be first-order poles. To see this, let us suppose that $f(p)/p$ has first-order poles p_k on the finite part of the line $\operatorname{Re} p = \alpha$, which line may further separate the strips of convergence I and II from each other (see fig. 38). If, moreover, $f(p)$ have the originals h_1 and h_2 with respect to I and II, respectively, then it follows (if necessary, in the sense of $(C, 1)$ limits):

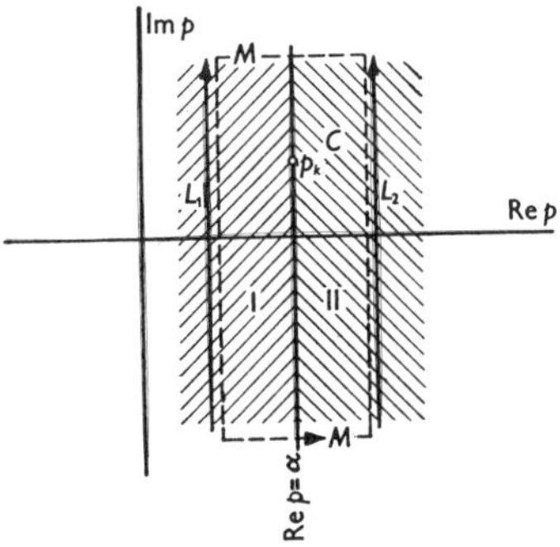

Fig. 38. Adjacent strips of convergence for one and the same image.

$$h_1(t) = \frac{1}{2\pi i} \int_{L_1} e^{pt} \frac{f(p)}{p}\, dp,$$

$$h_2(t) = \frac{1}{2\pi i} \int_{L_2} e^{pt} \frac{f(p)}{p}\, dp.$$

† See E. T. Whittaker and G. N. Watson, *A Course of Modern Analysis*, Cambridge, 1940, p. 115.

‡ See Doetsch, loc. cit. p. 126.

When subtracting we come to an integral with a closed contour M, that is,

$$h_2 - h_1 = \frac{1}{2\pi i}\int_M e^{pt}\frac{f(p)}{p}\,dp.$$

This transformation is allowed if $f(p)/p$ tends to zero uniformly for $|\operatorname{Im} p| \to \infty$ inside the strips of convergence; it is sufficient† for this if the definition integral is uniformly convergent in both I and II. Now the preceding contour integral can easily be evaluated on account of Cauchy's theorem of residues, the only singularities of $f(p)/p$ in the interior of M being the poles p_k (with corresponding residues R_k). We therefore can express $h_1(t)$ into $h_2(t)$ so as to have simultaneously

$$f(p) \doteqdot \begin{cases} h_2(t) & p \text{ in II}, \\ h_2(t) - \Sigma R_k e^{p_k t} & p \text{ in I}. \end{cases} \tag{40}$$

We have thus arrived at a simple rule to survey the change in original when passing a line containing a number of first-order poles.

A different point of view is as follows. Instead of keeping the image function constant, we may also introduce a new image by subtracting $p\Sigma \dfrac{R_k}{p-p_k}$ from $f(p)$, to make the image free from singularities at $\operatorname{Re} p = \alpha$. If $f(p)/p$ is subjected to the same restrictions as before, we now find the operational relation

$$f(p) - p\Sigma\frac{R_k}{p-p_k} \doteqdot h_2(t) - U(t)\,\Sigma R_k e^{p_k t}, \tag{41}$$

which is valid in both strips of convergence I and II.

Rules (40) and (41) will now be illustrated by three examples involving the ζ-function of Riemann.

Example 1. As to the operational relation (11), that is,

$$\zeta(p) \doteqdot [e^t], \quad 1 < \operatorname{Re} p < \infty,$$

the only singularity on the boundary $\operatorname{Re} p = 1$ is the pole at $p = 1$ with residue 1. The next pole of $f(p)/p$ is situated at $p = 0$; accordingly from (40) we first have

$$\zeta(p) \doteqdot [e^t] - e^t, \quad 0 < \operatorname{Re} p < 1. \tag{42}$$

Next the function $f(p)/p$ shows a pole at $p = 0$ with residue $\zeta(0) = -\frac{1}{2}$; in crossing the new boundary at $\operatorname{Re} p = 0$, again from (40):

$$\zeta(p) \doteqdot [e^t] - e^t + \tfrac{1}{2}, \quad -1 < \operatorname{Re} p < 0, \tag{43}$$

whose domain of convergence cannot be extended beyond $\operatorname{Re} p = -1$.

The difficulty in practice is that it may sometimes be doubtful whether the behaviour of $f(p)/p$ at infinity actually allows the crossing of the boundary. However, we may verify afterwards the operational relation obtained, namely, by means of

† See Doetsch, loc. cit. p. 51.

the definition integral itself. Accordingly, in the case of (42) and (43), if the upper limit of integration tends to infinity along the discontinuities of the function $[e^t]$, the definition integral yields the limits

$$\lim_{N\to\infty} p\int_{-\infty}^{\log N} e^{-ps}([e^s]-e^s)\,ds = \lim_{N\to\infty}\left\{\sum_{n=1}^{N}\frac{1}{n^p}-\frac{1}{1-p}\frac{1}{N^{p-1}}\right\} = \zeta(p), \quad 0<\mathrm{Re}\,p<1,$$
$$\lim_{N\to\infty} p\int_{-\infty}^{\log N} e^{-ps}([e^s]-e^s+\tfrac{1}{2})\,ds = \lim_{N\to\infty}\left\{\sum_{n=1}^{N}\frac{1}{n^p}-\frac{1}{1-p}\frac{1}{N^{p-1}}-\frac{1}{2N^p}\right\} = \zeta(p), \quad -1<\mathrm{Re}\,p<0, \tag{44}$$

which really converge, in the regions indicated, to the analytic continuation of the series $\sum_{1}^{\infty}\frac{1}{n^p}$, that is, to $\zeta(p)$. On the other hand, the last expression has no finite limit

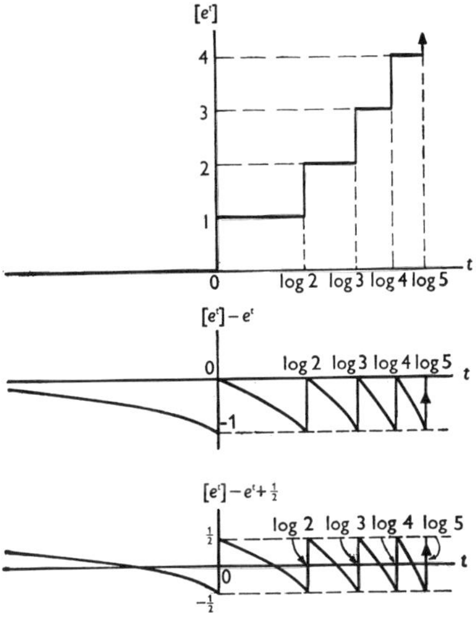

Fig. 39. Originals of $\zeta(p)$ in different strips of convergence.

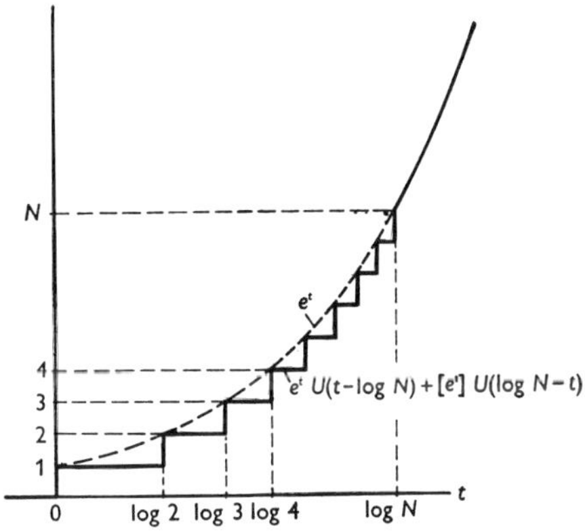

Fig. 40. The original of $\sum_{n=1}^{N}\frac{1}{n^p}-\frac{1}{1-p}\frac{1}{N^{p-1}}$ for $\mathrm{Re}\,p>1$.

for $\mathrm{Re}\,p<-1$; therefore, (43) is not valid at the left of the line $\mathrm{Re}\,p=-1$, though in crossing this line no pole of $\frac{1}{p}f(p)$ would be passed. For the rest also (20) will lead to the line $\mathrm{Re}\,p=-1$ as left boundary of the strip of convergence with respect to the original of (43). In fig. 39 the three different originals, which thus lead to the single image $\zeta(p)$, have been drawn. Moreover, fig. 40 shows the original corresponding to a finite value of N in (44) for the strip $\mathrm{Re}\,p>1$ instead of $0<\mathrm{Re}\,p<1$; the transition at $t=N$ from the step curve to the exponential curve clearly indicates how only for $N\to\infty$ the complete step function $[e^t]$ results.

We further remark that, again starting from (11) and crossing the pole at $p=1$, rule (41) leads to the operational relation

$$\zeta(p)-\frac{p}{p-1} \fallingdotseq [e^t]-e^t U(t), \quad 0<\mathrm{Re}\,p<\infty. \tag{45}$$

Example 2. Another application of the theory given above refers to the operational relation (III, 40). As is well known, the function $\frac{1}{p}\Pi(p)\zeta(p)$ has poles at $p = 1, 0, -1, -3, -5, \ldots$ with residues $1, -\frac{1}{2}, \frac{B_2}{2!}, \frac{B_4}{4!}, \frac{B_6}{6!}, \ldots$, respectively, in which B_n denotes Bernoulli's numbers, which can most easily be defined by

$$\frac{z}{e^z - 1} = \sum_{n=0}^{\infty} \frac{B_n}{n!} z^n \quad (|z| < 2\pi). \tag{46}$$

In applying rule (40) we find successively

$$\Pi(p)\zeta(p) \risingdotseq \begin{cases} \dfrac{1}{e^{e^{-t}} - 1}, & 1 < \operatorname{Re} p < \infty, \\ \dfrac{1}{e^{e^{-t}} - 1} - e^t, & 0 < \operatorname{Re} p < 1, \\ \dfrac{1}{e^{e^{-t}} - 1} - e^t + \frac{1}{2}, & -1 < \operatorname{Re} p < 0, \\ \dfrac{1}{e^{e^{-t}} - 1} - \sum\limits_{n=0}^{2N} \dfrac{B_n}{n!} e^{-(n-1)t}, & -2N-1 < \operatorname{Re} p < -2N+1 \quad (N \geqslant 1), \end{cases} \tag{47}$$

the right-hand members of which are just the remainders of the following infinite series:

$$\frac{1}{e^{e^{-t}} - 1} = \sum_{n=0}^{\infty} \frac{B_n}{n!} e^{-(n-1)t} \quad \left(t > \log \frac{1}{2\pi}\right).$$

Example 3. In this example we shall deal with the function

$$\xi(p) = (p-1)\pi^{-\frac{1}{2}p}\,\Pi(\tfrac{1}{2}p)\,\zeta(p),$$

which is important in Riemann's functional equation of the ζ-function, viz.

$$\zeta(1-p) = \frac{2}{(2\pi)^p} \cos\left(\frac{\pi p}{2}\right) \Gamma(p)\,\zeta(p),$$

which is equivalent to $\xi(p+\frac{1}{2})$ being an even function of p. The simplest operational deduction of this relation runs as follows. First from (24) we have with the help of the shift rule

$$\frac{\sin(\frac{1}{2}\pi p)\,\Pi(p)}{(2\pi n)^{p+1}} \risingdotseq \frac{\sin(2\pi n\, e^{-t})}{2\pi n} \quad (-1 < \operatorname{Re} p < 1),$$

which, when the sum over n is taken, leads to

$$\frac{\sin(\frac{1}{2}\pi p)\,\Pi(p)}{(2\pi)^{p+1}}\,\zeta(p+1) \risingdotseq \sum_{n=1}^{\infty} \frac{\sin(2\pi n\, e^{-t})}{2\pi n} \quad (0 < \operatorname{Re} p < 1).$$

The right-hand member is simply the Fourier series of $\frac{1}{2}\,\mathrm{Sa}\,(e^{-t}) = \frac{1}{2}([e^{-t}] - e^{-t} + \frac{1}{2})$ (cf. (XII, 6) and (XII, 10)); thus, according to (43), after p is replaced by $-p$, it has the image $-\frac{1}{2}\zeta(-p)$. If this image is put equal to the left-hand side of the operational relation above, Riemann's functional equation is found.

An alternative procedure for arriving at the functional equation, revealing at the same time the connexion existing between the ζ-function and the θ_3-function, is the following. If the shift and similarity rules are applied to (III, 16), we obtain

$$\frac{\Pi(\frac{1}{2}p)}{\pi^{\frac{1}{2}p}} \frac{1}{n^p} \risingdotseq e^{-\pi n^2 e^{-2t}} \quad (\operatorname{Re} p > 0),$$

and, after carrying out the summation over n,

$$\frac{\Pi(\frac{1}{2}p)}{\pi^{\frac{1}{2}p}}\zeta(p) \doteqdot \tfrac{1}{2}\{\theta_3(0,\mathrm{e}^{-2t})-1\}, \quad 1<\operatorname{Re}p<\infty. \tag{48}$$

The image divided by p is singular only at the poles $p = 1$ and $p = 0$, since the poles of the factor $\Pi(\frac{1}{2}p)$ are cancelled out by the zeros of $\zeta(p)$. From (41) we thus get

$$\frac{\Pi(\frac{1}{2}p)}{\pi^{\frac{1}{2}p}}\zeta(p) - \frac{1}{2(p-1)} \doteqdot \tfrac{1}{2}\{\theta_3(0,\mathrm{e}^{-2t})-1\} - \tfrac{1}{2}(\mathrm{e}^{t}-1)\,U(t),$$

of which the strip of convergence may possibly cover the whole complex plane of p. This actually appears to be true, as can easily be seen after simplifying the right-hand member by means of a well-known functional equation of the θ-functions. If then the ξ-function is introduced we finally obtain the relation

$$\frac{p}{p^2-\frac{1}{4}}\{\xi(p+\tfrac{1}{2})-\tfrac{1}{2}\} \doteqdot \tfrac{1}{2}\mathrm{e}^{\frac{1}{2}|t|}\{\theta_3(0,\mathrm{e}^{2|t|})-1\}, \quad -\infty<\operatorname{Re}p<\infty.$$

From the θ-series the convergence of the definition integral for any p follows at once. It is observed that in this relation the original is an even function of t. Thus, the original being unchanged by the substitution $t \to -t$, the image will only change its sign by the corresponding transformation $p \to -p$. Therefore $\xi(p+\frac{1}{2})$ is even in p, which statement is equivalent with Riemann's relation. Incidentally, the last operational relation is somewhat simplified by eliminating the factor $1/(p^2-\frac{1}{4})$; we then arrive at

$$p\xi(p+\tfrac{1}{2}) \doteqdot \frac{1}{2}\left(\frac{d^2}{dt^2}-\frac{1}{4}\right)\{\mathrm{e}^{\frac{1}{2}t}\theta_3(0,\mathrm{e}^{2t})\}, \quad -\infty<\operatorname{Re}p<\infty, \tag{49}$$

in which the variable t rather than $|t|$ occurs, the function between { } being even in t.

13. Operational relations having a line of convergence

The operational relations thus far discussed had a strip of convergence of non-vanishing width. Sometimes, however, the strip of convergence happens to reduce to a single line parallel to the imaginary axis, thus representing a transition between a strip of finite width and a non-existent strip. As an example we may quote the relation (III, 11)

$$\tfrac{1}{2}\mathrm{e}^{-\alpha|t|} \doteqdot \frac{\alpha p}{\alpha^2-p^2}, \quad -\operatorname{Re}\alpha<\operatorname{Re}p<\operatorname{Re}\alpha,$$

which, for α real, is valid only if $\alpha>0$. In the limiting case $\alpha = 0$, having performed the limit at both left and right sides of the relation itself, we obtain

$$1 \doteqdot 2\pi p\,\delta(ip), \quad \operatorname{Re}p = 0. \tag{50}$$

The validity of this operational relation can otherwise be verified by means of the definition integral; for $p = i\omega$ it becomes

$$p\int_{-\infty}^{\infty}\mathrm{e}^{-i\omega t}\,dt = 2p\int_{0}^{\infty}\cos(\omega t)\,dt,$$

which, according to (37), actually leads to the image required as a (C,r) limit with $r>0$. As is apparent from this example, the definition integral is not necessarily assumed to be convergent in the elementary sense whenever a line of convergence occurs. On the contrary, a large number of possibilities exists as in § 10, where we have discussed the character of the convergence on boundaries of non-vanishing strips. In addition, $f(p)/p$ is usually singular somewhere on the line of convergence, whereby the type of singularity may vary largely; this can most conveniently be surveyed by means of examples, which will now be given.

(1) $$\frac{\epsilon}{\pi(t^2+\epsilon^2)} \doteqdot p\,\mathrm{e}^{-\epsilon|p|}, \quad \mathrm{Re}\,p = 0 \ (\epsilon > 0). \tag{51}$$

The only singularity of $f(p)/p$ on the line of convergence lies at $p = 0$. Excluding this point the convergence of the definition integral is ordinary, as is seen by suitably transforming the path of integration, that is, closing it either in the upper t-plane (if $p/i < 0$) or in the lower t-plane (if $p/i > 0$), and then applying the theorem of residues with respect to the poles at $t = \pm i\epsilon$. The operational relation (51) is that of Cauchy's approximating function (v, 32).

(2) $$\frac{\sin(\lambda t)}{\pi t} \doteqdot pU(p^2+\lambda^2), \quad \mathrm{Re}\,p = 0 \ (\lambda > 0). \tag{52}$$

In this case $f(p)/p$ has two singularities, namely the discontinuities at $p = \pm i\lambda$. Otherwise this Dirichlet function (see (v, 15)) has an ordinarily convergent definition integral.

(3) $$\frac{1}{t} \doteqdot \pi\,|\,p\,|, \quad \mathrm{Re}\,p = 0. \tag{53}$$

Outside the singularity $p = 0$ the corresponding definition integral converges in the sense of a principal value.

(4) $$\mathrm{sgn}\,t \doteqdot \begin{cases} 2 & (p \neq 0) \\ 0 & (p = 0) \end{cases}, \quad \mathrm{Re}\,p = 0. \tag{54}$$

Outside the singularity at $p = 0$ the definition integral is summable (C, r) for $r > 0$, as follows from (38).

(5) $$t^n \doteqdot 2\pi i^n p\,\delta^{(n)}\left(\frac{p}{i}\right), \quad \mathrm{Re}\,p = 0.$$

Outside the singular point $p = 0$ the definition integral should be taken as a Cesàro limit of order r with $r > n$.

It is to be remarked that the ordinary rules remain valid in applications involving a line of convergence. So, for instance, with the attenuation rule from (50)

$$\mathrm{e}^{\lambda t} \doteqdot 2\pi\lambda\,\delta(ip - i\lambda), \quad \mathrm{Re}\,p = \mathrm{Re}\,\lambda,$$

and after combining the formulae for $\lambda = i$ and $\lambda = -i$:

$$\sin t \doteqdot 2\pi\,\delta(p^2+1), \quad \mathrm{Re}\,p = 0, \tag{55}$$

$$\cos t \doteqdot 2\pi p\,\delta(p^2+1), \quad \mathrm{Re}\,p = 0. \tag{56}$$

Also the important rule of the composition product may be applied, even if the definition integral of either factor converges only as a Cesàro limit; more precisely, if for some value of p outside the singularities the initial operational relations involved are summable (C, r_1), (C, r_2), respectively, then the operational relation for the composition product, if the latter exists, will be summable (C, r_1+r_2+1), whilst the corresponding image remains equal to $\frac{1}{p}f_1(p)\,f_2(p)$. This theorem, which is valid independently of the width of the strip of convergence, was proved by Doetsch† for one-sided originals.

It will be evident from the preceding that the introduction of operational relations possessing a line of convergence admits the possibility of operationally transforming two-sided originals that could only be treated before in their corresponding one-sided form. As to the images, a definite original may also correspond to a p-function that is not everywhere analytic on some line parallel to the imaginary axis. In particular, many relations may be found whose line of convergence coincides with the transition boundary of two different strips of convergence related to a single p-function. If on this line the singularities are merely poles of the first order (see the preceding section) we have in addition to (40) the general relation

$$f(p) \doteqdot h_2(t) - \tfrac{1}{2}\Sigma R_k e^{p_k t}, \quad \mathrm{Re}\, p = \alpha,$$

which indicates that on the line in question the original is simply the arithmetic mean of the values at the left and the right of it. So for (42) we get

$$\zeta(p) \doteqdot [e^t] - \tfrac{1}{2}e^t, \quad \mathrm{Re}\, p = 1.$$

Finally, any operational relation possessing a single line as domain of convergence can be transformed into a relation that has the imaginary p-axis as line of convergence; that is, by means of the attenuation rule. Both the definition integral and the inversion integral then reduce to the corresponding formula of (II, 5); as a consequence, the theory of the special operational relations above is equivalent to that of the Fourier identity.

14. Symmetry between the integrals of definition and inversion

The starting-point of our operational calculus was formed by the pair of integrals (II, 5) which emerged from the Fourier identity. The initial symmetry between both integrals was lost later on, since in the definition integral the integration was always fixed along the real axis, whilst the inversion integral in general admitted a considerable variation of its path of integration. As a consequence, the p-function had to be analytic. The lost

† *Ann. Scu. norm. sup. Pisa* (2), IV, 83, 1935.

symmetry is restored when operational relations with a line of convergence are introduced, as will be clear from what follows. If in the definition integral

$$\frac{f(p)}{p} = \int_{-\infty}^{\infty} e^{-ps} h(s)\, ds, \quad \alpha < \operatorname{Re} p < \beta, \tag{57}$$

the functions f and h are formally replaced by new functions, namely,

$$f(x) \to 2\pi x\, h^*(ix), \quad h(x) \to \frac{f^*(ix)}{ix}, \tag{58}$$

and, moreover, the following substitutions are made:

$$p = -it, \quad s = -ip,$$

we obtain the relation:

$$h^*(t) = \frac{1}{2\pi i} \int_{-i\infty}^{i\infty} e^{pt} \frac{f^*(p)}{p}\, dp, \quad \alpha' < \operatorname{Im} t < \beta'. \tag{59}$$

We at once recognize the expression above as the inversion integral corresponding to the new operational relation

$$h^*(t) \doteqdot f^*(p), \quad \operatorname{Re} p = 0,$$

that is, when expressed in terms of the initial functions

$$\frac{i}{t} f\left(\frac{t}{i}\right) \doteqdot 2\pi p\, h\left(\frac{p}{i}\right), \quad \operatorname{Re} p = 0. \tag{60}$$

The validity of this operational relation can be verified in a straightforward manner from (57). Since in so doing it turns out that the definition integral of (60) is simply the Fourier identity for $h(t)$, provided (57) is applicable for $\operatorname{Re} p = 0$, we arrive at the conclusion that any operational relation whose strip of convergence contains the imaginary axis (or, alternatively, reduces to this line) will also lead, according to (60), to a relation with a single line of convergence. Consequently, there are always two coexisting operational relations; in this way (50) is the counterpart of $p \doteqdot \delta(t)$. We may further draw attention to a new kind of symmetry between the image and the original. Whereas, concerning the image, the function $f(p)/p$ is necessarily analytic inside a non-vanishing strip of convergence, it follows likewise, from the integral representation (59), that the original of (60) is analytic in the range $\alpha' < \operatorname{Im} t < \beta'$ of the complex t-plane. Accordingly, originals that can be continued analytically outside $\operatorname{Im} t = 0$ possess a strip extending in a direction parallel to the real axis of t (see fig. 41), in contrast with the strip of the image functions which parallels the imaginary axis of the p-plane. Just as the definition integral represents the p-function in the whole strip of convergence, the inversion integral represents the original everywhere in the interior of the t-strip. As an example showing a t-strip of non-vanishing width we may mention the relation (40) with inversion integral as follows:

$$\frac{1}{e^{e^{-t}}-1} = \frac{1}{2\pi i}\int_{c-i\infty}^{c+i\infty} e^{pt}\,\Gamma(p)\,\zeta(p)\,dp \quad (c>1;\ -\tfrac{1}{2}\pi<\operatorname{Im} t<\tfrac{1}{2}\pi).$$

In every respect a complete symmetry thus obtains between the image and the original. Just as $f(p)/p$ has at least one singularity on the boundary of the p-strip, the original must have a singular point on the boundary of the t-strip (though perhaps at infinity) or on the real axis if the strip has zero width. In the example just mentioned the original is singular on the boundaries $\operatorname{Im} t = \pm\frac{1}{2}\pi$ at $t = \log(2k\pi)\pm\frac{1}{2}i\pi$ (k positive integral). Though in the way indicated any operational relation virtually involves two different kinds of strips of convergence, there does not seem to exist any definite relationship between them, since according to § 4 the p-strip is determined solely by the behaviour of $h(t)$ as $|t|\to\infty$, whilst, on the other hand, the

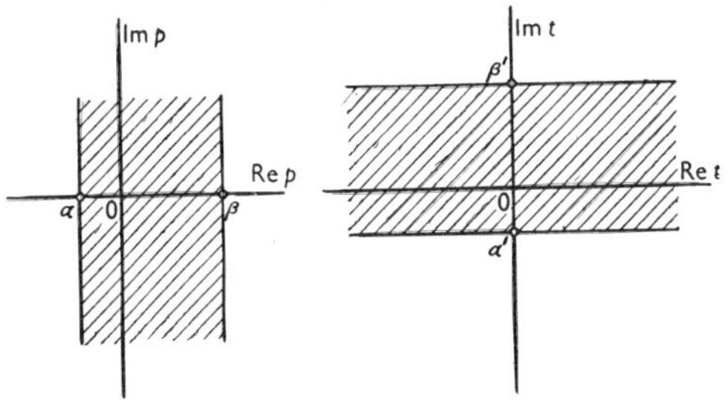

Fig. 41. The strips of convergence for p and t.

t-strip depends, amongst other things, upon the singularities of $h(t)$ in the finite part of the t-plane. It is further possible that either strip reduces to a single line simultaneously, as can be seen from

$$\sum_{n=-\infty}^{\infty} \delta(t-n) \fallingdotseq p \sum_{n=-\infty}^{\infty} \delta\left(\frac{p}{2\pi i}-n\right) \quad (\operatorname{Re} p = 0;\ \operatorname{Im} t = 0).$$

The symmetry between image and original also becomes apparent from many other features not treated here. We conclude the present section by remarking that the rules of the operational calculus may be considered in pairs such that corresponding rules change into one another by the substitution (58); for instance, the attenuation and the shift rule, or the rule of the composition product together with that for the product of two originals.

15. Uniqueness of the operational relations

As is obvious—from the fact that in this book the definition integral is taken as the basis of the calculus—at best only one image belongs to a given original. On the other hand, it is of special importance in practical applications to know whether the original is uniquely determined by the image given in some strip of convergence, since this knowledge is required in order to be able to infer the identity of the originals from the equality of the corresponding images. It has already been seen in § 11 that the inversion integral, if necessary as a $(C, 1)$ limit, always represents the original whenever the latter is normalized and of limited total fluctuation; in other words, there is at most one original that is everywhere normalized and of limited total fluctuation, and belonging to a given image in a given strip of convergence. Consequently, there does exist a one-to-one correspondence between image and original if the latter is subjected to the conditions of normalization and limited total fluctuation. As to practical applications, the restriction with regard to the originals is of little, if any, importance. If, however, other types of originals are considered as well, the question whether some image given in a certain strip of convergence may have different originals can be reduced to a simpler one, namely, whether the image $f(p) = 0$ may belong to non-vanishing originals. These originals $N(t)$ do indeed exist; after Doetsch† they are called null functions. They are highly discontinuous and satisfy

$$\int_{t_1}^{t_2} N(t)\,dt = 0 \tag{61}$$

for arbitrary limits t_1 and t_2.

It is evident that the addition of null functions to a given original does not effect the image in any way, that is, the original is determined uniquely but for null functions‡. Henceforth we shall refer to this as the uniqueness theorem. From (61) it follows further that null functions are different from zero only on some point set of zero Lebesgue measure. A simple example of a null function is drawn in fig. 42; it is zero everywhere except at the integral points where it has unit value, thus giving a true idea of the 'emptiness' of null functions. Just as in the case of the unit and of the δ-function, the null function of fig. 42 can be obtained as the limit of a continuous function, viz.

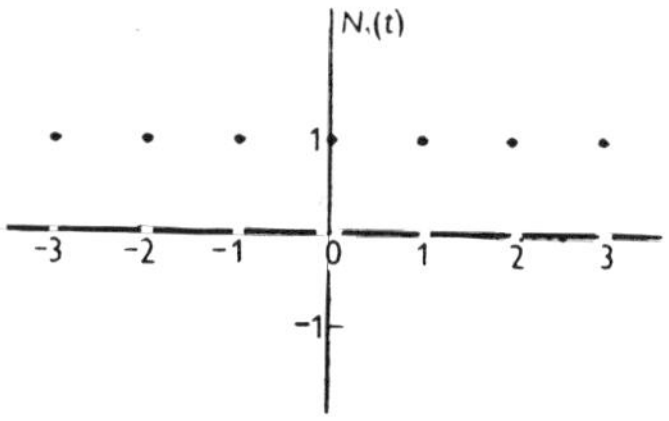

Fig. 42. Example of a null function.

$$\operatorname*{Lim}_{\epsilon \to 0} \frac{\epsilon}{\sin(\pi t) + \epsilon}.$$

It is almost unnecessary to emphasize that in physical and technical applications of the operational calculus the null functions can be wholly ignored.

In concluding this chapter we will briefly indicate the general trend followed in the proof of the uniqueness theorem, mainly because a remarkable property of the image function then becomes more vivid. Usually the

† Loc. cit. p. 34.

‡ See Doetsch, loc. cit. p. 52.

proof is first restricted to one-sided originals; in this respect we would mention that Bremekamp† showed that at most one original satisfying Dirichlet's conditions of Fourier series and integrals belongs to a given image. Lerch‡, however, based his proof of the theorem for one-sided originals on the problem of moments (cf. example 4 of v, § 8). First the definition integral for $\mathrm{Re}\,p > \alpha$ is transformed by an integration by parts as follows:

$$\frac{f(p)}{p-p_1} = p\int_0^\infty dt\, \mathrm{e}^{-(p-p_1)t}\int_0^t du\, \mathrm{e}^{-p_1 u} h(u),$$

in which p_1 denotes an arbitrary point inside the strip of convergence. If $f(p)$ has to be zero identically for $\mathrm{Re}\,p > \alpha$, the integral above must certainly vanish at the points

$$p = p_1 + (n+1)\,q_1 \quad (n = 0, 1, 2, \ldots)\ (q_1 > 0). \tag{62}$$

The substitution $s = \mathrm{e}^{-q_1 t}$ then leads to an infinite number of equations, viz.

$$\int_0^1 \psi(s)\, s^n\, ds = 0 \quad (n = 0, 1, 2, \ldots),$$

where
$$\psi(x) \equiv \int_0^{-\log x/q_1} \mathrm{e}^{-p_1 u} h(u)\, du. \tag{63}$$

Consequently the requirement $f(p) = 0$ implies the vanishing of all the moments of $\psi(x)$. Further, since p_1 lies inside the strip of convergence, the function $\psi(x)$ of (63) is continuous in the interval $0 < x < 1$. It is well known that the latter condition necessarily leads to a vanishing $\psi(x)$, whence it further follows that $\mathrm{e}^{-p_1 t} h(t)$ as well as $h(t)$ itself has a vanishing integral between arbitrary limits t_1 and t_2, which is (61). If so the uniqueness theorem has been proved for one-sided originals, then the extension for two-sided functions§ is not difficult. It is to be noticed that a strip of non-vanishing width was prescribed; thus the uniqueness theorem is not yet proved for operational relations that have a line of convergence. It should be kept in mind, however, that in this special case the theory of the Fourier integral is applicable as was already mentioned at the end of § 13; if the originals are then confined to functions that are absolutely integrable in $(-\infty, +\infty)$, the uniqueness proves true.

Though only the general lines of the proof of the uniqueness theorem were indicated, it may be observed from the preceding that the knowledge of the vanishing of the image function, $f(p) = 0$, is used only in part; in the case of one-sided originals we merely accounted for the vanishing in the sequence of points shown in (62). This illustrates the general property that a non-vanishing image having an infinite number of equidistant zeros on

† *Proc. K. Akad. Wet. Amst.* XL, 691, 1937.
‡ *Acta Math.* XXVII, 339, 1903.
§ Doetsch, loc. cit. p. 52.

a line parallel to the real axis can never have a one-sided original. Indeed, if such an original did exist, it could be nothing but a null function, whose image would then, however, be identically zero. The theorem under consideration was extended by Doetsch†, who proved that the image of a one-sided original is never zero at an infinite number of points p_n of the real axis, provided that the series $\sum_1^\infty \frac{1}{p_n}$ is divergent in the elementary sense. On the other hand, it is quite possible to construct one-sided originals that have an infinite number of equidistant zeros of the image on a line parallel to the imaginary axis, as can be seen from the rectangle function $U(t) - U(t-\alpha)$ whose image $1 - e^{-\alpha p}$ shows such a sequence of zeros on the imaginary axis itself. Furthermore, there do exist *two-sided* originals with an infinite number of equidistant zeros of the image on the real p-axis all lying inside the strip of convergence. The relation

$$\sin(\alpha p)\cos^p\alpha\,\Pi(p) \doteq e^{-e^{-t}}\sin(\tan\alpha\, e^{-t}), \quad -1 < \operatorname{Re} p < \infty \qquad (64)$$

which can be obtained with Euler's integral for the Γ-function, may serve as an example.

Finally, summarizing our general considerations concerning the uniqueness theorem, we may state that in practice the null functions are of no importance with respect to the *originals*, since they can be added at will. It should be borne in mind, however, that null functions used as *images* may sometimes lead to non-trivial operational relations. So the image of (50) is proportional to $x\delta(x)$ $(x = p/i)$, which is a null function in the sense of a Stieltjes integral.

† Loc. cit. p. 38.

CHAPTER VII

ASYMPTOTIC RELATIONS AND OPERATIONAL TRANSPOSITION OF SERIES

1. Introduction

Whereas the preceding chapter dealt with fundamental rather than numerical questions, we shall now summarize the theory in which numerical relationship between image and original is paramount. In so doing we shall confine ourselves to one-sided originals, though all the theorems to be discussed presently might easily be extended so as to include two-sided originals as well, though at the cost of simplicity. The whole basis is provided by a number of theorems showing the relationship between the behaviour of the original in the neighbourhood of $t = 0(\infty)$ and that of the image at $p = \infty(0)$, respectively. These theorems are called *Abel theorems* when, starting with a given property of the original, a conclusion is stated with respect to the corresponding image. In the other case, when the conclusion refers to the original, they are called *Tauber theorems*. As to these names, the theorems in question may be considered as extensions of the well-known Abel and Tauber theorems of ordinary power series.

In §§ 2–5 the general Abel and Tauber theorems are surveyed. In § 6 they are applied to derive general operational identities, whilst in §§ 7–11 they are studied in connexion with term-by-term transposition of series. That in this respect one should be very careful is seen from the series

$$\sum_{n=1}^{\infty} e^{-n^2 t^2},$$

which by means of (II, 24) would formally lead to the following representation of the image:

$$\sqrt{\pi}\, p \sum_{n=1}^{\infty} \frac{e^{p^2/4n^2}}{n},$$

which, however, is divergent for all values of p. The theory of the present chapter will reveal the in general asymptotic character of those series obtained by operationally transposing term by term. For this purpose the concept of asymptotic series is recalled in § 7. Further, if the series development of the original is given, then the validity of the series for the image will be shown to depend upon some Abel theorem, whilst the legitimacy of the transposition of a given series for the image depends upon a Tauber theorem. Since Tauber theorems are the more complicated, it will in general be more difficult to prove the legitimacy of the term-by-term interpretation of the series for the image than that for an original. In the last section of the present chapter we shall discuss Widder's inversion formula, which, in contrast to the Abel and Tauber theorems, yields numerical relations between image and original at any point.

2. Abel and Tauber theorems

As already mentioned, these theorems reveal the relationship between the values of the original at $t = 0(\infty)$ and that of the corresponding image at $p = \infty(0)$, respectively. Of most simple character are the rules

$$h(\infty) = f(0), \tag{1a}$$

$$h(0) = f(\infty), \tag{1b}$$

or, more precisely,

$$\operatorname*{Lim}_{t\to\infty} h(t) = \operatorname*{Lim}_{p\to+0} f(p), \quad \operatorname*{Lim}_{t\to+0} h(t) = \operatorname*{Lim}_{p\to\infty} f(p).$$

Either of these formulae is valid for one-sided originals provided both left- and right-hand members exist, whilst it is further to be supposed that p as well as t runs through real positive values. The validity of (1 a, b) is made plausible by first substituting $u = pt$ in the definition integral

$$f(p) = \int_0^\infty e^{-u} h\left(\frac{u}{p}\right) du,$$

and then taking the limit under the sign of integration†. It should be kept in mind that (1 a), for instance, states only the equality of left- and right-hand members; we should know in advance in some way or other that both limits do, in fact, exist. Of far more importance is the theorem that states the existence of the limit involving f, and the equality of both limits, given the existence of the limit involving h. These theorems are written down as follows:

$$\operatorname*{Lim}_{t\to\infty} h(t) \Rightarrow f(+0), \tag{2a}$$

$$h(+0) \Rightarrow \operatorname*{Lim}_{p\to\infty} f(p). \tag{2b}$$

According to the definitions in § 1 the statements above are typical Abel theorems; if the arrows pointed in the opposite direction the theorems would be of the Tauber character. To justify the terminology we shall discuss these theorems operationally, as they are originally formulated for power series $\sum_0^\infty a_n z^n$ (radius of convergence $\rho = 1$). In this respect we have

(1) *Abel's theorem:* If the series above is known to converge for $z = 1$, then its sum at this point is equal to the limit value for $z \to 1-0$ (this limit

† See Balth. van der Pol, *Phil. Mag.* VIII, 864, 1929.

in fact exists) of the function defined by the power series for $|z| < 1$. With the help of the arrow notation we may thus write

$$\sum_{n=0}^{\infty} a_n \Rightarrow \operatorname*{Lim}_{z \to 1} \sum_{n=0}^{\infty} a_n z^n. \tag{3a}$$

(2) *Tauber's theorem:* Given that the function defined by the power series tends to some limit for z approaching 1 from below through real values, then the power series converges at $z = 1$ with sum equal to the limit value above, provided $\operatorname*{Lim}_{\to \infty} (n a_n) = 0$. Therefore

$$\operatorname*{Lim}_{z \to 1} \sum_{n=0}^{\infty} a_n z^n \Rightarrow \sum_{n=0}^{\infty} a_n \quad \text{if} \quad \operatorname*{Lim}_{n \to \infty} (n a_n) = 0. \tag{3b}$$

By putting $z = e^{-p}$ it is quite easy to formulate both theorems in the language of operational calculus, if use is also made of the operational relation (VI, 14), that is,

$$\sum_{n=0}^{\infty} a_n e^{-pn} \doteqdot \sum_{n=0}^{[t]} a_n \quad (\operatorname{Re} p > 0).$$

If the image and original are denoted by $f(p)$ and $h(t)$, respectively, the theorems $(3a, b)$ become equivalent to

$$\operatorname*{Lim}_{t \to \infty} h(t) \Rightarrow \operatorname*{Lim}_{p \to +0} f(p) \quad \text{(Abel)},$$

$$\operatorname*{Lim}_{p \to +0} f(p) \Rightarrow \operatorname*{Lim}_{t \to \infty} h(t) \quad \text{(Tauber)},$$

if

$$\operatorname*{Lim}_{t \to \infty} t\{h(t) - h(t-1)\} = 0.$$

Consequently, the restricted theorem of Abel (Tauber) concerning ordinary power series leads to definite conclusions as to the image (original) when something is known about the corresponding original (image); the general Abel and Tauber theorems of the operational calculus are extensions of these restricted theorems. In addition, the simple example above already indicates that Abel theorems are in general the more simple, since Tauber theorems are valid only under special supplementary conditions. On the other hand, the Tauber theorems provide functional properties of much more profound character, with the result that they are extremely useful in the solution of intricate problems, such as the determination of the distribution of prime numbers. It can generally be said that any Tauber theorem supplies sufficient conditions to reverse the corresponding Abel theorem. In view of the symmetry existing between original and image (see VI, § 14) it would not be anticipated that Tauber theorems are essentially more difficult to prove than Abel theorems. The asymmetry occurring in the set of theorems is easily understood, however, by remembering that the data concerning the original can be woven into the definition integral which is the actual basis of the operational calculus. In the other case, that of the Tauber theorem, something has to be concluded about the original itself, whilst the data referring to the image will yield only an integral property of the original through the definition integral. To be sure, one could alternatively start with the inversion integral, but this is not fruitful since the inversion integral in general only exists as a Cesàro limit.

In discussing the general significance of (1) and (2) we observe that the behaviour of the image for large values of p is mainly determined by the character of the original at small values of t; and likewise, the image for small values of p is related to the original for t large. This is of particular importance from the point of view of physics; accordingly, the numerical behaviour of transients in electrical networks, immediately after switching on the exterior source, is largely correlated to the trend of the network admittance for high frequencies, whilst the transient at later times is closely related to the admittance for low frequencies (see VIII, § 6).

After discussing the general Abel theorems in the next section we shall deal with Tauber theorems in §§ 4 and 5, since it proves useful to treat real and complex Tauber theorems separately. In the first case the image is considered for real values of p (§ 4), whilst in the second the p-variable is allowed to approach the line $\operatorname{Re} p = 0$ from an arbitrary direction (§ 5).

3. Abel theorems

As some of the simplest Abel theorems we mention the following extensions of (2*a*) and (2*b*), which are both easily deduced with the help of appropriate sifting integrals:

THEOREM I. *For any one-sided operational relation* $h(t)\,U(t) \doteqdot f(p)$ *we have*

$$\operatorname*{Lim}_{t\to+0} \frac{h(t)}{t^\nu} \Rightarrow \operatorname*{Lim}_{p\to+\infty} \frac{p^\nu f(p)}{\Pi(\nu)} \quad (\nu > -1), \tag{4}$$

and, moreover, if the strip of convergence extends at least as far as the imaginary axis,

$$\operatorname*{Lim}_{t\to\infty} \frac{h(t)}{t^\nu} \Rightarrow \operatorname*{Lim}_{p\to+0} \frac{p^\nu f(p)}{\Pi(\nu)} \quad (\nu > -1). \tag{5}$$

As to (4), it is almost obvious that the strip of convergence is not necessarily required to extend to the line $\operatorname{Re} p = 0$, since in this case only the behaviour of the image for large values of p is significant. The first part of theorem I, that is, (4), can be proved directly by means of the theory of the sifting integrals as discussed in V, § 7; to this end the left-hand side of (4) is first rewritten as follows:

$$A = \operatorname*{Lim}_{t\to 0} \frac{h(t)}{t^\nu} = \operatorname*{Lim}_{x\to 0} \frac{h(x^{1/(\nu+1)})}{x^{\nu/(\nu+1)}} = 2 \operatorname*{Lim}_{\lambda\to\infty} \int_0^\infty \frac{h(\tau^{1/(\nu+1)})}{\tau^{\nu/(\nu+1)}}\, \delta(-\tau, \lambda)\, d\tau, \tag{6}$$

in which according to (v, 33), the approximate δ-function is taken equal to

$$\delta(t,\lambda) = \frac{\lambda}{2\Pi(\nu+1)} \exp\{-\,|\,\lambda t\,|^{(\nu+1)^{-1}}\} \quad (\nu > -1).$$

Returning to the initial variable t by putting $\tau = t^{\nu+1}$ and then identifying λ with $p^{\nu+1}$, we find with the aid of the definition integral corresponding to the operational relation under consideration that A is equal to the right-hand member of (4). Similarly, the second part (5) of theorem I is deduced by using the sifting property of the function (v, 35)

$$\delta(t,\lambda) = \frac{\exp\{-(\lambda\,|\,t\,|)^{-1}\}}{2\Pi(\nu)\,\lambda^{\nu+1}\,|\,t\,|^{\nu+2}},$$

which shows the peculiarity, already mentioned, that it is no approximation of the δ-function as it does not become infinitely large at $t = 0$.

The simple Abel theorems above thus correlate numerical values of image and original at the special points 0 and ∞. To obtain a similar relationship between the values at other points we first apply the attenuation rule and the integration rule for $\operatorname{Re} p > 0$ to $f(p) \doteqdot h(t)\,U(t)$. The result is

$$U(t)\int_0^t e^{-as}\,h(s)\,ds \doteqdot \frac{f(p+a)}{p+a}, \quad \operatorname{Re} p > \max(0, -\operatorname{Re} a),$$

which again is a one-sided relation. When rule (5) is applied we obtain from it

$$\lim_{t\to\infty} \frac{1}{t^{\nu}} \int_0^t e^{-as}\,h(s)\,ds \Rightarrow \lim_{p\to a+0} \frac{(p-a)^{\nu}}{\Pi(\nu)} \frac{f(p)}{p} \quad (\nu > -1;\ \operatorname{Re} a > 0).$$

This formula becomes particularly simple for $\nu = 0$; it then states that the definition integral, if convergent at $p = a$, is equal to the limit of the image as $p \to a$. The last is self-evident if the point $p = a$ lies inside the strip of convergence. If, on the other hand, $\operatorname{Re} p = \operatorname{Re} a$ acts as boundary of the strip of convergence, the formula states that at the boundary the definition integral, in so far as it is convergent, assumes values in accordance with the analytic continuation of the image function from the inside of the strip of convergence. It can generally be said that the knowledge provided by the Abel and Tauber theorems when approaching the boundary of convergence always proves very useful, whilst their results become trivial when reference is made to points inside the strip of convergence. Moreover, by a simple displacement the point $p = a$ can always be shifted to the origin, which makes the formulation of the theorems as simple as possible.

For a better understanding we shall give two non-trivial applications of (5), both in connexion with the Riemann ζ-function:

(1) From (VI, 11) it follows that

$$\frac{p}{p+1}\zeta(p+1) \doteqdot e^{-t}[e^t], \quad 0 < \operatorname{Re} p < \infty.$$

In this case the original tends to unity as $t \to \infty$. Thus we infer the existence of the limit of the image ($p \to 0$), this also being equal to 1; or, alternatively,

$$\operatorname*{Lim}_{p \to 1+0} \{(p-1)\zeta(p)\} = 1,$$

which states that the residue of $\zeta(p)$ at its pole $p = 1$ is necessarily equal to unity.

(2) Starting from

$$\zeta(p+1) = \sum_{n=1}^{\infty} \frac{1}{n n^p} \doteqdot \sum_{n=1}^{[e^t]} \frac{1}{n} \quad (\operatorname{Re} p > 0),$$

we further find
$$\zeta(p+1) - \frac{1}{p} \doteqdot U(t) \left\{ \sum_{n=1}^{[e^t]} \frac{1}{n} - t \right\} \quad (\operatorname{Re} p > 0).$$

As t tends to infinity the original tends to Euler's constant

$$C = \operatorname*{Lim}_{N \to \infty} \left(\sum_{n=1}^{N} \frac{1}{n} - \log N \right),$$

as is easily seen by putting $N = e^t$. By Abel's theorem we then get

$$C = \operatorname*{Lim}_{p \to 0} \left\{ \zeta(p+1) - \frac{1}{p} \right\}.$$

Thus, simply by applying the Abel theorem twice, the behaviour of $\zeta(p)$ in the vicinity of $p = 1$ is found to be

$$\zeta(p) = \frac{1}{p-1} + C + \dots. \tag{7}$$

To enlarge the field of application we may generalize the Abel theorems (4) and (5) by replacing ν by $\nu + n + 1$, $f(p)$ by $\frac{1}{p^{n+1}} f(p)$, and the original by the corresponding expression obtained in accordance with the integration rule (IV, 35). We then get the following theorem:

THEOREM II. *Given the one-sided operational relation*

$$h(t)\, U(t) \doteqdot f(p),$$

then it follows that

$$\operatorname*{Lim}_{t \to +0} \frac{\int_0^t h(s) \left(1 - \frac{s}{t}\right)^n ds}{n!\, t^{\nu+1}} \Rightarrow \operatorname*{Lim}_{p \to \infty} \frac{p^\nu f(p)}{\Pi(\nu+n+1)} \quad (n = 0, 1, 2, \dots)\ (\nu > -n-2) \tag{8}$$

and, moreover, if the strip of convergence extends at least as far as the imaginary axis,

$$\operatorname*{Lim}_{t \to \infty} \frac{\int_0^t h(s) \left(1 - \frac{s}{t}\right)^n ds}{n!\, t^{\nu+1}} \Rightarrow \operatorname*{Lim}_{p \to +0} \frac{p^\nu f(p)}{\Pi(\nu+n+1)} \quad (n = 0, 1, 2, \dots)\ (\nu > -n-2). \tag{9}$$

These formulae allow a simple interpretation; for instance, if (9) is rewritten as

$$\operatorname*{Lim}_{t\to\infty}\frac{1}{t^\nu}\frac{\int_0^t d\tau_0\int_0^{\tau_0}d\tau_1\ldots\int_0^{\tau_{n-1}}d\tau_n h(\tau_n)}{\int_0^t d\tau_0\int_0^{\tau_0}d\tau_1\ldots\int_0^{\tau_{n-1}}d\tau_n}\Rightarrow\operatorname*{Lim}_{p\to+0}\frac{(n+1)!}{\Pi(\nu+n+1)}p^\nu f(p)$$

$$(n = 0, 1, 2, \ldots)\ (\nu > -n-2), \qquad (10)$$

it states that, in order to be able to tell something about the image, the original itself need not tend to a limit; on the contrary, the existence of the limit of the mean value in (10) is already sufficient, no matter how large the order n may be. In the electrical networks just mentioned the current, as a function of time, is often taken as the original; if in transient phenomena the current is represented by a Fourier series, then its first-order mean value as $t\to\infty$ tends to the constant term of the series, this constant merely being the direct-current component. According to the Abel theorem (9) or (10) for $\nu = n = 0$, this constant is equal to $f(0)$; in physical terms this means (see (VIII, 17)) that Ohm's law remains valid with respect to the direct-current component in the case of oscillating currents.

Returning to (10) it should be remarked that this formula does not produce the n of the minimum order that is required to get a definite result for the left-hand member. As to the form of the left-hand member of (9), there is a close relationship to the Cesàro limits of VI, § 9, since in the underlying definitions (VI, 36) just the same integral occurs. In view of this for $\nu = -1$ we may also write equation (9) as follows:

$$(C, n)\int_0^\infty h(s)\,ds \Rightarrow \operatorname*{Lim}_{p\to+0}\frac{f(p)}{p}. \qquad (11)$$

An example showing that (9) can be applied for $\nu = 0$, but not the original Abel theorem 5, is provided by the operational relation

$$\sin t\,U(t) \risingdotseq \frac{p}{p^2+1} \quad (\operatorname{Re} p > 0).$$

For, by putting $n = 1$, $\nu = 0$ in (9) and taking the limit

$$\operatorname*{Lim}_{t\to\infty}\frac{1}{t}\int_0^t \sin s\left(1-\frac{s}{t}\right)ds = 0,$$

it follows that for $p\to 0$ the image must tend to zero. Another non-trivial application of (9) again refers to Riemann's ζ-function in the one-sided operational relation (VI, 45),

$$\zeta(p) - \frac{p}{p-1} \risingdotseq [e^t] - e^t\,U(t), \quad 0 < \operatorname{Re} p < \infty.$$

In this case the limit $h(\infty)$ does not exist either; a non-vanishing limit is present for the first-order mean value, namely,

$$\lim_{t\to\infty}\frac{1}{t}\int_0^t ([e^s]-e^s)\,ds = \lim_{t\to\infty}\frac{t[e^t]-\log([e^t]!)-e^t+1}{t},$$

of which the right-hand side is easily shown to be $-\frac{1}{2}$ after using Stirling's approximation (see (28)); from (9) or (10) (if $\nu = n = 0$) it further follows that $\zeta(0) = -\frac{1}{2}$.

4. Real Tauber theorems

As already indicated by the arrow notation, it is not certain in advance whether the preceding Abel theorems may be reversed so as to obtain the corresponding Tauber theorems; that is, from a limiting property of the image for real values of p one cannot infer corresponding properties of the original. The very simple example

$$\sin t\, U(t) \doteqdot \frac{p}{p^2+1} \quad (\operatorname{Re} p > 0),$$

for which $\lim_{p\to 0} f(p) = 0$, already shows that applying $f(0) \Rightarrow h(\infty)$ would lead to the fallacious property $\lim_{t\to\infty} \sin t = 0.$

This is also evident from the general formula (11) according to which the existence of the Cesàro limit of some arbitrary order n of the integral on the left implies the existence of, and equality to, the right-hand limit. If the right-hand member is given, without any further data, it is impossible to learn the order n actually required. Therefore, in the case of Tauber theorems, some non-trivial supplementary conditions are usually given that prove sufficient to guarantee some property or other of the original. Of course, one would possibly prefer a set of necessary *and* sufficient conditions so as to be able to reverse the Abel theorems in the Tauber sense. In the case of simple originals, however, it is usually more convenient to know sufficient conditions. An example of this is provided by (9) for $n = 0$ and $\nu \geqslant -1$, leading to the following theorem†:

THEOREM III. *Given the operational relation*

$$f(p) \doteqdot h(t)\, U(t) \quad (\operatorname{Re} p > 0),$$

then a sufficient condition for the validity of

$$\lim_{p\to 0+} p^\nu f(p) \Rightarrow \lim_{t\to\infty} \Pi(\nu+1)\frac{\int_0^t h(s)\,ds}{t^{\nu+1}} \quad (\nu \geqslant -1) \tag{12}$$

is that a positive constant K can be found so as to have

$$Kt^\nu + h(t) \geqslant 0 \quad (t > 0). \tag{13}$$

† G. Doetsch, *Theorie und Anwendung der Laplace-Transformation*, Berlin, 1939, and New York, 1943, p. 210.

This theorem is easily applied to originals that are positive for $t > 0$. A further simplification is possible if $h(t)$ happens to be monotonic for $t > 0$; it is then permissible† to change the right-hand member of (12) by means of De l'Hospital's rule; thus

$$\operatorname*{Lim}_{t\to\infty} \Pi(\nu+1)\frac{\int_0^t h(s)\,ds}{t^{\nu+1}} \Rightarrow \operatorname*{Lim}_{t\to\infty} \Pi(\nu)\frac{h(t)}{t^\nu}. \tag{14}$$

Therefore we have the following theorem:

THEOREM IV. *Given the operational relation*

$$f(p) \doteq h(t)\,U(t) \quad (\operatorname{Re} p > 0),$$

then a sufficient condition for the validity of

$$\operatorname*{Lim}_{p\to 0+} p^\nu f(p) \Rightarrow \operatorname*{Lim}_{t\to\infty} \Pi(\nu)\frac{h(t)}{t^\nu} \quad (\nu \geqslant -1) \tag{15}$$

is that $h(t)$ should be monotonic for $t > 0$.

There exist many other conditions concerning the legitimacy of reversing Abel theorems into Tauber theorems; further details will not be given, however, because of their complicated nature.

Example. To study the number-theoretical function $d(n)$, which denotes the total number of divisors of n (1 and n included), let us start with the series (III, 36) of the ζ-function. After squaring this series it follows that

$$\zeta^2(p) = \sum_{n=1}^{\infty} \frac{d(n)}{n^p} \quad (\operatorname{Re} p > 1). \tag{16}$$

The Dirichlet series is first rewritten as

$$\zeta^2(p+1) = \sum_{n=1}^{\infty} \frac{d(n)}{n}\frac{1}{n^p} \quad (\operatorname{Re} p > 0),$$

and then (VI, 10) is applied, leading to the following operational relation

$$\zeta^2(p+1) \doteq \sum_{n=1}^{[e^t]} \frac{d(n)}{n} \quad (\operatorname{Re} p > 0). \tag{17}$$

Its domain of convergence extends on the left to $\operatorname{Re} p = 0$. Since the residue of $\zeta(p)$ at its pole $p = 1$ is known to be 1, the image function of (17) obeys

$$\operatorname*{Lim}_{p\to 0} p^2 f(p) = 1.$$

Further, since the original is monotonic we obtain from (15) when applied with $\nu = 2$:

$$1 = 2 \operatorname*{Lim}_{t\to\infty} \frac{1}{t^2} \sum_{n=1}^{[e^t]} \frac{d(n)}{n},$$

or, when x replaces e^t,

$$\operatorname*{Lim}_{x\to\infty} \frac{\sum_{n=1}^{[x]} \frac{d(n)}{n}}{(\log x)^2} = \frac{1}{2}.$$

Finally, it is usual to write the above statement about the function $d(n)$ in the alternative form

$$\sum_{n=1}^{[x]} \frac{d(n)}{n} \sim \tfrac{1}{2}(\log x)^2 \quad (x \to \infty). \tag{18}$$

† See Doetsch, loc. cit. p. 209.

5. Complex Tauber theorems

So far we have discussed Tauber theorems in which only real values of p were important, since the assumption was made that the point $p = 0$ was approached along the real p-axis. It will be evident that further properties of the originals are obtained if limiting properties of the image are also known for complex values of p. Restricting ourselves again to strips of convergence extending on the left up to the line $\operatorname{Re} p = 0$, we mention the theorem of Ikehara†, which even involves the behaviour of the image in the approach to an arbitrary point on the boundary of convergence. In continuing the enumeration of the theorems in the preceding section we then have

THEOREM V. *In order that, with reference to the operational relation*

$$f(p) \doteqdot h(t)\, U(t) \quad (\operatorname{Re} p > 0),$$

the following limiting formula should be valid:

$$\operatorname*{Lim}_{p \to +0} f(p) = \operatorname*{Lim}_{t \to \infty} h(t),$$

it is sufficient that, simultaneously,

(1) $e^t h(t)$ *is positive and non-decreasing for* $t \geqslant 0$;

(2) *a constant* A *exists such that, for* $\operatorname{Re} p$ *tending to* 0†,

$$\operatorname*{Lim}_{\operatorname{Re} p \to +0} \left\{\frac{f(p)}{p} - \frac{A}{p}\right\}$$

tends to some bounded function g *(of argument* $\operatorname{Im} p$*) uniformly in any finite interval* $-a < \operatorname{Im} p < a$.

The condition of uniform convergence in (2) amounts to the requirement that, given $-a < y = \operatorname{Im} p < a$, a number δ can be found (depending perhaps on ϵ and a, but not on y), such that

$$\left|\frac{f(x+iy)}{x+iy} - \frac{A}{x+iy} - g(y)\right| < \epsilon$$

for any x subject to $0 \leqslant x = \operatorname{Re} p \leqslant \delta$. This manner of approaching the imaginary axis is illustrated in fig. 43 by means of shading.

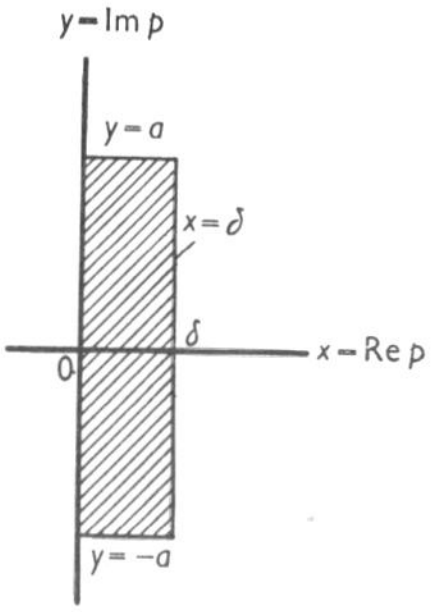

Fig. 43. Illustrating the approach to the imaginary axis mentioned in the complex Tauber theorem.

In particular, if the function $f(p)/p$ on the boundary is singular only at $p = 0$, this being a first-order pole, condition (2) is satisfied. For a better understanding we may treat two examples; the first does not allow the application of theorem V, whereas the second example does.

† *J. Math. Phys.* x, 1, 1931.

We first consider the operational relation

$$\frac{1}{p^2+1} \fallingdotseq (1-\cos t)\, U(t) \quad (\operatorname{Re} p > 0),$$

which does not fulfil condition (2) of theorem V, since the function

$$\frac{1}{p} f(p) - \frac{A}{p}$$

may remain finite at the pole $p = 0$—if A is chosen equal to 1—yet be infinitely large at the additional poles at $p = \pm i$. It is therefore reasonable that to $f(0) = 1$ there should not correspond $h(\infty) = 1$; for the rest, the (trivial) result

$$\lim_{t \to \infty} \frac{1}{t} \int_0^t (1-\cos s)\, ds = 1$$

can be obtained with the help of (12) and (13).

The second example to be discussed is

$$\frac{1}{p+1} \fallingdotseq (1-e^{-t})\, U(t) \quad (\operatorname{Re} p > 0).$$

In this case the function $\dfrac{f(p)}{p} - \dfrac{1}{p}$ is regular for $\operatorname{Re} p > -1$, and thus it possesses the required property as $\operatorname{Re} p \to 0$; $f(0) = h(\infty) = 1$ as it should.

It may be remarked that theorem V is usually written so as to be applicable to $p = 1$, or even to an arbitrary value† of p, instead of to our $p = 0$ which was chosen to facilitate comparison with the real Tauber theorems already given.

The preceding complex Tauber theorem is particularly useful in the derivation of the law concerning the distribution of prime numbers (see XII, § 5). That it is equally well suited for less complicated problems of number theory may be seen in the discussion of the example below. The difficulty in those applications generally amounts to our not knowing whether $e^t h(t)$ is monotonic. If necessary, the latter condition may be replaced, however, by the requirement that the function

$$\phi(p) = \frac{p}{p-1} f(p-1)$$

must be 'completely monotonic' for $p > 1$, that is, $\phi(p)$ should satisfy

$$\phi(p) \geqslant 0, \quad \phi'(p) \leqslant 0, \quad \phi''(p) \geqslant 0, \quad \phi'''(p) \leqslant 0, \quad \text{etc.}$$

This can be shown by application of the attenuation rule and a theorem of Bernstein‡, which states that the one-sided original with domain of convergence $\operatorname{Re} p > 0$ is a non-decreasing function of t if, and only if, the image is completely monotonic for $0 < p < \infty$.

† See Widder, loc. cit. p. 233. ‡ *Acta Math.* LI, 56, 1928.

Example. Referring to the example of the foregoing section, we shall derive a second property of the function $d(n)$. Let us apply (VI, 10) to (16); it gives

$$\zeta^2(p) \doteqdot \sum_{n=1}^{[e^t]} d(n) \quad (\mathrm{Re}\,p > 1).$$

In order to make the operational relation suited for application of theorem V, we first eliminate the second-order pole at $p = 1$. This is accomplished by subtraction of another relation, namely,

$$\frac{p^2}{(p-1)^2} - \zeta^2(p) \doteqdot \left\{(t+1)\,e^t - \sum_{n=1}^{[e^t]} d(n)\right\} U(t) = M(e^t)\,U(t) \quad (\mathrm{Re}\,p > 1). \tag{19}$$

Now it is important to notice that the function

$$M(x) \equiv x(\log x + 1) - \sum_{n=1}^{[x]} d(n)$$

is positive for $x > 1$, as is easily seen in the plot in fig. 44. Since for any divisor of the number n ($< x$) there correspond two integers m_1, m_2 whose product is less than x, the function $\sum_1^{[x]} d(n)$ is equal to the total number of lattice points (that is, of integral co-ordinates m_1, m_2) that lie in the first quadrant of the (m_1, m_2)-plane, and 'inside' the hyperbola $m_1 m_2 = x$. Moreover, this total number is equal to the shaded area in fig. 44, whence it easily follows that $M(x)$ is the dashed area between the hyperbole and the step curve

$$m_2 = \left[\frac{x}{[m_1]+1}\right];$$

thus $M(x)$ is necessarily positive. If, further, the integration and attenuation rules are applied to (19) we find

$$\frac{1}{p} - \frac{p}{(p+1)^2}\,\zeta^2(p+1) \doteqdot U(t)\,e^{-t}\int_0^t M(e^\tau)\,d\tau \quad (\mathrm{Re}\,p > 0). \tag{20}$$

Since M is positive the original multiplied by e^t is obviously monotonically increasing with t. On the other hand, the left-hand member of (20) divided by p has only one singularity on the boundary of convergence $\mathrm{Re}\,p = 0$, namely, the pole of the first order at $p = 0$. Hence theorem V is applicable, leading with the aid of (7) to

$$2 - 2C = \lim_{t\to\infty} \left\{e^{-t}\int_0^t M(e^\tau)\,d\tau\right\},$$

or, after introducing the definition of $M(x)$,

$$\frac{1}{x}\int_1^x \sum_{n=1}^{[s]} d(n)\,\frac{ds}{s} \sim \log x - 2(1-C) \quad (x \to \infty),$$

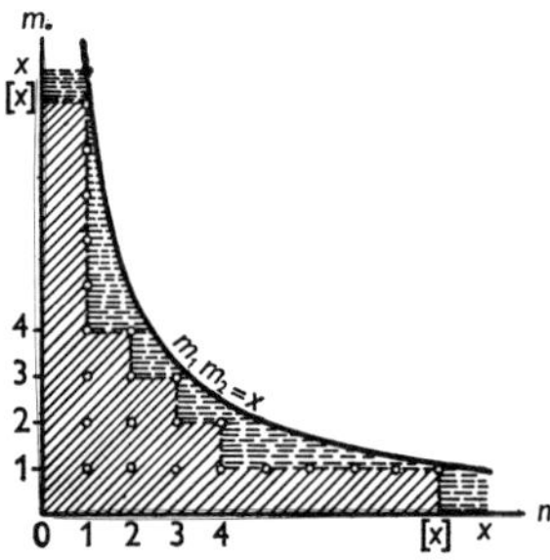

Fig. 44. The lattice related to the function $\sum_1^{[x]} d(n)$.

this being an approximation to the first-order mean value of the function

$$\frac{1}{x}\sum_{n=1}^{[x]} d(n).$$

It should finally be noticed that the real Tauber theorem III, when applied to the operational relation (20), would produce an expression for the second-order mean value.

6. Operational equalities

The preceding Abel and Tauber theorems can be applied not only to special operational relations but also to the many operational rules as treated in chapter IV. Then the corresponding results are equalities containing both original and image. Let us first discuss some suitable examples to make things clear.

(1) From the rule (IV, 41 *a*) for one-sided relations possessing a strip of convergence $\operatorname{Re} p > 0$:

$$\int_p^\infty \frac{f(s)}{s}\,ds \doteqdot U(t)\int_0^t \frac{h(s)}{s}\,ds,$$

it follows† with the help of Abel's theorem

$$\int_0^\infty \frac{h(s)}{s}\,ds \Rightarrow \int_0^\infty \frac{f(s)}{s}\,ds. \tag{21}$$

In this way it is easy to evaluate Dirichlet's integral by applying (21) to $h(t) = \sin t\, U(t)$. The arrow in (21) may be reversed inasmuch as the restrictions of the Tauber theorems are satisfied.

(2) From the rule (IV, 29) involving the product of two originals, which we now suppose to be one-sided with strip of convergence $\operatorname{Re} p > 0$, we obtain after Abel's theorem (11), provided $\alpha_2 + c \leqslant 0$,

$$\int_0^\infty h_1(s)\,h_2(s)\,ds \Rightarrow \frac{i}{2\pi}\int_{c-i\infty}^{c+i\infty} f_1(s)\,f_2(-s)\frac{ds}{s^2} \quad (c>0). \tag{22}$$

This equation is closely related to Parseval's well-known theorem of Fourier series; the latter is obtained by taking for h_1 and h_2 the following cut-off functions:

$$h_1(t) = \{U(t) - U(t-2\pi)\}\sum_{n=-\infty}^{\infty} a_n\,e^{int},$$

$$h_2(t) = \{U(t) - U(t-2\pi)\}\sum_{n=-\infty}^{\infty} b_n\,e^{int},$$

because then, on account of the orthogonality of the trigonometric functions, the left-hand side of (22) becomes

$$\int_0^{2\pi} h_1(s)\,h_2(s)\,ds = 2\pi\sum_{n=-\infty}^{\infty} a_n b_{-n}.$$

(3) From a rule that is closely related to (IV, 33 *b*), namely,

$$\left(e^{-t}\frac{d}{dt}\right)^n h(t) \doteqdot p(p+1)(p+2)\ldots(p+n-1)\,f(p+n),$$

† The relation (21) was derived by Balth. van der Pol, *Phil. Mag.* VIII, 864, 1929, by an integration with respect to λ from 0 to ∞ of the similarity rule (IV, 1 *a*).

it follows again from (11) for any one-sided operational relation that

$$\int_0^\infty \left(e^{-s}\frac{d}{ds}\right)^n h(s)\,ds \Rightarrow (n-1)!\,f(n).$$

For instance, in the case of the function $f(p) = p^{-\nu}$ an extension of Euler's second Γ-function integral is obtained, viz.

$$\frac{(n-1)!}{n^\nu}\Pi(\nu) = \int_0^\infty \left(e^{-s}\frac{d}{ds}\right)^n (s^\nu)\,ds,$$

in which n is a positive integer and ν fractional > -1.

7. Asymptotic series

Since in practical application of the operational calculus we often encounter asymptotic series it may be useful to recall first the underlying definitions. After Poincaré it is said that the series $\sum_{n=0}^{\infty} u_n(x)$ approximates the function $S(x)$ at $x = a$ asymptotically if the difference between $S(x)$ and the finite sum

$$\sum_{n=0}^{N} u_n(x),$$

after dividing by the last term taken into account, tends to zero for x tending to a, where a may be finite or infinite. A conventional notation for the asymptotic equality is the following:

$$S(x) \approx \sum_{n=0}^{\infty} u_n(x) \quad (x \to a).$$

Whereas the asymptotic expansion in mathematical symbols is characterized by

$$\operatorname*{Lim}_{x\to a} \frac{S(x) - \sum_{n=0}^{N} u_n(x)}{u_N(x)} = 0, \tag{23}$$

the analogue with respect to the ordinary sum definition of a convergent series is

$$\operatorname*{Lim}_{N\to\infty} \left\{S(x) - \sum_{n=0}^{N} u_n(x)\right\} = 0.$$

The difference between the two definitions is obviously the following: in the ordinary sum the number of terms N is variable, x being held constant whilst in the asymptotic expansion x is varied while N remains constant. Power series, being ordinarily convergent within their circle of convergence, are at the same time asymptotic at $x = 0$.

Of special interest are the asymptotic series $(x \to 0)$ corresponding to those power series for which the error is less than the absolute value of the last term $u_N(x)$ whenever the series is cut off at this term. It may even be supposed that the 'power series' contains non-integral powers of x. Let $\sum_0^\infty u_n(x)$ denote such a 'power series' in which the exponents of x shall increase steadily with n. If then

$$\left| S(x) - \sum_{n=0}^{N} u_n(x) \right| < | u_N(x) |,$$

the asymptotic character at $x = 0$ follows immediately; first we have

$$\left| S(x) - \sum_{n=0}^{N} u_n(x) \right| = \left| \left\{ S(x) - \sum_{n=0}^{N+1} u_n(x) \right\} + u_{N+1}(x) \right| < 2 \, | u_{N+1}(x) |,$$

and then

$$\frac{\left| S(x) - \sum_{n=0}^{N} u_n(x) \right|}{| u_N(x) |} < 2 \frac{| u_{N+1}(x) |}{| u_N(x) |}.$$

The right-hand member of this inequality vanishes in the limit as $x \to 0$, since the quotient u_{N+1}/u_N is a positive power of x except for a constant factor. In view of (23) the 'power series' is indeed asymptotic for $x \to 0$.

As to numerical evaluation, those series for which the error is less than the absolute value of the last term taken into account are favourably broken off after the term of smallest absolute value. This does not mean, however, that it would be impossible to obtain still better approximations (for instance, by properly accounting for the first term neglected, by means of a weight factor). A well-known example is provided by the exponential integral. By repeated integration by parts it is found that

$$\begin{aligned} -e^p \operatorname{Ei}(-p) &= \int_0^\infty \frac{e^{-u}}{p+u} \, du \\ &= \frac{0!}{p} - \frac{1!}{p^2} + \frac{2!}{p^3} - \dots + \frac{(-1)^{N-1}(N-1)!}{p^N} + (-1)^N N! \int_0^\infty \frac{e^{-u}}{(p+u)^{N+1}} \, du. \end{aligned}$$

If for the moment we confine ourselves to $p > 0$, the term with the integral at the right is easily shown to be less in absolute value than $(N-1)!/p^N$, by replacing the numerator of the integrand by unity. Thus we get the asymptotic series

$$-e^p \operatorname{Ei}(-p) \approx \sum_{n=0}^{\infty} (-1)^n \frac{n!}{p^{n+1}} \quad (p \to \infty), \tag{24}$$

which, at least for $p > 0$, is of the special character that the error is always less than the last term taken into account. Finally, a function is never uniquely determined by any asymptotic expansion. This is clearly shown by the relation

$$\lim_{p \to \infty} p^{N+1} \left\{ f(p) - \sum_{n=0}^{N} (-1)^n \frac{n!}{p^{n+1}} \right\} = 0$$

which is not only satisfied by $f(p) = -e^p \operatorname{Ei}(-p)$ but also by

$$-e^p \operatorname{Ei}(-p) + A \, e^{-\alpha p} \quad \text{if} \quad \alpha > 0.$$

8. Operational transposition of power series in p^{-1} and in t

In § 1 of the present chapter an example was given showing that transposition of series term by term is not in general legitimate. This applies to both original and image. Even power series that are convergent for all finite values of t may lead to divergent series for the image function. So, for instance, the power series of the function $e^{-t^2}U(t)$, when transposed term by term for $\operatorname{Re} p > 0$, produces the series

$$1-\frac{2!}{1!}\frac{1}{p^2}+\frac{4!}{2!}\frac{1}{p^4}-\frac{6!}{3!}\frac{1}{p^6}+\dots, \tag{25}$$

which is divergent throughout. Yet, according to (II, 25), the original under consideration certainly has an image. Only in a few general cases is it always possible to obtain a convergent series for the *original* from a convergent series for the *image*. One of these cases was treated in § 3 of the preceding chapter; as to power series in $1/p$ with radius of convergence ρ we have the operational relation (VI, 15), that is,

$$\sum_{n=0}^{\infty}\frac{a_n}{p^n} \doteq U(t)\sum_{n=0}^{\infty}a_n\frac{t^n}{n!} \quad \left(\operatorname{Re} p > \frac{1}{\rho}\right), \tag{26}$$

in which the legitimacy of transposing term by term was shown with the help of Borel's method of summing. In this example the t-series converges everywhere; thus the original is an integral function of t except for the unit-function factor. Also in the case of non-integral exponents n the method using the Borel sum is suited for proving (26). It is possible that the strip of convergence of (26) extends farther to the left than is indicated by $\operatorname{Re} p = 1/\rho$. For, if p_0 denotes (on the circle of convergence) the singularity of $f(p)$ that is as far to the right as possible, the left-hand boundary of convergence will be given by the line $\operatorname{Re} p = \operatorname{Re} p_0$ (see fig. 45).

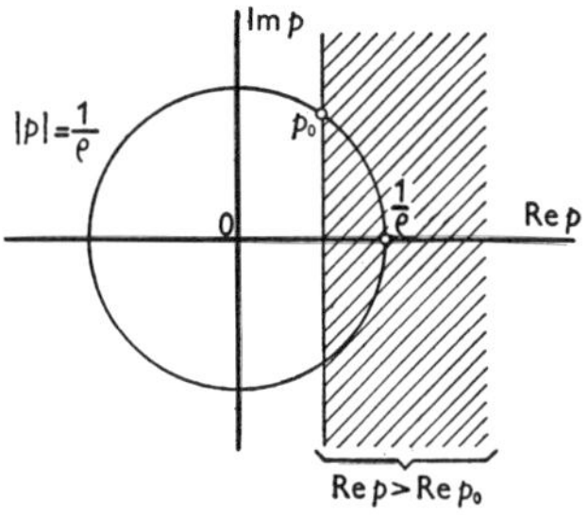

Fig. 45. The strip of convergence of a power series in p^{-1}.

The transposition of power series in p^{-1} thus leads to power series in t that are everywhere convergent. On the other hand, if a power series in t is given (as a one-sided function), then the probability that the corresponding series in p will also be convergent is greatly diminished, since the new coefficients are obtained from the old by multiplying the latter by $n!$, as is seen from (26). Yet the theory of asymptotic series in combination with some appropriate Abel theorem allows of a general treatment of power series in t, whether convergent or only asymptotic. The following theorem can be stated:

THEOREM VI. *If a one-sided original is represented asymptotically as $t \to 0^+$ by some power series of not necessarily integral exponents exceeding -1, then the p-series obtained by transposing the original term by term represents the corresponding image asymptotically as $p \to \infty$.*

The proof of theorem VI is very simple indeed. Writing the original in the form

$$h(t)\, U(t) \approx U(t) \sum_{n=0}^{\infty} c_n t^{m_n} \quad (t \to 0)\ (m_n > -1),$$

then, according to (23), the asymptotic character of this series is expressed by

$$\operatorname*{Lim}_{t \to 0} \frac{h(t) - \sum_{n=0}^{N} c_n t^{m_n}}{t^{m_N}} = 0.$$

Further, since $m_N > -1$, with the help of Abel's theorem (4),

$$0 = \operatorname*{Lim}_{p \to \infty} p^{m_N} \left\{ f(p) - \sum_{n=0}^{N} c_n \frac{\Pi(m_n)}{p^{m_n}} \right\},$$

which in turn is just the condition expressing the asymptotic character of $f(p)$:

$$f(p) \approx \sum_{n=0}^{\infty} c_n \frac{\Pi(m_n)}{p^{m_n}} \quad (p \to \infty).$$

This proves the theorem under consideration.

Either of the series involved in theorem VI may be convergent as well as asymptotic, in whole or in part. More precisely, the following cases are possible:

(1) The t-series as well as the p-series converges everywhere. An example is provided by the relation

$$\sin(2\sqrt{t})\, U(t) \doteqdot \sqrt{\frac{\pi}{p}}\, e^{-1/p}, \quad 0 < \operatorname{Re} p < \infty, \tag{27}$$

which can be obtained by transposing the power series of $\sin(2\sqrt{t})$ term by term.

(2) The t-series converges everywhere; the p-series only in some part of the complex p-plane containing $p = \infty$. This situation is encountered in the example

$$\frac{p^2}{p^2+1} = 1 - \frac{1}{p^2} + \frac{1}{p^4} - \ldots \doteqdot U(t)\left(1 - \frac{t^2}{2!} + \frac{t^4}{4!} - \ldots\right) = \cos t\, U(t) \quad (\operatorname{Re} p > 0).$$

(3) The t-series is convergent throughout; the p-series is asymptotic as $p \to \infty$ without being convergent. See example (25).

(4) The t-series is convergent only in some finite part of the t-plane containing $t = 0$; the p-series is asymptotic as $p \to \infty$. As an example we may quote

$$-p\, e^{p}\, \operatorname{Ei}(-p) \doteqdot \frac{U(t)}{t+1} \quad (\operatorname{Re} p > 0),$$

which is obtained from (III, 34) by a shift; the power series in t converges only for $|t| < 1$, whilst the corresponding series in p leads to the asymptotic (24).

(5) Both the t-series and the p-series diverge throughout. An example of this is given by the original coming from (24) when p is changed into $1/t$, viz.

$$-e^{1/t}\,\mathrm{Ei}\left(-\frac{1}{t}\right) \approx \sum_{n=0}^{\infty}(-1)^n n!\,t^{n+1} \quad (t \to 0).$$

This original certainly possesses an image for $\mathrm{Re}\,p > 0$, since the left-hand side behaves like $\log t$ as $t \to \infty$. The asymptotic series for the image is then

$$f(p) \approx \sum_{n=0}^{\infty}(-1)^n \frac{n!(n+1)!}{p^{n+1}} \quad (p \to \infty).$$

Consequently both series are asymptotic here.

Example. Stirling's series. This famous asymptotic series is extremely useful in approximating the function $\Pi(\nu)$. It can be derived operationally as follows. Let us put

$$\Pi(p) = p^p e^{-p} \sqrt{(2\pi p)}\, e^{\mu(p)}. \tag{28}$$

Then Stirling's approximation can be reduced to the statement $\mu(\infty) = 0$. A simple relation involving $\mu(p)$ is obtained when the logarithm of (28) is taken and then the functional equation of the Γ-function is applied, namely,

$$p\{\mu(p+1) - \mu(p)\} = p - (p^2 + \tfrac{1}{2}p)\log\left(1 + \frac{1}{p}\right). \tag{29}$$

The right-hand side of (29) can easily be translated into the t-language with the help of the following relation:

$$\frac{1-e^{-t}}{t}\,U(t) \doteqdot p\log\left(1+\frac{1}{p}\right) \quad (\mathrm{Re}\,p > 0), \tag{30}$$

which in turn may be verified by applying the rule for multiplying the original by t. If it is supposed that $p\mu(p)$ has some original $h(t)$, one further finds, upon transposing the left-hand side of (29) with the attenuation rule and the right-hand side by means of (30), an equation for $h(t)$. After some simplification the result is found to be

$$p\mu(p) \doteqdot \frac{1}{t}\left(\frac{1}{e^t - 1} - \frac{1}{t} + \frac{1}{2}\right) U(t), \quad 0 < \mathrm{Re}\,p < \infty. \tag{31}$$

The Stirling series in question is obtained from this relation simply by transposing the power series for the original term by term. The latter is, on account of (VI, 46),

$$U(t) \sum_{n=2}^{\infty} \frac{B_n}{n!} t^{n-2}. \tag{32}$$

Therefore
$$\mu(p) \approx \sum_{n=2}^{\infty} \frac{B_n}{n(n-1)} \frac{1}{p^{n-1}} = \frac{1}{12p} - \frac{1}{360p^3} + \frac{1}{1260p^5} - \cdots \quad (p \to \infty). \tag{33}$$

A closer investigation will show that the asymptotic development (33) is valid not only for real values of $p \to \infty$ but also for $|p| \to \infty$ in any sector $-\pi + \delta < \arg p < \pi - \delta$ of the complex p-plane, thus excluding the negative real axis of p.

9. Transposition of series with ascending powers of p

In the preceding section only series with descending powers of p were discussed. This is quite natural, since the individual terms, being proportional to p^{-N}, always have an original, however large N may be. It is remarkable that ascending power series of p also exist, thus containing p^{ν} with $\nu > 1$, that may be transposed term by term with the help of (III, 3). Though the latter relation is not valid for these values of ν, yet the resulting series may be true in the asymptotic sense. We shall confine ourselves to series that contain integral powers of $p^{\frac{1}{2}}$; hence the image may be written as

$$f(p) = \sum_{n=0}^{\infty} c_n p^{\frac{1}{2}n}, \tag{34}$$

which is supposed to have a non-vanishing radius of convergence. Let us apply (III, 3) formally; we must investigate whether the resulting series for $t > 0$ can actually represent the original asymptotically, that is,

$$h(t) \approx \sum_{n=0}^{\infty} c_n \frac{t^{-\frac{1}{2}n}}{\Pi(-\frac{1}{2}n)} \quad (t \to \infty). \tag{35}$$

A sufficient condition as to the validity of (35) is that $t^N R_N(t)$ must be a monotonic function of t for $N > 0$, where $R_N(t)$ denotes the remainder

$$R_N(t) = h(t) - \sum_{n=0}^{2N-1} c_n \frac{t^{-\frac{1}{2}n}}{\Pi(-\frac{1}{2}n)}.$$

In order to prove this we first differentiate the power series for $f(p)/p$ inside its circle of convergence N times in succession; this easily leads to

$$\operatorname*{Lim}_{p \to 0} \sqrt{p}\left\{p\left(-\frac{d}{dp}\right)^N \frac{f(p)}{p} - (-1)^N \sum_{n=0}^{2N-1} c_n \frac{\Pi(\frac{1}{2}n-1)}{\Pi(\frac{1}{2}n-N-1)} p^{\frac{1}{2}n-N}\right\} = 0.$$

Next, on account of the monotonic character of $t^N R_N(t)$, the Tauber theorem IV can be applied with $\nu = \frac{1}{2}$, giving

$$\operatorname*{Lim}_{t \to \infty} t^{N-\frac{1}{2}}\left\{h(t) - \sum_{n=0}^{2N-1} c_n \frac{t^{-\frac{1}{2}n}}{\Pi(-\frac{1}{2}n)}\right\} = 0, \tag{36}$$

which is the very definition of asymptotic development; hence (35) follows. It is further striking that in (35) the terms with even values of n disappear because $\Pi(-\frac{1}{2}n)$ is infinitely large for n even. These terms would come from integral powers of p which individually would lead to derivatives of the δ-function. This kind of transposition of power series, dropping the even terms, was already applied by Heaviside in transient phenomena of cables, though without a proper proof. Carson† has also made extensive investigations concerning these series.

† *Electric Circuit Theory and the Operational Calculus*, New York, 1926, pp. 50–84.

Example 1. Starting from the relation

$$\sqrt{p} \doteq \frac{U(t)}{\sqrt{(\pi t)}} \quad (\operatorname{Re} p > 0),$$

after multiplying by e^{-t} and integrating, we obtain

$$\frac{1}{\sqrt{(p+1)}} \doteq \operatorname{erf}(\sqrt{t})\, U(t), \quad 0 < \operatorname{Re} p < \infty. \tag{37}$$

From this relation it is easy to get

$$\frac{\sqrt{p}}{\sqrt{p}+1} \doteq e^t \operatorname{erfc}(\sqrt{t})\, U(t), \quad 0 < \operatorname{Re} p < \infty, \tag{38}$$

to which the preceding theory may be applied. Term-by-term transposition of the power series

$$\frac{\sqrt{p}}{\sqrt{p}+1} = p^{\frac{1}{2}} - p + p^{\frac{3}{2}} - p^2 + \dots$$

leads to

$$e^t \operatorname{erfc}(\sqrt{t}) \approx \frac{1}{\sqrt{(\pi t)}}\left(1 - \frac{1}{2t} + \frac{3}{4t^2} - \frac{3.5}{8t^3} + \frac{3.5.7}{16t^4} - \dots\right); \tag{39}$$

suitable integration by parts of the error function directly proves (39). The image of (38) represents the frequency admittance of an infinite cable consisting per unit length of a series resistance and a shunt capacitance (of equal magnitudes, in suitably chosen units of course), the whole in series with a unit resistance (see XV, § 4). According to the theory in § 5 of the next chapter the original here represents the current as a function of time after a unit voltage has been applied to this cable. The expansion (39) is very useful in determining the current a long time after the electromotive force was switched on.

Let us omit the resistance in series with the foregoing cable, and let a voltage $\{1 - e^{-t}\}\, U(t)$ be applied. This will give rise to a current equal to the original in the following operational relation

$$\frac{\sqrt{p}}{p+1} \doteq -i\, e^{-t} \operatorname{erf}(i\sqrt{t})\, U(t) = U(t) \int_0^t \frac{e^{-u}}{\sqrt{\{\pi(t-u)\}}}\, du \quad (\operatorname{Re} p > 0).$$

Again transposing term by term the series

$$\frac{\sqrt{p}}{p+1} = p^{\frac{1}{2}} - p^{\frac{3}{2}} + p^{\frac{5}{2}} - p^{\frac{7}{2}} + \dots,$$

we find the asymptotic expansion

$$\int_0^t \frac{e^{-u}}{\sqrt{\{\pi(t-u)\}}}\, du \approx \frac{1}{\sqrt{(\pi t)}}\left(1 + \frac{1}{2t} + \frac{3}{4t^2} + \frac{3.5}{8t^3} + \dots\right) \quad (t \to \infty).$$

On the other hand, if the voltage applied is equal to $(1 - \cos t)\, U(t)$, we are led to the relation

$$\frac{\sqrt{p}}{p^2+1} \doteq \frac{U(t)}{\sqrt{\pi}} \int_0^t \frac{\sin u}{\sqrt{(t-u)}}\, du \quad (\operatorname{Re} p > 0),$$

which does not allow us to derive an asymptotic series for the original from the corresponding power series of the image. This is due to the oscillating character of the original $h(t)$ as $t \to \infty$, which in turn is due to the oscillating voltage applied.

Example 2. It is even possible that the series for the original is convergent rather than asymptotic. This can be seen from the example

$$e^{-\sqrt{p}} \doteq \operatorname{erfc}\left(\frac{1}{2\sqrt{t}}\right) U(t), \quad 0 < \operatorname{Re} p < \infty. \tag{40}$$

This relation can be derived as follows. In the integrand of the corresponding definition integral we substitute for the error function its integral representation. If the order of integration in the corresponding repeated integral is changed the second integral can easily be solved explicitly, whence for the image

$$\sqrt{\frac{2}{\pi}}\, p^{\frac{1}{2}} \int_0^\infty \exp\left\{-\tfrac{1}{2}\sqrt{p}\left(w^2+\frac{1}{w^2}\right)\right\} dw.$$

Taking the arithmetic mean of this integral and that obtained from it by the substitution $w \to w^{-1}$, a new integral is found which is readily evaluated.

When the image series of (40)

$$\mathrm{e}^{-\sqrt{p}} = 1-\frac{p^{\frac{1}{2}}}{1!}+\frac{p}{2!}-\frac{p^{\frac{3}{2}}}{3!}+\ldots$$

is transposed term by term the result is for $t>0$

$$1-\frac{1}{\sqrt{(\pi t)}}\left(1-\frac{1}{12t}+\frac{1}{160t^2}-\ldots\right),$$

which is the *convergent* development in powers of $1/t$ of

$$1-\frac{1}{\sqrt{(\pi t)}}\int_0^1 \mathrm{e}^{-w^2/4t}\, dw,$$

this being the original of the relation (40).

One should be very careful in applying (35) to the transposition of power series with ascending powers of p, as is observed from the analogous series

$$\mathrm{e}^{\sqrt{p}} = 1+\frac{p^{\frac{1}{2}}}{1!}+\frac{p}{2!}+\frac{p^{\frac{3}{2}}}{3!}+\ldots,$$

which would yield the function

$$1+\frac{1}{\sqrt{(\pi t)}}\left\{1-\frac{1}{12t}+\frac{1}{160t^2}-\ldots\right\} = \left\{2-\operatorname{erfc}\left(\frac{1}{2\sqrt{t}}\right)\right\} U(t)$$

as original of $\mathrm{e}^{\sqrt{p}}$. This t-function, however, has the image $2-\mathrm{e}^{-\sqrt{p}}$, which is quite different from $\mathrm{e}^{\sqrt{p}}$; indeed, the latter has no original at all, as can be made plausible from the absolute divergence of the corresponding inversion integral, even if taken in Abel's sense. In addition, it is easily verified that the (sufficient) condition mentioned above for the validity of (35) is not satisfied either.

Whereas in applying the rule (35) the positive integral powers of p automatically disappear, it is still possible that power series containing only positive integral powers have a definite operational meaning. For instance, if it is known that the one-sided original $h(t)\, U(t)$ possesses moments of any order

$$M_n = \int_0^\infty h(s)\, s^n\, ds \quad (n = 0, 1, 2, \ldots),$$

and further that $f(p)/p$ has a power series of non-vanishing radius of convergence, then for the image

$$f(p) = p\sum_{n=0}^{\infty}\frac{(-1)^n}{n!}M_n p^n. \tag{41}$$

This Maclaurin series can be proved by first applying (IV, 38), then the integration rule (IV, 34a), and finally Abel's theorem I. So, in the case of the relation

$$\{\theta_3(0,t)-1\}\, U(t) \risingdotseq \sqrt{(\pi p)}\coth\{\sqrt{(\pi p)}\}-1 = f(p) \quad (\operatorname{Re} p>0),$$

derivable from (51) to be discussed later on, the corresponding moments are found to be

$$\int_0^\infty \{\theta_3(0,s)-1\}s^n\,ds = (-1)^n n!(4\pi)^{n+1}\frac{B_{2n+2}}{(2n+2)!}.$$

In accordance with (41) the power series for the image is

$$f(p) = 4\pi p \sum_{n=0}^{\infty} \frac{B_{2n+2}}{(2n+2)!}(4\pi p)^n \quad (|\,p\,|<\pi).$$

It is almost unnecessary to emphasize that merely knowing the power series of the image (corresponding to a one-sided original) does not imply the existence of the moments. So, the image $1/(e^p-1)$ of (v, 8) can be expanded into powers of p if $|\,p\,|<2\pi$, yet the original $[t]\,U(t)$ does not have any moment integral.

10. Expansion in rational fractions (Heaviside's expansion theorem II)

This section is devoted to operational treatment of images for which $f(p)/p$ can be expanded as a sum of rational fractions. A rational function is the most simple; it is the quotient of two polynomials: $N(p)/D(p)$. Moreover, if it is supposed that the degree of $N(p)$ is less than that of $D(p)$, we may write

$$\frac{f(p)}{p} = \frac{N(p)}{D(p)} = \sum_{n=1}^{N}\sum_{k=1}^{m_n}\frac{a_{n,k}}{(p-p_n)^k}, \tag{42}$$

thus without an additional polynomial. It is evident that the finite sum in (42) may be transposed term by term. The original of any separate term is

$$\frac{p}{(p-p_n)^k} \doteq \frac{e^{p_n t}\,t^{k-1}}{(k-1)!}U(t) \quad (\operatorname{Re}p > \operatorname{Re}p_n).$$

Let us enumerate the poles p_n according to decreasing values of $\operatorname{Re}p_n$; thus all poles lie on the left of (some perhaps on) the vertical line $\operatorname{Re}p = \operatorname{Re}p_1$. Consequently

$$f(p) \doteq U(t)\sum_{n=1}^{N} e^{p_n t}\sum_{k=1}^{m_n} a_{n,k}\frac{t^{k-1}}{(k-1)!}, \quad \operatorname{Re}p_1 < \operatorname{Re}p < \infty. \tag{43}$$

It is also an easy matter to determine the original belonging to $f(p)$ in other domains, such as $\operatorname{Re}p_2 < \operatorname{Re}p < \operatorname{Re}p_1$, but they are of less importance. The transposition of rational functions considered above only for the strip $\operatorname{Re}p > \operatorname{Re}p_1$—that lying as far to the right as possible—is called Heaviside's (second) expansion theorem, in honour of Heaviside who frequently applied it†. This theorem is useful in the operational solution of linear differential equations with constant coefficients (VIII, § 3), and in the theory of transient phenomena in electrical networks.

† Heaviside's first expansion theorem will be treated in VIII, § 8.

On the other hand, if the degree of $N(p)$ exceeds (or is equal to) that of $D(p)$, a polynomial containing positive integral powers of p must be added to (42), which will lead to the δ-function and its derivatives. For instance, if all the poles are of the first order, the result is

$$p\sum_{n=1}^{N}\frac{a_n}{p-p_n}+\sum_{m=1}^{k}b_m p^m \risingdotseq U(t)\sum_{n=1}^{N}a_n\,\mathrm{e}^{p_n t}+\sum_{m=1}^{k}b_m\,\delta^{(m-1)}(t), \quad \mathrm{Re}\,p_1<\mathrm{Re}\,p<\infty.$$

We would like to draw special attention to a property that characterizes the importance of expanding the image in rational fractions. Instead of taking the complete image function we consider, in succession, first the fraction corresponding to the pole p_1, then the sum of the fractions belonging to p_1, p_2, next the sum of those corresponding to p_1, p_2, p_3, etc.†. The successive operational relations are

$$p\frac{a_1}{p-p_1} \risingdotseq U(t)\,a_1\,\mathrm{e}^{p_1 t},$$

$$p\left(\frac{a_1}{p-p_1}+\frac{a_2}{p-p_2}\right) \risingdotseq U(t)\,(a_1\,\mathrm{e}^{p_1 t}+a_2\,\mathrm{e}^{p_2 t}),$$

$$p\left(\frac{a_1}{p-p_1}+\frac{a_2}{p-p_2}+\frac{a_3}{p-p_3}\right) \risingdotseq U(t)\,(a_1\,\mathrm{e}^{p_1 t}+a_2\,\mathrm{e}^{p_2 t}+a_3\,\mathrm{e}^{p_3 t}),$$

etc.

Obviously the right-hand members represent the complete original approximately as $t\to\infty$. The more poles taken into account, the better the approximation. It is of great importance that this method of approximation is not restricted to rational functions, as will be seen below.

Let the one-sided original $h(t)\,U(t)$ have a strip of convergence $\mathrm{Re}\,p>\mathrm{Re}\,p_1$, and let its image be such that in the remaining part of the p-plane the only singularities of $\frac{1}{p}f(p)$ are first-order poles $p_1, p_2, \ldots$, with residues $a_1, a_2, \ldots$. In this case it will often be possible to obtain better and better approximations of the original ($t\to\infty$) by transposing more and more principal parts $a_1/(p-p_1)$, $a_2/(p-p_2)$, ...‡. For instance, suppose that the procedure (VI, 41) of widening the strip of convergence (by suitably neutralizing the poles) is applicable to the operational relation under consideration. Then it follows

$$f(p)-p\sum_{n=1}^{N}\frac{a_n}{p-p_n} \risingdotseq h(t)-U(t)\sum_{n=1}^{N}a_n\,\mathrm{e}^{p_n t}, \tag{44}$$

in which the left-hand abscissa of convergence is equal to $\alpha_N=\mathrm{Re}\,p_N-b_N$, where b_N is some positive constant. On the other hand, this abscissa of

† We suppose that any vertical line in the p-plane contains at most one pole. As before, these poles are enumerated according to decreasing values of $\mathrm{Re}\,p_n$.

‡ To keep the reasoning as simple as possible we shall assume that the poles all lie on the left of the imaginary axis.

convergence is obtainable from the general formula (VI, 20). As soon as α_N becomes negative we thus have

$$\overline{\operatorname{Lim}_{x\to\infty}} \frac{1}{x}\log\left|\int_x^\infty \left\{h(s)-\sum_{n=1}^{N} a_n\, e^{p_n s}\right\} ds\right| = \operatorname{Re} p_N - b_N.$$

From this limiting relation we infer that, for ϵ sufficiently small and positive, there can be found a number $x_0(\epsilon)$ such that the expression under the limit sign is less than $\operatorname{Re} p_N-\epsilon$; this can be written equivalently as

$$\left|\frac{\int_x^\infty \left\{h(s)-\sum_{n=1}^{N} a_n\, e^{p_n s}\right\} ds}{e^{p_N x}}\right| < e^{-\epsilon x} \quad \text{if} \quad x > x_0(\epsilon),$$

whence it further follows that

$$\operatorname{Lim}_{x\to\infty} \frac{\int_x^\infty \left\{h(s)-\sum_{n=1}^{N} a_n\, e^{p_n s}\right\} ds}{e^{p_N x}} = 0. \tag{45}$$

If, moreover, De l'Hospital's rule may be applied (for which it is sufficient that the expression in braces is a monotonic function of s) the result is

$$\operatorname{Lim}_{t\to\infty} \frac{h(t)-\sum_{n=1}^{N} a_n\, e^{p_n t}}{e^{p_N t}} = 0, \tag{46}$$

which shows that for large values of t we have, under the conditions indicated, the approximation

$$h(t) \approx \sum_{n=1}^{N} a_n\, e^{p_n t}.$$

Finally we would emphasize that the larger N taken, the better the approximation.

As an example may be cited (VI, 11)

$$\zeta(p) \doteq [e^t], \quad 1 < \operatorname{Re} p < \infty.$$

In this case $f(p)/p$ has simple poles at $p = 1$ and $p = 0$ with residues 1 and $-\frac{1}{2}$, respectively. Accordingly, the original $[e^t]$ has approximations corresponding to

$$\frac{p}{p-1} \doteq e^t\, U(t), \qquad \frac{p}{p-1}-\frac{1}{2} \doteq (e^t-\tfrac{1}{2})\, U(t),$$

of which the last is the better, since its mean value as $t \to \infty$ is exactly equal to that of the true original.

Even in the case of higher-order poles the corresponding principal parts may lead to useful approximations for the original. To see this, consider the example of § 5; after applying the attenuation rule it is found that

$$\frac{p}{p+1}\, \zeta^2(p+1) \doteq e^{-t} \sum_{n=1}^{[e^t]} d(n) \quad (\operatorname{Re} p > 0).$$

Here the function $f(p)/p$ has a pole of the second order with principal part

$$\frac{1}{p^2}+\frac{(2C-1)}{p}.$$

The corresponding operational relation, that is,

$$\frac{1}{p}+(2C-1) \doteqdot (t+2C-1)\,U(t),$$

leads to the following approximation:

$$e^{-t}\sum_{n=1}^{[e^t]} d(n) \sim t+2C-1 \quad (t\to\infty).$$

This is completely in accordance with the result of § 5, where by the use of the complex Tauber theorem it was shown that

$$\int_1^x \frac{ds}{s}\sum_{n=1}^{[s]} d(n) \sim x\log x - 2x(1-C) \quad (x\to\infty);$$

when differentiated through with respect to x it just gives the approximation found above.

In the foregoing, where the original was approximated by taking a number of principal parts of $f(p)/p$, it was left undecided whether the process had to be stopped somewhere owing, for example, to singularities different from poles. If such singularities are not present we may continue the procedure *ad infinitum*, provided, of course, that there are an infinite number of poles and, what is more, that the strip of convergence can continually be widened by way of the successive relations (44). The equation (45) then holds good for all values of N; in the sense of Poincaré it obviously implies the asymptotic expansion

$$\int_\infty^x h(s)\,ds \approx \sum_{n=1}^{\infty} a_n \frac{e^{p_n x}}{p_n} \quad (x\to\infty).$$

If, again, the rule of De l'Hospital is applicable, the original itself is asymptotically represented by

$$h(x) \approx \sum_{n=1}^{\infty} a_n\, e^{p_n x} \quad (x\to\infty).$$

Instead of asymptotic there may often be ordinary equality, such that in the case of the relation

$$p\sum_{n=1}^{\infty}\frac{a_n}{p-p_n} \doteqdot U(t)\sum_{n=1}^{\infty} a_n\, e^{p_n t} \quad (\operatorname{Re} p > \operatorname{Re} p_1), \tag{47}$$

the original is simply obtained by a term-by-term transposition of the infinite series in rational fractions (a so-called Mittag-Leffler series). For instance, the result so obtained will certainly be true if the path of integration in the inversion integral can be closed at the left (see end of VI, § 11), since then the right-hand side of (47) is found by means of Cauchy's theorem of residues (see also fig. 37). Furthermore, if the poles p_n are real the original is a Dirichlet series, which type of series was studied before as an image function (see VI, § 3). The following three examples are cited:

$$pP(p,1) = p\sum_{n=0}^{\infty}\frac{(-1)^n}{n!}\frac{1}{p+n} \risingdotseq U(t)\sum_{n=0}^{\infty}\frac{(-1)^n}{n!}\mathrm{e}^{-nt} \quad (\mathrm{Re}\,p>0),$$

$$\psi(p)+C = p\sum_{n=1}^{\infty}\frac{1}{n(p+n)} \risingdotseq U(t)\sum_{n=1}^{\infty}\frac{\mathrm{e}^{-nt}}{n} \quad (\mathrm{Re}\,p>-1), \tag{48}$$

$$\surd(\pi p)\coth\{\surd(\pi p)\} = p\sum_{n=-\infty}^{\infty}\frac{1}{p+\pi n^2} \risingdotseq U(t)\sum_{n=-\infty}^{\infty}\mathrm{e}^{-\pi n^2 t} \quad (\mathrm{Re}\,p>0). \tag{49}$$

The first of them has already been dealt with in (III, 19, 20) in the discussion of the Prym functions. The second example (48) gives, after the convergent series at the right has been summed,

$$\psi(p)+C \risingdotseq -U(t)\log(1-\mathrm{e}^{-t}), \quad -1<\mathrm{Re}\,p<\infty, \tag{50}$$

which is also found by a straightforward reduction of its definition integral (by means of integration by parts) to (III, 14) involving the ψ-function. Finally, the original (49) is merely a θ-function; thus

$$\surd(\pi p)\coth\{\surd(\pi p)\} \risingdotseq U(t)\,\theta_3(0,t), \quad 0<\mathrm{Re}\,p<\infty. \tag{51}$$

The series for the image function in (49) is based upon the well-known expansion of the cotangent function in rational fractions, namely,

$$\pi\cot(\pi p) = \frac{1}{p}+2\sum_{n=1}^{\infty}\frac{p}{p^2-n^2} = (C,0)\sum_{n=-\infty}^{\infty}\frac{1}{p+n}, \tag{52}$$

which itself can also be transposed term by term, though its poles stretch to infinity at both sides of the origin. By first differentiating (III, 11) so as to obtain

$$\mathrm{sgn}\,t\,\mathrm{e}^{-n|t|} \risingdotseq \frac{2p^2}{p^2-n^2} \quad (-n<\mathrm{Re}\,p<n),$$

the term-by-term transposition leads to

$$\pi p\cot(\pi p)-1 \risingdotseq \frac{\mathrm{sgn}\,t}{\mathrm{e}^{|t|}-1}, \quad -1<\mathrm{Re}\,p<1. \tag{53}$$

In any of the examples above the expansion in rational fractions of the image function converges everywhere outside the poles. Meromorphic functions of this kind are of genus zero. It is also possible that the expansion of (47) in rational fractions, as it stands, does not converge owing to the special behaviour of p_n and a_n for large values of n. Convergence may be maintained, however, by subtracting from any term the first n terms of its corresponding power-series development:

$$\frac{a_n}{p-p_n} = -a_n\left(\frac{1}{p_n}+\frac{p}{p_n^2}+\frac{p^2}{p_n^3}+\dots\right).$$

If at least N such terms are necessary to guarantee the convergence, the function under consideration is of genus N. Therefore, the development of a meromorphic function whose poles are all of the first order and whose genus is N, is generally

$$\frac{f(p)}{p} = \sum_{n=1}^{\infty} a_n\left(\frac{1}{p-p_n}+\frac{1}{p_n}+\frac{p}{p_n^2}+\frac{p^2}{p_n^3}+\ldots+\frac{p^{N-1}}{p_n^N}\right). \tag{54}$$

There are some difficulties in transposing these image functions term by term, unless $N = 0$. For instance, the term $\delta(t)\sum_1^{\infty}\frac{a_n}{p_n}$ of the original in general diverges. Yet, new operational relations are readily obtained, namely, by first differentiating (54) N times in succession, which can be inferred from the following example. After writing the series for the image of (48) in the form

$$\psi(p)+C = \sum_{n=1}^{\infty}\left(\frac{1}{n}-\frac{1}{p+n}\right), \tag{55}$$

the function $\psi(p)+C$ appears to be of genus 1, whilst the function $\frac{1}{p}\{\psi(p)+C\}$ is of genus 0 on account of (48). Straightforward transposition of (55) term by term is impossible, unlike the case of

$$p\psi'(p) = p\sum_{n=1}^{\infty}\frac{1}{(p+n)^2},$$

which is obtained after differentiating. The last series leads to the new relation

$$p\psi'(p) \risingdotseq U(t)\,t\sum_{n=1}^{\infty}e^{-nt} = \frac{t}{e^t-1}\,U(t) \quad (\mathrm{Re}\,p > -1). \tag{56}$$

11. Transposition of other series

In the preceding sections we studied term-by-term transposition of several types of series, particularly power series and expansions in rational fractions. The question arises whether still other series exist that admit term-by-term transposition, leading to convergent or asymptotic developments.

In confining ourselves again to one-sided originals, one of the few statements that will generally hold is the following. Term-by-term transposition of any series

$$U(t)\sum_{n=0}^{\infty}h_n(t)$$

for $\mathrm{Re}\,p > 0$ is allowed if it is uniformly convergent for $t \geqslant 0$. For, if we put

$$h(t)\,U(t) = \sum_{n=0}^{N}\{h_n(t)+R_N(t)\}\,U(t),$$

and supposing
$$h_n(t)\,U(t) \doteqdot f_n(p),$$
we have
$$f(p) = \sum_{n=0}^{N} f_n(p) + p\int_0^\infty e^{-pt} R_N(t)\,dt.$$

Further, on account of the uniform convergence for $t \geqslant 0$, $|R_N(t)| < \epsilon$ if $N > N_0(\epsilon)$, where $N_0(\epsilon)$ is independent of t. Therefore
$$\left| f(p) - \sum_{n=0}^{N} f_n(p) \right| \leqslant |p|\,\epsilon \int_0^\infty e^{-\mathrm{Re}\,pt}\,dt = \epsilon \frac{|p|}{\mathrm{Re}\,p},$$
and consequently
$$f(p) = \sum_{n=0}^{\infty} f_n(p).$$

Particularly, it is often desired to transpose term by term those series which are expansions in orthogonal functions. As an example we may mention the functions $e^{-\frac{1}{2}t} L_n(t)$ in which $L_n(t)$ denotes the Laguerre polynomial of degree n. According to (VI, 17) its image becomes
$$e^{-\frac{1}{2}t} L_n(t)\,U(t) \doteqdot n!\,p\frac{(p-\frac{1}{2})^n}{(p+\frac{1}{2})^{n+1}} \quad (\mathrm{Re}\,p > -\tfrac{1}{2}).$$
Therefore, the operational relation
$$p\sum_{n=0}^{\infty} c_n \frac{(p-\frac{1}{2})^n}{(p+\frac{1}{2})^{n+1}} \doteqdot U(t)\,e^{-\frac{1}{2}t} \sum_{n=0}^{\infty} \frac{c_n}{n!} L_n(t) \quad (\mathrm{Re}\,p > 0) \tag{57}$$
is certainly valid if the right-hand side, being an expansion in orthogonal functions of the original, is known to converge uniformly for $t \geqslant 0$.

12. A real inversion formula

Until now the inversion integral has been considered as a general means of producing the original corresponding to a given image function. The inversion integral is of no direct value if $f(p)$ is given only for real p, as we would first have to calculate it—by means of analytic continuation—at the path of integration $\mathrm{Re}\,p = c$. In principle the analytic continuation is always possible, though often difficult. On the other hand, the original can be expressed in a form that involves only the values of the image at the real axis of p. Best known in this respect is Widder's formula†, which in the case of one-sided originals reads
$$h(t) = \lim_{N\to\infty} \left\{ \frac{p^{N+1}}{N!} \left(-\frac{d}{dp}\right)^N \frac{f(p)}{p} \right\}_{p=N/t} \quad (t > 0). \tag{58}$$

In this formula $h(t)$ should be replaced by $\frac{1}{2}\{h(t+0)+h(t-0)\}$ at points of discontinuity. The expression (58) can be derived from a sifting integral. If for $f(p)/p$ the definition integral is inserted, and the differentiation is

† 'The inversion of the Laplace integral and the related moment problem', *Trans. Amer. Math. Soc.* XXXVI, 107–200, 1934.

performed under the sign of integration, the right member of (58) becomes the sifting integral for the mean value of $h(t)$ that corresponds to the approximation function

$$\delta(t-1) = \lim_{N\to\infty} \frac{N^{N+1}}{N!} e^{-Nt} t^N U(t),$$

which, on account of Stirling's formula, is equivalent to (v, 37). Incidentally, (58) may be written in a more elegant form as follows:

$$h(t) = \lim_{N\to\infty} \left\{ \frac{\dfrac{d^N}{dp^N}\dfrac{f(p)}{p}}{\dfrac{d^N}{dp^N}\dfrac{1}{p}} \right\}_{p=N/t}. \tag{59}$$

It may be expected that, in considering Widder's expression for finite values of N rather than for $N\to\infty$, we obtain a set of approximations for $h(t)$. The larger N chosen, the better the approximation will be. We obtain successively

$$N = 0:\quad f(0),$$

$$N = 1:\quad f\left(\frac{1}{t}\right) - \frac{1}{t} f'\left(\frac{1}{t}\right),$$

$$N = 2:\quad f\left(\frac{2}{t}\right) - \frac{2}{t} f'\left(\frac{2}{t}\right) + \frac{2}{t^2} f''\left(\frac{2}{t}\right),$$

etc.

The first of them just delivers the exact value at $t = \infty$ because of Abel's theorem.

It is important to notice that, in the case of any known one sided operational relation, the limit relation (58) may lead to interesting properties. The following example due to Tricomi† may be cited. Starting from

$$e^{-p} \doteqdot U(t-1) \quad (\mathrm{Re}\, p > 0),$$

and putting in (58) $f(p) = e^{-p}$, we find

$$U(t-1) = \lim_{N\to\infty} \frac{\sum\limits_{n=0}^{N} \dfrac{(N/t)^n}{n!}}{e^{N/t}} \quad (t>0),$$

or, putting $t = x^{-1}$,

$$U(1-x) = \lim_{N\to\infty} \frac{\sum\limits_{n=0}^{N} \dfrac{(Nx)^n}{n!}}{e^{Nx}} \quad (x>0).$$

For $x = 1$ this formula becomes

$$\frac{1}{2} = \lim_{N\to\infty} \frac{\sum\limits_{n=0}^{N} \dfrac{N^n}{n!}}{e^N},$$

which shows that the exponential series of e^x is approximately halved if, for large integral values of x, it is broken off at the term $n = x$.

† *R.C. Accad. Lincei*, xxv, 416, 1936.

Whereas Widder's formula applies to any point where $h(t)$ has limited total fluctuation, different expressions may be found of less generality, though also relating the one-sided original to the trend of the image function along the real axis. Examples are provided by

(1) If $h(t)$ can be developed as a Taylor series in the neighbourhood of $t = 0$, it follows that

$$h(t) = \sum_{n=0}^{\infty} \frac{t^n}{(n!)^2} \left\{ \frac{d^n}{dx^n} f\left(\frac{1}{x}\right) \right\}_{x=0}. \tag{60}$$

To verify this statement, the definition integral is written as

$$f\left(\frac{1}{x}\right) = \int_0^\infty \mathrm{e}^{-u} h(xu)\, du.$$

Performing the differentiation under the sign of integration, we obtain further

$$\left\{ \frac{d^n}{dx^n} f\left(\frac{1}{x}\right) \right\}_{x=0} = h^{(n)}(0) \int_0^\infty \mathrm{e}^{-u} u^n\, du = n!\, h^{(n)}(0)$$

leading to (60), this being the Maclaurin series of $h(t)$.

(2) If there is a strip of absolute convergence, the integral of the original for $t > 0$ obeys

$$\int_0^t h(s)\, ds = \lim_{\lambda \to \infty} \sum_{n=1}^{\infty} \frac{(-1)^{n+1}}{n!} \frac{f(\lambda n)}{\lambda n} \mathrm{e}^{\lambda n t}.$$

It should be noticed that in this formula† two limits occur; one corresponding to $\lambda \to \infty$, the other with respect to the order of the partial sums.

(3) If $h(t)$ can be expanded in a Taylor series, it follows that

$$h(t) = \lim_{p \to \infty} [[\mathrm{e}^{pt} f(p)]], \tag{61}$$

in which $[[\phi(x)]]$ has the same meaning as before, namely, the part of the Laurent expansion

$$\phi(x) = \sum_{n=-\infty}^{\infty} c_n x^n$$

that contains only the negative powers of x. Formula (61) is obtained by applying Abel's theorem (4) to (IV, 4a).

In this connexion we mention further the following equation:

$$\int_{t-0}^{t+0} h(s)\, ds = \lim_{N \to \infty} \left(-\frac{e}{t}\right)^N \left\{ \frac{d^N}{dp^N} \frac{f(p)}{p} \right\}_{p=N/t} \quad (t > 0), \tag{62}$$

by means of which it becomes possible to determine those values of t for which the original is impulsive, for elsewhere the limit is automatically zero. Again, (62) can be deduced from a suitable sifting integral; inserting

† For a proof, see Doetsch, loc. cit. p. 133.

the definition integral and integrating by parts we obtain for the right-hand side of (62)

$$\lim_{N\to\infty} N\int_0^\infty ds\left(1-\frac{1}{s}\right)(s\,e^{1-s})^N\int_0^{ts} du\,h(u).$$

Finally, the result required is found after using the limit

$$\operatorname{sgn} t\,\delta(t) = \lim_{N\to\infty}\frac{Nt}{2}e^{-Nt}(t+1)^{N-1}U(t+1),$$

which in turn follows from (v, 38); the left-hand side is equal to the even δ-function multiplied by $\operatorname{sgn} t$. The limit (62), sieving out the impulse functions occurring in the original, is particularly useful when applied to the operational relation (vi, 9) involving a Dirichlet series:

$$h(t) = \sum_{n=1}^\infty a_n\,\delta(t-\lambda_n) \risingdotseq p\sum_{n=1}^\infty a_n\,e^{-p\lambda_n}.$$

The coefficients of the Dirichlet series

$$\phi(p) = \sum_{n=1}^\infty a_n\,e^{-p\lambda_n}$$

are then given by $\quad a_n = \lim\limits_{N\to\infty}\left(-\frac{e}{\lambda_n}\right)^N\phi^{(N)}\left(\frac{N}{\lambda_n}\right).$

In concluding this chapter we remark that, just as in the case of Widder's inversion integral, it is possible to replace the *definition* integral by expressions that are completely free of exponential functions, which otherwise are so characteristic in the operational calculus. Whereas in Widder's formula the limit as $N\to\infty$ of the Nth-order differential quotient is important, the analogous expression for the image function contains a limit of a repeated integral. Thus, for one-sided originals,

$$f(p) = \lim_{N\to\infty}(N+1)!\left(\frac{p}{N}\right)^{N+1}\int_0^{N/p}d\tau_0\int_0^{\tau_0}d\tau_1\int_0^{\tau_1}d\tau_2\ldots\int_0^{\tau_{N-1}}d\tau_N\,h(\tau_N), \qquad (63)$$

which is easily proved after showing it to be equal to

$$\lim_{N\to\infty} p\int_0^{N/p}h(t)\left(1-\frac{pt}{N}\right)^N dt.$$

Moreover, similarly to the corresponding (59), we have

$$f(p) = \lim_{N\to\infty}\frac{\int_0^{N/p}d\tau_0\int_0^{\tau_0}d\tau_1\int_0^{\tau_1}d\tau_2\ldots\int_0^{\tau_{N-1}}d\tau_N\,h(\tau_N)}{\int_0^{N/p}d\tau_0\int_0^{\tau_0}d\tau_1\int_0^{\tau_1}d\tau_2\ldots\int_0^{\tau_{N-1}}d\tau_N},$$

indicating clearly how the image function is related to the higher-order mean values of the corresponding original.

CHAPTER VIII

LINEAR DIFFERENTIAL EQUATIONS WITH CONSTANT COEFFICIENTS

1. Introduction

Whereas the preceding chapters were devoted to discussing fundamental principles and establishing general rules, the rest of the book will be aimed substantially at problems that are most rapidly solved by means of the operational calculus. A simple example is provided by the linear differential equations with constant coefficients. These equations are of utmost importance in the description of electrical and mechanical systems whose basic elements, such as inductances, resistances, moments of inertia, do not depend on time nor on currents, voltages, displacements, and the like. First of all it is shown that the solution of the given differential equation is reducible to an algebraic problem. For systems initially at rest the treatment is as simple as possible (§§ 2, 3). Also in the case of general initial conditions (at $t = 0$, say) the operational calculus readily leads to a solution if the required function is replaced by zero for $t < 0$ (§ 4). As in further discussions the concepts of admittance and impedance play an important part, we shall treat them in some detail in § 5. Next, in §§ 6 and 7, they are shown to be particularly useful in the operational treatment of transient phenomena occurring in electrical networks when switched on or off. Heaviside's first expansion theorem is discussed in § 8; it is of primary importance in the description of the response immediately after the electromotive force has been switched on. Finally, a classification of different types of admittance is given in § 9.

2. Inhomogeneous equations with the unit function at the right

The most simple linear differential equation with constant coefficients, having the unit function on the right-hand side, is the following:

$$\left(a_0\frac{d^n}{dt^n}+a_1\frac{d^{n-1}}{dt^{n-1}}+\dots+a_{n-1}\frac{d}{dt}+a_n\right)h(t) = U(t),$$

or, by way of abbreviation,

$$P\left(\frac{d}{dt}\right)h(t) = U(t). \tag{1}$$

In physical problems leading to differential equations it is customary to place on the right the terms that are independent of the function required, which usually represent the exterior forces acting upon the system under consideration. In the case of equation (1) such influences from the outside start to work at $t = 0$, and are completely absent for $t < 0$. For instance, equation (1) is obtained for the charge q of a condenser C if at the time $t = 0$ an electromotive force of unit magnitude comes into action. More precisely, if the condenser is connected in series with an inductance L and a resistance R (see fig. 46), the charge on the condenser satisfies the differential equation

$$L\frac{d^2q}{dt^2}+R\frac{dq}{dt}+\frac{1}{C}q = U(t), \tag{2}$$

which is a special case of (1). To solve the general equation (1) operationally, it is natural to identify $h(t)$ with the original of the calculus†. Then the equation is transposed term by term, which comes down to multiplying it by $p\,e^{-pt}$ and integrating from $-\infty$ to $+\infty$; in other words, the complete equation is subjected to the operator

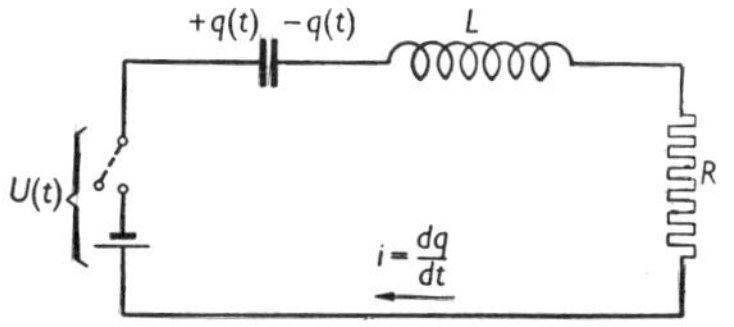

Fig. 46. An example of an electric circuit.

$$p\int_{-\infty}^{\infty} dt\, e^{-pt}.$$

It is to be remarked that in the general method of Laplace the solution of the differential equation is assumed to be of the form

$$h(t) = \int e^{pt}\phi(p)\,dp,$$

where the function ϕ and the path of integration have to be suitably chosen afterwards. If some line parallel to the imaginary axis happens to be an adequate path of integration the substitution above reduces to introducing the image of $h(t)$ by means of the inversion integral. The direct term-by-term transposition of (1) is easily performed if the differentiation rule is applicable; at any rate this is allowed inside the strip of convergence corresponding to the nth derivative of $h(t)$. Let us assume that such a strip exists; moreover, restricting ourselves to the region $\operatorname{Re} p > 0$, the image of the right-hand side of (1) is equal to 1. Therefore from (1)

$$(a_0 p^n + a_1 p^{n-1} + \ldots + a_{n-1}p + a_n)\,f(p) = P(p)\,f(p) = 1.$$

The image of the required solution is thus determinable from a mere algebraic equation; upon dividing we obtain

$$f(p) = \frac{1}{P(p)} = \frac{1}{a_0 p^n + a_1 p^{n-1} + \ldots + a_{n-1}p + a_n}, \tag{3}$$

whose original, if existing for $\operatorname{Re} p > 0$, is a particular solution of the differential equation (1). In particular, if the polynomial $P(p)$ has no

† For instance, see Balth. van der Pol, *Phil. Mag.* VII, 1153, 1929.

multiple zeros the original is easily found. By expansion in rational fractions we first have

$$\frac{1}{P(p)}=\frac{p}{pP(p)}=\frac{1}{P(0)}+\sum_{k=1}^{n}\frac{1}{p_k P'(p_k)}\frac{p}{p-p_k}. \tag{4}$$

If, again, p_1 is the zero farthest to the right, the original of any individual term is determined by

$$\frac{p}{p-p_k} \risingdotseq e^{p_k t}\,U(t)$$

for $\operatorname{Re} p > \operatorname{Re} p_1$. Therefore, as the original of (3), and thus as a particular solution of (1), we find

$$h(t)=U(t)\left\{\frac{1}{P(0)}+\sum_{k=1}^{n}\frac{e^{p_k t}}{p_k P'(p_k)}\right\}. \tag{5}$$

The method used above is obviously based upon Heaviside's second expansion theorem, the theory of which was given in § 10 of the preceding chapter. It may not be superfluous to emphasize that the particular solution (5) fulfils equation (1) in the *whole* t-range (including $t = 0$). It is that solution of (1) which is subjected to the supplementary condition that, before the exterior force has come into action, that is, for $t < 0$, the system is completely at rest; this implies that at the point of transition $t = 0$ the function and its derivatives of orders up to $n-1$ are continuous, their values being zero. On the other hand, it is equally possible to construct n other particular solutions of (1), namely, by transposing (1) with respect to the strips $\operatorname{Re} p_2 < \operatorname{Re} p < \operatorname{Re} p_1$, $\operatorname{Re} p_3 < \operatorname{Re} p < \operatorname{Re} p_2$, etc. For practical applications, however, these solutions are of little importance, since they correspond to less simple boundary conditions. Further, it is evident how to deal with equations for which $P(p)$ has multiple zeros; in this case originals of the type (VII, 43) with $k > 1$ are obtained. Although of minor importance in practice, it may be remarked that even the eigenfunctions $e^{p_k t}$ of the homogeneous equation can be found similarly from relations having a line of convergence. For transposition of the corresponding homogeneous equation leads to $P(p)f(p) = 0$, which is satisfied, amongst others, by the image of the operational relations

$$f_k(p)=2\pi p\,\delta\left(\frac{p-p_k}{i}\right) \risingdotseq e^{p_k t} \quad (\operatorname{Re} p = \operatorname{Re} p_k),$$

as, according to (V, 27), we have

$$P(p)\,f_k(p)=P(p_k)\,f_k(p)=0\times f_k(p)=0.$$

In practical applications it will usually be preferred to join any pair of rational fractions corresponding to conjugate zeros, at least if the coefficients of (1) are real throughout, in order to obtain (5) immediately in a real form.

Example. The transposition of the preceding equation (2) for the charge on the condenser, excited by the unit-function voltage, leads to

$$f(p) = \frac{1}{Lp^2+Rp+\dfrac{1}{C}}.$$

Introducing the following abbreviations

$$\alpha = \frac{R}{2L}, \quad \omega_1 = \sqrt{\left(\alpha^2-\frac{1}{CL}\right)},$$

we obtain for the expansion in rational fractions

$$f(p) = C+\frac{1}{2L\omega_1}\left\{\frac{1}{\alpha+\omega_1}\,\frac{p}{p+\alpha+\omega_1}-\frac{1}{\alpha-\omega_1}\,\frac{p}{p+\alpha-\omega_1}\right\},$$

and thus for the original

$$q(t) = U(t)\left\{C+\frac{e^{-\alpha t}}{2\omega_1 L}\left(\frac{e^{-\omega_1 t}}{\alpha+\omega_1}-\frac{e^{\omega_1 t}}{\alpha-\omega_1}\right)\right\}.$$

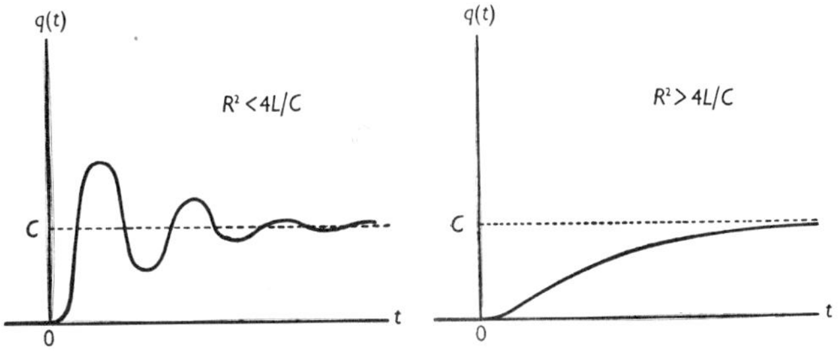

Fig. 47. The charge of the condenser of fig. 46, the e.m.f. being $U(t)$.

This expression for the charge is in a real form if and only if $R^2 \geqslant 4L/C$. An alternative solution, which is in a real form if $R^2 \leqslant 4L/C$, can be given after first having rewritten

$$f(p) = \frac{1}{Lp}\,\frac{p}{\{(p+\alpha)^2+\omega_0^2\}} \qquad \left[\omega_0 = \sqrt{\left(\frac{1}{CL}-\alpha^2\right)}\right],$$

whence, by application of some elementary rules to (III, 22), it easily follows that

$$q(t) = \frac{U(t)}{L\omega_0}\int_0^t e^{-\alpha\tau}\sin(\omega_0\tau)\,d\tau = \frac{U(t)}{L\omega_0(\omega_0^2+\alpha^2)}[\omega_0-e^{-\alpha t}\{\alpha\sin(\omega_0 t)+\omega_0\cos(\omega_0 t)\}].$$

The different character of $q(t)$ according as $R^2<4L/C$ or $R^2>4L/C$, is illustrated by fig. 47.

3. Inhomogeneous equations with arbitrary right-hand member

The special equation (1), which had the unit function $U(t)$ in its right-hand member, was treated in some detail because it is simple and therefore very useful to explain the general operational procedure of solving linear differential equations with constant coefficients. We now proceed to equations having an arbitrary force function at the right:

$$P\left(\frac{d}{dt}\right)h(t) = \left(a_0\frac{d^n}{dt^n}+a_1\frac{d^{n-1}}{dt^{n-1}}+\ldots+a_n\right)h(t) = \varphi(t). \tag{6}$$

Let us assume, once more, that the nth-order derivative of $h(t)$ has a non-vanishing strip of convergence. Then from (6) it follows that $P(p)f(p) = \phi(p)$, if $\phi(p)$ is the image corresponding to $\varphi(t)$ in the said strip. Again, the image of the solution in question is found by a simple division:

$$f(p) = \frac{\phi(p)}{P(p)}.$$

Furthermore, if $1/P(p)$ is expanded in rational fractions, that is to say

$$\frac{1}{P(p)} = \sum_{k=1}^{n} \frac{1}{P'(p_k)} \frac{1}{p-p_k},$$

the image becomes

$$f(p) = \sum_{k=1}^{n} \frac{1}{P'(p_k)} \frac{1}{p} \frac{p}{p-p_k} \phi(p). \tag{7}$$

Finally, after applying the composition-product rule, the original is easily written down for $\operatorname{Re} p > \operatorname{Re} p_1$, in so far as the relation $\varphi(t) \risingdotseq \phi(p)$ is valid there. The result is

$$h(t) = \sum_{k=1}^{n} \frac{e^{p_k t}}{P'(p_k)} \int_{-\infty}^{t} e^{-p_k \tau} \varphi(\tau)\, d\tau. \tag{8}$$

This again is a particular solution of the differential equation. Just as in the preceding section other particular solutions can be obtained with respect to the strips $\operatorname{Re} p_{i+1} < \operatorname{Re} p < \operatorname{Re} p_i$ lying between the remaining zeros of $P(p)$; they are unimportant in practice. Concerning multiple zeros of $P(p)$, the situation is similar to that of the foregoing section; any of the rational fractions can be transposed with the help of the composition-product rule, irrespective of the order of the zeros.

Example. The charge $q(t)$ on the condenser of the example of the foregoing section, under the influence of an electromotive force e^{-t^2} (which is thus working for $t < 0$ too), satisfies the differential equation

$$L\frac{d^2q}{dt^2} + R\frac{dq}{dt} + \frac{1}{C}q = e^{-t^2}.$$

The transposition of this equation into the p-language is

$$\left(Lp^2 + Rp + \frac{1}{C}\right) f(p) = \sqrt{\pi}\, p\, e^{\frac{1}{4}p^2},$$

whence it follows that

$$f(p) = \frac{\sqrt{\pi}}{L} \frac{1}{p} \times p\, e^{\frac{1}{4}p^2} \times \frac{p}{(p+\alpha)^2 + \omega_0^2},$$

in which ω_0 and α have the same meaning as before. Upon applying the composition-product rule we obtain

$$q(t) = \frac{1}{L\omega_0} \int_{0}^{\infty} e^{-(t-\tau)^2 - \alpha\tau} \sin(\omega_0 \tau)\, d\tau.$$

This expression is in real form if $R^2 \leqslant 4L/C$; it can be reduced to error functions of complex arguments. Moreover, it is easily verified that the particular solution $q(t)$ above corresponds to the condenser being uncharged, and the circuit being current-free, at $t = -\infty$. If $R^2 > 4L/C$ we had better expand $f(p)$ into rational fractions.

4. Differential equations with boundary conditions

In the preceding two sections a particular solution of the differential equation was found, which was examined afterwards to see to what special supplementary conditions it corresponded. It is also possible to treat operationally the more general problem of constructing the solution $h(t)$ that satisfies *given* boundary conditions at any given point (for simplicity we shall choose the point $t = 0$). This is achieved by means of a differential equation for the one-sided function $h^*(t) = h(t)\,U(t)$; in the new differential equation the prescribed values of the derivatives of $h(t)$ at $t = 0$ occur as parameters in the right-hand member. First the derivatives of the function $h^*(t)$ are calculated:

$$\left.\begin{aligned} \frac{d}{dt}h^*(t) &= h'(t)\,U(t)+h(t)\,\delta(t) = h'(t)\,U(t)+h(0)\,\delta(t), \\ \frac{d^2}{dt^2}h^*(t) &= h''(t)\,U(t)+h'(t)\,\delta(t)+h(0)\,\delta'(t) \\ &= h''(t)\,U(t)+h'(0)\,\delta(t)+h(0)\,\delta'(t), \\ \text{etc.,} & \end{aligned}\right\} \tag{9}$$

whence the differential equation for $h^*(t)$ follows readily:

$$P\left(\frac{d}{dt}\right)h^*(t) = U(t)\,P\left(\frac{d}{dt}\right)h(t)+\{a_{n-1}h(0)+a_{n-2}h'(0)+\ldots\}\,\delta(t)$$
$$+\{a_{n-2}h(0)+a_{n-3}h'(0)+\ldots\}\,\delta'(t)+\ldots.$$

Consequently it is expected that, to obtain the solution $h(t)$ of

$$P\left(\frac{d}{dt}\right)h(t) = \varphi(t) \tag{10}$$

for $t > 0$ whose derivatives assume given values $h^{(k)}(0)$ at $t = 0$, we may take the part $t > 0$ of the particular solution of

$$P\left(\frac{d}{dt}\right)h^*(t) = \varphi(t)\,U(t)+\delta(t)\sum_{k=0}^{n-1}a_{n-k-1}h^{(k)}(0)$$
$$+\delta'(t)\sum_{k=0}^{n-2}a_{n-k-2}h^{(k)}(0)+\ldots+\delta^{(n-1)}(t)\,a_0h(0) \tag{11}$$

that vanishes, together with its derivatives, for $t < 0$. The new equation is always inhomogeneous, even if the equation for $h(t)$ is homogeneous ($\varphi = 0$). The required particular solution of (11) is easily obtained by applying the theory of the preceding section. After transposition of (11) we find for the image of $h^*(t)$

$$f^*(p) = f_1^*(p)+f_2^*(p), \tag{12}$$

in which
$$f_1^*(p) = \frac{\phi^*(p)}{P(p)}$$

and
$$f_2^*(p) = \frac{p\sum\limits_{k=0}^{n-1} a_{n-k-1}h^{(k)}(0)+p^2\sum\limits_{k=0}^{n-2} a_{n-k-2}h^{(k)}(0)+\ldots+p^n a_0 h(0)}{a_n+pa_{n-1}+p^2a_{n-2}+\ldots+p^n a_0}, \quad (13)$$

if $\phi^*(p)$ denotes the image of $\varphi(t)\,U(t)$. The corresponding original $h^*(t)$ is obviously wholly independent of the values of $\varphi(t)$ at $t<0$; this is not astonishing, since, once the initial conditions at $t = 0$ are given, the earlier values of $\varphi(t)$ have ceased to interfere.

We shall now prove that the solution of (10) for $t>0$, subject to the given boundary conditions at $t = 0$, is the right-hand part $(t>0)$ of the one-sided original of (12) inside the strip $\operatorname{Re} p > \operatorname{Re} p_0$, where p_0 is the zero farthest to the right of $P(p)$. To do this, consider the expansion of $f^*(p)$ in powers of $1/p$; provided $\varphi(0)$ is finite (so that, according to Abel's theorem, $\phi^*(\infty)$ is finite too), the part $f_1^*(p)$ yields $(1/p)^n$ as the lowest power of $1/p$. The lower powers of $1/p$ occurring in $f^*(p)$ thus originate from the other part, $f_2^*(p)$. It is then easily verified from (13) that

$$f^*(p) = h(0)+\frac{h'(0)}{p}+\frac{h''(0)}{p^2}+\ldots+\frac{h^{(n-1)}(0)}{p^{n-1}}+\frac{c_n}{p^n}+\ldots.$$

Finally, term-by-term transposition for $\operatorname{Re} p>0$ leads to a power series for $h^*(t)$ whose first n terms just yield the prescribed values of the derivatives at $t = +0$.

The separate parts of $f^*(p)$ and their corresponding contributions to $h^*(t)$ admit of a simple interpretation. For the first term $f_1^*(p) \doteqdot h_1^*(t)$ is independent of the given boundary values at $t = 0$ and therefore represents the system for $t>0$ if it is supposed to be completely at rest for $t<0$. In other words, $h_1^*(t)$ is characteristic for the switch-on phenomena, that is, the response due to suddenly applying the impressed force $\varphi(t)$ to the system at rest at $t = 0$. On the other hand, the second term, $f_2^*(p) \doteqdot h_2^*(t)$, is independent of $\varphi(t)$, and is determined solely by the initial conditions at $t = 0$. The function $h_2^*(t)$ would be the exact solution if, for $t>0$, $\varphi(t)$ was zero though the values of $\varphi(t)$ for $t<0$ may affect the initial values of $h(t)$, $h'(t)$, etc. at $t = 0$. Therefore, $h_2^*(t)$ describes the system after the exterior force has become inactive; in other words, it represents the typical switch-off phenomena, corresponding to certain specified initial conditions. Both kinds of phenomena can be discussed on fig. 48, where some electrical system is excited by an electromotive force $e(t)$. If the switch I is on and II is off, the differential equation governing the current through the system is of the form $P(d/dt)\,i(t) = e(t)$. The switch-on current $i_1(t)$ is obtained when the electrical system is assumed to be completely at

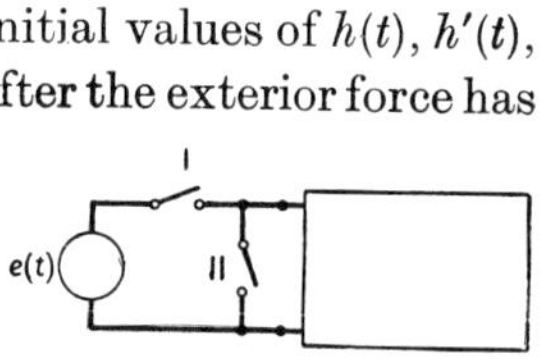

Fig. 48. Illustrating switch-on and switch-off phenomena.

rest before $t = 0$, while at $t = 0$ the switch I is suddenly closed, the switch II being kept off, so that the impressed voltage is thus $e(t)\,U(t)$. On the other hand, the switch-off current $i_2(t)$ is generated if II is switched on and I is switched off at $t = 0$, so as to make the exterior source inactive for $t > 0$. We may just call $i_2(t)$ the short-circuit current.

We return to these transient phenomena in subsequent sections; we may remark, however, that $h_2^*(t)$ can be found by expanding $f_2^*(p)$ into rational fractions, whilst the rule of the composition product is used in the determination of $h_1^*(t)$. Concerning the short-circuit phenomenon $h_2^*(t)$ it is observed that according to (13) the image $f_2^*(p)$ is a rational function of p with the special properties that its numerator and denominator in general are of equal degree, and the numerator contains a factor p. This makes Heaviside's second expansion theorem easily applicable, since $f_2^*(p)/p$ will not contain terms other than the rational fractions. We may therefore put

$$f_2^*(p) = \sum_{k=1}^{n} c_k \frac{p}{p-p_k},$$

whence it follows that $$h_2^*(t) = U(t) \sum_{k=1}^{n} c_k \mathrm{e}^{p_k t}.$$

Consequently, the short-circuit current is a linear combination of the eigenfunctions $\mathrm{e}^{p_k t}$, that is, of the non-vanishing solutions of the homogeneous equation $P(d/dt)\,h(t) = 0$, valid for all values of t (see the end of § 2). The weight factors c_k depend upon the given initial conditions.

Example 1. Let us once more return to the circuit of fig. 46; we now suppose the electromotive force $\mathrm{e}^{-\beta t}$ to be active for $t > 0$, whilst at $t = 0$ the charge $q(t)$ and the current $i(t) = dq/dt$ are prescribed. We have thus to construct that solution of

$$L\frac{d^2q}{dt^2} + R\frac{dq}{dt} + \frac{1}{C}q = \mathrm{e}^{-\beta t},$$

which obeys the boundary conditions $q(t) = q(0)$ and $dq/dt = i(0)$, at $t = 0$. According to the theory given above we may equally well study the differential equation for $q^*(t) = q(t)\,U(t)$, which reads

$$L\frac{d^2q^*}{dt^2} + R\frac{dq^*}{dt} + \frac{1}{C}q^* = \mathrm{e}^{-\beta t}U(t) + \{Li(0) + Rq(0)\}\,\delta(t) + Lq(0)\,\delta'(t),$$

and transpose it operationally. If this is done we arrive at the following relation

$$q^*(t) \doteqdot \frac{\dfrac{p}{p+\beta} + \{Li(0) + Rq(0)\}\,p + Lq(0)\,p^2}{Lp^2 + Rp + \dfrac{1}{C}}.$$

The charge $q(t)$ for $t > 0$ is equal to the sum of two different components:

(1) The switch-on part:

$$\frac{p}{(p+\beta)\left(Lp^2 + Rp + \dfrac{1}{C}\right)} \doteqdot U(t)\,\frac{\mathrm{e}^{-\beta t} - \mathrm{e}^{-\alpha t}\left\{\cos(\omega_0 t) + \dfrac{\alpha-\beta}{\omega_0}\sin(\omega_0 t)\right\}}{L\{(\alpha-\beta)^2 + \omega_0^2\}}.$$

The right-hand function would describe the charge on the condenser after switching on the electromotive force $e^{-\beta t}$ suddenly at $t = 0$, provided both the charge on and the current to the condenser vanish at $t = 0$.

(2) The switch-off part:

$$\frac{\{Li(0)+Rq(0)\}p+Lq(0)p^2}{Lp^2+Rp+\dfrac{1}{C}} \doteqdot U(t)\,e^{-\alpha t}\left(q_0\cos(\omega_0 t)+\{i(0)+\alpha q(0)\}\frac{\sin(\omega_0 t)}{\omega_0}\right).$$

In this case the t-function represents, for $t > 0$, the charge on the condenser if at $t = 0$ it is suddenly short-circuited.

Example 2. Another example, not dealing with network theory, of a differential equation reducible to an inhomogeneous one is provided by Humbert's trigonometric functions of the third order†. These functions are solutions of the differential equation

$$\frac{d^3\lambda}{dt^3}+\lambda = 0,$$

that can be determined as follows. Let functions λ_1, λ_2 and λ_3 be defined by first expanding $e^{-\rho^2 t}$ into powers of ρ and then letting $\rho^3 = 1$. Thus

$$e^{-\rho^2 t} = 1-\rho^2 t+\rho\frac{t^2}{2!}-\frac{t^3}{3!}+\rho^2\frac{t^4}{4!}-\rho\frac{t^5}{5!}+\ldots = \lambda_1(t)+\rho\lambda_2(t)+\rho^2\lambda_3(t). \tag{14}$$

Consequently

$$\left.\begin{aligned}\lambda_1(t) &= 1-\frac{t^3}{3!}+\frac{t^6}{6!}-\ldots = \tfrac{1}{3}e^{-t}+\tfrac{2}{3}e^{\frac{1}{2}t}\cos(\tfrac{1}{2}\sqrt{3}\,t),\\ \lambda_2(t) &= \frac{t^2}{2!}-\frac{t^5}{5!}+\frac{t^8}{8!}-\ldots = \tfrac{1}{3}e^{-t}-\tfrac{2}{3}e^{\frac{1}{2}t}\cos(\tfrac{1}{2}\sqrt{3}\,t+\tfrac{1}{3}\pi),\\ \lambda_3(t) &= -t+\frac{t^4}{4!}-\frac{t^7}{7!}+\ldots = \tfrac{1}{3}e^{-t}-\tfrac{2}{3}e^{\frac{1}{2}t}\cos(\tfrac{1}{2}\sqrt{3}\,t-\tfrac{1}{3}\pi).\end{aligned}\right\} \tag{15}$$

Apparently these functions are extensions of the conventional trigonometric functions of the second order, which may be defined similarly by means of

$$e^{it} = \cos t+i\sin t,$$

if use is made of $i^2 = -1$.

The one-sided trigonometric functions of the third order satisfy the inhomogeneous equation

$$\left(\frac{d^3}{dt^3}+1\right)h^*(t) = \lambda''(0)\,\delta(t)+\lambda'(0)\,\delta'(t)+\lambda(0)\,\delta''(t),$$

from which it is easily seen (ignoring null functions) that

$$\left(\frac{d^3}{dt^3}+1\right)h_1^*(t) = \delta''(t),\quad \left(\frac{d^3}{dt^3}+1\right)h_2^*(t) = \delta(t),\quad \left(\frac{d^3}{dt^3}+1\right)h_3^*(t) = -\delta'(t).$$

From these equations it is not difficult to deduce the following operational relations:

$$\lambda_1(t)\,U(t) \doteqdot \frac{p^3}{p^3+1},\quad 0<\mathrm{Re}\,p<\infty,$$

$$\lambda_2(t)\,U(t) \doteqdot \frac{p}{p^3+1},\quad 0<\mathrm{Re}\,p<\infty,$$

$$\lambda_3(t)\,U(t) \doteqdot -\frac{p^2}{p^3+1},\quad 0<\mathrm{Re}\,p<\infty,$$

by means of which many properties of the λ-functions may be derived.

† P. Humbert, *Le calcul symbolique*, Paris, 1934, p. 29.

5. Admittances and impedances

With a view to application of the operational calculus to electrical networks it may be useful to discuss the concepts of admittance and impedance in some detail. The former was introduced by Heaviside† who defined the admittance as the mathematical operator $A(d/dt)$ that has to be applied to the electromotive force $e(t)$ in order to get the current $i(t)$, entering (leaving) the network at $P(Q)$ (see fig. 49), in response to the said electromotive force. Furthermore, the impedance $Z(d/dt)$ is defined as being the inverse operator of the admittance $A(d/dt)$; thus

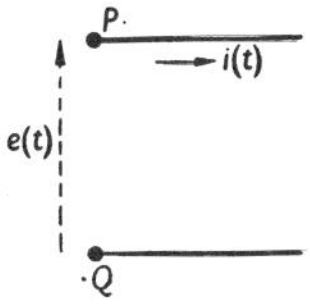

Fig. 49. The terminals of a network.

$$i(t) = A\left(\frac{d}{dt}\right) e(t), \quad e(t) = Z\left(\frac{d}{dt}\right) i(t).$$

While thus in Heaviside's definition the given voltage $e(t)$ may be wholly arbitrary, the present-day definitions of admittance and impedance are concerned only with cisoidal time dependence, $e^{i\omega t}$, where thus the applied voltage is alternating and of frequency $\nu = \omega/2\pi$. In the quasi-stationary theory of electrical networks, that is to say, when the networks are governed by linear differential equations with constant coefficients, the response current will also vary in time according to $e^{i\omega t}$. The operators $A(d/dt)$ and $Z(d/dt)$ then become equal to the ratios i/e and e/i respectively; they are independent of the time and are simply functions of the frequency or, alternatively, of $i\omega$. To distinguish the general concepts of admittance and impedance in the sense of Heaviside from the restricted analogues the latter will be called the frequency admittance and the frequency impedance, respectively, such as to have for the special $e(t) = e^{i\omega t}$:

$$i(t) = A(i\omega)\, e^{i\omega t} = \frac{e^{i\omega t}}{Z(i\omega)}.$$

The functions $A(i\omega)$ and $Z(i\omega)$ may further be defined for complex values of ω, in such sense that, if we put $p = i\omega$, the current corresponding to the impressed voltage $e(t) = E_0\, e^{pt}$ is determined by

$$i(t) = A(p)\, E_0\, e^{pt} = \frac{E_0\, e^{pt}}{Z(p)}. \tag{16}$$

It thus follows that the frequency admittance and the frequency impedance are functions of the complex variable p, which determine the ratio between current and voltage both for damped vibrations ($\operatorname{Re} p < 0$) and for increasing vibrations ($\operatorname{Re} p > 0$), whilst for undamped vibrations $\operatorname{Re} p = 0$.

In addition to these frequency functions we shall introduce the response current $i_{\mathsf{J}}(t)$ due to the unit-function voltage $U(t)$. After Barnes and Pren-

† *Proc. Roy. Soc.* LII, 504, 1892–3.

dergast‡, this function of time (per unit of voltage) will be called the time admittance of the network; it is identical with Carson's indicial admittance. The knowledge of either the time admittance or the frequency admittance proves sufficient to determine the response of the network to any arbitrary voltage $e(t)$. As a consequence, there must exist a certain definite relationship between both kinds of admittance functions, which is most easily surveyed in the following operational manner. Suppose $E(p)$ and $I(p)$ to be the images for $\operatorname{Re} p > 0$ of the applied voltage $e(t)$ and the current $i(t)$ in question, respectively. Now, the inversion integral states that

$$e(t) = \frac{1}{2\pi i}\int_{c-i\infty}^{c+i\infty} e^{pt}\frac{E(p)}{p}\,dp \quad (c > 0).$$

Therefore, since any component e^{pt} of the voltage $e(t)$ causes a current of amount $A(p)\,e^{pt}$, and the system is linear, the total voltage $e(t)$ produces a current equal to

$$i_e(t) = \frac{1}{2\pi i}\int_{c-i\infty}^{c+i\infty} e^{pt}A(p)\,E(p)\frac{dp}{p} \quad (c > 0).$$

The right-hand side, however, is nothing but the original corresponding to the image $A(p)\,E(p)$; consequently

$$I(p) = A(p)\,E(p) \quad (\operatorname{Re} p > 0), \tag{17}$$

which is clearly the operational formulation of Ohm's law. In particular, if $e(t)$ is chosen equal to $U(t)$, we have to put $E(p) = 1$. Since in this particular case $I(p)$ is the image of the time admittance $i_{\Gamma}(t)$, we have consequently

$$i_{\Gamma}(t) \fallingdotseq A(p) \quad (\operatorname{Re} p > 0). \tag{18}$$

In words, the frequency admittance, as a function of p, and the time admittance, as a function of t, stand to each other as image and original in the operational sense.

Simple quasi-stationary systems, to which the theory above is applicable, are networks that are built up by means of ideal resistors, capacitors, self-inductances and mutual inductances, since then linear differential equations with constant coefficients obtain. The network of fig. 50 may serve as an example; its behaviour is determined by the following set of equations:

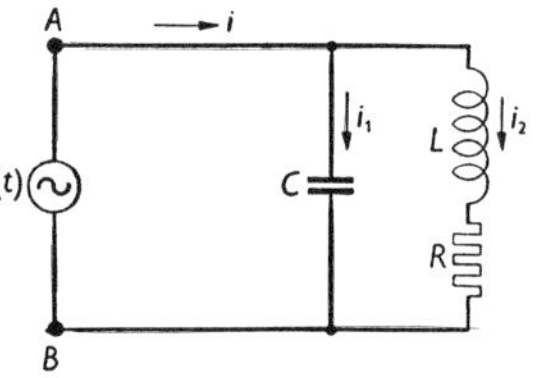

Fig. 50. Example of an electric network with two meshes.

$$e(t) = \frac{1}{C}\int_{-\infty}^{t} i_1(\tau)\,d\tau = Ri_2(t) + L\frac{di_2}{dt}, \quad i = i_1 + i_2.$$

After eliminating i_1 and i_2 we get a relation between i and e, namely,

$$e = \left(R + L\frac{d}{dt}\right)\left(i - C\frac{de}{dt}\right). \tag{19}$$

‡ *J. Math. Phys.* XI, 27, 1932.

Choosing $e = E_0 e^{pt}$, $i = I_0 e^{pt}$, we have

$$E_0 = (R+Lp)(I_0 - CpE_0).$$

Consequently for the frequency admittance of the system

$$A(p) = \frac{I_0}{E_0} = Cp + \frac{1}{R+Lp}. \tag{20}$$

The time admittance is found as the original of (20); therefore

$$i_{\text{ʃ}}(t) = C\delta(t) + \frac{U(t)}{R}(1 - e^{-Rt/L}),$$

which function at the same time represents the current due to an electromotive force of amount $U(t)$.

The treatment of more complicated networks will be postponed till the next chapter, where we will show how to determine the admittance between the terminals P and Q from the complete set of current and voltage equations. There is often a far more simple method, however. If mutual inductances are absent, the admittance between P and Q can easily be found by first giving separate impedances R, $1/Cp$, Lp, to the individual resistors, capacitors and self-inductances, respectively, and then applying Ohm's law, according to which in any intermediate stage either the impedances or the admittances should be added, depending on whether the individual elements are connected in series or in parallel. For instance, in fig. 50 the admittance (20) is the sum of the admittances Cp and $1/(R+Lp)$ occurring in the branches i_1 and i_2 respectively. The second part in turn is the inverse of the sum of impedances R and Lp, whose elements are connected in series in the i_2 branch. In this way it generally turns out that both the impedance and the admittance of any network containing only a finite number of resistors, capacitors and self-inductances, can be written as the quotient of two polynomials in p with positive coefficients, whilst the absolute difference between the degrees of numerator and denominator never exceeds 1. So, in the example cited above, from (20),

$$A(p) = \frac{LCp^2 + RCp + 1}{Lp + R}. \tag{21}$$

It may already be remarked here that later on (IX, § 3) the concept of admittance is somewhat generalized; by so doing it is possible that the difference in degree of numerator and denominator becomes other than -1, 0 or 1. Moreover, for non-quasi-stationary systems, such as cables (XV, § 4), the admittance is often a more complicated transcendent. In other parts of physics and engineering the concepts of admittance and impedance also serve a useful purpose. For instance, in acoustics, where the frequency impedance at a point is defined by the ratio of a sound pressure $P = P_0 e^{pt}$ to the corresponding velocity of a membrane, a column of air, etc. There is a time admittance as well, which describes the response to a suddenly impinging force, represented by the unit function $U(t)$ on the right-hand side of the underlying differential equation. A closer examination of (18) shows that it is only valid if the function $A(p)$ has no singularities in the half-plane $\mathrm{Re}\, p > 0$, since otherwise the strip of convergence would not extend to the imaginary axis. In particular, poles p_r subject to $\mathrm{Re}\, p_r > 0$ are forbidden; this actually applies to all passive systems, that is to say, systems without sources of energy. For, if it were possible to have a pole p_1 on the right of the imaginary axis, then from $A(p_1) = \infty$ according to (16) a current $a e^{p_1 t}$ of non-vanishing amplitude might exist without any e.m.f. being present; further, this current would gradually increase ($\mathrm{Re}\, p_1 > 0$), which is inconsistent with there being no sources of energy. The poles corresponding to a passive network thus all lie to the left of the imaginary axis, or at best on the axis itself (the point at infinity included). The latter circumstance, however, never occurs if the network is not only passive but also dissipative, that is to say, if the existing currents transform a part of the electric energy into heat. To see this, suppose p_1 is a pole of $A(p)$ on the imaginary axis. The above current $a e^{p_1 t}$ would then represent an undamped oscillation; but this is contradictory to the fact that in dissipative systems the amplitude of the current should decrease. Dissipative systems have only damped eigenfunctions $e^{p_k t}$ ($\mathrm{Re}\, p_k < 0$), though in the case of passive but non-dissipative systems undamped eigenfunctions may occur. Obviously these definitions are independent of whether the system under consideration is electrical, mechanical or acoustical.

We now discuss some general properties of frequency and time admittances that are valid irrespective of the particular structure of a system. Of special importance thereby is the admittance A for imaginary values of p $(= i\omega)$ whose behaviour is decisive for undamped oscillations. The corresponding function $A(i\omega)$ depends on *two* real functions of ω, e.g. the modulus and phase of $A(i\omega)$, whereas the time admittance $i_{\Gamma}(t)$ is determined by a *single* real function of t. The asymmetry occurring here is only apparent, since $i_{\Gamma}(t)$ may also be compared with the single function $A(p)$ for real p; the latter function represents the response of the system to aperiodic oscillations e^{pt}. In practice, however, the behaviour of undamped oscillations is far more important, so that usually the two real functions determining $A(i\omega)$ are considered side by side with the real function $i_{\Gamma}(t)$ (see fig. 51 illustrating these functions for the network of fig. 50 if $R^2C < L$).

In the following P and Q will denote the real and imaginary part of $A(iw)$, respectively, so as to have for real values of ω

$$A(i\omega) = P(\omega) + iQ(\omega). \tag{22}$$

As general properties† we may mention:

(1) From the requirement that (in the case of an electrical network) any real e.m.f. causes a current that is real too, it follows that $P(\omega)$ is an even function of ω and $Q(\omega)$ an odd function of ω. For, the real e.m.f.

$$\cos(\omega t) = \tfrac{1}{2}e^{i\omega t} + \tfrac{1}{2}e^{-i\omega t}$$

leads to a current $\qquad \tfrac{1}{2}A(i\omega)\,e^{i\omega t} + \tfrac{1}{2}A(-i\omega)\,e^{-i\omega t}$,

provided that the network is assumed to be a linear system. On account of (22) this yields as imaginary part of the current:

$$\frac{i}{2}\{P(\omega) - P(-\omega)\}\sin(\omega t) + \frac{i}{2}\{Q(\omega) + Q(-\omega)\}\cos(\omega t),$$

which has to vanish for all values of t. This is possible only if the coefficients of $\sin(\omega t)$ and $\cos(\omega t)$ are both zero. Thus P is even and Q is odd in ω. For the network of fig. 50 these functions are, on account of (20),

$$P(\omega) = \frac{R}{R^2 + L^2\omega^2}, \quad Q(\omega) = \frac{(R^2C - L)\,\omega + L^2C\omega^3}{R^2 + L^2\omega^2}.$$

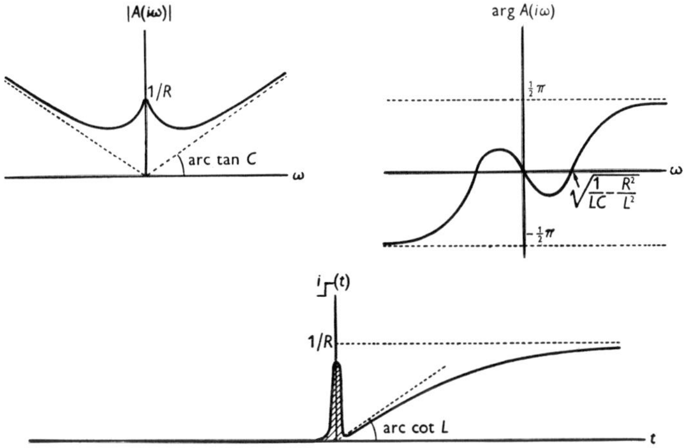

Fig. 51. Frequency and time admittance of the network of fig. 50.

(2) From the fact that the influence of the applied voltage $U(t)$ is felt only *after* the moment $t = 0$ (effect comes after cause) it follows that the admittance for $\operatorname{Re} p > 0$ of any dissipative system is uniquely determined by either of its components $P(\omega)$, $Q(\omega)$ for real values of ω; viz.

$$A(p) = \frac{2p}{\pi}\int_0^\infty \frac{P(s)}{p^2 + s^2}\,ds = \frac{1}{\pi}\int_{-\infty}^\infty \frac{P(s)}{p + is}\,ds \quad (\operatorname{Re} p > 0), \tag{23}$$

$$A(p) = 2A(0) + \frac{2p^2}{\pi}\int_0^\infty \frac{Q(s)}{s(p^2 + s^2)}\,ds = A(\infty) - \frac{i}{\pi}\int_{-\infty}^\infty \frac{Q(s)}{p + is}\,ds \quad (\operatorname{Re} p > 0). \tag{24}$$

† An exhaustive treatment is given by M. Bayard, 'Relations entre les parties réelles et imaginaires des impédances', *Rev. gén. Élect.* XXXVII, 659, 1935. Cf. also J. H. Schouten, *Tijdschr. Ned. Radiogenootschap*, XI, 129, 1945.

To prove these formulae we first observe that for any dissipative system the integrand of the inversion integral of (18) is singular on the line $\mathrm{Re}\,p = 0$ only at the pole $p = 0$ where the residue is $A(0)$. Therefore the operational relation (18) can be extended by applying the method of (VI, 40). It then covers a strip $\mathrm{Re}\,p_0 < \mathrm{Re}\,p < \infty$, where p_0 is the first singularity to the left of the imaginary axis:

$$A(p) \doteqdot i_{\Gamma}(t) - A(0) \quad (\mathrm{Re}\,p > \mathrm{Re}\,p_0). \tag{25}$$

If in the inversion integral we choose $c = 0$, putting at the same time $p = is$, we obtain

$$i_{\Gamma}(t) - A(0) = \frac{1}{2\pi i}\int_{-\infty}^{\infty} \frac{e^{ist}}{s} A(is)\, ds,$$

which by means of (22), upon using the even character of $P(s)$ and the odd character of $Q(s)$, is transformed into

$$i_{\Gamma}(t) - A(0) = \frac{1}{\pi}\int_0^{\infty} \frac{\sin(ts)}{s} P(s)\, ds + \frac{1}{\pi}\int_0^{\infty} \frac{\cos(ts)}{s} Q(s)\, ds. \tag{26}$$

Further, since $i_{\Gamma}(t)$ must be zero for negative values of t (when the voltage $U(t)$ has not yet become effective) the right-hand side of (26) is equal to $-A(0)$ for $t = -t'$ $(t' > 0)$; thus for $t' > 0$:

$$A(0) = \frac{1}{\pi}\int_0^{\infty} \frac{\sin(t's)}{s} P(s)\, ds - \frac{1}{\pi}\int_0^{\infty} \frac{\cos(t's)}{s} Q(s)\, ds.$$

Consequently, in combining this with (26) it is found possible to express the time admittance for $t' > 0$ in either P- or Q-values:

$$i_{\Gamma}(t') = \frac{2}{\pi}\int_0^{\infty} \frac{\sin(t's)}{s} P(s)\, ds = 2A(0) + \frac{2}{\pi}\int_0^{\infty} \frac{\cos(t's)}{s} Q(s)\, ds \quad (t' > 0). \tag{27}$$

After multiplying these expressions by $U(t')$ and transposing them for $\mathrm{Re}\,p > 0$, we arrive at the identities (23) and (24). In addition to this we remark that it is easy to verify the equivalence of (27) to the respective inversion integrals (having a line of convergence) of the following relations:

$$\operatorname{sgn} t\, i_{\Gamma}(|t|) \doteqdot 2P(ip) \quad (\mathrm{Re}\,p = 0), \tag{28}$$

$$i_{\Gamma}(|t|) - 2A(0) \doteqdot -2iQ(ip) \quad (\mathrm{Re}\,p = 0). \tag{29}$$

(3) For dissipative systems the mutual relationship between the real and imaginary parts of the frequency admittance $A(i\omega)$ for real frequencies ω is expressed by the formulae

$$Q(\omega) = \frac{2\omega}{\pi}\int_0^{\infty} \frac{P(s) - P(\omega)}{s^2 - \omega^2}\, ds = \frac{1}{\pi}\int_{-\infty}^{\infty} \frac{P(s)}{s - \omega}\, ds, \tag{30}$$

$$P(\omega) - P(\infty) = \frac{2}{\pi}\int_0^{\infty} \frac{sQ(s) - \omega Q(\omega)}{\omega^2 - s^2}\, ds = -\frac{1}{\pi}\int_{-\infty}^{\infty} \frac{Q(s) - Q(\omega)}{s - \omega}\, ds. \tag{31}$$

The first of them may be obtained by first subtracting from (23) the identity

$$P(\omega) = \frac{2p}{\pi}\int_0^\infty \frac{P(\omega)}{p^2+s^2}\,ds \quad (\operatorname{Re} p > 0),$$

giving
$$A(p) - P(\omega) = \frac{2p}{\pi}\int_0^\infty \frac{P(s)-P(\omega)}{p^2+s^2}\,ds \quad (\operatorname{Re} p > 0).$$

Unlike (23) the last integral is convergent even for $\operatorname{Re} p = 0$. If then (22) is taken into account, we are easily led to (30) for $p = i\omega$. In quite the same manner (31) follows from (24).

(4) The following formulae express $P(\omega)$ and $Q(\omega)$ in terms of the time admittance:

$$P(\omega) = \omega\int_0^\infty \sin(\omega t)\,\{i_\Gamma(t) - A(0)\}\,dt, \quad Q(\omega) = \omega\int_0^\infty \cos(\omega t)\,\{i_\Gamma(t) - A(0)\}\,dt.$$

They can be proved by separating the definition integral of (25) with the help of (22) into its real and imaginary parts for $p = i\omega$.

It will be obvious that the above-mentioned properties may still hold for other functions $A(p)$, not necessarily restricted to the ratio of two quantities such as current and voltage, either of which is caused by the other. In fact, the proofs above are based on very general properties of complex functions. In this connexion it should be noted that in Kramers's dispersion theory† the set of formulae (30, 31) were derived for the real part ξ and the imaginary part η of a certain coefficient of polarization, the combination $\xi + i\eta$ simply being an analytic function of the frequency $\omega = p/i$.

Finally, the formulae given above may gain in elegance if a suitable change of variables is made. For instance, (23) is transformed by $s = p\tan\phi$ into

$$A(p) = \frac{2}{\pi}\int_0^{\frac{1}{2}\pi} P(p\tan\phi)\,d\phi \quad (\operatorname{Re} p > 0).$$

6. Transient phenomena

Discussing below the response of a system to a sudden change we conveniently fix our mind on electrical systems. The simplest case is then furnished by introducing a unit voltage at $t = 0$, whereby the potential across the terminals is raised instantaneously from zero to unit value. As has already been seen in § 2, the underlying differential equation may have $U(t)$ on its right-hand side only; it is then an easy matter to solve the equation operationally. It is less simple, however, if the e.m.f. is also involved in the differential operator, as in equation (19) concerning the network of fig. 50. The concept of admittance then proves very useful, since in the first place the frequency admittance follows easily from the structure of the system

† *Atti Congr. Int. Fisisi*, XI, 22, 1927.

(see formula (20)), and secondly, the behaviour of the system under the influence of a unit-function e.m.f. is determined at once from the original of the frequency admittance, thus leading to the time admittance.

We need not particularly emphasize the importance of the relation (18), which we shall now examine in greater detail. From VII, § 2, we know that the behaviour of the original as $t \to 0(\infty)$ is closely connected to that of the image as $p \to \infty(0)$. Therefore the relation (18) expresses that in the very beginning the current is determined by the behaviour of the admittance at very high frequencies, whilst later on it mainly depends on the behaviour of the admittance in the low-frequency range. In particular, by means of the Abel theorems (VII, 4 and 5) for $\nu = 0$ we have

$$A(\infty) = i_{\Gamma}(0), \quad A(0) = i_{\Gamma}(\infty),$$

provided $i_{\Gamma}(t)$ has limiting values as $t \to 0$ and $t \to \infty$. It is further known that the existence of the left-hand members does not imply that of the right. Yet the possible applications of Tauber theorems to finite passive electrical networks can be surveyed easily by considering the general form of the time admittance $i_{\Gamma}(t)$. Since in this case $A(p)$ is always the quotient of two polynomials in the variable p, differing in degree by at most ± 1, the following expansion in rational fractions is valid generally:

$$A(p) = \alpha p + \beta + \frac{\gamma}{p} + \sum_{k=1}^{n} c_k \frac{p}{p-p_k} \quad (\operatorname{Re} p_k \leqslant 0;\ p_k \neq 0). \tag{32}$$

The original of this function then gives for the corresponding time admittance the expression

$$i_{\Gamma}(t) = \alpha\delta(t) + U(t)\left\{\beta + \gamma t + \sum_{k=1}^{n} c_k e^{p_k t}\right\}. \tag{33}$$

In this expression the exponential terms represent damped oscillations in so far as $\operatorname{Re} p_k < 0$. Only in the case of passive, non-dissipative, networks can γ be different from zero; in this case $A(p)$ may also have some poles different from zero on the imaginary axis ($\operatorname{Re} p_k = 0$) leading to undamped oscillations in $i_{\Gamma}(t)$. Moreover, if $\gamma = 0$ the condition (VII, 13) with respect to the Tauber theorem III for $\nu = 0$ is satisfied; we therefore obtain

$$A(0) \Rightarrow \operatorname*{Lim}_{t\to\infty} \frac{1}{t}\int_0^t i_{\Gamma}(\tau)\, d\tau. \tag{34}$$

In words: after a long time the mean current (= the direct-current component per unit voltage) becomes equal to the admittance $A(0)$ for ordinary direct current ($p = 0$). Thus Ohm's law with reference to the mean current for $t \to \infty$ is the physical equivalent of some mathematical Tauber theorem. For dissipative networks ($\operatorname{Re} p_k < 0$) the mean current on the right of (34) may be replaced by the actual value $i_{\Gamma}(\infty)$.

It is now easy to study transient phenomena due to electromotive forces of quite general character, completely different from the unit-function voltage above. To this purpose the relation (17) is of utmost importance. Of particular simplicity is the δ-function voltage $e(t) = \delta(t)$ for which $E(p) = p$. If $i_\delta(t)$ denotes the current due to the impulse e.m.f. $\delta(t)$ it follows immediately that

$$i_\delta(t) \risingdotseq pA(p) \quad (\operatorname{Re} p > \operatorname{Re} p_0). \tag{35}$$

For instance, in the case of the network of fig. 50 we obtain from (20)

$$i_\delta(t) = C\delta'(t) + \frac{1}{L}\,\mathrm{e}^{-Rt/L}\,U(t).$$

The equation (17) provides us with many relations between transients corresponding to different electromotive forces $e(t)$. First we infer from (18) and (35) by applying the differentiation rule that

$$i_\delta(t) = \frac{d}{dt} i_\mathit{\Gamma}(t). \tag{36}$$

Furthermore, the current i_e due to any arbitrary e.m.f. $e(t)$ can always be obtained if either the time admittance $i_\mathit{\Gamma}(t)$ or its derivative $i_\delta(t)$ is known. To see this we first rewrite (17) as follows:

$$i_e(t) \risingdotseq \frac{1}{p} \times A(p) \times pE(p) = \frac{1}{p} \times pA(p) \times E(p).$$

Then, by means of the composition-product rule,

$$i_e(t) = \int_{-\infty}^{\infty} i_\mathit{\Gamma}(\tau)\, e'(t-\tau)\, d\tau = \int_{-\infty}^{\infty} i_\delta(\tau)\, e(t-\tau)\, d\tau. \tag{37}$$

It is thus also a matter of indifference whether we know either the time admittance $i_\mathit{\Gamma}(t)$, or the function $i_\delta(t)$, or the frequency admittance $A(p)$, because the current is always uniquely determined by each of them whenever the e.m.f. applied is given.

In concluding this section, let us take the alternating e.m.f.

$$e(t) = \sin(\omega t)\, U(t),$$

switched on at $t = 0$. The image of this function is $E(p) = \omega p/(p^2+\omega^2)$. According to (17) the response current is given by

$$i(t) \risingdotseq \frac{\omega}{p} A(p) \frac{p^2}{p^2+\omega^2},$$

which upon applying the rule of the composition product yields

$$i(t) = \omega \int_{-0}^{t} i_\mathit{\Gamma}(\tau) \cos\{\omega(t-\tau)\}\, d\tau.$$

Although $i_{\Gamma}(t)$ is zero for $t<0$ we may not replace the lower limit of integration $(-\infty)$ by zero. The notation -0 indicates that the integration should start somewhat below zero, so as to include, if necessary, any impulse term $\delta(t)$ in $i_{\Gamma}(t)$.

7. Time impedances

In the theory of electrical networks there exists a high degree of symmetry between the properties of currents on the one hand and electromotive forces on the other. Numerous properties remain valid if currents are replaced by e.m.f.'s, and vice versa, provided the admittance $A(p)$ is at the same time replaced by its inverse, the impedance $Z(p)$. Consequently, just as one can ask for the current if the e.m.f. is given (this problem was in fact treated above), one may deduce formulae expressing the voltage in terms of the input current; the role played by the admittance is then taken over by the impedance. For example, if the current at the terminals is given by $i(t) \risingdotseq I(p)$, the corresponding input voltage obeys

$$e(t) \risingdotseq Z(p)\, I(p) \quad (\mathrm{Re}\, p>0), \tag{38}$$

since the network is supposed to be linear and since the current e^{pt} produces a voltage $Z(p)\,e^{pt}$; for the rest (38) may be derived in exactly the same manner as (17). In particular, corresponding to the unit current $i(t) = U(t)$ we have the following electromotive force:

$$e_{\Gamma}(t) \risingdotseq Z(p) \quad (\mathrm{Re}\, p>0), \tag{39}$$

which we shall call the *time impedance* by analogy to the time admittance $i_{\Gamma}(t)$. In the case of the network of fig. 50 the frequency impedance is found from (21) to be

$$Z(p) = \frac{1}{A(p)} = \frac{Lp+R}{LCp^2+RCp+1}.$$

After expanding into rational fractions and transposing term by term the time impedance is as follows:

$$e_{\Gamma}(t) = U(t)\left\{R - \frac{Lp_1+R}{RCp_1+2}\,e^{p_1 t} - \frac{Lp_2+R}{RCp_2+2}\,e^{p_2 t}\right\},$$

where
$$p_{1,2} = -\frac{R}{2L} \pm i\sqrt{\left(\frac{1}{LC}-\frac{R^2}{4L^2}\right)}$$
are the poles of $Z(p)$.

The relation (39) will also be valid for $\mathrm{Re}\, p > \mathrm{Re}\, p_0$ if, now, p_0 is that pole of $Z(p)/p$ which is as far to the right as possible; in the case of passive networks, $Z(p)$, like $A(p)$, never has a pole on the right of the imaginary axis. Furthermore, the time admittance and the time impedance can be directly related with each other by means of the rules of the operational calculus. In this way the identity $Z(p)\,A(p) = 1$, if first rewritten as

$$\frac{1}{p}\times Z(p)\times pA(p) = \frac{1}{p}\times pZ(p)\times A(p) = 1$$

yields at once

$$\int_{-0}^{t+0} e_{\Gamma}(t-\tau)\, i'_{\Gamma}(\tau)\, d\tau = \int_{-0}^{t+0} e'_{\Gamma}(t-\tau)\, i_{\Gamma}(\tau)\, d\tau = U(t), \tag{40}$$

in which the limits of integration originating from the composition product $(-\infty, +\infty)$ are properly replaced by zero and t, or, more precisely, by -0 and $t+0$, in order to account for possible terms involving δ-functions.

Though in practice the time admittance is preferred, since usually the voltages rather than the currents are known, it is advantageous to use them together, especially in combinations of switching-on and switching-off phenomena. For instance, let us assume a system completely at rest for $t<0$. Let further at $t = 0$ the unit voltage $U(t)$ come into action; at time $t = t_1$, however, the voltage source is cut out. In this special case the switching-on current is cut out at $t = t_1$ and the question of determining the remaining voltage may arise. The operational solution runs as follows. For all values of the time t the current is represented by

$$i(t) = i_{\Gamma}(t)\, U(t_1-t) \doteqdot I(p),$$

whilst (38) may be written in the following manner:

$$e(t) \doteqdot \frac{1}{p}\times pZ(p)\times I(p).$$

If now the rule of the composition product is applied, the answer to the question above is found in the form

$$e(t) = \int_{-\infty}^{\infty} e'_{\Gamma}(t-\tau)\, i_{\Gamma}(\tau)\, U(t_1-\tau)\, d\tau = \int_{-0}^{t_1} e'_{\Gamma}(t-\tau)\, i_{\Gamma}(\tau)\, d\tau. \tag{41}$$

It is most striking that (41) applies to all three intervals $t<0$, $0<t<t_1$, $t>t_1$, where the course of $e(t)$ is quite different. With respect to the first two intervals the upper limit of integration t_1 may be replaced by t, since in the interval $t<\tau<t_1$ the argument of e'_{Γ} becomes negative and thus implies a vanishing e'. Thus, if $t<t_1$ the integral (41) reduces to $U(t)$ on account of (40), as it should; for $t>t_1$, however, the function (41) is far from trivial.

8. Series for small values of t. Heaviside's first expansion theorem

In the preceding sections we have repeatedly dealt with images that are rational functions of p, whose numerator and denominator were different in degree by at most unity. In particular this applies to the frequency admittance $A(p)$ and the frequency impedance $Z(p)$ of electrical circuits.

Further the image (13) of the short-circuit current (see § 4) had numerator and denominator of equal degree. In their operational transposition these three quantities led to the time admittance, the time impedance and the short-circuit current, whilst the respective originals could be determined rigorously by first expanding the images into rational fractions and then applying Heaviside's second expansion theorem. But for small values of t (e.g. immediately after the source has been switched on) these expansions are of little value. It is then better to use a series in powers of t, to be obtained by developing the image into powers of $1/p$. In order to include all the above possibilities we shall assume the image to be a rational function of the form

$$f(p) = \frac{a_0+a_1p+a_2p^2+\ldots+a_np^n+a_{n+1}p^{n+1}}{b_0+b_1p+b_2p^2+\ldots+b_np^n}$$

$$= \frac{\dfrac{a_0}{p^n}+\dfrac{a_1}{p^{n-1}}+\ldots+\dfrac{a_{n-1}}{p}+a_n+a_{n+1}p}{\dfrac{b_0}{p^n}+\dfrac{b_1}{p^{n-1}}+\ldots+\dfrac{b_{n-1}}{p}+b_n} \tag{42}$$

in which b_n is supposed different from zero.

Let this function be developed into powers of $1/p$, viz.

$$f(p) = \frac{\Delta_1}{b_n}p-\frac{\Delta_2}{b_n^2}+\frac{\Delta_3}{b_n^3}\frac{1}{p}-\frac{\Delta_4}{b_n^4}\frac{1}{p^2}+\frac{\Delta_5}{b_n^5}\frac{1}{p^3}-\ldots, \tag{43}$$

where the coefficients are easily determined as follows:

$$\Delta_1 = a_{n+1},$$

$$\Delta_2 = \begin{vmatrix} a_{n+1} & b_n \\ a_n & b_{n-1} \end{vmatrix},$$

$$\Delta_3 = \begin{vmatrix} a_{n+1} & b_n & 0 \\ a_n & b_{n-1} & b_n \\ a_{n-1} & b_{n-2} & b_{n-1} \end{vmatrix},$$

etc.

Obviously the series (43) converges outside any circle enclosing all the poles of $f(p)$ and having its centre at the origin $p = 0$. It is then legitimate to transpose the series term by term (see (VII, 26)); confining ourselves, moreover, to $\operatorname{Re} p > 0$, we obtain for the original the following convergent series:

$$h(t) = \frac{\Delta_1}{b_n}\delta(t)-U(t)\left\{\frac{\Delta_2}{b_n^2}-\frac{\Delta_3}{b_n^3}\frac{t}{1!}+\frac{\Delta_4}{b_n^4}\frac{t^2}{2!}-\frac{\Delta_5}{b_n^5}\frac{t_5}{3!}+\ldots\right\}.$$

This is nothing but Heaviside's first expansion theorem, which is of great importance in the calculation of transient phenomena.

Example. Returning once more to the network of fig. 50, we have for its admittance (21)

$$A(p) = \frac{LCp^2 + RCp + 1}{Lp + R} = Cp + \frac{1}{Lp} - \frac{R}{L^2 p^2} + \dots;$$

consequently, for the time admittance,

$$i_{\Gamma}(t) = C\delta(t) + U(t)\left\{\frac{t}{L} - \frac{R^2}{2L^2}t^2 + \dots\right\}.$$

9. Classification of admittances

The preceding discussions enable us to divide the possible types of electrical admittances into different classes. Three distinct cases can be distinguished as to the difference in degree of the numerator (degree m) and the denominator (degree n), since $|m-n| \leqslant 1$:

(1) $m - n = 1$:

In this case the coefficient a_{n+1} in (42) is different from zero; therefore the development (43) starts as follows:

$$A(p) = c_0 p + c_1 + \frac{c_2}{p} + \dots,$$

where $c_0 \neq 0$. For high frequencies the network behaves like a capacitor of capacitance c_0. Further, the first few terms of the time admittance are given by

$$i_{\Gamma}(t) = c_0 \delta(t) + U(t)(c_1 + c_2 t + \dots).$$

Thus an impulse function of amplitude c_0 at $t = 0$ is present, followed by a continuous function having c_1 as initial value (see fig. 52*a*).

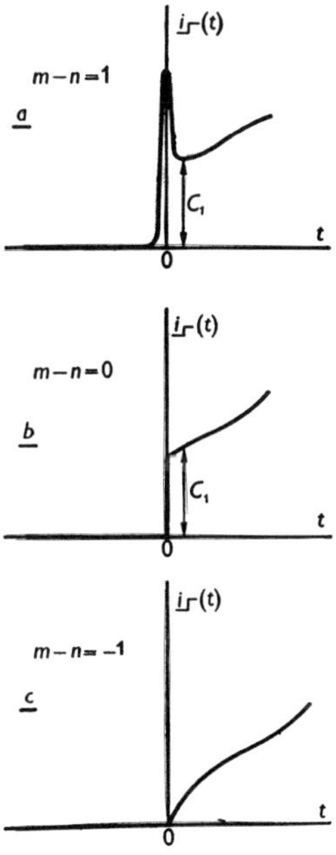

Fig. 52. The time admittance for $t \to 0$.

(2) $m - n = 0$:

In (42) we may take $a_{n+1} = 0$. Therefore the development of $A(p)$ starts with a constant term c_1; for high frequencies the network behaves as a resistor. The corresponding development of $i_{\Gamma}(t)$ begins with the term $c_1 U(t)$. This leads to the curve of fig. 52*b*, showing a sudden jump at $t = 0$.

(3) $m - n = -1$:

We now may suppose in (42) $a_{n+1} = a_n = 0$; hence the first term of $A(p)$ is c_2/p. Accordingly, the network has the character of a self-inductance $1/c_2$ for high frequencies. The expansion for $i_{\Gamma}(t)$ starts with the term $c_2 t\, U(t)$, so as to make the curve of fig. 52*c* continuous at $t = 0$; its slope there is determined by the constant c_2.

A different classification of the electrical networks is still possible, namely, with reference to their behaviour for low frequencies. This comes down to the properties of $i_\Gamma(t)$ for *large* values of t. If p^μ and p^ν are the lowest powers actually occurring in the numerator and denominator of $A(p)$ respectively, then the absolute difference of μ and ν, too, cannot exceed 1. Therefore, again, we have three different cases according as $\mu-\nu = 1, 0,$ or -1. In the low-frequency range the corresponding behaviour of the network is, in the same order, that of a capacitor, that of a resistor, or that of a self-inductance. As $p \to 0$ the respective functions $A(p)/p$, $A(p)$, $pA(p)$, have a finite non-vanishing limit. Accordingly, in (32) $\beta = \gamma = 0$, $\gamma = 0$, $\gamma \neq 0$, respectively. Consequently, the time admittance $i_\Gamma(t)$ given by (33) is schematically represented for $t \to \infty$ by the three diagrams of fig. 53, provided we confine ourselves to systems without undamped eigenfunctions $e^{p_k t}$.

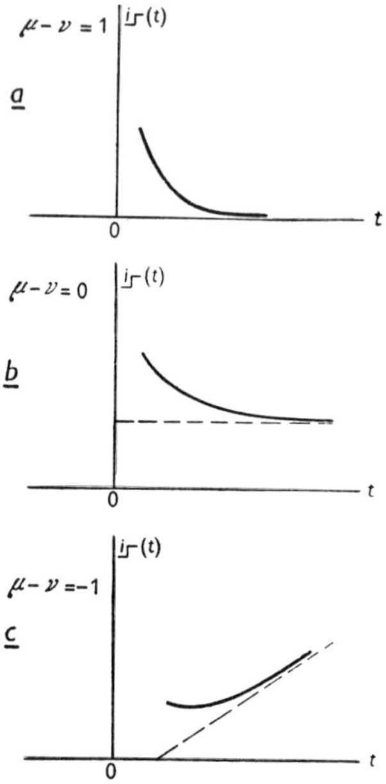

Fig. 53. The time admittance for $t \to \infty$.

Finally, in combining the two distinct classifications we are led to a total number of nine different types of networks. In the next chapter we shall classify the mutual admittances analogously. The latter classification as well as that above has been discussed previously by Nijenhuis and Stumpers†.

† 'On some properties of electrical networks', *Physica*, *'s-Grav.*, VIII, 289, 1941.

CHAPTER IX

SIMULTANEOUS LINEAR DIFFERENTIAL EQUATIONS WITH CONSTANT COEFFICIENTS; ELECTRIC-CIRCUIT THEORY

1. Introduction

Just as in the preceding chapter we applied the operational method to a *single* linear differential equation with constant coefficients, thus obtaining an algebraic equation of the first degree for the image, we can apply it to a *set* of differential equations, thereby reducing its solution to that of a system of first-degree algebraic equations. Such systems are encountered, for instance, in the quasi-stationary theory of electrical networks, at least if we are interested in a complete survey of all the currents rather than in the current flowing through some particular branch. The disadvantage of getting a set of differential equations is neutralized by them being of the second order only.

We shall briefly deduce the general form of the equations in § 2, whereupon they will be approached operationally in § 3. Though our terminology and the accompanying examples will always refer to electrical systems, the general outlines equally apply to other fields, where, if necessary, the maximum order of the individual differential equations may be greater than two. In §§ 4–6 we shall discuss in succession the time admittance, some general applications, questions of boundary conditions, and transient phenomena. The body of the theory below is substantially the same as that in the preceding chapter except that in the following a whole *set* of simultaneous equations is taken as starting-point. Finally, filter networks of the ladder and lattice types are treated in §§ 7 and 8, these being special cases of general electrical networks.

2. Equations of electric-circuit theory

In the quasi-stationary treatment the electrical networks are supposed to consist of a finite number of branches which are connected to one another at a certain number of vertices. Each separate branch contains a resistance R, a capacitance C and a self-inductance L in series. For the present we shall not account for mutual inductance between different branches. The electrical behaviour of such a system is governed by the two laws of Kirchhoff. According to the first law the algebraic sum of currents flowing into any vertex is equal to zero. The second law of Kirchhoff states that the total voltage drop around a closed loop is equal to the total electromotive force therein impressed.

We shall consider a network with N independent meshes. Let us make a definite choice out of the great many possibilities of taking a system of N independent meshes, and number them by $1, 2, \ldots, N$. Let further $i_1, i_2, \ldots, i_N$ denote the independent mesh currents and $e_1, e_2, \ldots, e_N$ the corresponding set of independent electromotive forces. It is then not difficult to write down the system of N differential equations in question, by properly accounting for the actual branch currents (i.e. linear combinations with coefficients $+1$ or -1 of the mesh currents) and the various electrical elements themselves, even if we had not ignored mutual inductances. We shall not give the general equations, however. We rather prefer to study the particular case of planar networks (an assembly of branches and vertices that can be drawn in a plane without crossings) where the coefficients of the equations can easily be interpreted in terms of the real network elements, owing to the fact that the N meshes can be chosen such that any branch current is the algebraic sum of at most two mesh currents. We simply take the N *adjacent* meshes, none of which has parts of the network in its interior (see fig. 54 with $N = 4$). In this special case it is also easy to fix the positive directions of the mesh currents adequately, for instance, counter-clockwise as in fig. 54.

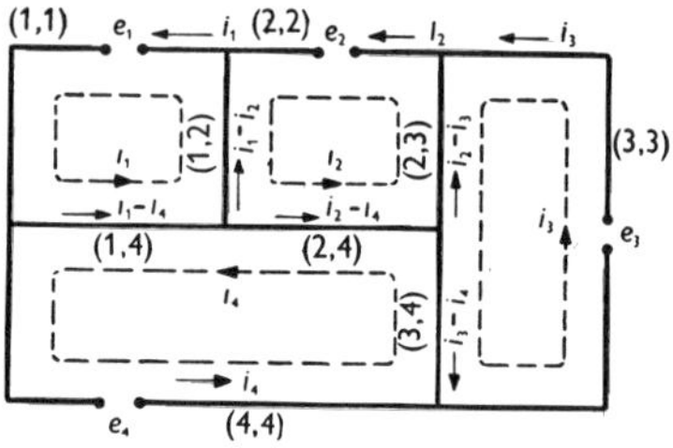

Fig. 54. An electrical network with four meshes.

Let (n, m) indicate the branch that is common to the nth and the mth mesh; in addition, (n, n) denotes the branch that belongs to the nth mesh only. Let L^*_{nm}, R^*_{nm} and C^*_{nm} represent the self-inductance, the resistance and the capacitance of the branch (n, m), all connected in series. The second law of Kirchhoff applied to the nth mesh then yields

$$e_n = \left(L^*_{nn}\frac{d}{dt} + R^*_{nn} + \frac{1}{C^*_{nn}}\int_{-\infty}^{t} dt\right) i_n$$
$$+ \sum_{m \neq n}\left(L^*_{nm}\frac{d}{dt} + R^*_{nm} + \frac{1}{C^*_{nm}}\int_{-\infty}^{t} dt\right)(i_n - i_m) \quad (n = 1, 2, \ldots, N). \quad (1)$$

Upon properly arranging the currents i_m we can write this in the following manner:

$$e_n = \sum_{m=1}^{N}\left(L_{nm}\frac{d}{dt} + R_{nm} + \frac{1}{C_{nm}}\int_{-\infty}^{t} dt\right) i_m \quad (n = 1, 2, \ldots, N), \quad (2)$$

if the new coefficients involved are defined by

(1) if $n = m$:

$$L_{nn} = \sum_{m=1}^{N} L^*_{nm}, \quad R_{nn} = \sum_{m=1}^{N} R^*_{nm}, \quad \frac{1}{C_{nn}} = \sum_{m=1}^{N} \frac{1}{C^*_{nm}}; \quad (3)$$

(2) if $n \neq m$: $\quad L_{nm} = -L^*_{nm}, \quad R_{nm} = -R^*_{nm}, \quad C_{nm} = -C^*_{nm}. \quad (4)$

The canonical equations (2) are basic for electric-circuit theory. It should further be emphasized that the coefficients in (2) are positive for $n = m$, negative for $n \neq m$, at least if the currents have been oriented as in fig. 54. These properties need not necessarily hold if mutual inductances are admitted, though of course the general form of (2) remains. Moreover, there are two other ways of expressing the equations (2):

(1) Upon introducing the charges

$$q_n(t) = \int_{-\infty}^{t} i_n(\tau)\, d\tau,$$

we obtain a set of equations without integral signs, viz.

$$e_n = \sum_{m=1}^{N} \left(L_{nm} \frac{d^2}{dt^2} + R_{nm} \frac{d}{dt} + \frac{1}{C_{nm}} \right) q_m \quad (n = 1, 2, \ldots, N). \tag{5}$$

(2) If the currents and charges are introduced simultaneously, the system (2) can be written as follows:

$$e_n = \frac{d}{dt} \left(\sum_{m=1}^{N} L_{nm} i_m \right) + \sum_{m=1}^{N} R_{nm} i_m + \sum_{m=1}^{N} \frac{q_m}{C_{nm}} \quad (n = 1, 2, \ldots, N). \tag{6}$$

The latter expressions immediately remind us of the well-known Lagrangian equations of mechanics. Defining the magnetic energy T, the electric energy V, and the heat dissipation per unit time F, by means of

$$T = \frac{1}{2} \sum_{n=1}^{N} \sum_{m=1}^{N} L_{nm} i_n i_m, \tag{7}$$

$$V = \frac{1}{2} \sum_{n=1}^{N} \sum_{m=1}^{N} \frac{q_n q_m}{C_{nm}}, \tag{8}$$

$$F = \frac{1}{2} \sum_{n=1}^{N} \sum_{m=1}^{N} R_{nm} i_n i_m, \tag{9}$$

we easily show that (6) is equivalent to

$$\frac{d}{dt} \frac{\partial T}{\partial i_n} + \frac{\partial V}{\partial q_n} + \frac{\partial F}{\partial i_n} = e_n \quad (n = 1, 2, \ldots, N). \tag{10}$$

3. Transposition of the circuit equations; mutual impedances and admittances

We now proceed to the operational transposition of the general network equations (2). For this purpose we only need the differentiation and integration rules. It is further sufficient to assume that the following $2N$ operational relations

$$i_n(t) \fallingdotseq I_n(p), \quad e_n(t) \fallingdotseq E_n(p) \quad (n = 1, 2, \ldots, N),$$

together with those corresponding to the first-order derivatives of $i_n(t)$ and $e_n(t)$ have a common strip of convergence situated at $\operatorname{Re} p > 0$. If so, then from (2)

$$\sum_{m=1}^{N} Z_{nm}(p)\, I_m(p) = E_n(p) \quad (n = 1, 2, \ldots, N), \tag{11}$$

in which the new coefficients Z_{nm} are defined by

$$Z_{nm}(p) = L_{nm} p + R_{nm} + \frac{1}{C_{nm} p}. \tag{12}$$

Though we have established equations (12) only for a special kind of network, that is to say, for planar networks, they apply equally to non-planar networks. For planar networks, however, it is easy to interpret Z_{nm}; after substituting (3) and (4) in (12) we obtain

(1) if $n = m$: $$Z_{nn}(p) = p \sum_{m=1}^{N} L^*_{nm} + \sum_{m=1}^{N} R^*_{nm} + \frac{1}{p} \sum_{m=1}^{N} \frac{1}{C^*_{nm}},$$

(2) if $n \neq m$: $$Z_{nm}(p) = -\left(pL^*_{nm} + R^*_{nm} + \frac{1}{pC^*_{nm}}\right),$$

whence it follows that Z_{nn} is equal to the total impedance of the nth mesh, whilst $-Z_{nm}$ for $n \neq m$ is equal to the impedance of the branch that is common to the nth and the mth mesh.

In order to calculate the currents due to the given sources of voltage we have first to solve the system (11) for the images. This solution may be written in the following way:

$$I_n(p) = \sum_{m=1}^{N} A_{nm}(p)\, E_m(p), \tag{13}$$

in which the coefficients A_{nm} are expressible as the quotients of two determinants involving the known parameters Z_{nm}. Thereupon the currents themselves are found by forming the originals of the right-hand sides of (13), which can be done by transposing them term by term, and with the help of the composition-product rule. Herewith we may conclude the general discussion of the operational method.

We shall now discuss the significance of the functions $A_{nm}(p)$, which turn out to be the natural extensions of the admittances treated in the foregoing chapter. To see this we assume that all the currents and voltages vary with time like e^{pt}. Then the system (13) proves to be valid for the amplitudes $i_n(0)$, $e_n(0)$, of the currents and voltages respectively, without any operational transposition. Further, a special coefficient such as $A_{12}(p)$ equals the ratio of the current $i_1(0)\, e^{pt}$ to the electromotive force $e_2(0)\, e^{pt}$, if all the other sources of voltage are short-circuited, that is, $e_1 = e_3 = e_4 = \ldots = e_N = 0$. In this case we have obviously a four-terminal network (see fig. 55); across one pair of terminals the electromotive force e_2

is applied, whilst $A_{12}e_2$ is the current at some other pair of terminals supposed to be short-circuited. Therefore, again, the function A_{12} represents a frequency admittance, though it differs from the former definition in that it relates the current and voltage at different places. Generally A_{nm} $(n \neq m)$ is called the *transfer admittance*; A_{nn} is the *driving-point admittance* at the nth pair of terminals, in which the current and voltage refer to the same place. Similar names exist for the impedances; e.g. the transfer impedance $Z_{21}(p)$ of the $2N$-terminal network is the ratio of the voltage across the second pair of terminals, $e_2(0)\,e^{pt}$, to the current through the first pair of terminals, $i_1(0)\,e^{pt}$, if the remaining $N-1$ pairs of terminals are open-circuited. In this case $i_2 = i_3 = \ldots = i_N = 0$; here again the system acts as a four-terminal network.

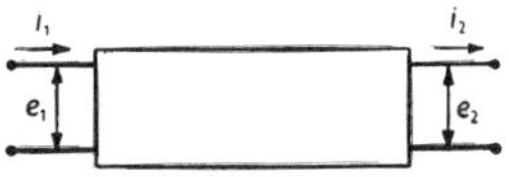

Fig. 55. A four-terminal network.

Before discussing the significance of the transfer admittances in their relation to the operational calculus it may be worth while to give a review of some important properties:

(1) The symmetry property $A_{nm} = A_{mn}$; this relation is evident for the special networks considered above since $Z_{nm} = Z_{mn}$. The latter symmetry is true on account of Z_{nm} being the impedance of the branch common to the nth and the mth mesh.

(2) $A_{nm}(p)$ is a rational function of p. If μ and ν are the degrees of numerator and denominator respectively, then†

$$\mu-\nu \leqslant 1 \quad \text{for} \quad n \neq m, \qquad |\mu-\nu| \leqslant 1 \quad \text{for} \quad n = m.$$

(3) In the case of passive networks none of the functions $A_{nm}(p)$ has poles to the right of the imaginary axis; this does not hold for the zeros, however, unlike the case of driving-point admittances.

(4) The following relationship exists between the admittances and the impedances:

$$\sum_{r=1}^{N} A_{nr}(p)\,Z_{rm}(p) = \delta_{nm}, \tag{14}$$

in which $\delta_{nm} = 1$ if $n = m$, and $\delta_{nm} = 0$ of $n \neq m$; (14) is a consequence of (13) being the solution of (11).

Example. Consider the simple network of fig. 56, in which there are two mesh currents, i_1 and i_2. The equations (1) reduce to

$$e_1 = Ri_1 + \frac{1}{C}\int_{-\infty}^{t} (i_1 - i_2)\,d\tau, \quad e_2 = L\frac{di_2}{dt} + \frac{1}{C}\int_{-\infty}^{t} (i_2 - i_1)\,d\tau.$$

According to (11) the operational transformation leads to

$$E_1 = \left(R + \frac{1}{Cp}\right) I_1 - \frac{I_2}{Cp}, \quad E_2 = -\frac{I_1}{Cp} + \left(Lp + \frac{1}{Cp}\right) I_2,$$

† See, for instance, W. Nijenhuis and F. L. H. M. Stumpers, 'On some properties of electrical networks', *Physica*, 's-*Grav.*, VIII, 289, 1941.

of which the solution is given by

$$I_1 = \frac{(LCp^2+1)\,E_1+E_2}{RLCp^2+Lp+R}, \quad I_2 = \frac{E_1+(RCp+1)\,E_2}{RLCp^2+Lp+R}.$$

We consequently obtain for the driving-point admittances

$$A_{11} = \frac{LCp^2+1}{RLCp^2+Lp+R}, \quad A_{22} = \frac{RCp+1}{RLCp^2+Lp+R}, \tag{15a}$$

and for the transfer admittances

$$A_{12} = A_{21} = \frac{1}{RLCp^2+Lp+R}. \tag{15b}$$

In the last expression the difference in degree of numerator and denominator is two; this would be impossible in the case of a driving-point admittance. The different character of the three admittances mentioned is clear from fig. 57 showing $|A(i\omega)|$ as a function of ω.

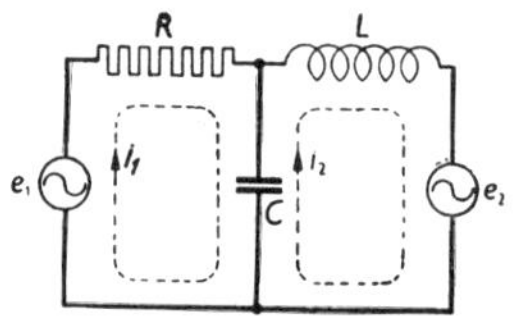

Fig. 56. Example of an electric network with two meshes.

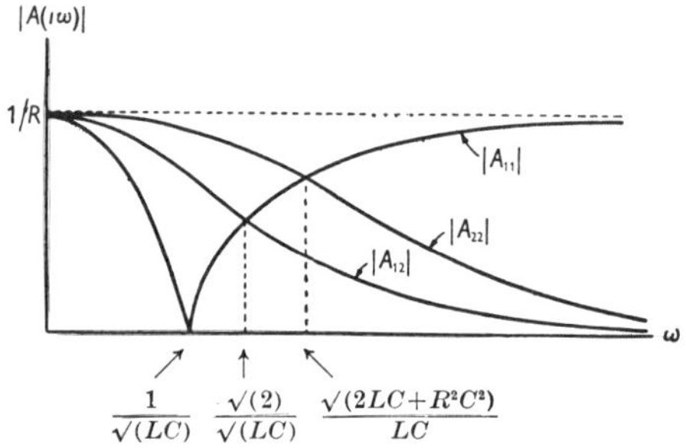

Fig. 57. The modulus of the driving point and transfer admittances of the network of fig. 56.

4. Time admittances of electric circuits

The simplest kind of electric-network transient phenomenon is obtained if all sources of voltage except one are short-circuited, the remaining electromotive force being assumed equal to the unit function $U(t)$. Therefore, let $E_k = 1$ and $E_m = 0$ $(m \neq k)$; then (13) reduces to

$$I_n(p) = A_{nk}(p).$$

Thus, if the original of $A_{nk}(p)$ is denoted by $i_{nk\Gamma}(t)$, the latter time function represents the current through the nth pair of terminals in response to the unit voltage impressed at the kth pair of terminals, just as in the preceding chapter (§ 5) the function $i_{\Gamma}(t)$ was the original of the driving-point admittance $A(p)$. Instead of the single operational relation (VIII, 18) we now have, for k fixed, N relations

$$i_{nk\Gamma}(t) \doteqdot A_{nk}(p), \quad 0 < \operatorname{Re} p < \infty \quad (n = 1, 2, \ldots, N), \tag{16}$$

in which the left-hand functions for $n \neq k$ are suitably called *transfer-time admittances* corresponding to the kth source of voltage $e_k(t) = U(t)$. Moreover, from the symmetry property $A_{nk} = A_{kn}$ it follows that $i_{nk\Gamma}(t) = i_{kn\Gamma}(t)$.

The generalized time admittances have similar properties as the restricted time admittances of the foregoing chapter, in particular in their relationship to the frequency admittances. Accordingly, the generalized admittances can be divided into several types as to their behaviour for large or small values of t, corresponding to the character of the frequency admittance for low or high frequencies respectively†. There is one important difference, however. Owing to the fact that the degrees μ and ν of the numerator and denominator respectively of the transfer frequency admittances only satisfy $\mu-\nu \leqslant 1$, rather than $|\mu-\nu| \leqslant 1$, there are now more than three classes; we must now distinguish according to

$$\mu-\nu = 1, 0, -1, -2, \ldots.$$

That is to say, the expansion in powers of $1/p$ of the frequency admittance may start with any of p^1, p^0, p^{-1}, p^{-2}, ..., such that the main term in the corresponding time admittance, as $t \to 0$, is proportional to $\delta(t)$, t^0, t^1, t^2, ..., respectively. Completely independent of this is that the frequency admittance can equally well be written as the quotient of two polynomials in the variable $1/p$ whose degrees (μ' numerator, ν' denominator) also satisfy $\mu'-\nu' \leqslant 1$. From this again it is possible to arrive at certain conclusions as to the behaviour of the time admittances for $t \to \infty$. Just as was described in § 9 of the foregoing chapter, in this case also the combination of properties for t small (p large) and t large (p small) furnishes a natural classification of the transfer impedances. Furthermore, the generalized time admittances can be calculated with the help of either an expansion in rational fractions (Heaviside's second expansion theorem) or a power series (Heaviside's first expansion theorem) for the frequency admittance. Finally, more complicated transient phenomena can also be calculated through by means of the composition-product rule, once the time admittance is known.

Example. Let us again consider the circuit of fig. 56. To the source of voltage e_1 there corresponds the following pair of time admittances:

$$i_{11\Gamma}(t) \doteqdot A_{11}(p), \quad i_{12\Gamma}(t) \doteqdot A_{12}(p).$$

From the originals of (15) for $\operatorname{Re} p > 0$ we then obtain, if $R > \frac{1}{2}\sqrt{(L/C)}$,

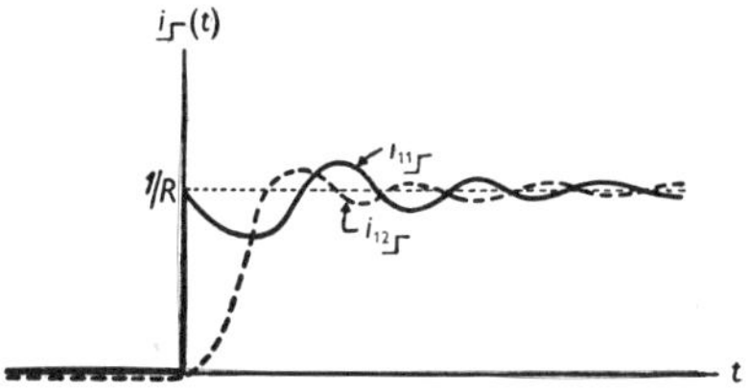

Fig. 58. Two time admittances of the network of fig. 56.

$$i_{11\Gamma}(t) = \frac{U(t)}{R}\left\{1-\frac{e^{-t/2RC}}{RC\omega_0}\sin(\omega_0 t)\right\},$$

$$i_{12\Gamma}(t) = \frac{U(t)}{R}\left[1-e^{-t/2RC}\left\{\cos(\omega_0 t)+\frac{\sin(\omega_0 t)}{2RC\omega_0}\right\}\right],$$

in which ω_0 is an abbreviation for

$$\omega_0 = \sqrt{\left(\frac{1}{CL}-\frac{1}{4R^2C^2}\right)}.$$

These two time admittances are shown schematically in fig. 58.

† See W. Nijenhuis and F. L. H. M. Stumpers, loc. cit.

5. Applications of the general circuit theory

The operational method discussed in the preceding two sections, apart from giving the solutions of some special problems, is also useful in deducing general network properties. This will be demonstrated in the following example:

$$A'_{kk}(0) = 2\{V(\infty) - T(\infty)\}, \tag{17}$$

which states a definite relationship between the derivative of the driving-point admittance at $p = 0$ and the difference in electric and magnetic energy of the whole system at $t = \infty$, delivered by an electromotive force $e_k(t) = U(t)$ switched on at $t = 0$ to the terminals k, if it is understood that before $t = 0$ the system was completely at rest.

To prove (17) we start with an identity that is easily derived from (14), namely,

$$A_{kk}(p) - A_{kk}(-p) = \sum_{m=1}^{N} \sum_{n=1}^{N} \{Z_{mn}(-p) - Z_{mn}(p)\} A_{km}(-p) A_{kn}(p).$$

Then we apply (12); the resistance terms automatically disappear, whence it follows that

$$\frac{A_{kk}(p) - A_{kk}(-p)}{p}$$
$$= -\frac{2}{p} \sum_{m=1}^{N} \sum_{n=1}^{N} \left\{ L_{mn} \times pA_{km}(-p) \times A_{kn}(p) + \frac{1}{C_{mn}} \times A_{km}(-p) \times \frac{A_{kn}(p)}{p} \right\}. \tag{18}$$

To transpose this into the t-language, we assume the network dissipative with respect to the source e_k, so as to have the first pole p_0 of $A_{kk}(p)$ (counted from the right) lying to the left of the imaginary axis (see VIII, § 5). On account of (16) we then have

$$A_{kn}(p) \risingdotseq i_{kn\Gamma}(t) = q'_{kn\Gamma}(t) \quad (\mathrm{Re}\, p > \mathrm{Re}\, p_0),$$
$$pA_{kn}(p) \risingdotseq i'_{kn\Gamma}(t) \quad (\mathrm{Re}\, p > \mathrm{Re}\, p_0).$$

Moreover, if the variable p is replaced by $-p$ the new relations have a strip in common with the old; we thus have simultaneously for $0 < \mathrm{Re}\, p < -\mathrm{Re}\, p_0$

$$A_{kn}(p) \risingdotseq i_{kn\Gamma}(t), \quad pA_{km}(-p) \risingdotseq i'_{km\Gamma}(-t),$$
$$\frac{A_{kn}(p)}{p} \risingdotseq q_{kn}(t), \quad A_{km}(-p) \risingdotseq -q'_{km}(-t).$$

All this enables us to transpose (18) operationally with the help of the composition-product rule. The result is found to be

$$\frac{A_{kk}(p) - A_{kk}(-p)}{p} \risingdotseq 2 \sum_{m=1}^{N} \sum_{n=1}^{N}$$
$$\times \left\{ -L_{mn} \int_{-\infty}^{t+0} i_{kn\Gamma}(t-\tau)\, i'_{km\Gamma}(-\tau)\, d\tau + \frac{1}{C_{mn}} \int_{-\infty}^{t+0} q_{kn}(t-\tau)\, q'_{km}(-\tau)\, d\tau \right\}, \tag{19}$$

valid in the aforementioned strip of convergence. Let us now determine the limit of the original as $t \to \infty$. By replacing the first factor in either of the integrands by $i_{kn\mathcal{J}}(\infty)$ and $q_{kn\mathcal{J}}(\infty)$ respectively, and then integrating the second factors, we obtain as a result $4V(\infty) - 4T(\infty)$, if the definitions (7) and (8) are accounted for. Since we assumed the system to be passive this limit necessarily exists, and it then follows from the Abel theorem that the limit above is equal to the limit of the left-hand side of (19) as $p \to 0$, that is, $2A'_{kk}(0)$. This completes the proof of (17).

The result just obtained can also be interpreted somewhat differently as follows. With the help of the elementary rules it is easy to derive from

$$A_{kk}(p) \doteqdot i_{kk\mathcal{J}}(t) \quad (\operatorname{Re} p > \operatorname{Re} p_0),$$

the relations

$$A'_{kk}(p) \doteqdot -\int_{-\infty}^{t} \tau\, i'_{kk\mathcal{J}}(\tau)\, d\tau = \int_{-\infty}^{t} i_{kk\mathcal{J}}(\tau)\, d\tau - t i_{kk\mathcal{J}}(t) \quad (\operatorname{Re} p > 0).$$

If to this Abel's theorem is applied then (17) is found equal to

$$\lim_{t \to \infty} \left\{ \int_{-\infty}^{t} i_{kk\mathcal{J}}(\tau)\, d\tau - t i_{kk\mathcal{J}}(t) \right\},$$

of which the first term represents the total energy $\mathfrak{A}$ actually delivered to the system by the unit voltage $e_k(t) = U(t)$ up to the time t, whilst the second term indicates the fictitious work $\mathfrak{B}$ that would have been done since $t = 0$ if the system were always in its final situation, corresponding to $t = \infty$ (this second term comes down to heat dissipation). Consequently, (17) may also be formulated as follows:

$$\mathfrak{A} - \mathfrak{B} = 2V(\infty) - 2T(\infty) = A'_{kk}(0). \tag{20}$$

The above theorem was first stated, without proof, by Heaviside†. It was later on derived by Lorentz‡ in a direct manner from Maxwell's equations, thus without introducing the principles of quasi-stationary theory. Van der Pol§ stated the connexion with the frequency admittance A_{kk}.

If, instead of (18), the analogous identity for

$$\frac{1}{p}\{A_{kk}(p) + A_{kk}(-p)\}$$

is taken as starting-point, then only the resistance terms pertain, while the influence of both self-inductance and capacitance disappears. By similar reasoning we find

$$A_{kk}(+0) = i_{kk\mathcal{J}}(\infty) = 2F(\infty). \tag{21}$$

† *Electrical Papers*, London, 1892, vol. II, 412.

‡ *Collected Works*, The Hague, 1936, vol. III, 331; *Proc. Nat. Acad. Sci., Wash.*, VIII, 333, 1922.

§ 'A new theorem on electrical networks', *Physica*, 's-Grav., IV, 585, 1937.

6. Circuit equations with initial conditions

In § 4 of the preceding chapter we discussed how to account for the initial conditions at $t = 0$ to which the required solution of a differential equation with constant coefficients may be subjected. A new differential equation was derived, namely, that for $h^*(t) = h(t)\,U(t)$. The situation is quite the same in the case of a *set* of such equations. We shall confine ourselves again to the equations of an electric circuit. Thus, let us ask for those solutions of (2) or (5) for which the currents $i_n(t) = q_n'(t)$, and the charges $q_n(t)$ have prescribed values at $t = 0$. Introducing the new functions

$$i_n^*(t) = i_n(t)\,U(t) = q_n'(t)\,U(t),$$

we then obtain, in view of (2),

$$\sum_{m=1}^{N}\left(L_{nm}\frac{d}{dt}+R_{nm}+\frac{1}{C_{nm}}\int_{-\infty}^{t}dt\right)i_m^*$$
$$= e_n\,U(t)+\delta(t)\sum_{m=1}^{N}L_{nm}i_m(0)-U(t)\sum_{m=1}^{N}\frac{q_m(0)}{C_{nm}} \quad (n = 1,2,\ldots,N). \tag{22}$$

To solve (22) operationally we assume that the following relations

$$i_n^*(t) = i_n(t)\,U(t) \fallingdotseq I_n^*(p), \quad e_n(t)\,U(t) \fallingdotseq E_n^*(p),$$

together with those corresponding to the first-order derivatives of the first set, have a common strip of convergence. We then find, instead of (11), that

$$\sum_{m=1}^{N}Z_{nm}(p)\,I_m^*(p) = E_n^*(p)+p\sum_{m=1}^{N}L_{nm}i_m(0)-\sum_{m=1}^{N}\frac{q_m(0)}{C_{nm}} \quad (n = 1,2,\ldots,N), \tag{23}$$

or, more briefly, and remembering the definitions (7) and (8),

$$\sum_{m=1}^{N}Z_{nm}(p)\,I_m^*(p) = E_n^*(p)+p\left(\frac{\partial T}{\partial i_n}\right)_{t=0}-\left(\frac{\partial V}{\partial q_n}\right)_{t=0},$$

which has been given by Carson† in a slightly different notation. The last equations have left-hand sides completely equal to those of (11); they can therefore be solved in terms of the former admittance coefficients $A_{nm}(p)$ of § 3. So we find that

$$I_n^*(p) = \sum_{m=1}^{N}A_{nm}(p)\left\{E_m^*(p)+p\left(\frac{\partial T}{\partial i_m}\right)_0-\left(\frac{\partial V}{\partial q_m}\right)_0\right\}. \tag{24}$$

We have now arrived at the same situation as in VIII, § 4, inasmuch as the operational solution splits up into two parts, viz.

(*a*) $$i_n^{(1)}(t)\,U(t) \fallingdotseq f_n^{(1)}(p) = \sum_{m=1}^{N}A_{nm}(p)\,E_m^*(p).$$

† *Bell. Syst. Tech. J.* xv, 340, 1936.

This part is characteristic for the switching-on phenomena, since it represents the situation as time goes on for a system that was at rest before $t = 0$.

$$(b) \qquad i_n^{(2)}(t)\,U(t) \doteq f_n^{(2)}(p) = \sum_{m=1}^{N} A_{nm}(p)\left\{p\left(\frac{\partial T}{\partial i_m}\right)_0 - \left(\frac{\partial V}{\partial q_m}\right)_0\right\}.$$

This second part describes the typical switching-off phenomenon, which occurs if at $t = 0$ all sources of voltage are short-circuited, the further course of the system being fully determined by the values of $i_m(0)$ and $q_m(0)$. The last phenomenon is, in general, according to Heaviside's second expansion theorem, a linear combination of 'eigenstates' r, each of which represents a situation where all the currents and voltages have a common time factor $e^{p_r t}$. The constants p_r are the zeros of the determinant $|Z_{nm}(p)|$ of (11); the characteristic frequencies are $\omega_r = p_r/i$.

7. Ladder networks; filters

In the operational form of Kirchhoff's network equations, (11), all the mesh currents are present simultaneously. One usually confines oneself to special types of networks, however; in particular, to the *ladder networks* and the *lattice networks*, both satisfying the condition that any mesh has never more than two neighbours. Consequently, any equation involves at most three mesh currents. The scheme of a general ladder network is presented in fig. 59, in which the elements Z_n and $Q_{n+\frac{1}{2}}$ denote arbitrary impedances.

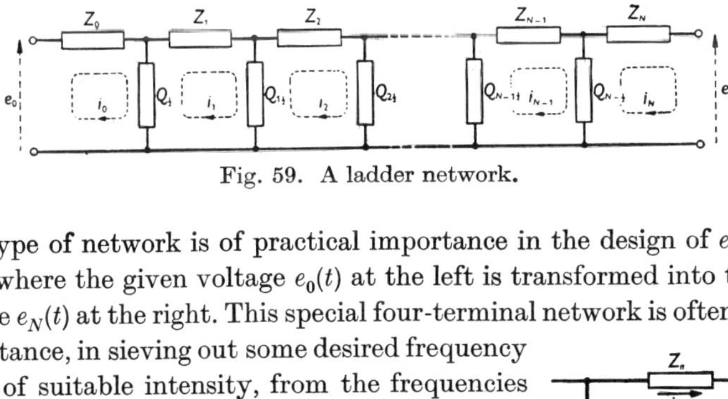

Fig. 59. A ladder network.

This type of network is of practical importance in the design of *electrical filters* where the given voltage $e_0(t)$ at the left is transformed into the new voltage $e_N(t)$ at the right. This special four-terminal network is often useful, for instance, in sieving out some desired frequency band, of suitable intensity, from the frequencies occurring in the original input signal.

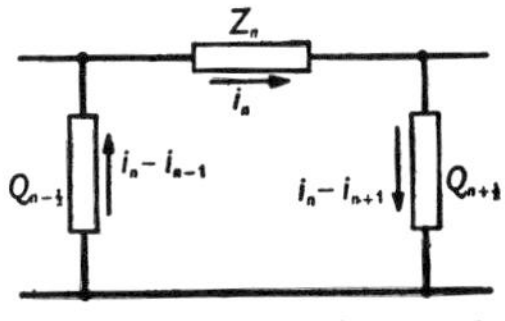

Fig. 60. An interior mesh of a ladder network.

In contrast with the custom in filter theory, which is based on the concepts of image impedance, iterative impedance, and the like, we shall here derive our results in a straightforward manner, directly from the Kirchhoff equations. Concerning the interior meshes (see fig. 60), the equations (11) reduce to

$$Z_n I_n + Q_{n+\frac{1}{2}}(I_n - I_{n+1}) + Q_{n-\frac{1}{2}}(I_n - I_{n-1}) = 0 \quad (n = 1, 2, \ldots, N-1), \qquad (25)$$

whilst the zeroth and the Nth mesh lead to different equations containing the images $E_0(p)$ and $E_N(p)$ of the electromotive forces $e_0(t)$ and $e_N(t)$, if, for the sake of symmetry, we assume electromotive forces impressed at both sides of the circuit. The images of the mesh currents $I_n(p)$ are thus to be obtained in terms of E_0, E_N, Z_n, $Q_{n+\frac{1}{2}}$, from the following system of equations:

$$\left.\begin{aligned}
&(Z_0+Q_{\frac{1}{2}})\,I_0 \quad -Q_{\frac{1}{2}}\,I_1 &&= E_0,\\
&-Q_{\frac{1}{2}}\,I_0+(Q_{\frac{1}{2}}+Q_{1\frac{1}{2}}+Z_1)\,I_1-Q_{1\frac{1}{2}}\,I_2 &&= 0,\\
&\qquad -Q_{1\frac{1}{2}}\,I_1+(Q_{1\frac{1}{2}}+Q_{2\frac{1}{2}}+Z_2)\,I_2-Q_{2\frac{1}{2}}\,I_3 &&= 0,\\
&\qquad\vdots\\
&-Q_{N-1\frac{1}{2}}\,I_{N-2}+(Q_{N-1\frac{1}{2}}+Q_{N-\frac{1}{2}}+Z_{N-1})\,I_{N-1}-Q_{N-\frac{1}{2}}\,I_N &&= 0,\\
&\qquad -Q_{N-\frac{1}{2}}\,I_{N-1}+(Q_{N-\frac{1}{2}}+Z_N)\,I_N &&= -E_N.
\end{aligned}\right\} \tag{26}$$

In order better to survey this set of $N+1$ equations for the $N+1$ unknowns, we may write them in the form of a single difference equation of the second order (cf. chapter XIII), viz.

$$\Delta_n\{Q_n(p).\Delta_n I_n(p)\}-Z_n(p)\,I_n(p) = -\phi_n(p) \quad (n = 0,1,2,\dots,N), \tag{27}$$

in which the operator Δ_n, frequently to be used later on, is defined by

$$\Delta_n f(n) = f(n+\tfrac{1}{2})-f(n-\tfrac{1}{2}). \tag{28}$$

The right-hand member, $-\phi_n$, of (27) is zero except at the end-points $n = 0$ and $n = N$, where it is equal to $-E_0$ and $+E_N$ respectively, if $Q_{-\frac{1}{2}}$ and $Q_{N+\frac{1}{2}}$ are put equal to zero.

Ladder networks are also closely connected with continued fractions. This is most easily seen from the practical example of a filter terminated by some impedance Z' so that $E_N = Z'I_N$. When viewed from the left, the circuit is seen to act as a two-terminal network with driving-point impedance

$$\frac{E_0}{I_0} = Z_0+\frac{1|}{|\dfrac{1}{Q_{\frac{1}{2}}}}+\frac{1|}{|Z_1}+\frac{1|}{|\dfrac{1}{Q_{1\frac{1}{2}}}}+\frac{1|}{|Z_2}+\dots+\frac{1|}{|Z_{N-1}}+\frac{1|}{|\dfrac{1}{Q_{N-\frac{1}{2}}}}+\frac{1|}{|Z_N+Z'}. \tag{29}$$

This formula can be derived either from (26) or more directly as follows. The circuit viewed from the terminals of e_0 is evidently a series connexion of Z_0 with the remainder; this remainder in turn is simply a parallel connexion of the admittance $1/Q_{\frac{1}{2}}$ and that formed by the series connexion of Z_1 and the new remainder of the network behind Z_1; and so on.

In this way any continued fraction with partial numerators equal to 1, and denominators that can be interpreted alternately as impedance and admittance, can represent the driving-point impedance of some ladder network which is easily constructed. In particular, a non-terminating ladder network may correspond to an infinite continued fraction. To exemplify we mention

$$\sqrt{\left(\frac{C}{R}p\right)}\tanh\{\sqrt{(RCp)}\} = \frac{1|}{\left|\frac{1}{Cp}\right.} + \frac{1|}{\left|\frac{3}{R}\right.} + \frac{1|}{\left|\frac{5}{Cp}\right.} + \frac{1|}{\left|\frac{7}{R}\right.} + \dots \doteqdot \frac{U(t)}{R}\theta_2\left(0, \frac{\pi t}{RC}\right), \quad (30)$$

whose p-function is the frequency admittance I_0/E_0 of a filter whose elements are the capacitances $1/pZ_n = C/(4n+1)$ and the resistances

$$Q_{n+\frac{1}{2}} = R/(4n+3).$$

The original of the continued fraction, involving the θ-function (see for its derivation formulae 4 and 11 of chapter XI), is just the time admittance; it therefore represents the current entering the circuit at the left if the unit voltage $U(t)$ is applied at the entrance.

The converse problem is also important. Starting with some rational function for a given admittance one may ask whether it is realizable by a two-terminal ladder network. The answer is often in the affirmative†, owing to the existence of identities such as

$$\frac{d_0 p^n + d_2 p^{n-2} + d_4 p^{n-4} + \dots}{d_1 p^{n-1} + d_3 p^{n-3} + d_5 p^{n-5} + \dots} = a_1 p + \frac{1|}{|a_2 p} + \frac{1|}{|a_3 p} + \dots + \frac{1|}{|a_n p},$$

$$\frac{d_0 p^{n/2} + d_2 p^{n/2-1} + d_4 p^{n/2-2} + \dots + d_n}{d_1 p^{n/2} + d_3 p^{n/2-1} + d_5 p^{n/2-2} + \dots + d_{n-1} p} = a_1 + \frac{1|}{|a_2 p} + \frac{1|}{|a_3} + \frac{1|}{|a_4 p} + \dots + \frac{1|}{|a_n p},$$

in which

$$a_1 = \frac{d_0}{d_1}, \quad a_k = \frac{D_{k-1}^2}{D_{k-2} D_k} \quad (k = 2, 3, \dots, n-1), \quad a_n = \frac{D_{n-1}}{d_n D_{n-2}};$$

$$D_0 = 1,$$

$$D_k = \begin{vmatrix} d_1 & d_0 & 0 & 0 & 0 & 0 & \\ d_3 & d_2 & d_1 & d_0 & 0 & 0 & \\ d_5 & d_4 & d_3 & d_2 & d_1 & d_0 & \\ & & & \ddots & & & \\ & & & & & & d_k \end{vmatrix},$$

n being an even number in the second identity.

For instance, the first identity leads to a ladder network with self-inductances corresponding to the series impedances $a_1 p$, $a_3 p$, $a_5 p$, ..., and capacitors corresponding to shunt impedances $1/(a_2 p)$, $1/(a_4 p)$, ... Whether or not such ladder networks are actually realizable depends on the coefficients a_k being positive or not.

† Cf. W. Bader, *Arch. Elektrotech.* XXXIV, 293, 1940.

In returning to the difference equation (27) we observe that this equation is linear in I_n with respect to n with coefficients in general depending on n. Of course, the solution is most easily found if both the series impedances Z_n and the shunt impedances $Q_{n+\frac{1}{2}}$ are independent of n (homogeneous filter). If, moreover, Z and Q denote these constant impedances respectively, the solution appears as simple as possible if the two boundary impedances Z_0 and Z_N are taken equal to $\frac{1}{2}Z$ instead of Z. In this case the solution of (26) or (27) becomes

$$I_n(p) = \frac{E_0 \cosh\{2(N-n)\,\alpha\} - E_N \cosh(2n\alpha)}{\sqrt{(QZ+\frac{1}{4}Z^2)}\sinh(2N\alpha)}, \tag{31}$$

in which

$$\alpha = \operatorname{arc\,sinh}\left\{\sqrt{\left(\frac{Z}{4Q}\right)}\right\} = i\operatorname{arc\,sin}\left\{\sqrt{\left(-\frac{Z}{4Q}\right)}\right\} \quad (\operatorname{Re}\alpha > 0). \tag{32}$$

In addition, if the filter is made a two-terminal network (observed from the left) by applying the terminal impedance $Z' = E_N/I_N$, E_N can be eliminated by means of (31) for $n = N$. The final expression for the image of the nth mesh current is then found to be

$$I_n(p) = \frac{E_0}{\sqrt{(QZ+\frac{1}{4}Z^2)}} \frac{\{e^{-2n\alpha} + R\,e^{-2(2N-n)\,\alpha}\}}{(1-R\,e^{-4N\alpha})}, \tag{33}$$

in which

$$R(p) = \frac{\sqrt{(QZ+\frac{1}{4}Z^2)} - Z'}{\sqrt{(QZ+\frac{1}{4}Z^2)} + Z'}. \tag{34}$$

As in the general case, $A(p) = I_n/E_0$ not only determines the ratio between the *images* of the current in the nth mesh and the voltage, but also the ratio between the quantities *themselves* if both are proportional to e^{pt}. If then p is replaced by $i\omega$, we obtain the familiar expression for the admittance of the homogeneous low-pass filter of alternating-current technique. It is also possible to consider the p-function I_n/E_0 as being a transfer frequency admittance.

Owing to its simplicity, formula (33) can be interpreted physically in two different ways, in virtue of two distinct expansions.

(1) We first consider the geometric series

$$I_n = \frac{E_0}{\sqrt{(QZ+\frac{1}{4}Z^2)}}\left\{\sum_{k=0}^{\infty} R^k\,e^{-2\alpha(n+2kN)} + \sum_{k=0}^{\infty} R^{k+1}\,e^{2\alpha(n-2kN-2N)}\right\}. \tag{35}$$

This development shows that the currents produced by the alternating electromotive force $E_0 = e_0\,e^{pt}$ can be looked upon as an assembly of travelling waves. For, if the exponential factors of (35) are multiplied by the time factor $e^{pt} = e^{i\omega t}$, the separate terms become proportional to

$$e^{-2\operatorname{Re}\alpha\,(n+2kN)}\,e^{i\{\omega t - 2\operatorname{Im}\alpha\,(n+2kN)\}},$$

$$e^{2\operatorname{Re}\alpha\,(n-2kN-2N)}\,e^{i\{\omega t + 2\operatorname{Im}\alpha\,(n-2kN-2N)\}}.$$

In the order given, these factors correspond to waves travelling through the filter to the right and to the left respectively. The velocity of propagation is, distances being measured in units of cell lengths, $v = \frac{1}{2}\omega/\mathrm{Im}\,\alpha$; the amplitude is decreased by a factor $e^{-2\,\mathrm{Re}\,\alpha}$ when the wave passes from one mesh to the next. Moreover, the successive terms of (35) have phase differences equal to $4N\,\mathrm{Im}\,\alpha = 2N\omega/v$, corresponding to the transit time required for a wave to travel twice the whole system of N cells, in the forward and backward directions successively. Obviously it is just as if in the filter a travelling wave had started at the left (travelling to the right; the term with $k = 0$ in the first series of (35)), had then been reflected at the right (producing a wave travelling to the left; $k = 0$ in the second series of (35)), followed by a reflexion at the left, and so on and on. In this picture the loss in amplitude after each reflexion at the right is accounted for by a factor R, since after k successive reflexions the amplitude is R^k. Therefore, R of (34) has the meaning of a coefficient of reflexion.

(2) Another interpretation is obtained according to Heaviside's second expansion theorem by means of a development in partial fractions (a finite value of $A(\infty)$ is assumed), viz.

$$\frac{A_n(p)}{p} = \frac{I_n}{pE_0} = \Sigma \frac{A_{n,r}}{p-p_r}.$$

The original of this series (whether or not the latter terminates) represents the current due to the unit voltage in the form of a combination of eigenfunctions $e^{p_r t}$. The corresponding characteristic frequencies $\omega_r = p_r/i$ are obtainable from the zeros of the denominator of (33); they therefore depend strongly upon the terminal impedance Z'.

There are, of course, many other systems performing small vibrations that allow of either of the physical interpretations given above, for instance, vibrations of a string. Our first interpretation in this case corresponds to D'Alembert's solution of the differential equation which describes the motion of the string in terms of travelling waves reflected forwards and backwards at the ends. The second interpretation corresponds to Bernoulli's description in which the solution is expanded into a Fourier series.

The solution of our filter problem is particularly simple if the coefficient of reflexion happens to be zero, since then only one wave of (35) remains, travelling to the right; in this case the filter acts as if it were infinitely long ($N \to \infty$). But this is only possible if

$$Z' = \sqrt{(QZ + \tfrac{1}{4}Z^2)},$$

that is, if the filter is terminated by some transcendental function, which is never realizable rigorously by means of a finite number of self-inductances, resistances and capacitances.

Let us now discuss an example of a homogeneous, and another of an inhomogeneous, filter of the ladder-type network.

Example 1. The theory of ladder networks is most conveniently illustrated by the common low-pass filter (fig. 61) in which both series and shunt impedances are constant except that the former is halved at the end-points. In this case the series impedances are inductances L, and the shunt impedances are capacitances C. Consequently, we have to take

$$Z = Lp, \quad Q = \frac{1}{Cp}, \quad \alpha = \operatorname{arc\,sinh}\left(\frac{p}{2}\sqrt{(LC)}\right).$$

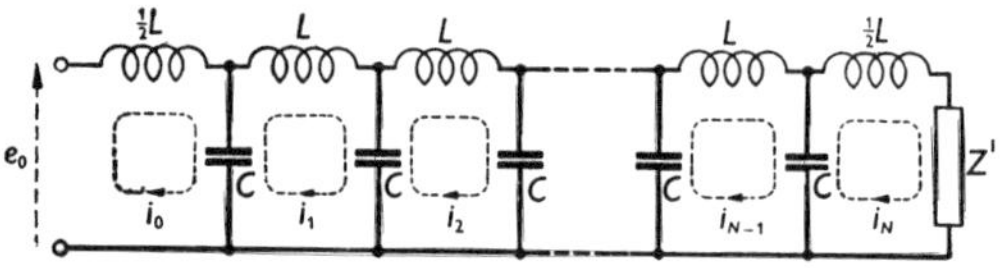

Fig. 61. The homogeneous low-pass filter.

If the filter is infinitely long, or terminated by the impedance

$$\sqrt{\left(QZ+\frac{Z^2}{4}\right)} = \sqrt{\left(\frac{L}{C}+\frac{L^2p^2}{4}\right)}, \tag{36}$$

we must take R equal to zero; the transfer admittance with respect to the nth mesh current then follows from

$$A_n = \frac{I_n}{E_0} = \frac{e^{-2n\alpha}}{\sqrt{(QZ+\frac{1}{4}Z^2)}} = \frac{\sqrt{\frac{C}{L}}}{\sqrt{\left(\frac{p^2}{\omega_0^2}+1\right)}}\left\{\sqrt{\left(\frac{p^2}{\omega_0^2}+1\right)}-\frac{p}{\omega_0}\right\}^{2n} \quad \left(\omega_0 = \frac{2}{\sqrt{(LC)}}\right). \tag{37}$$

The expression clearly indicates that the low-pass filter especially lets through the *low* frequencies occurring in the electromotive force impressed at the left. For $p/i = \omega < \omega_0$ the factor $\{\ \}^{2n}$ has unit absolute value and therefore the currents in different cells have equal absolute values. But if $\omega > \omega_0$ the absolute value of the current decreases exponentially with n, thus giving at the right a current that is almost zero for a sufficiently long filter. Consequently ω_0 may be taken as the cut-off frequency of the low-pass filter.

The original of the p-function (37) is the time admittance with respect to the nth current, that is to say, the current through the nth self-inductance caused by the unit voltage $e_0(t) = U(t)$ at the left side of the filter. Furthermore, upon multiplication of the image by p, the new original becomes equal to the current produced by the electromotive force $\delta(t)$. Anticipating the result (x, 12) of the next chapter, involving a Bessel function, we then find that

$$i_{n,\delta}(t) = \frac{2}{L} J_{2n}(\omega_0 t)\, U(t). \tag{38}$$

By a Fourier analysis of this Bessel function it is easily shown that it contains only frequencies between 0 and ω_0. Since any arbitrary voltage $e_0(t)$ can be built up by impulse functions, it is quite understandable why the frequencies between 0 and ω_0 predominate in the response current. As to the property that the low-pass filter does not suppress the frequencies $\omega > \omega_0$ completely, this is due to the fact that the latter frequencies, though absent in $J_{2n}(\omega_0 t)$, are present in the cut-off function $J_{2n}(\omega_0 t)\, U(t)$; however, the amplitudes of these frequencies are smaller than those for $\omega < \omega_0$. The functions $|A_n(i\omega)|$ and $i_{n,\delta}(t)$ are shown in fig. 62.

In passing we note that the homogeneous low-pass filter is easily transformed, by a limiting process, into a coaxial, loss-free, cable of self-inductance $\bar{L}$ and capacitance

$\overline{C}$, both per unit of length (xv, § 4). To see this we need only take the number of cells per unit of length in the filter equal to N, and then let N grow indefinitely. In order to get the current at the distance x from the left end of the cable, we have to put $L = \overline{L}/N$, $C = \overline{C}/N$, $n = Nx$, so as to obtain

$$\omega_0 = N\overline{\omega}_0 = \frac{2N}{\sqrt{(\overline{L}\overline{C})}},$$

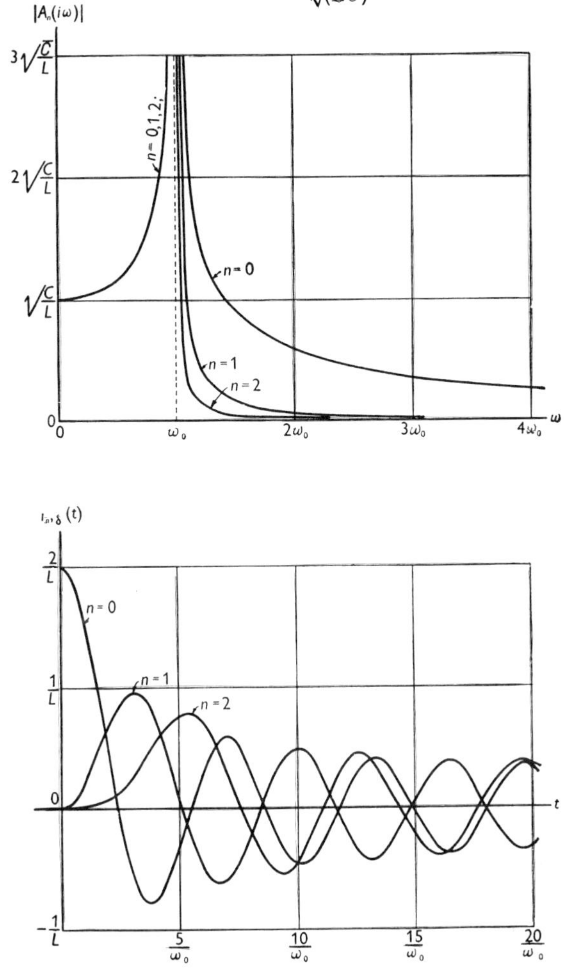

Fig. 62. Transfer admittances and responses to an impulse e.m.f. for the homogeneous low-pass filter.

and then from (37) as $N \to \infty$,

$$A(x) = \sqrt{\left(\frac{\overline{C}}{\overline{L}}\right)} \lim_{N\to\infty} \left\{\sqrt{\left(\frac{p^2}{N^2\overline{\omega}_0^2}+1\right)} - \frac{p}{N\overline{\omega}_0}\right\}^{2Nx} = \sqrt{\left(\frac{\overline{C}}{\overline{L}}\right)} e^{-2px/\overline{\omega}_0}.$$

Again, multiplication of this image by p and transposition into the t-language together lead to

$$i_\delta(t) = \sqrt{\left(\frac{\overline{C}}{\overline{L}}\right)} \delta\left(t - \frac{2x}{\overline{\omega}_0}\right), \tag{39}$$

which is the response current in the cable due to the impulsive voltage $\delta(t)$. Further details of cables will be postponed till later. The gradual transition for $N \to \infty$ of the low-pass filter into a cable is also evident from a consideration of (38) for small t. In fact, the numerical behaviour of J_{2n} shows that $i_{n,\delta}$ has no appreciable magnitude before the moment given about by $\omega_0 t \sim 2n$ or $t \sim 2n/\omega_0 = 2x/\varpi_0$ (see fig. 63). It signifies that the first observable effect of the impulse is moving along the filter with a velocity $\frac{1}{2}\varpi_0$; this is in complete agreement with the velocity of propagation of the sharply defined impulse existing according to (39) in the case of a cable.

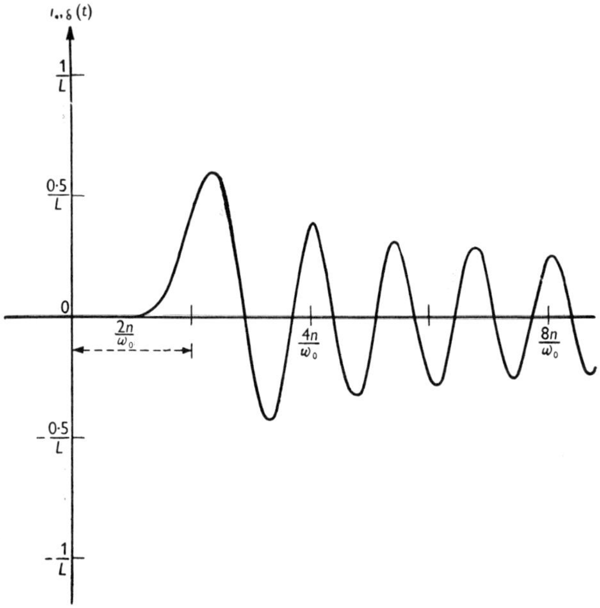

Fig. 63. Showing the retarded response to an impulse e.m.f. for the homogeneous low-pass filter.

We now return to the low-pass filter under consideration before. It is easy to obtain from (38) a relation fulfilled by the Bessel function. Since the image $I_n = pA_n(p)$ of the function (38) must satisfy the difference equation (27), we have for $n \neq 0$ that

$$\Delta_n^2 I_n - \frac{4p^2}{\omega_0^2} I_n = 0.$$

Consequently the Bessel function $J_{2n}(\omega_0 t) = J_{2n}(2x)$ of (38) itself is a solution of the corresponding t-equation; that means

$$\left(\Delta_n^2 - \frac{d^2}{dx^2}\right) J_{2n}(2x) = 0. \tag{40}$$

This difference-differential equation is characteristic† for the Bessel function $J_{2n}(2x)$ in its dependence upon the order n and the argument x.

Just as this property of Bessel functions was found by the special choice (36) as to the terminal impedance Z', we may derive other interesting properties by another

† Cf. Balth. van der Pol, 'Discontinuous phenomena in radio communication', *J. Inst. Elect. Engrs, London*, LXXXI, 381, 1937.

suitable function Z'. In this respect the Bessel-function approximations of Nijenhuis† should be mentioned. They are found by taking Z' equal to zero; it then follows from (33) that the impulse voltage $e_0(t) = \surd(L/C)\,\delta(t)$ causes a current in the nth mesh whose image is given by

$$i_{n,\delta}(t) \doteq \frac{p}{\surd(p^2+1)} \frac{\{\surd(p^2+1)+p\}^{2N-2n}+\{\surd(p^2+1)-p\}^{2N-2n}}{\{\surd(p^2+1)+p\}^{2N}-\{\surd(p^2+1)-p\}^{2N}}, \tag{41}$$

if for the sake of simplicity ω_0 is taken equal to 1. Let us now apply the preceding theory. The analogue of D'Alembert's method (development into a geometric series) leads to the image

$$i_{n,\delta}(t) \doteq \sum_{k=0}^{\infty} \frac{p}{\surd(p^2+1)}\{e^{-(2n+4kN)\operatorname{arc\,sinh} p}+e^{(2n-4kN-4N)\operatorname{arc\,sinh} p}\},$$

whence it follows with the aid of (x, 12) for the original

$$i_{n,\delta}(t) = U(t)\sum_{k=0}^{\infty}\{J_{2n+4kN}(t)+J_{-2n+4kN+4N}(t)\}. \tag{42}$$

As to Bernoulli's method we need the expansion in rational fractions whereby the zeros of the denominator of (41) play a part. In this case the zeros lie on the imaginary axis at

$$p_r = i\sin\left(\frac{\pi r}{2N}\right) \quad (r = \text{integer}).$$

Accordingly we easily obtain

$$i_{n,\delta}(t) \doteq \frac{(-1)^n}{4N}\left(\frac{p}{p+i}+\frac{p}{p-i}\right)+\frac{1}{2N}\sum_{r=-N+1}^{N-1}\cos\left(\frac{\pi n r}{N}\right)\frac{p}{p-i\sin(\pi r/2N)}.$$

Combining the fractions belonging to conjugate complex zeros the original can be written as follows:

$$i_{n,\delta}(t) = \frac{U(t)}{2N}\sum_{r=-N+1}^{N}\cos\left(\frac{\pi n r}{N}\right)\cos\left\{t\sin\left(\frac{\pi r}{2N}\right)\right\}. \tag{43}$$

Finally, identifying the right-hand members of (42) and (43) we arrive at the remarkable equality

$$J_{2n}(t) = \frac{1}{2N}\sum_{r=-N+1}^{N}\cos\left(\frac{\pi n r}{N}\right)\cos\left\{t\sin\left(\frac{\pi r}{2N}\right)\right\}$$
$$-J_{2n+4N}(t)-J_{2n+8N}(t)-\ldots-J_{-2n+4N}(t)-J_{-2n+8N}(t)-\ldots.$$

It is worth noting that for suitable values of n, N, t, the Bessel functions at the right are negligible in comparison with the sum, which is favourable in the numerical calculation of the Bessel function‡.

Example 2. Wigge § has discussed a non-homogeneous filter (Z_n and $Q_{n+\frac{1}{2}}$ depending on n) having the special property that the characteristic frequencies $\omega_r = p_r/i$ are those of an arithmetic progression: ω_0, $3\omega_0$, $5\omega_0$, ... The series impedances are again self-inductances but they now depend on n, viz.

$$\frac{Z_n}{p} = L_n = 2L_0\frac{(N+1)(N+2)\ldots(N+n)}{N(N-1)\ldots(N-n+1)}\frac{(2N+1)(2N-1)\ldots(2N-2n+3)}{(2N+3)(2N+5)\ldots(2N+2n+1)}$$
$$(1 \leqslant n \leqslant N),$$

† W. Nijenhuis, 'On transients in homogeneous ladder networks of finite length', *Physica*, *'s-Grav.*, IX, 817, 1942.

‡ Cf. Nijenhuis, loc. cit.

§ *Z. Hochfrequenztech. Elektroak.* LVII, 101, 1941.

whilst, for $n = 0$, $Z_0 = L_0 p$. The capacitances, occurring in the form of shunt impedances, are given by

$$\frac{1}{pQ_{n+\frac{1}{2}}} = C_{n+\frac{1}{2}}$$
$$= \frac{1}{L_0\omega_0^2(N+1)(2N+1)} \frac{N(N-1)\dots(N-n+1)}{(N+2)(N+3)\dots(N+n+1)} \frac{(2N+3)(2N+5)\dots(2N+2n+1)}{(2N-1)(2N-3)\dots(2N-2n+1)}$$
$$(0 \leqslant n \leqslant N).$$

Fig. 64 illustrates the network. It can be shown, using equation (26), that for this special filter network the ratio of the image of the current (I_0) to the image of the voltage (E_0), both taken at the left-hand side, is determined by the following driving-point admittance:

$$A(p) = \frac{1}{L_0 p} \frac{(p^2+0^2\omega_0^2)(p^2+2^2\omega_0^2)\dots\{p^2+(2N)^2\omega_0^2\}}{(p^2+1^2\omega_0^2)(p^2+3^2\omega_0^2)\dots\{p^2+(2N+1)^2\omega_0^2\}}. \tag{44}$$

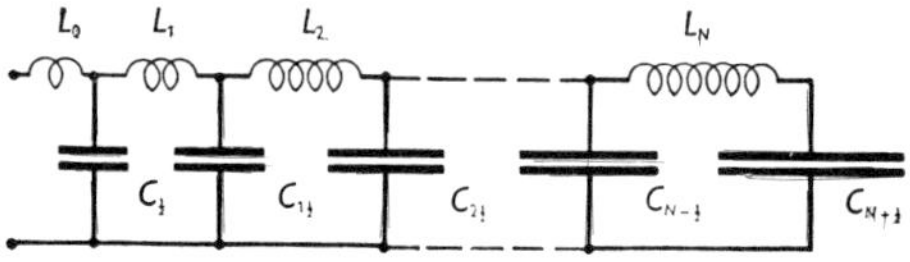

Fig. 64. An inhomogeneous low-pass filter.

As before, the original of $pA(p)$ provides us with the input current $i_\delta(t)$ in response to the impulse voltage $\delta(t)$. The operational relation (XIII, 37(b)) then easily leads to

$$i_\delta(t) = \frac{U(t)}{L_0} P_{2n+1}\{\cos(\omega_0 t)\}, \tag{45}$$

in which P denotes the Legendre polynomial. Although the elements of the filter under discussion depend on n in a rather complicated manner, yet it has some simple properties, e.g. the characteristic frequencies are harmonic and real, and the transients are simple expressions involving Legendre polynomials.

8. Lattice networks

A second type of network frequently encountered in practice is the *lattice network*, where again the Kirchhoff equations involve no more than three mesh currents, just as for ladder networks. A lattice network consists of N consecutive cells, each of which contains a pair of equal impedances Z_n and a second pair of equal impedances Z'_n that are interconnected

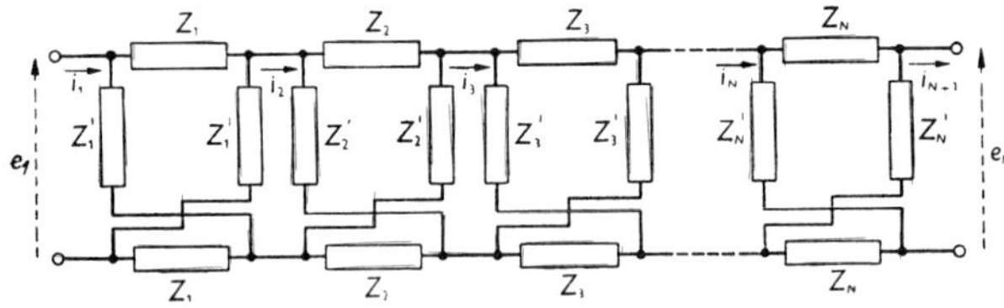

Fig. 65. A lattice network.

according to the scheme of fig. 65. Although the number of meshes of this network is $2N+1$, yet it follows from the symmetry that it is sufficient to account for $N+1$ independent mesh currents, of which the nth, i_n, flows through the connexion of two consecutive cells. To see this let us consider the nth and the $(n-1)$th cell, while supposing the other cells not present (see fig. 66). The current i_n through the connexions of both cells can be split up into two parts; the first part follows the single arrows, the second the double arrows. In their respective ways both parts meet the same impedances; they are therefore both equal to $\frac{1}{2}i_n$. If then the whole lattice network is built up by joining together the pairs formed by the first and the second cell, the second and the third cell, etc., then it is easily verified that the series impedances Z_n are all traversed by the current $\frac{1}{2}i_n+\frac{1}{2}i_{n+1}$, and similarly the shunt impedances Z'_n by the current $\frac{1}{2}i_n-\frac{1}{2}i_{n+1}$. Therefore the system of equations (11) for the individual meshes reads as follows:

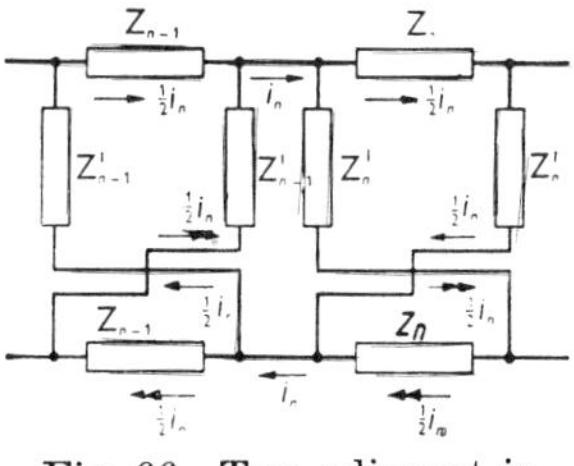

Fig. 66. Two adjacent interior meshes of a lattice network.

$$\left.\begin{aligned}
&(Z_1+Z'_1)\,I_1+(Z_1-Z'_1)\,I_2 &&= 2E_1,\\
&(Z_1-Z'_1)\,I_1+(Z_1+Z'_1+Z_2+Z'_2)\,I_2+(Z_2-Z'_2)\,I_3 &&= 0,\\
&\quad(Z_2-Z'_2)\,I_2+(Z_2+Z'_2+Z_3+Z'_3)\,I_3+(Z_3-Z'_3)\,I_4 &&= 0,\\
&\qquad\cdots\\
&\quad(Z_{N-1}-Z'_{N-1})\,I_{N-1}+(Z_{N-1}+Z'_{N-1}+Z_N+Z'_N)\,I_N\\
&\qquad\qquad\qquad\qquad+(Z_N-Z'_N)\,I_{N+1} &&= 0,\\
&\qquad\qquad(Z_N-Z'_N)\,I_N+(Z_N+Z'_N)\,I_{N+1} &&= -2E_{N+1}.
\end{aligned}\right\} \quad (46)$$

Again, the algebraic equations determining the image functions I_n can be summarized in a single difference equation, namely,

$$\Delta_n\{Z'_{n-\frac{1}{2}}(p)\,\Delta_n I_n(p)\}-\nabla_n\{Z_{n-\frac{1}{2}}(p)\,\nabla_n I_n(p)\} = -2\phi(n) \quad (n = 1, 2, \ldots, N+1), \quad (47)$$

where in addition to the difference operator Δ_n of (28), we also use the sum operator

$$\nabla_n f(n) = f(n+\tfrac{1}{2})+f(n-\tfrac{1}{2}). \quad (48)$$

As in the discussion of the ladder network, $\phi(n)$ is everywhere zero, except at the boundaries ($n = 1$ and $n = N+1$) where it assumes the values E_1 and $-E_{N+1}$ respectively, if it is understood that

$$Z_0 = Z'_0 = Z_{N+1} = Z'_{N+1} = 0.$$

The system (46) can be generally solved if, as usual, we assume that

$$Z_1 Z'_1 = Z_2 Z'_2 = \ldots = Z_N Z'_N = Z_0^2, \quad (49)$$

which in the conventional theory of filters means that the image impedance of any of the consecutive cells equals a fixed impedance Z_0. The simplest solution is obtained if in addition an extra impedance $\frac{1}{2}Z_0$ is added to the cells at the ends; the currents of the corresponding circuit, which is shown in fig. 67, are then determinable from

$$I_n = \frac{E_1}{2Z_0}\prod_{k=1}^{n-1}\left(\frac{Z_0-Z_k}{Z_0+Z_k}\right) - \frac{E_{N+1}}{2Z_0}\prod_{k=n}^{N}\left(\frac{Z_0-Z_k}{Z_0+Z_k}\right). \tag{50}$$

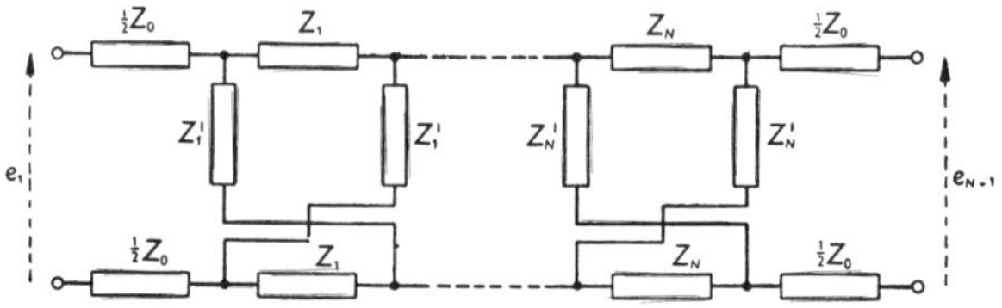

Fig. 67. A lattice network with suitably chosen end cells.

This formula can still be simplified if the source of voltage at the right is short-circuited ($E_{N+1} = 0$), which means that the filter at the right of the Nth cell is terminated by the common image impedance Z_0; the mutual admittance between the left end and the nth branch of the filter then becomes

$$A_n = \frac{I_n}{E_1} = \frac{1}{2Z_0}\prod_{k=1}^{n-1}\left(\frac{Z_0-Z_k}{Z_0+Z_k}\right) = \frac{1}{2Z_0}\prod_{k=1}^{n-1}\left(\frac{Z_k'-Z_0}{Z_k'+Z_0}\right). \tag{51}$$

The corresponding originals again represent the time admittances, that is to say, the currents in the successive branches as functions of the time, if at the left end of the filter a unit voltage is applied.

Whereas formula (51) enables us to build up lattice networks whose admittances consist of a *product* of simple functions, the latter networks are particularly useful in the realization of admittances given in the form of *series*. Moreover, by means of parallel connexions of the lattice networks discussed above it is possible to realize admittances that are represented by a sum of products of p-functions. It is even possible, by properly using an infinite number of elements, to arrive at the representation of convergent infinite products and series. In order to see how in this way many transcendental admittances can be realized by suitably chosen networks, it is worth while to consider the following examples.

Example 1. If we take, according to the diagram of fig. 67,

$$Z_0 = \sqrt{\frac{L}{C}}, \quad Z_k = \frac{k}{Cp}, \quad Z_k' = \frac{Lp}{k} \quad (k = 1, 2, \ldots, N),$$

the image impedance becomes a pure resistance; the series impedances are capacitances, decreasing with k and k^{-1}, and the shunt impedances are self-inductances, decreasing in just the same manner.

According to (51) the admittances of the corresponding circuit in which we assume $e_{N+1} = 0$ (see fig. 68) are given by

$$A_n = \frac{I_n}{E_1} = \frac{1}{2}\sqrt{\left(\frac{C}{L}\right)}\prod_{k=1}^{n-1}\left(\frac{p\sqrt{(LC)}-k}{p\sqrt{(LC)}+k}\right) \doteqdot \frac{\sqrt{C}}{2\sqrt{L}}\, P_{n-1}(-1+2e^{-t/\sqrt{(LC)}})\, U(t). \tag{52}$$

Obviously the network under consideration is dissipative; the admittance above has poles on the negative p-axis at $-1/\sqrt{(LC)}, -2/\sqrt{(LC)}, \ldots, -(n-1)/\sqrt{(LC)}$, and zeros on the positive p-axis at $1/\sqrt{(LC)}, 2/\sqrt{(LC)}, \ldots, (n-1)/\sqrt{(LC)}$. It is easily verified that the absolute value of A_n on the imaginary axis, that is to say for real frequencies ($p = i\omega$), is $\frac{1}{2}\sqrt{(C/L)}$. Consequently, though any undamped alternating signal in passing this filter undergoes a phase shift varying with frequency, the change in amplitude is independent of the frequency. The Legendre polynomials $P_1, P_2, \ldots, P_N$ are characteristic for the time admittances of the various branches. The right-hand member of (52), which may be derived from (XI, 35), is just the current in the nth branch due to the unit voltage $U(t)$ applied at the left of the filter.

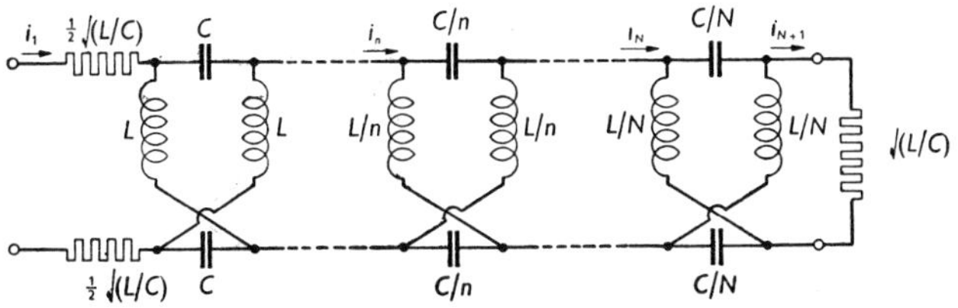

Fig. 68. Lattice network leading to Legendre polynomials.

Example 2. Another lattice network leads to Laguerre polynomials. Let us choose the elements so that

$$Z_1 = \frac{\lambda}{4p+1}, \quad Z_1' = \lambda(4p+1), \quad Z_0 = \lambda,$$

$$Z_2 = Z_3 = \ldots = Z_N = \frac{\lambda}{2p}, \quad Z_2' = Z_3' = \ldots = Z_N' = 2\lambda p.$$

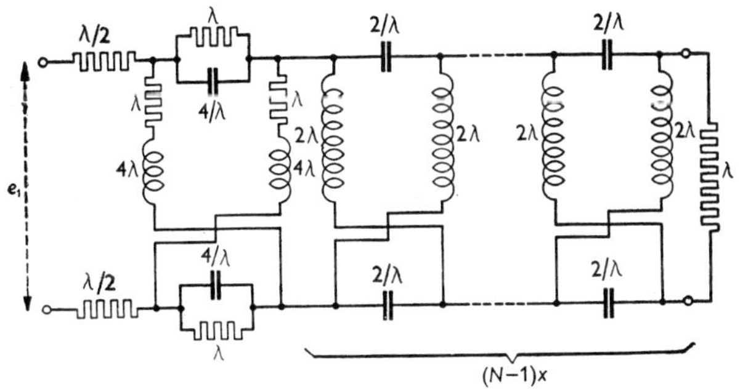

Fig. 69. Lattice network leading to Laguerre polynomials.

Then the series impedances consist of capacitances $2/\lambda$ and the shunt impedances of self-inductances 2λ, except those of the first cell which have values twice as large, whilst the latter are connected in parallel and in series, respectively, with a pure resistance λ. According to the scheme of fig. 69, which is another special case of fig. 67, the transfer admittance (51) becomes

$$A_n = \frac{p}{2\lambda} \frac{(p-\frac{1}{2})^{n-2}}{(p+\frac{1}{2})^{n-1}} \quad (n \neq 1).$$

The original of this p-function is the corresponding time admittance, which is found to be

$$i_{\Gamma}(t) = \frac{U(t)}{2\lambda(n-2)!} e^{-\frac{1}{2}t} L_{n-2}(t), \tag{53}$$

if use is made of (VI, 17) and the attenuation rule. Therefore, the currents in the successive branches are represented by the Laguerre polynomials $L_0, L_1, \ldots, L_{N-1}$, multiplied by the factor

$$U(t)\,e^{-\frac{1}{2}t}.$$

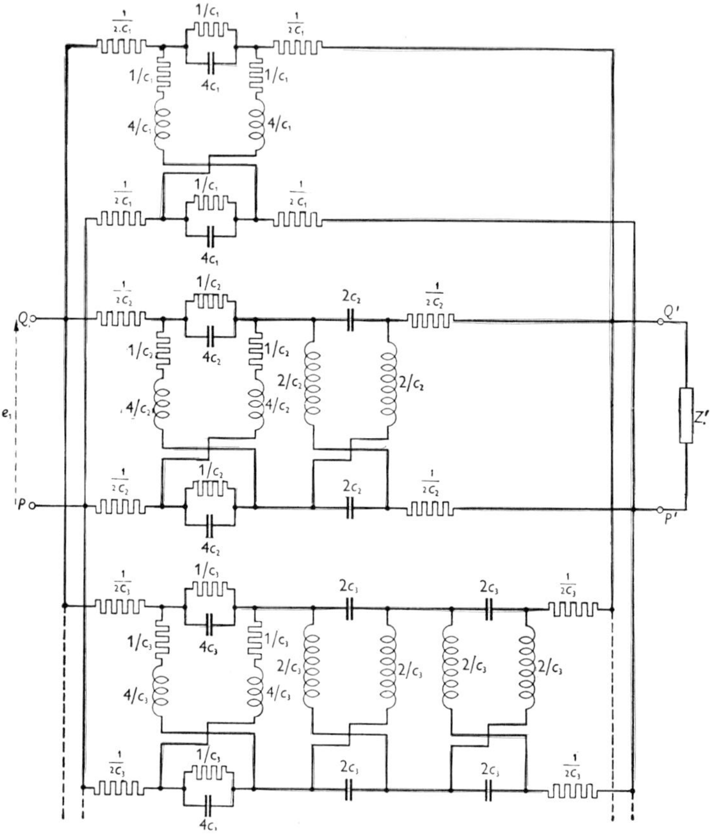

Fig. 70. Lattice network leading to a series expansion of Laguerre polynomials.

This is important with a view to the realization of networks having a prescribed time admittance. Since the very functions $e^{-\frac{1}{2}t}L_n(t)\,U(t)$, where n is a non-negative integer, are orthogonal for the interval $0 < t < \infty$, that is,

$$\int_0^\infty e^{-x} L_n(x)\,L_m(x)\,dx = \begin{cases} (n!)^2 & (n=m), \\ 0 & (n \neq m), \end{cases}$$

it is most convenient to assume the prescribed time admittance expanded as follows:

$$i_{\mathcal{J}}(t) = U(t) \sum_{N=1}^{\infty} \frac{c_N}{(N-1)!}\, e^{-\frac{1}{2}t} L_{N-1}(t),$$

in which the coefficients c_N are easily determined if $i_{\mathcal{J}}(t)$ is given. If now (see fig. 70) a parallel connexion is made† of the lattice filters above, which individually have

† A similar device was given by Y. W. Lee, *J. Math. Phys.* XI, 83, 1932, where, however, $i_\delta(t)$ instead of $i_{\mathcal{J}}(t)$ is expanded into orthogonal functions.

time admittances equal to $U(t)\, c_N\, \mathrm{e}^{-\frac{1}{2}t} L_{N-1}(t)$, then in the case of an arbitrary terminal impedance Z', the transfer time admittance (unit voltage across PQ, current through Z' between P' and Q') is determined by

$$i_{\Gamma}(t) = \mathrm{e}^{-\frac{1}{2}t} \frac{\dfrac{c_1}{0!} L_0(t) + \dfrac{c_2}{1!} L_1(t) + \dfrac{c_3}{2!} L_2(t) + \ldots}{2 + Z'(c_1 + c_2 + c_3 + \ldots)} \, U(t).$$

In terms of the p-language the device given above reduces to a possible realization of the transfer frequency admittance $A(p)$ if the latter is explicitly given in the form of

$$A(p) = \frac{p}{2 + Z'(c_1 + c_2 + \ldots)} \left\{ \frac{c_1}{(p+\frac{1}{2})} + c_2 \frac{(p-\frac{1}{2})}{(p+\frac{1}{2})^2} + c_3 \frac{(p-\frac{1}{2})^2}{(p+\frac{1}{2})^3} + \ldots \right\}.$$

The sum $c_1 + c_2 + c_3 + \ldots$ equals $i_{\Gamma}(0)$ as can be verified by expressing c_n as an integral according to its significance as the coefficient of the expansion with respect to the complete orthogonal functions $\mathrm{e}^{-t/2} L_{n-1}(t)$.

The diagram of fig. 70 is also suitable for negative c_n; the upper end at the right-hand side of the horizontal filter then has to be connected to P', and the lower end to Q'.

CHAPTER X

LINEAR DIFFERENTIAL EQUATIONS WITH VARIABLE COEFFICIENTS

1. Introduction

Just as linear differential equations with constant coefficients may often be solved by transposing them term by term, so also may the less simple differential equations with variable coefficients. But the resulting equation is not, as before, algebraic in p; on the contrary, it is also in general a linear differential equation with variable coefficients. Whether the latter is essentially simpler than the initial equation depends on the nature of the variable coefficients occurring in the original equation. All this is particularly easy to survey if, as below, we confine ourselves to coefficients that are rational functions of the independent variable. In this case the most general form of the differential equation is reducible to the following:

$$\sum_r \sum_s a_{rs} t^r \frac{d^s}{dt^s} h(t) = -\phi(t),$$

in which $a_{r,s}$ are constants, and r and s both run through a finite set of non-negative integers.

According to (IV, 38) and the differentiation rule, the equation above is transposed into

$$p \sum_r \sum_s a_{rs} \left(-\frac{d}{dp}\right)^r \{p^{s-1} f(p)\} = -\phi(p), \tag{1}$$

which is obviously of the same form as the initial differential equation. There is in general also a typical reciprocity. To explain this we first remind the reader that the maximum value of s in the original equation is usually referred to as the *order* of the differential equation. Secondly, let the maximum value of r therein be referred to as the 'degree' of the differential equation; this does not necessarily lead to confusion with the customary notion of this word, since the differential equations considered here are only of the linear type, i.e. of the first degree in the usual sense. With these definitions in mind, the reciprocity referred to above concerns a sufficient condition that by the process of operational transposition the degree and the order of the differential equations involved change into each other. The condition in question requires that a_{r0} should vanish for $r \neq 0$. This rule can easily be verified from the foregoing equations.

The operational method may be expected to be particularly useful if in the original equation the degree is smaller than the order, since then the p-equation is of lower order. It should be borne in mind, however, that the p-equation may admit solutions that have no original at all. On the other hand, in order that the method be successful the only requirement is that the t-solution in question should have an image.

It often happens that the solution $h(t)$ possesses an image only if first made one-sided: $h^*(t) = h(t)\,U(t)$. In this case we have first to derive a new equation, satisfied by $h^*(t)$, from that initially given to hold for $h(t)$; this procedure was carried out for differential equations with constant coefficients in VIII, § 4.

As to the determination of the constants of integration occurring in any solution of the p-equation fulfilling the side conditions, the Abel and Tauber theorems prove very useful.

To illustrate the efficiency of the operational calculus in dealing with the type of differential equations specified, we shall subsequently discuss the functions of Laguerre, Hermite, Bessel, Legendre, and the hypergeometric function which covers all these special functions. The general method is to deduce simple operational relations for the function itself when starting from the corresponding differential equation. If necessary, the function may be first multiplied by some other simple function. From these topics especially the heuristic value of the operational calculus is quite evident; not only is the given differential equation solved, whether formally or not, but also many properties of the solution become easily perceivable. It often happens that almost trivial relations between the images found represent corresponding relations between the originals themselves that are of essentially profound character; they could be deduced otherwise, but most probably only by means of a great amount of cumbersome calculation.

2. Laguerre polynomials

For these polynomials we already have the operational relation in (VI, 17). For a better understanding of the general operational procedure in the resolution of a differential equation, we shall derive the same relation once more but differently. We now only suppose that the underlying t-functions $L_n(t)$ are polynomials in t, and satisfy the differential equation

$$t\frac{d^2L_n}{dt^2}+(1-t)\frac{dL_n}{dt}+nL_n = 0. \tag{2}$$

We know in advance that the function L_n as polynomial has a simple image (that is, one without impulse functions) if and only if it is made one-sided. Therefore, we shall restrict ourselves to the differential equation for $L_n^*(t) = L_n(t)\,U(t)$. To this end we first calculate the individual derivatives, viz.

$$\frac{dL_n^*}{dt} = L_n'\,U(t)+L_n(0)\,\delta(t),$$

$$\frac{d^2L_n^*}{dt^2} = L_n''(t)\,U(t)+L_n'(0)\,\delta(t)+L_n(0)\,\delta'(t),$$

and then obtain for the new differential equation

$$t\frac{d^2L_n^*}{dt^2}+(1-t)\frac{dL_n^*}{dt}+nL_n^* = L_n(0)\delta(t)+\{L_n'(0)-L_n(0)\}t\delta(t)+L_n(0)t\delta'(t). \quad (3)$$

On the right-hand side, the term with $t\delta(t)$ may be omitted since it represents a null function and consequently does not contribute to the operational transposition (see VI, § 15). Moreover, $t\delta'(t)$ can be replaced by $-\delta(t)$ according to (V, 66). Therefore, the final conclusion is that the right-hand side of (3) may be ignored completely; both $L_n(t)$ and $L_n(t)\,U(t)$ satisfy the same homogeneous differential equation (2).

Now, the operational transposition leads to the following equation:

$$-p\frac{d}{dp}(pf_n^*)+pf_n^*+p\frac{d}{dp}f_n^*+nf_n^* = 0,$$

in which $f_n^*(p)$ stands for the image of $L_n^*(t)$. On rearranging we easily obtain that

$$(p-p^2)\frac{df_n^*}{dp}+nf_n^* = 0, \quad (4)$$

which is of the first order and of the second degree (in p). This differential equation is much simpler than (2), the latter being of the second order and of the first degree (in t). Owing to this the general solution of (4) can be given immediately; it then leads to the following operational relation:

$$f_n^*(p) = A\left(1-\frac{1}{p}\right)^n \risingdotseq L_n(t)\,U(t) \quad (\mathrm{Re}\,p>0),$$

where the constant of integration, A, has to be determined in accordance with the conventional normalization of the Laguerre polynomials, $L_n(0)=n!$. This is afforded by means of Abel's theorem, stating that $h(0)=f(\infty)$; it gives $A = L_n(0) = n!$. We have thus found the relation (VI, 17) anew, namely,

$$L_n(t)\,U(t) \fallingdotseq n!\left(1-\frac{1}{p}\right)^n \quad (\mathrm{Re}\,p>0).$$

With this relation at hand it is quite easy to write down the successive Laguerre polynomials, viz.

$$\left.\begin{aligned}
L_0(t)\,U(t) &= U(t) &&\fallingdotseq 1,\\
L_1(t)\,U(t) &= (1-t)\,U(t) &&\fallingdotseq 1-\frac{1}{p},\\
L_2(t)\,U(t) &= (2-4t+t^2)\,U(t) &&\fallingdotseq 2\left(1-\frac{1}{p}\right)^2,\\
L_3(t)\,U(t) &= (6-18t+9t^2-t^3)\,U(t) &&\fallingdotseq 6\left(1-\frac{1}{p}\right)^3,
\end{aligned}\right\} \quad \mathrm{Re}\,p>0.$$

etc.

These low-order polynomials are represented, together with their images, in fig. 71.

Before concluding the present section, just a few remarks which apply not only to the differential equation of the Laguerre polynomials but also to other equations to be discussed in due course.

(1) The differential equation (2) possesses a second solution, linearly independent of $L_n(t)$, in the form

$$w_n(t) = L_n(t)\int_a^t \frac{e^s}{sL_n^2(s)}\,ds.$$

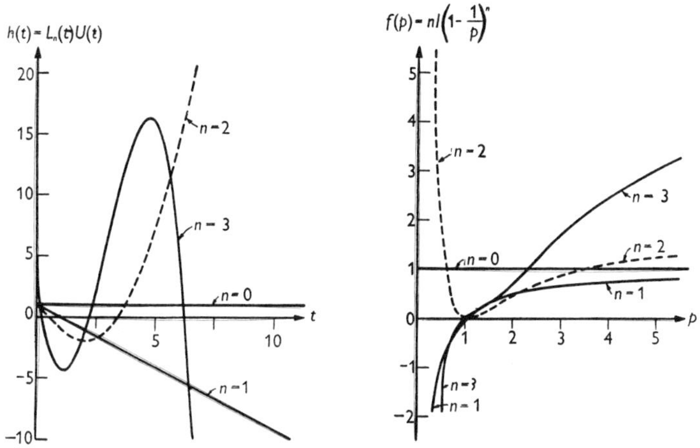

Fig. 71. One-sided Laguerre polynomials with their images.

Though the corresponding one-sided function $W_n(t)\,U(t)$ certainly has an image, the latter is not found as a solution of (4). As a matter of fact, the differential equation for the one-sided function $W_n(t)\,U(t)$ is different from that for $L_n(t)\,U(t)$; the former function does not satisfy the homogeneous equation (2) but it is a solution of (3) when the right-hand member is replaced by $\frac{1}{n!}\delta(t)$. This can be verified when properly accounting for the logarithmic singularity of $W_n(t)$ at $t = 0$.

(2) Let us multiply the differential equation (2) by the factor t; the resulting equation then becomes

$$t^2h''(t) + t(1-t)\,h'(t) + nt\,h(t) = 0.$$

It is very important to note that both $L_n(t)\,U(t)$ and $W_n(t)\,U(t)$ satisfy this new equation; in other words, the operational transform

$$p\frac{d}{dp}\left\{(1-p)\frac{df}{dp} + \frac{n}{p}f\right\} = 0,$$

which is nothing but (4) multiplied by the operator $-p\dfrac{d}{dp}\left(\dfrac{1}{p}\right)\ldots$, is satisfied by the images of both $L_n(t)\,U(t)$ and $W_n(t)\,U(t)$. Therefore multiplication of a differential equation by some power of t increases the number of independent solutions that can be found operationally; this is a consequence of the simultaneous increase of the order of the differential equation for the image function.

3. Hermite polynomials

The Hermite polynomials were encountered in the operational relation (v, 65). Let us next discuss them in connexion with their differential equation

$$\frac{d^2\mathrm{He}_n}{dt^2} - 2t\frac{d\mathrm{He}_n}{dt} + 2n\mathrm{He}_n = 0. \tag{5}$$

From the point of view of the operational calculus, the above equation is somewhat less simple than the differential equation (2) of the Laguerre polynomials, because the one-sided function $\mathrm{He}_n^*(t) = \mathrm{He}_n(t)\,U(t)$ does not satisfy the homogeneous equation (5) but the inhomogeneous equation

$$\left(\frac{d^2}{dt^2} - 2t\frac{d}{dt} + 2n\right)\mathrm{He}_n^* = \mathrm{He}_n'(0)\,\delta(t) + \mathrm{He}_n(0)\,\delta'(t). \tag{6}$$

Since either $\mathrm{He}_n'(0)$ or $\mathrm{He}_n(0)$ is zero, only one term of the right-hand side actually remains. Let us confine ourselves for the moment to even values of n by putting $n = 2m$; then $\mathrm{He}_{2m}'(0) = 0$ as $\mathrm{He}_n(t)$ is even or odd according as n is even or odd. We thus find that

$$\left(\frac{d^2}{dt^2} - 2t\frac{d}{dt} + 4m\right)\mathrm{He}_{2m}^* = \mathrm{He}_{2m}(0)\,\delta'(t),$$

which upon transformation leads to the following image equation:

$$\left\{2p\frac{d}{dp} + (4m+p^2)\right\}f_{2m}^* = \mathrm{He}_{2m}(0)\,p^2. \tag{7}$$

A particular solution of the inhomogeneous equation (7) is easily obtained from the knowledge that $Cp^{-2m}\,\mathrm{e}^{-\frac{1}{4}p^2}$ is the general solution of the corresponding homogeneous equation. When the latter function is expanded into a Laurent series, and then subjected to the operator on the left-hand side of (7), then all the powers of p generated will of course cancel out. In like manner it is easily shown that only one term, namely, that proportional to p^2, remains when the same operator is acted upon

$$\alpha\left[\left[\frac{\mathrm{e}^{-\frac{1}{4}p^2}}{p^{2m}}\right]\right] = \alpha\left[\left[\frac{1}{p^{2m}}\left(1 - \frac{p^2}{4} + \frac{p^4}{32} - \ldots\right)\right]\right]$$
$$= \alpha\left\{\frac{1}{p^{2m}} - \frac{1}{4p^{2m-2}} + \frac{1}{32p^{2m-4}} - \ldots + \frac{(-1)^m}{4^m m!}\right\}.$$

The symbol [[ ]] has the same meaning here as before in (IV, $4a$); it suppresses the terms of positive powers of p.

To obtain the proper coefficient of p^2 occurring on the right-hand side of (7) we have to choose α equal to

$$(-4)^m m!\,\mathrm{He}_{2m}(0) = 4^m(2m)!.$$

Having thus found a particular solution of the differential equation (7), we can at once write down the general solution, namely,

$$f^*_{2m} = Ap^{-2m}\,e^{-\frac{1}{4}p^2} + 4^m(2m)!\left[\left[\frac{e^{-\frac{1}{4}p^2}}{p^{2m}}\right]\right].$$

Now, it turns out that the inversion integral with respect to the p-function above diverges unless $A = 0$. Consequently, as the one-sided Hermite polynomial does have a convergent inversion integral, we necessarily arrive at $A = 0$ so that

$$\mathrm{He}_{2m}(t)\,U(t) \risingdotseq 4^m(2m)!\left[\left[\frac{e^{-\frac{1}{4}p^2}}{p^{2m}}\right]\right] \quad (\mathrm{Re}\,p > 0),$$

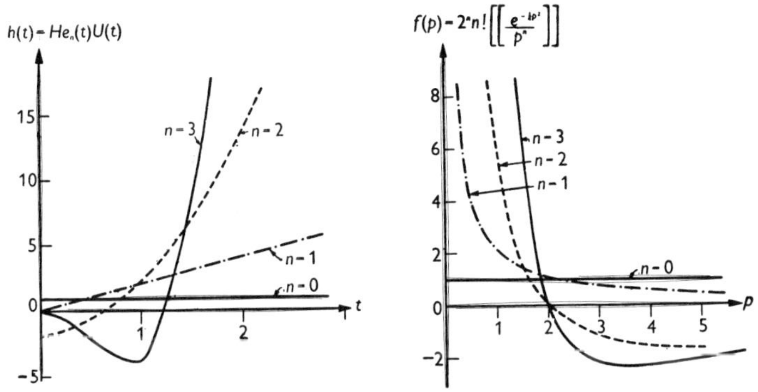

Fig. 72. One-sided Hermite polynomials with their images.

or somewhat simplified, after replacing $2m$ by n, and t by $\frac{1}{2}t$,

$$\mathrm{He}_n(\tfrac{1}{2}t)\,U(t) \risingdotseq n!\left[\left[\frac{e^{-p^2}}{p^n}\right]\right], \quad 0 < \mathrm{Re}\,p < \infty. \tag{8}$$

It is to be remarked that this relation also holds for the Hermite polynomials of odd orders. Finally, the simple relation enables us to write down at once the successive polynomials, viz.

$$\left.\begin{aligned}
\mathrm{He}_0(t)\,U(t) &= U(t) && \risingdotseq 1, && \\
\mathrm{He}_1(t)\,U(t) &= 2tU(t) && \risingdotseq \frac{2}{p} && = 2[[p^{-1}e^{-\frac{1}{4}p^2}]], \\
\mathrm{He}_2(t)\,U(t) &= (4t^2-2)\,U(t) && \risingdotseq \frac{8}{p^2} - 2 && = 8[[p^{-2}e^{-\frac{1}{4}p^2}]], \\
\mathrm{He}_3(t)\,U(t) &= (8t^3-12t)\,U(t) && \risingdotseq \frac{48}{p^3} - \frac{12}{p} && = 48[[p^{-3}e^{-\frac{1}{4}p^2}]],
\end{aligned}\right\}\ \mathrm{Re}\,p > 0.$$

etc.

These originals and images are shown in fig. 72.

Other simple operational relations may be derived after first introducing a suitable change of variables. As an example consider the function $y_n(t) = \mathrm{He}_n(\sqrt{t})$ whose differential equation is readily obtained from (5), viz.

$$\left\{2t\frac{d^2}{dt^2}+(1-2t)\frac{d}{dt}+n\right\}y_n = 0.$$

That of the corresponding one-sided function, $y_n^*(t) = y_n(t)\,U(t)$, turns out to be

$$\left\{2t\frac{d^2}{dt^2}+(1-2t)\frac{d}{dt}+n\right\}y_n^* = -\mathrm{He}_n(0)\,\delta(t),$$

whence it follows for the image $\phi_n^*(p)$ of $y_n^*(t)$ that

$$\left\{2p(1-p)\frac{d}{dp}+(n-p)\right\}\phi_n^* = -\mathrm{He}_n(0)\,p.$$

For odd values of n especially, the solution is easily obtainable since then the right-hand member vanishes. In this case the solution is uniquely determined except for a constant factor; it is thus found that

$$\mathrm{He}_{2m+1}(\sqrt{t})\,U(t) \doteqdot \frac{A}{\sqrt{p}}\left(\frac{1}{p}-1\right)^m \quad (\mathrm{Re}\,p>0).$$

The required value of A is most easily calculated with the help of Abel's theorem applied in the form

$$\lim_{t\to\infty} \Pi(m+\tfrac{1}{2})\frac{h(t)}{t^{m+\frac{1}{2}}} = \lim_{p\to 0} p^{m+\frac{1}{2}} f(p).$$

We may just state the final result to be

$$\mathrm{He}_{2m+1}(\sqrt{t})\,U(t) \doteqdot \frac{2^{2m+1}\Pi(m+\frac{1}{2})}{\sqrt{p}}\left(\frac{1}{p}-1\right)^m, \quad 0<\mathrm{Re}\,p<\infty. \tag{9}$$

The usefulness of such operational relations becomes particularly evident from the following example, in which starting from (9) and (VI, 17) a connexion is found between the Laguerre polynomials on the one hand, and the odd-order Hermite polynomials on the other. For this purpose the image of (9) is split up into factors according to

$$(-1)^m\,2^{2m+1}\Pi(m+\tfrac{1}{2})\times\frac{1}{p}\times\sqrt{p}\times\left(1-\frac{1}{p}\right)^m,$$

which upon application of the composition-product rule and (4, 18) immediately leads to

$$\mathrm{He}_{2m+1}(\sqrt{t}) = (-1)^m\frac{(2m+1)!}{(m!)^2}\int_0^t \frac{L_m(\tau)}{\sqrt{(t-\tau)}}\,d\tau \quad (t>0). \tag{10}$$

4. Bessel functions (operational relations)

In this section a number of operational relations for the Bessel function $J_\nu(t)$ will be derived through the operational transposition of its differential equation

$$J''_\nu(t)+\frac{1}{t}J'_\nu(t)+\left(1-\frac{\nu^2}{t^2}\right)J_\nu(t)=0, \tag{11}$$

either directly or after introducing some new variables.

As is well known, the ordinary Bessel function of the first kind, $J_\nu(t)$, is the solution of (11) that can be developed into ascending powers of t when the first term is chosen equal to $(\frac{1}{2}t)^\nu/\Pi(\nu)$. It is obvious that the two-sided Bessel function is not likely to have a simple image; however, the one-sided function $J^*_\nu(t) = J_\nu(t)\,U(t)$ will possess an image for $\operatorname{Re} p > 0$, provided $\operatorname{Re}\nu > -1$. Since $J_\nu(t)$ is proportional to t^ν as $t\to 0$, the homogeneous differential equation (11) also applies to the one-sided function $J^*_\nu(t)$ for $\operatorname{Re}\nu > -1$; this may be verified in like manner as before in other cases. Upon operational transposition the equation for the corresponding image is obtained in the following form:

$$\left\{(p^2+1)\frac{d^2}{dp^2}+\left(p-\frac{2}{p}\right)\frac{d}{dp}+\left(\frac{2}{p^2}-\nu^2\right)\right\}f^*_\nu(p)=0.$$

Though the last differential equation, like that shown in (11), is still of the second order, its general solution is algebraic, namely,

$$f^*_\nu(p)=\frac{p}{\sqrt{(p^2+1)}}[A\{\sqrt{(p^2+1)}-p\}^\nu+B\{\sqrt{(p^2+1)}-p\}^{-\nu}].$$

The constants of integration, A and B, are again determined by means of Abel's theorem, at least for $\operatorname{Re} p > 0$. For large values of p we have first, approximately, that

$$p^\nu f^*_\nu(p)\sim\frac{A}{2^\nu}+B2^\nu p^{2\nu}\quad(|p|\to\infty).$$

Next, according to Abel's theorem, the right-hand member must tend in the limit to the finite number

$$\operatorname*{Lim}_{t\to 0}\Pi(\nu)\frac{h(t)}{t^\nu}=\Pi(\nu)\operatorname*{Lim}_{t\to 0}\frac{J_\nu(t)}{t^\nu}=\frac{1}{2^\nu}.$$

This is actually possible, but only if $B = 0$, $A = 1$, provided we confine ourselves to values of ν with $\operatorname{Re}\nu > 0$. The final result then becomes

$$J_\nu(t)\,U(t)\doteqdot\frac{p}{\sqrt{(p^2+1)}}\{\sqrt{(p^2+1)}-p\}^\nu=\frac{p}{\sqrt{(p^2+1)}}\,e^{-\nu\operatorname{arc\,sinh}p},\quad 0<\operatorname{Re}p<\infty. \tag{12}$$

A closer examination will show that this operational relation is valid even when $\operatorname{Re}\nu > -1$. The original and image for the orders $\nu = 0, 1, 2$ are shown in fig. 73.

The many-valued exponent $-\nu \operatorname{arc\,sinh} p$ just accounts for the individual branches of the many-valued Bessel function, for which it is well known that $J_\nu(x)$ is multiplied by $e^{2\pi i\nu}$ when the argument x describes a complete contour around the origin in the counter-clockwise direction.

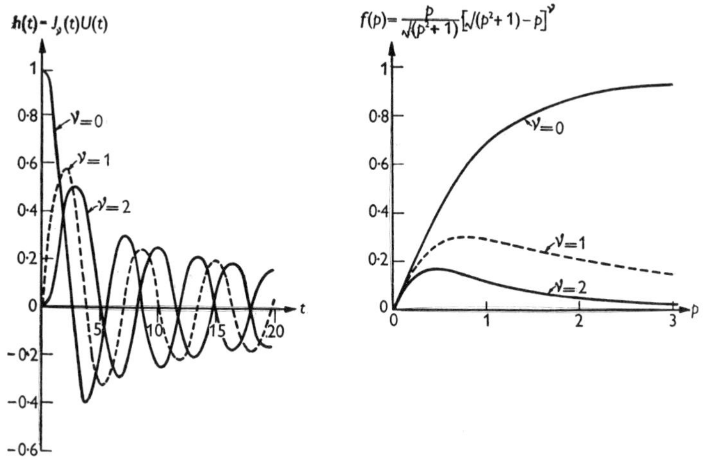

Fig. 73. One-sided Bessel functions with their images.

As a direct application of the important relation (12) we mention the corresponding inversion integral, namely,

$$J_\nu(t)\, U(t) = \frac{1}{2\pi i}\int_{c-i\infty}^{c+i\infty} \frac{e^{pt-\nu \operatorname{arc\,sinh} p}}{\sqrt{(p^2+1)}}\, dp \quad (c>0;\ \operatorname{Re}\nu > -1). \tag{13}$$

We have thus obtained an integral representation of the one-sided Bessel function valid for $\operatorname{Re}\nu > -1$ which may be compared with that of the two-sided function, viz.

$$J_\nu(t) = \frac{1}{2\pi i}\int_L \frac{e^{pt-\nu \operatorname{arc\,sinh} p}}{\sqrt{(p^2+1)}}\, dp,$$

valid for *all* complex values of ν and t, provided $\operatorname{Re} t > 0$; the path of integration L is shown in fig. 74 together with that of (13), K. Apparently the simplicity of the path of integration in (13) causes the integral to vanish automatically when $t < 0$.

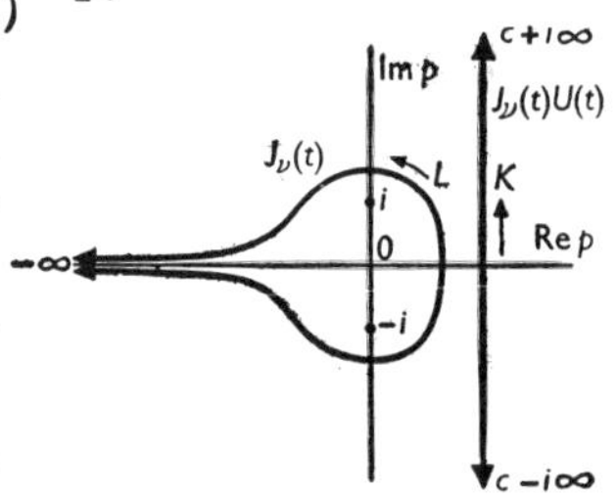

Fig. 74. Paths of integration of the integrals for $J_\nu(t)$ and $J_\nu(t)\, U(t)$.

In general the operational relation (12) applies only to one of the two linearly independent solutions J_ν and $J_{-\nu}$ of (11). Only if $-1 < \operatorname{Re}\nu < 1$ does (12) represent the images both of $J_\nu(t)$ and $J_{-\nu}(t)$. In this case, any solution of (11) can be transposed operationally; in particular, the Hankel functions which are often required in physical applications because of their specific asymptotic behaviour for large values of t. The two existing types of Hankel functions are most conveniently defined as the following combinations of J_ν and $J_{-\nu}$:

$$H_\nu^{\substack{(1)\\(2)}}(t) = \pm\frac{i}{\sin(\nu\pi)}\{e^{\mp\nu\pi i}J_\nu(t) - J_{-\nu}(t)\}. \tag{14}$$

It is not difficult to derive the corresponding images with the aid of (12) for $-1 < \operatorname{Re}\nu < 1$, but we shall only explicitly mention the limiting case $\nu = 0$, which reads

$$H_0^{\substack{(1)\\(2)}}(t)\,U(t) \doteqdot \frac{p}{\sqrt{(p^2+1)}}\left(1 \mp \frac{2i}{\pi}\operatorname{arc\,sinh} p\right), \quad 0 < \operatorname{Re} p < \infty. \tag{15}$$

In concluding the discussion of the operational relation (12) we remark that, after an application of the rule (IV, 42), it reduces to the simple relation

$$\frac{J_\nu(t)}{t}U(t) \doteqdot \frac{p}{\nu}e^{-\nu\operatorname{arc\,sinh} p}, \quad 0 < \operatorname{Re} p < \infty \;\; (\operatorname{Re}\nu > 0). \tag{16}$$

In order that the operational calculus should supply us with a rapid approach to perhaps unusual and intricate properties of Bessel functions, it is desirable to have other operational relations at hand also, which may be obtained after adequate change of variables. Of the great variety of possibilities we shall treat only two examples; in both cases the differential equation for the original is of the first degree (in t) and, accordingly, that of the corresponding image is of the first order.

(1) The differential equation for $y_\nu(t) = t^\nu J_\nu(t)$ can easily be obtained from (11); it is found that

$$t\frac{d^2y_\nu}{dt^2} + (1-2\nu)\frac{dy_\nu}{dt} + ty_\nu = 0, \tag{17}$$

of which a second independent solution is $t^\nu J_{-\nu}(t)$. When $\operatorname{Re}\nu > 0$, the same homogeneous differential equation (17) is satisfied by the one-sided function $t^\nu J_\nu(t)\,U(t)$, but not by the analogue $t^\nu J_{-\nu}(t)\,U(t)$.

Let $\phi_\nu(p)$ denote the image of $t^\nu J_\nu(t)\,U(t)$. Then, upon transposing (17), it is found that

$$\left(p+\frac{1}{p}\right)\phi_\nu'(p) + \left(2\nu - \frac{1}{p^2}\right)\phi_\nu(p) = 0,$$

of which the general solution is easily given; we then get the operational relation

$$t^\nu J_\nu(t)\,U(t) \doteqdot A\frac{p}{(p^2+1)^{\nu+\frac{1}{2}}}.$$

Again, the constant of integration, A, can be obtained by applying Abel's theorem. The final result may be shown to hold even when $\operatorname{Re}\nu > -\frac{1}{2}$; thus

$$t^\nu J_\nu(t)\, U(t) \doteqdot \frac{2^\nu \Pi(\nu-\frac{1}{2})}{\sqrt{\pi}} \frac{p}{(p^2+1)^{\nu+\frac{1}{2}}}, \quad 0<\operatorname{Re} p<\infty \quad (\operatorname{Re}\nu > -\tfrac{1}{2}). \tag{18}$$

Fig. 75 shows the original and image for $\nu = 1$ and $\nu = 2$.

In particular, for non-negative integral values of ν the relation (18) is reducible to the following form:

$$t^n J_n(t)\, U(t) \doteqdot p\left(-\frac{1}{p}\frac{d}{dp}\right)^n \left(\frac{1}{\sqrt{(p^2+1)}}\right), \quad 0<\operatorname{Re} p<\infty. \tag{19}$$

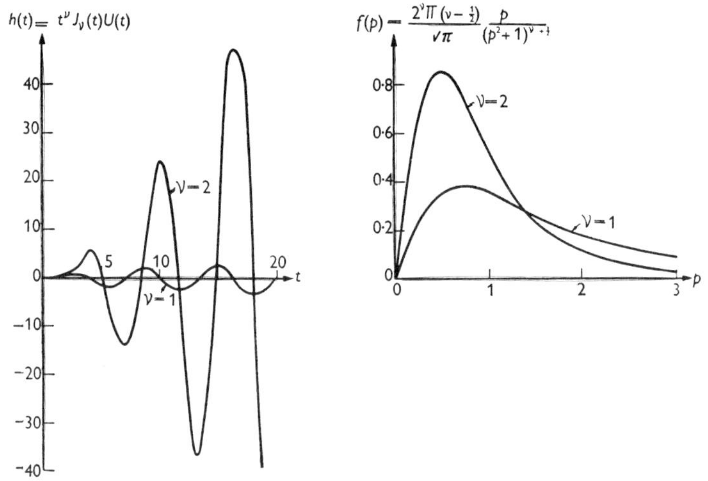

Fig. 75. Examples showing the original and image of the operational relation (18).

(2) By changing both the dependent and independent variables according to

$$z_\nu(t) = t^{\frac{1}{2}\nu} J_\nu(2\sqrt{t}),$$

the equation (11) is transformed into

$$t\frac{d^2 z_\nu}{dt^2} + (1-\nu)\frac{dz_\nu}{dt} + z_\nu = 0,$$

of which a second solution is $t^{\frac{1}{2}\nu} J_{-\nu}(2\sqrt{t})$. The corresponding image equation is again of the first order, namely,

$$pf'(p) + \left(\nu - \frac{1}{p}\right) f(p) = 0.$$

Proceeding in the same manner as in the first example we finally obtain the following operational relation:

$$t^{\frac{1}{2}\nu} J_\nu(2\sqrt{t})\, U(t) \doteqdot \frac{e^{-1/p}}{p^\nu}, \quad 0 < \operatorname{Re} p < \infty \ (\operatorname{Re} \nu > -1). \tag{20}$$

The original and image for $\nu = 1$ and $\nu = 2$ are plotted in fig. 76.

The last relation at once furnishes the conventional expansion of the Bessel function into powers of its argument, viz.

$$J_\nu(z) = \sum_{n=0}^{\infty} \frac{(-1)^n (\frac{1}{2}z)^{\nu+2n}}{n!\, \Pi(\nu+n)}, \tag{21}$$

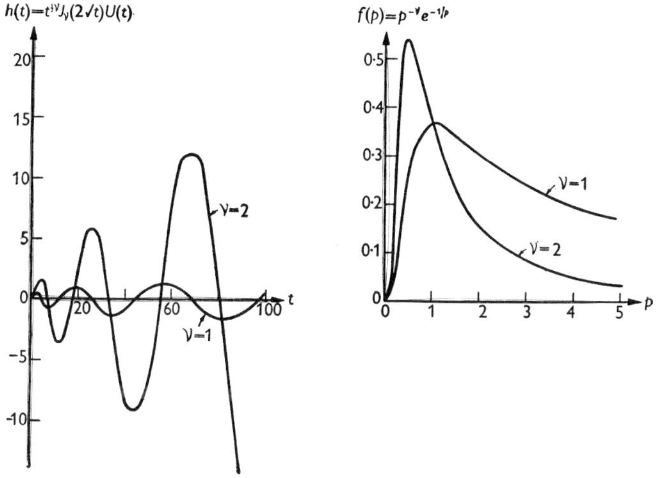

Fig. 76. Examples showing the original and image of the operational relation (20).

which may be derived from (20) by term-by-term transposition of the image, the latter being expanded into powers of $1/p$, and putting $z = 2\sqrt{t}$. As is well known, the Bessel function of the first kind of arbitrarily complex order ν and argument z is customarily defined by the very series (21).

Another operational relation for non-negative integral values of ν is the following:

$$\frac{J_n(2\sqrt{t})}{t^{\frac{1}{2}n}}\, U(t) \doteqdot [[(-p)^n e^{-1/p}]], \quad 0 < \operatorname{Re} p < \infty. \tag{22}$$

It can be deduced by first expanding the Bessel function on the left into powers of t, using (21), and then transposing the result obtained term by term. It should be noticed that here, in contrast with the image of (8), the series in powers of $1/p$ of the function [[ ]] involves an infinite number of terms.

Still another interesting formula: from the definition integral corresponding to (22) it follows after the substitutions $p = 1/x$, $t = x^2 s$ that

$$\frac{x^n}{n!} - \frac{x^{n+1}}{(n+1)!} + \frac{x^{n+2}}{(n+2)!} - \ldots = x\int_0^\infty \frac{e^{-xs} J_n\{2x\sqrt{s}\}}{s^{\frac{1}{2}n}}\, ds \quad (x > 0),$$

in which the left-hand function is just the 'remainder' of the exponential e^{-x}, apart from sign.

It may further be emphasized that the variables occurring in (20) are also useful in connexion with the function $\mathbf{Y}_n(t)$, which for integral values (n) of ν represents a second solution of (11), linearly independent of the ordinary Bessel function $J_n(t)$. Though in text-books explicit expressions for $\mathbf{Y}_n$ are usually quite complicated, yet we have the comparatively simple formula $(n = 1, 2, 3, \ldots)$

$$x^{\frac{1}{2}n}\mathbf{Y}_n(2\sqrt{x}) = (-1)^n \sum_{k=0}^{\infty} (-1)^k \frac{d}{dk}\left\{\frac{x^k}{\Pi(k)\,\Pi(k-n)}\right\}.$$

Term-by-term transposition leads to the following operational relation:

$$t^{\frac{1}{2}n}\mathbf{Y}_n(2\sqrt{t})\,U(t) \risingdotseq (-1)^n \sum_{k=0}^{\infty} (-1)^k \frac{d}{dk}\left\{\frac{1}{\Pi(k-n)\,p^k}\right\}.$$

Finally, on account of (VI, 4a), the terms with $k \geqslant n$ can be expressed by an exponential integral; we thus find that

$$t^{\frac{1}{2}n}\mathbf{Y}_n(2\sqrt{t})\,U(t) \risingdotseq \frac{1}{p^n}\left\{e^{-1/p}\,\overline{\mathrm{Ei}}\left(\frac{1}{p}\right) - \sum_{k=1}^{n} (k-1)!\,p^k\right\}, \quad 0 < \mathrm{Re}\,p < \infty$$
$$(n = 1, 2, 3, \ldots). \quad (23)$$

Thus far we have exclusively dealt with Bessel functions of real arguments, but the methods developed above apply equally to Bessel functions of complex argument. Of particular importance is the 'Bessel function of imaginary argument', defined by

$$I_\nu(x) = e^{-\frac{1}{2}\pi i\nu} J_\nu(ix), \quad (24)$$

from which it is evident that $I_\nu(x)$ is real for real values of x, just as $J_\nu(x)$ itself. The differential equation satisfied by $I_\nu(t)$ is slightly different from (11), in that $(1 - \nu^2/t^2)$ is replaced by $(-1 - \nu^2/t^2)$. The further investigation proceeds on the same lines; omitting all details we state only the relations analogous to (18) and (20), viz.

$$t^\nu I_\nu(t)\,U(t) \risingdotseq \frac{2^\nu}{\sqrt{\pi}}\Pi(\nu - \tfrac{1}{2})\frac{p}{(p^2-1)^{\nu+\frac{1}{2}}}, \quad 1 < \mathrm{Re}\,p < \infty \ (\mathrm{Re}\,\nu > -\tfrac{1}{2}), \quad (25)$$

$$t^{\frac{1}{2}\nu} I_\nu(2\sqrt{t})\,U(t) \risingdotseq \frac{e^{1/p}}{p^\nu}, \quad 0 < \mathrm{Re}\,p < \infty \ (\mathrm{Re}\,\nu > -1). \quad (26)$$

In addition, there also exist important relations that involve I_ν in the image rather than in the original. To obtain a simple example, consider the following formula:

$$\frac{I_\nu(p)}{p^\nu} = \frac{1}{2^\nu \sqrt{\pi}\,\Pi(\nu-\frac{1}{2})} \int_{-1}^{1} e^{-pu}(1-u^2)^{\nu-\frac{1}{2}}\,du \quad (\operatorname{Re}\nu > -\tfrac{1}{2}), \tag{27}$$

which may be verified by comparing the expansions of either side into powers of p, after applying Euler's integral of the first kind to the coefficients on the right. But the right-hand member of (27) is simply the definition integral of a certain cut-off function; accordingly, we are led to a new operational relation, viz.

$$\frac{I_\nu(p)}{p^{\nu-1}} \doteq \frac{(1-t^2)^{\nu-\frac{1}{2}}}{\sqrt{\pi}\,2^\nu \Pi(\nu-\frac{1}{2})} U(1-t^2), \quad -\infty < \operatorname{Re} p < \infty \quad (\operatorname{Re}\nu > -\tfrac{1}{2}). \tag{28}$$

Another kind of Bessel function is represented by the well-known K-function, defined by

$$K_\nu(z) = \frac{\pi i}{2} e^{\frac{1}{2}\nu\pi i} H_\nu^{(1)}(iz) = \frac{\pi}{2\sin(\nu\pi)}\{I_{-\nu}(z) - I_\nu(z)\}. \tag{29}$$

It, too, has a simple integral representation, completely analogous to that of $I_\nu(p)$, that is, (27). We may just write down its operational interpretation, that is to say, the relation

$$\frac{K_\nu(p)}{p^{\nu-1}} \doteq \frac{\sqrt{\pi}}{2^\nu \Pi(\nu-\frac{1}{2})}(t^2-1)^{\nu-\frac{1}{2}}\,U(t-1), \quad 0 < \operatorname{Re} p < \infty \quad (\operatorname{Re}\nu > -\tfrac{1}{2}), \tag{30}$$

which we further transform by means of the shift rule into

$$\frac{e^p K_\nu(p)}{p^{\nu-1}} \doteq \frac{\sqrt{\pi}}{2^\nu \Pi(\nu-\frac{1}{2})}(t^2+2t)^{\nu-\frac{1}{2}}\,U(t), \quad 0 < \operatorname{Re} p < \infty \quad (\operatorname{Re}\nu > -\tfrac{1}{2}). \tag{31}$$

The last operational relation is of primary importance in Bessel-function theory, since it is basic for the investigation concerning the asymptotic behaviour of the general Bessel function. It may therefore be worth while to sketch an independent proof of (31). We first observe that the one-sided original on the right-hand side of (31) satisfies the following homogeneous differential equation ($\operatorname{Re}\nu > -\frac{1}{2}$):

$$(t^2+2t)\,h'(t) - 2(\nu-\tfrac{1}{2})(t+1)\,h(t) = 0.$$

Then, after operational transposition, the resulting p-equation is easily shown to be satisfied by the left-hand member of (31). A second solution of the p-equation, however, is proportional to $e^p p^{1-\nu} I_\nu(p)$. Consequently, the image corresponding to the original of (31) is at any rate expressible as follows:

$$\frac{e^p}{p^{\nu-1}}\{\alpha K_\nu(p) + \beta I_\nu(p)\}.$$

Next, the constants α and β remain to be chosen properly. Confining ourselves to $\mathrm{Re}\,\nu > \frac{1}{2}$, it is seen that the original under consideration is finite at $t = 0$; the same must therefore hold for the image as $p \to \infty$, whence it follows that $\beta = 0$. The constant of normalization, α, can be calculated from Abel's theorem, once the dominant term of $K_\nu(p)$ is assumed known.

The complete asymptotic expansion of K_ν is obtainable by first expanding the original of (31) into ascending powers of t, and then transposing term by term. The result is the well-known series

$$K_\nu(p) \approx \sqrt{\left(\frac{\pi}{2p}\right)}\, \mathrm{e}^{-p} \sum_{n=0}^{\infty} \frac{\Pi(\nu+n-\frac{1}{2})}{\Pi(\nu-n-\frac{1}{2})} \frac{1}{n!(2p)^n} \quad (p \to \infty), \tag{32}$$

which terminates if, and only if, ν is an odd multiple of $\frac{1}{2}$; in the latter case the expansion is not only an asymptotic, but also the rigorous expression, i.e. an equality for all values of p.

An alternative expression for $K_{n+\frac{1}{2}}$, where n is a positive integer, is the following:

$$K_{n+\frac{1}{2}}(p) = \sqrt{\left(\frac{\pi}{2}\right)}\, p^{n+\frac{1}{2}} \left(-\frac{1}{p}\frac{d}{dp}\right)^n \left(\frac{\mathrm{e}^{-p}}{p}\right). \tag{33}$$

In concluding this section we may mention the functions $\mathrm{ber}\,(z)$ and $\mathrm{bei}\,(z)$, which for $\nu = 0$ were introduced by Lord Kelvin in his electromagnetic theory of the skin-effect in wires of circular cross-section. The functions of unrestricted order and argument can be defined in terms of the Bessel function of the first kind by means of

$$\mathrm{e}^{\pm i\pi\nu} J_\nu(\mathrm{e}^{\mp\frac{1}{4}i\pi} x) = \mathrm{ber}_\nu\,(x) \pm i\,\mathrm{bei}_\nu\,(x).$$

Then, analogous to (20), we have

$$\mathrm{e}^{\pm i\pi\nu}\, t^{\frac{1}{2}\nu} J_\nu(2\,\mathrm{e}^{\mp\frac{1}{4}i\pi}\sqrt{t})\, U(t) \doteqdot \mathrm{e}^{\pm\frac{3}{4}i\pi\nu} \frac{\mathrm{e}^{\pm i/p}}{p^\nu} \quad (\mathrm{Re}\,p > 0;\ \mathrm{Re}\,\nu > -1),$$

which can be found either from the differential equation or by a term-by-term transposition from the corresponding series (21). It consequently follows that

$$\left.\begin{aligned} t^{\frac{1}{2}\nu}\,\mathrm{ber}_\nu\,(2\sqrt{t})\, U(t) &\doteqdot \frac{\cos\left(\frac{1}{p} + \frac{3}{4}\pi\nu\right)}{p^\nu} \\ t^{\frac{1}{2}\nu}\,\mathrm{bei}_\nu\,(2\sqrt{t})\, U(t) &\doteqdot \frac{\sin\left(\frac{1}{p} + \frac{3}{4}\pi\nu\right)}{p^\nu} \end{aligned}\right\} \quad 0 < \mathrm{Re}\,p < \infty \quad (\mathrm{Re}\,\nu > -1). \tag{34}$$

5. Bessel functions (applications)

The operational relations derived in the preceding section form an adequate starting-point either for proving results already known, or for establishing essentially new expressions, all involving Bessel functions.

Possible hints as to what relation is most useful in a certain case cannot be given; now one is the better, then another. The majority of the examples given below are also to be found elsewhere†. We shall start the discussion with some applications of the simple operational relation (20), because it yields the elementary properties of the Bessel functions quite easily.

(1) When in (20) the original is differentiated n times in succession with respect to t, then the right-hand member must be multiplied by p^n. This new image obviously corresponds to $t^{\frac{1}{2}(\nu-n)}J_{\nu-n}(2\sqrt{t})\,U(t)$. Consequently, after putting $x = 2\sqrt{t}$, we immediately find that

$$\left(\frac{d}{x\,dx}\right)^n \{x^\nu J_\nu(x)\} = x^{\nu-n}J_{\nu-n}(x). \tag{35}$$

(2) Upon multiplying the original of (20) by t, and applying to the image the corresponding operator $-p\dfrac{d}{dp}\left(\dfrac{1}{p}\ldots\right.$, we obtain

$$t^{\frac{1}{2}\nu+1}J_\nu(2\sqrt{t})\,U(t) \doteqdot (\nu+1)\frac{e^{-1/p}}{p^{\nu+1}} - \frac{e^{-1/p}}{p^{\nu+2}}.$$

Applying once more the relation (20), to either of the two terms of the image above, and then replacing $2\sqrt{t}$ by x and ν by $\nu-1$, we arrive at the well-known recurrence formula

$$J_{\nu+1}(x) + J_{\nu-1}(x) = \frac{2\nu}{x}J_\nu(x). \tag{36}$$

(3) In like manner, by application of the operator

$$t\frac{d}{dt} \doteqdot -p\frac{d}{dp}$$

to (20), it is found that $\quad J'_\nu(x) = \dfrac{\nu}{x}J_\nu(x) - J_{\nu+1}(x);$

elimination of $J_\nu(x)$ from this and (36) then leads to

$$J'_\nu(x) = \tfrac{1}{2}\{J_{\nu-1}(x) - J_{\nu+1}(x)\}. \tag{37}$$

The last identity can be written in the following way:

$$\left(\frac{d}{dx} + \tfrac{1}{2}\Delta_\nu\right)J_{2\nu}(x) = 0,$$

from which it is not difficult to derive again the difference-differential equation (IX, 40).

(4) By application of the composition-product rule to the simple identity

$$\frac{1}{p}\frac{1}{p^{\mu-\nu}}\frac{e^{-1/p}}{p^{\nu-1}} = \frac{e^{-1/p}}{p^\mu},$$

† Balth. van der Pol, 'On the operational solution of linear differential equations and an investigation of the properties of these solutions', *Phil. Mag.* LXXX, 861, 1929.

and using (20), it follows that

$$t^{\frac{1}{2}\mu}J_\mu(2\sqrt{t})\,U(t) = \frac{U(t)}{\Pi(\mu-\nu)}\int_0^t (t-\tau)^{\mu-\nu}\tau^{\frac{1}{2}(\nu-1)}J_{\nu-1}(2\sqrt{\tau})\,d\tau$$

$$(0<\operatorname{Re}\nu<\operatorname{Re}\mu+1).$$

Consequently, after putting $s = 2\sqrt{\tau}$ and $x = 2\sqrt{t}$,

$$\int_0^x J_{\nu-1}(s)\,(x^2-s^2)^{\mu-\nu}s^\nu\,ds = 2^{\mu-\nu}\,\Pi(\mu-\nu)\,x^\mu J_\mu(x).$$

(5) The definition integral corresponding to (20), after some simple transformations, yields the following formula:

$$\int_0^\infty e^{-x^2}x^{\nu+1}J_\nu(2\alpha x)\,dx = \tfrac{1}{2}\alpha^\nu e^{-\alpha^2} \quad (\operatorname{Re}\nu > -1).$$

A rather different class of properties is obtained when the investigation is based on the operational relation (18). For instance, consider the following factorization:

$$\frac{1}{\cos(\nu\pi)}\frac{p}{p^2+1} = \frac{1}{p}\times\frac{2^\nu\Pi(\nu-\frac{1}{2})}{\sqrt{\pi}}\frac{p}{(p^2+1)^{\nu+\frac{1}{2}}}\times\frac{2^{-\nu}\Pi(-\nu-\frac{1}{2})}{\sqrt{\pi}}\frac{p}{(p^2+1)^{-\nu+\frac{1}{2}}}.$$

When the rule of the composition product is applied, and (18) is used, we easily arrive at the result that

$$\cos(\nu\pi)\int_0^x J_\nu(x-\xi)\,J_{-\nu}(\xi)\left(\frac{x}{\xi}-1\right)^\nu d\xi = \sin x \quad (-\tfrac{1}{2}<\operatorname{Re}\nu<\tfrac{1}{2};\ x>0).$$

Another application of (18) is the proof that Bessel functions of half-integral orders ($\nu = n+\frac{1}{2}$, n integral) are expressible in finite terms of x, $\sin x$ and $\cos x$. To see this we first derive with the aid of the similarity rule

$$t^{n+\frac{1}{2}}J_{n+\frac{1}{2}}(xt)\,U(t) \doteqdot \frac{(2x)^{n+\frac{1}{2}}n!}{\sqrt{\pi}}\frac{p}{(p^2+x^2)^{n+1}}$$

$$= \sqrt{\left(\frac{2}{\pi}\right)}x^{n+\frac{1}{2}}\left(-\frac{1}{x}\frac{d}{dx}\right)^n\left(\frac{p}{p^2+x^2}\right) \quad (\operatorname{Re}p>0;\ n\geqslant 0;\ x>0),$$

and then, after transposing the right-hand member and putting $t = 1$,

$$J_{n+\frac{1}{2}}(x) = \sqrt{\left(\frac{2}{\pi}\right)}x^{n+\frac{1}{2}}\left(-\frac{1}{x}\frac{d}{dx}\right)^n\left(\frac{\sin x}{x}\right). \tag{38}$$

Let us next turn to the operational relation (12), which is very useful in evaluating integrals.The following examples may be given for reference.

(1) From the identity

$$\frac{1}{p}\times\frac{p}{\sqrt{(p^2+1)}}e^{-\nu\operatorname{arc\,sinh}p}\times\frac{p}{\sqrt{(p^2+1)}}e^{\nu\operatorname{arc\,sinh}p} = \frac{p}{p^2+1},$$

we immediately find upon application of the composition-product rule

$$\int_0^x J_\nu(s)\, J_{-\nu}(x-s)\, ds = \sin x \quad (-1 < \operatorname{Re}\nu < 1;\ x > 0). \tag{39}$$

(2) In like manner, from the identity

$$\frac{1}{p} \times \frac{p}{\sqrt{(p^2+1)}}\, e^{-\nu \operatorname{arc\,sinh} p} \times \frac{p}{\mu}\, e^{-\mu \operatorname{arc\,sinh} p} = \frac{1}{\mu} \frac{p}{\sqrt{(p^2+1)}}\, e^{-(\mu+\nu) \operatorname{arc\,sinh} p},$$

and using (12) and (16), we find that

$$\mu \int_0^x J_\mu(s)\, J_\nu(x-s) \frac{ds}{s} = J_{\mu+\nu}(x) \quad (\operatorname{Re}\mu > 0;\ \operatorname{Re}\nu > -1). \tag{40}$$

(3) In order to prove the well-known formula of Frullani, that is,

$$\int_0^\infty \{J_0(as) - \cos(bs)\} \frac{ds}{s} = \log\left(\frac{2b}{a}\right), \tag{41}$$

we first apply the rule (IV, 41 a) to the following operational relation:

$$\{J_0(at) - \cos(bt)\}\, U(t) \doteqdot p \left\{ \frac{1}{\sqrt{(p^2+a^2)}} - \frac{p}{p^2+b^2} \right\} \quad (\operatorname{Re} p > 0).$$

The result is found to be

$$U(t) \int_0^t \{J_0(as) - \cos(bs)\} \frac{ds}{s} \doteqdot \log\left(\frac{2\sqrt{(p^2+b^2)}}{\sqrt{(p^2+a^2)}+p} \right) \quad (\operatorname{Re} p > 0),$$

from which (41) follows on account of Abel's theorem, $h(\infty) = f(0)$.

Hitherto we have only discussed formulae involving a finite number of Bessel functions, if not only one. We shall now treat a few series containing an infinite number; they, too, can readily be got from the operational relations deduced in § 4.

(1) With the help of (12), and using the identity

$$\frac{p}{\sqrt{(p^2+1)}} (1 + 2\alpha^2 + 2\alpha^4 + 2\alpha^6 + \ldots) = 1 \quad (|\alpha| < 1),$$

in which $\alpha = \sqrt{(p^2+1)} - p$,

it is found that $J_0(x) + 2J_2(x) + 2J_4(x) + 2J_6(x) + \ldots = 1.$ (42)

Moreover, since for integral values of n we know that

$$J_n(-x) = (-1)^n J_n(x),$$

(42) may be written alternatively

$$\sum_{n=-\infty}^{\infty} J_{2n}(x) = \sum_{n=-\infty}^{\infty} J_n(x) = 1.$$

(2) From (12) together with

$$\frac{p}{\sqrt{(p^2+1)}}(1-2\alpha^2+2\alpha^4-2\alpha^6+\ldots) = \frac{p^2}{p^2+1} \quad (|\alpha|<1),$$

we similarly find that

$$J_0(x)-2J_2(x)+2J_4(x)-2J_6(x)+\ldots = \sum_{n=-\infty}^{\infty}(-1)^n J_{2n}(x) = \cos x. \tag{43}$$

(3) Again from (12), but now using

$$\frac{p}{\sqrt{(p^2+1)}}(\alpha-\alpha^3+\alpha^5-\alpha^7+\ldots) = \frac{p}{2(p^2+1)} \quad (|\alpha|<1),$$

it follows that

$$J_1(x)-J_3(x)+J_5(x)-J_7(x)+\ldots = \sum_{n=0}^{\infty}(-1)^n J_{2n+1}(x) = \tfrac{1}{2}\sin x. \tag{44}$$

(4) Next, (12) and the following identity

$$\frac{\alpha^\nu}{\sqrt{(p^2+1)}} = \frac{2p}{\sqrt{(p^2+1)}}\sum_{n=0}^{\infty}\alpha^{\nu+2n+1} \quad (|\alpha|<1),$$

lead to

$$\int_0^x J_\nu(s)\,ds = 2\sum_{n=0}^{\infty}J_{\nu+2n+1}(x). \tag{45}$$

(5) Let us differentiate (IV, 45) and transpose term by term the corresponding series

$$\frac{\sin t}{t}U(t) \risingdotseq p\operatorname{arc\,cot} p = p\arcsin\left(\frac{1}{\sqrt{(p^2+1)}}\right)$$

$$= \frac{p}{\sqrt{(p^2+1)}}\left\{1+\frac{1}{2}\frac{1}{3}\frac{1}{(p^2+1)}+\frac{1.3}{2.4}\frac{1}{5}\frac{1}{(p^2+1)^2}+\ldots\right\} \quad (\operatorname{Re} p>0),$$

with the aid of (18); the result is easily found to be

$$\sum_{n=0}^{\infty}\frac{(\frac{1}{2}x)^n}{(2n+1)\,n!}J_n(x) = \frac{\sin x}{x}. \tag{46}$$

(6) Finally, let us apply (20) to the following series:

$$\frac{e^{-(1-\alpha)/p}}{p^\nu} = \frac{e^{-1/p}}{p^\nu}+\frac{\alpha}{1!}\frac{e^{-1/p}}{p^{\nu+1}}+\frac{\alpha^2}{2!}\frac{e^{-1/p}}{p^{\nu+2}}+\ldots.$$

We then obtain:

$$\frac{J_\nu(2\sqrt{\{t(1-\alpha)\}})}{(1-\alpha)^{\frac{1}{2}\nu}} = \sum_{n=0}^{\infty}\frac{(\alpha\sqrt{t})^n}{n!}J_{\nu+n}(2\sqrt{t}) \quad (\operatorname{Re}\nu>-1;\ \alpha<1).$$

This formula is nothing but the well-known multiplication theorem of Bessel functions; after putting $x = 2\sqrt{t}$ and $y = \sqrt{(1-\alpha)}$, we get

$$\frac{J_\nu(yx)}{y^\nu} = \sum_{n=0}^{\infty}\{\tfrac{1}{2}x(1-y^2)\}^n\frac{J_{\nu+n}(x)}{n!} \quad (\operatorname{Re}\nu>-1), \tag{47}$$

which is actually valid for arbitrarily complex x, y.

Whereas the results so far obtained undoubtedly belong to the elementary part of Bessel-function theory, we shall now discuss some formulae not so well known.

Example 1. *Extension of an integral discussed by Sommerfeld.* We shall derive from (47) an integral formula that is an extension of one given by Sommerfeld, who applied it to describe the electromagnetic field of a Hertzian dipole, placed upon the surface of the earth assumed plane. The generalization can be obtained in the following manner. To begin with, the series expansion

$$\frac{p}{\sqrt{(p^2+1)}}\{\sqrt{(p^2+1)}-p\}^\nu e^{-a\{\sqrt{(p^2+1)}-p\}} = \sum_{n=0}^{\infty}\frac{(-a)^n}{n!}\frac{p}{\sqrt{(p^2+1)}}\{\sqrt{(p^2+1)}-p\}^{\nu+n}$$

is transposed term by term with the aid of (12). Then, the resulting series is summed by the use of (47). Next, the new result is subjected to the shift rule which gives

$$\frac{p}{\sqrt{(p^2+1)}}\{\sqrt{(p^2+1)}-p\}^\nu e^{-a\sqrt{(p^2+1)}} \doteqdot \left(\frac{t-a}{t+a}\right)^{\frac{1}{2}\nu} J_\nu\{\sqrt{(t^2-a^2)}\}\, U(t-a), \quad 0<\operatorname{Re} p<\infty$$
$$(\operatorname{Re}\nu > -1). \quad (48)$$

Finally, the corresponding definition integral, when the substitutions $p = z/\rho$, $a = -ik\rho$, $t = \sqrt{(\lambda^2\rho^2+a^2)}$ are performed, yields the identity

$$\left\{\sqrt{\left(\frac{z^2}{\rho^2}+1\right)}-\frac{z}{\rho}\right\}^\nu \frac{e^{ik\sqrt{(\rho^2+z^2)}}}{\sqrt{(\rho^2+z^2)}} = (\tan\tfrac{1}{2}\theta)^\nu\frac{e^{ikr}}{r}$$
$$= \int_0^\infty J_\nu(\lambda\rho)\, e^{-z\sqrt{(\lambda^2-k^2)}}\left\{\frac{\sqrt{(\lambda^2-k^2)}+ik}{\sqrt{(\lambda^2-k^2)}-ik}\right\}^{\frac{1}{2}\nu}\frac{\lambda\, d\lambda}{\sqrt{(\lambda^2-k^2)}}, \quad (49)$$

in which r and θ may be interpreted as spherical co-ordinates, and z and ρ as cylindrical co-ordinates. This formula is valid for all k (the square root being defined with positive real part), though in its derivation an imaginary value of k (a being real) has been assumed. Sommerfeld's integral mentioned above is obtained when ν is taken zero; it proves to be of great importance in the resolution of partial differential equations of the second order (see chapters XV and XVI).

Example 2. The special case of the operational relation (23) when $n = 0$, that is,

$$\mathbf{Y}_0(2\sqrt{t})\, U(t) \doteqdot e^{-1/p}\,\overline{\operatorname{Ei}}\left(\frac{1}{p}\right) \quad (\operatorname{Re} p > 0), \quad (50)$$

leads to a relationship between the functions $\mathbf{Y}_0$ and J_0. In the first place it follows from (VI, 3) that the image can be written as follows:

$$e^{-1/p} \int_{\infty}^{-1/p} e^{-s}\frac{ds}{s} = \int_0^\infty \frac{e^{-u/p}}{1-u}\, du,$$

and, secondly, when the integrand is transposed with the help of (20), we immediately arrive at the result in question, namely,

$$\mathbf{Y}_0(2\sqrt{t}) = \int_0^\infty \frac{J_0(2\sqrt{\tau})}{t-\tau}\, d\tau \quad (t>0).$$

Example 3. *Expressions for the derivative of the Bessel function with respect to its order.* Properties of these derivatives easily follow from the three operational relations below, which can be found by differentiation of (12), (18) and (20) with respect to ν. They are, respectively,

$$\frac{\partial}{\partial\nu} J_\nu(t)\, U(t) \doteqdot -\frac{p}{\sqrt{(p^2+1)}} \operatorname{arc\,sinh} p\, e^{-\nu \operatorname{arc\,sinh} p} \quad (\operatorname{Re} p>0;\ \operatorname{Re}\nu>-1), \tag{51}$$

$$t^\nu\left(\log t+\frac{\partial}{\partial\nu}\right) J_\nu(t)\, U(t) \doteqdot \frac{2^\nu \Pi(\nu-\frac{1}{2})}{\sqrt{\pi}} \frac{p}{(p^2+1)^{\nu+\frac{1}{2}}}\left\{\log\left(\frac{2}{p^2+1}\right)+\psi(\nu-\tfrac{1}{2})\right\}$$
$$(\operatorname{Re} p>0;\ \operatorname{Re}\nu>-\tfrac{1}{2}), \tag{52}$$

$$t^{\frac{1}{2}\nu}\left\{\tfrac{1}{2}\log t+\frac{\partial}{\partial\nu}\right\} J_\nu(2\sqrt{t})\, U(t) \doteqdot -\frac{\log p}{p^\nu} e^{-1/p} \quad (\operatorname{Re} p>0;\ \operatorname{Re}\nu>-1), \tag{53}$$

in which $\psi(x)$ is the logarithmic derivative of $\Pi(x)$.

To formulate, with the aid of (51), a certain property of $\frac{\partial}{\partial\nu} J_\nu$ as simply as possible we introduce the definition of the integral-Bessel function according to

$$\operatorname{Ji}(x) = \int_\infty^x \frac{J_0(s)}{s}\, ds.$$

After applying the rule (IV, 40) to (12) with $\nu = 0$ we then obtain the following operational relation:

$$\operatorname{Ji}(t)\, U(t) \doteqdot \log\{\sqrt{(p^2+1)}-p\} = -\operatorname{arc\,sinh} p \quad (\operatorname{Re} p>0). \tag{54}$$

The composition-product rule is now applied to the image of (51) according to the factorization

$$-\frac{1}{p} \times \frac{p^2}{\sqrt{(p^2+1)}} e^{-\nu \operatorname{arc\,sinh} p} \times \operatorname{arc\,sinh} p.$$

The result is of striking simplicity, viz.

$$\frac{\partial}{\partial\nu} J_\nu(t) = \int_0^t \operatorname{Ji}(t-\tau)\, J_\nu'(\tau)\, d\tau \quad (\operatorname{Re}\nu>0).$$

Two alternative expressions for the same differential quotient can be derived by factorizing (52) and (53) analogously and using (IV, 44), (III, 26), and (35) for $n = 1$; they are

$$\frac{\partial}{\partial\nu} J_\nu(t) = \{\psi(\nu-\tfrac{1}{2})-\log\tfrac{1}{2}t\}\, J_\nu(t)+2\int_0^t \left(\frac{\tau}{t}\right)^\nu J_{\nu-1}(\tau)\, \operatorname{Ci}(t-\tau)\, d\tau \quad (\operatorname{Re}\nu>0),$$

$$\frac{\partial}{\partial\nu} J_\nu(t) = (C+\log\tfrac{1}{2}t)\, J_\nu(t)+\int_0^t \log\left(1-\frac{\tau^2}{t^2}\right)\left(\frac{\tau}{t}\right)^\nu J_{\nu-1}(\tau)\, d\tau \quad (\operatorname{Re}\nu>0).$$

6. Legendre functions (operational relations)

Whereas Bessel functions are easily approached operationally because of the simplicity of their various images only involving algebraic or simple transcendental functions, the Legendre functions in turn are operationally mapped by the Bessel functions.

The term *Legendre function* applies to any solution of the following differential equation:

$$(1-x^2)\, y''-2xy'+\nu(\nu+1)\, y = 0. \tag{55}$$

In particular, the solutions that are equal to unity at the regular singularity $x = 1$ are called *Legendre functions of the first kind and of degree* ν, henceforth denoted by $P_\nu(x)$. Most generally known are the *Legendre polynomials*, $P_n(x)$, obtained when ν is taken integral $(=n)$. In order to give an operational treatment of these functions for unrestricted values of ν, bearing in mind that they are normalized according to $P_\nu(1) = 1$, it is simplest to consider the images of $h^*(t) = P_\nu(t)\,U(t-1)$. That is to say, they are not cut off at $t = 0$ but at $t = 1$. Proceeding as in the case of other differential equations treated before, we easily verify that the homogeneous equation (55) remains valid for $h^*(t)$, provided we make use of some well-known properties of the Legendre functions under consideration, according to which both $P_\nu(1)$ and $P'_\nu(1)$ are finite whatever ν may be†. Moreover, actual transposition of (55) leads to the following differential equation for the image, $f_\nu(p)$, of $h^*(t)$:

$$p^2 f''_\nu - \{p^2 + \nu(\nu+1)\} f_\nu = 0.$$

When in this equation the substitution $f_\nu = \sqrt{p}\, g_\nu$ is made, it reduces to the differential equation (11) of the Bessel functions of order $\nu + \frac{1}{2}$ and imaginary argument. Consequently, $f_\nu(p)/\sqrt{p}$ must be some linear combination of $I_{\nu+\frac{1}{2}}(p)$ and $K_{\nu+\frac{1}{2}}(p)$. Thus

$$f_\nu(p) = \sqrt{p}\,\{A I_{\nu+\frac{1}{2}}(p) + B K_{\nu+\frac{1}{2}}(p)\}. \tag{56}$$

The constants of integration, A and B, are uniquely determined by the condition $P_\nu(1) = 1$. This can be seen from the Abel theorem $h(0) = f(\infty)$ when applied to the shifted one-sided relation

$$P_\nu(t+1)\,U(t) \doteqdot e^p f_\nu(p).$$

This theorem then leads to

$$1 = \lim_{p\to\infty} \sqrt{p}\, e^p \{A I_{\nu+\frac{1}{2}}(p) + B K_{\nu+\frac{1}{2}}(p)\}.$$

However, since $\sqrt{p}\, e^p I_{\nu+\frac{1}{2}}(p)$ does not remain finite as $p \to \infty$ (compare the derivation of (31)), A is necessarily equal to zero. The value of B is next found with the help of (32). The final result then becomes

$$P_\nu(t)\,U(t-1) \doteqdot \sqrt{\left(\frac{2p}{\pi}\right)}\, K_{\nu+\frac{1}{2}}(p), \quad 0 < \operatorname{Re} p < \infty. \tag{57}$$

This operational relation, stating a relationship between the Legendre functions and the Hankel functions of imaginary argument, holds for all values of ν, even for $\operatorname{Re}\nu < \frac{1}{2}$, since both P_ν and $K_{\nu+\frac{1}{2}}$ are not changed when ν is replaced by $-\nu-1$. Moreover, for integral values n of ν the image on the right-hand side of (57) is an elementary function (see (33)), thus leading to

$$P_n(t)\,U(t-1) \doteqdot p^{n+1}\left(-\frac{1}{p}\frac{d}{dp}\right)^n \left(\frac{e^{-p}}{p}\right), \quad 0 < \operatorname{Re} p < \infty \;\; (n \geqslant 0). \tag{58}$$

† Cf. E. W. Hobson, *Spherical and Ellipsoidal Harmonics*, Cambridge, 1931, p. 189.

It is easily seen that $P_n(x)$ is a polynomial in x of degree n. For that purpose we apply the shift rule and (IV, 4a) to (58) by which it is found that for Legendre polynomials cut off at $t = 0$ it follows that

$$P_n(t)\,U(t) \doteqdot \left[\left[p^{n+1}\left(-\frac{1}{p}\frac{d}{dp}\right)^n \left(\frac{e^{-p}}{p}\right)\right]\right], \quad 0 < \operatorname{Re} p < \infty \quad (n \geqslant 0). \tag{59}$$

The first few of the complete operational relations involving the Legendre polynomials are now easily constructed:

$$P_0(t)\,U(t) = U(t) \doteqdot 1,$$

$$P_1(t)\,U(t) = t\,U(t) \doteqdot \frac{1}{p} = \left[\left[p^2\left(-\frac{1}{p}\frac{d}{dp}\right)\left(\frac{e^{-p}}{p}\right)\right]\right],$$

$$P_2(t)\,U(t) = \tfrac{1}{2}(3t^2-1)\,U(t) \doteqdot \frac{3}{p^2}-\frac{1}{2} = \left[\left[p^3\left(-\frac{1}{p}\frac{d}{dp}\right)^2\left(\frac{e^{-p}}{p}\right)\right]\right],$$

$$P_3(t)\,U(t) = \tfrac{1}{2}(5t^3-3t)\,U(t) \doteqdot \frac{15}{p^3}-\frac{3}{2p} = \left[\left[p^4\left(-\frac{1}{p}\frac{d}{dp}\right)^3\left(\frac{e^{-p}}{p}\right)\right]\right],$$

etc. $(\operatorname{Re} p > 0)$.

Summarizing the above we see that the operational relation for the Legendre polynomials emerges as a particular case of a relation that is valid for unrestricted values of the degree ν. This was made possible by means of cutting off $P_\nu(t)$ at $t = 1$, where both the function and its derivative are finite throughout. A similar procedure cannot be performed at $t = -1$, since $P_\nu(-1)$ is infinite, unless ν is integral. This, however, does not mean that the analogue of (58) is missing; on the contrary, executing the necessary operations while restricting ourselves to integral values of ν, we also find that

$$P_n(t)\,U(t+1) \doteqdot p^{n+1}\left(-\frac{1}{p}\frac{d}{dp}\right)^n\left(\frac{e^{p}}{p}\right) \quad (\operatorname{Re} p > 0).$$

Finally, by subtraction from (58), we obtain

$$P_n(t)\,U(1-t^2) \doteqdot 2p^{n+1}\left(-\frac{1}{p}\frac{d}{dp}\right)^n\left(\frac{\sinh p}{p}\right) = (-1)^n\sqrt{(2\pi p)}\,I_{n+\frac{1}{2}}(p),$$
$$-\infty < \operatorname{Re} p < \infty. \tag{60}$$

In accordance with the original at the left being a function cut off at both sides (vanishing non-identically only in that region over which the Legendre polynomials are orthogonal), the convergence of (60) holds for unrestricted values of p.

Let us now turn to a second solution of (55) and, in particular, to that solution which is commonly labelled as the *Legendre function of the second kind*, usually designated by $Q_\nu(x)$. A simple operational relation in this respect, not restricted to integral values of ν, can be established by admitting it as the image rather than as the original ($Q_\nu(x)$ is singular at $x = 1$). It turns out that, upon transposition of the differential equation for

$$\frac{I_{\nu+\frac{1}{2}}(t)}{\sqrt{t}}\, U(t) \quad (\mathrm{Re}\,\nu > -1), \tag{61}$$

we arrive at an equation that, according to (55), can be solved generally by means of the following linear combination:

$$p\{AP_\nu(p) + BQ_\nu(p)\}. \tag{62}$$

But, in order that (62) should represent the image of (61), we have to choose A equal to zero if $\mathrm{Re}\,\nu > -\frac{1}{2}$; in fact, since the original under discussion behaves as t^ν for $t \to 0$, the image (Abel's theorem!) should be proportional to $p^{-\nu}$ as $p \to \infty$, whilst (again applying Abel's theorem) it easily follows from (57) that $P_\nu(p)$ for $p \to \infty$ increases proportionally to p^ν, if $\mathrm{Re}\,\nu > -\frac{1}{2}$. Thus with this restriction on ν, A must be zero in (62), the image otherwise increasing at least like $p^{\nu+1}$ instead of $p^{-\nu}$. Finally, the constant B has to be determined from the normalization of Q_ν.

An alternative and more direct procedure starts with the following formula:

$$Q_\nu(z) = \frac{1}{2^{\nu+1}} \int_{-1}^{1} \frac{(1-s^2)^\nu}{(z-s)^{\nu+1}}\, ds, \tag{63}$$

which defines the Legendre function of the second kind when $\mathrm{Re}\,\nu > -1$ for all values of z not lying on the cut along the real z-axis at the left of $z = -1$. For, identifying z with p and transposing the integrand for $\mathrm{Re}\,p > s$, we immediately obtain

$$pQ_\nu(p) \doteqdot \frac{t^\nu U(t)}{2^{\nu+1}\Pi(\nu)} \int_{-1}^{1} e^{ts}(1-s^2)^\nu\, ds \quad (\mathrm{Re}\,p > 1),$$

whence it follows in connexion with (27) that

$$pQ_\nu(p) \doteqdot \sqrt{\left(\frac{\pi}{2t}\right)}\, I_{\nu+\frac{1}{2}}(t)\, U(t), \quad 1 < \mathrm{Re}\,p < \infty \ (\mathrm{Re}\,\nu > -1). \tag{64}$$

Furthermore, for non-negative integral values of n, $Q_n(z)$ is elementary; in this case the cut may be reduced to the finite interval $-1 \leqslant x \leqslant 1$ of the real axis. However, since in this range of the independent variable an appropriate second solution of Legendre's differential equation is also often required, an appropriate definition of $Q_n(x)$ is then

$$Q_n(x) = \tfrac{1}{2}Q_n(x+i0) + \tfrac{1}{2}Q_n(x-i0),$$

which, in addition, has a simple operational image on a single line of convergence. This image can be found by deriving, with the aid of (63), the following expression for the mean value:

$$Q_n(t) = \frac{(-1)^n}{2^{n+1}n!} \frac{d^n}{dt^n} \int_{-1}^{1} \frac{(1-s^2)^n}{t-s}\, ds \quad (-1 < t < 1).$$

Then this is easily transposed with the help of

$$\frac{1}{t-s} \doteqdot \pi \mid p \mid e^{-ps} \quad (\operatorname{Re} p = 0),$$

which relation, in its turn, may be deduced from (VI, 53), and whose definition integral should also be taken as a principal value. The resulting image is found to be, using (27),

$$Q_n(t) = (-1)^n \pi \sqrt{\left(\frac{\pi}{2p}\right)} \mid p \mid I_{n+\frac{1}{2}}(p) = \pi \mid p \mid p^n \left(-\frac{1}{p}\frac{d}{dp}\right)^n \left(\frac{\sinh p}{p}\right), \qquad \operatorname{Re} p = 0, \quad (65)$$

in which the Q-function thus occurs in the original. For $n = 0$ and $n = 1$ we have

$$Q_0(t) = \tfrac{1}{2}\log\left|\frac{1+t}{1-t}\right| \doteqdot \pi \sin(|p|), \quad \operatorname{Re} p = 0,$$

$$Q_1(t) = \tfrac{1}{2}t\log\left|\frac{1+t}{1-t}\right| - 1 \doteqdot -\pi \mid p \mid \frac{d}{dp}\left(\frac{\sinh p}{p}\right), \quad \operatorname{Re} p = 0.$$

7. Legendre functions (applications)

The operational relations discussed in the preceding section all contain a Legendre function and either a Bessel or a Hankel function. Consequently, numerous relations between these functions can be established. The corresponding definition integrals already furnish such expressions; for instance, (57), (60) and (64) lead to

$$\int_1^\infty e^{-pt} P_\nu(t)\,dt = \sqrt{\left(\frac{2}{\pi p}\right)} K_{\nu+\frac{1}{2}}(p) \quad (\operatorname{Re} p > 0),$$

$$\int_{-1}^1 e^{-pt} P_n(t)\,dt = 2p^n \left(-\frac{1}{p}\frac{d}{dp}\right)^n \left(\frac{\sinh p}{p}\right),$$

$$\int_0^\infty e^{-pt} I_{\nu+\frac{1}{2}}(t)\frac{dt}{\sqrt{t}} = \sqrt{\left(\frac{2}{\pi}\right)} Q_\nu(p) \quad (\operatorname{Re} p > 1;\ \operatorname{Re}\nu > -1).$$

The same holds, of course, with respect to the respective inversion integrals; that of (65), for example, will give after some reduction

$$Q_n(t) = \sqrt{(\tfrac{1}{2}\pi)} \int_0^\infty \frac{J_{n+\frac{1}{2}}(\omega)}{\sqrt{\omega}} \sin(\omega t - \tfrac{1}{2}n\pi)\,d\omega.$$

We will just take a few properties of Legendre functions as representative of a great number, which are found to be substantially equivalent with properties of Bessel functions already known.

(1) The recurrence relation (36) applied to the K-function becomes

$$K_{\nu+1}(p) - K_{\nu-1}(p) = \frac{2\nu}{p} K_\nu(p).$$

After multiplication through by $\sqrt{p}$, and transposition with the aid of (57), we obtain that

$$P_{\nu+1}(t)-P_\nu(t) = (2\nu+1)\int_0^t P_\nu(\tau)\,d\tau.$$

(2) Let us first derive from (60) by means of the similarity rule

$$P_n(yt)\,U(1-y^2t^2) \doteqdot (-1)^n\sqrt{\left(\frac{2\pi p}{y}\right)}\,I_{n+\frac{1}{2}}\left(\frac{p}{y}\right) \quad (y>0),$$

and then, with the help of (47), express the Bessel function of imaginary argument in the form of a series of Bessel functions. Now, again by (60), we may determine the original of the individual terms, and thus find

$$P_n(yt)\,U(1-y^2t^2) = \frac{1}{y^{n+1}}\sum_{k=0}^{\infty}\frac{1}{k!}\left(\frac{y^2-1}{2y^2}\right)^k\frac{d^k}{dt^k}\{P_{n+k}(t)\,U(1-t^2)\},$$

or, alternatively, in trigonometric functions $(-1<t<1)$,

$$P_n(\cos\theta\,\cos\theta') = \frac{1}{\cos^{n+1}\theta'}\sum_{k=0}^{\infty}\frac{(-\frac{1}{2}\tan^2\theta')^k}{k!}P^{(k)}_{n+k}(\cos\theta).$$

(3) The asymptotic series (32) is obtained anew, but with coefficients that are simply expressible in terms of Legendre functions, when one starts with

$$P_\nu(t+1)\,U(t) \doteqdot \sqrt{\left(\frac{2p}{\pi}\right)}\,e^p\,K_{\nu+\frac{1}{2}}(p) \quad (\mathrm{Re}\,p>0),$$

which follows from (57). Let the Legendre function occurring in the original be expanded into ascending powers of t, and the new original be transposed term by term. The resulting identity is only asymptotic, that is,

$$K_\nu(p) \approx \sqrt{\left(\frac{\pi}{2p}\right)}\,e^{-p}\sum_{n=0}^{\infty}\frac{P^{(n)}_{\nu-\frac{1}{2}}(1)}{p^n} \quad (p\to\infty).$$

(4) By first expanding the original of (64) into a power series with the help of (21), and then transposing the result term by term, the following convergent series is obtained:

$$Q_\nu(p) = \frac{\sqrt{\pi}}{2^{\nu+1}}\sum_{n=0}^{\infty}\frac{\Pi(\nu+2n)}{4^n n!\,\Pi(\nu+n+\frac{1}{2})}\frac{1}{p^{\nu+2n+1}} \quad (|\,p\,|>1;\ \mathrm{Re}\,\nu>-1).$$

8. Hypergeometric functions

The differential equations treated thus far can be considered as special cases of the general hypergeometric differential equation, either directly or after introduction of suitable variables:

$$Ly \equiv \left\{\left(x\frac{d}{dx}+\alpha_1\right)\left(x\frac{d}{dx}+\alpha_2\right)\cdots\left(x\frac{d}{dx}+\alpha_r\right) - \frac{d}{dx}\left(x\frac{d}{dx}+\gamma_1-1\right)\cdots\left(x\frac{d}{dx}+\gamma_s-1\right)\right\}y = 0. \quad (66)$$

The simplest solution of the differential equation (66) is that in the form of a hypergeometric series, viz.

$$
\begin{aligned}
{}_rF_s(\alpha_1, \ldots, \alpha_r; \gamma_1, \ldots, \gamma_s; x) \\
= \frac{\Gamma(\gamma_1)\,\Gamma(\gamma_2)\ldots\Gamma(\gamma_s)}{\Gamma(\alpha_1)\,\Gamma(\alpha_2)\ldots\Gamma(\alpha_r)} \sum_{n=0}^{\infty} \frac{\Gamma(\alpha_1+n)\,\Gamma(\alpha_2+n)\ldots\Gamma(\alpha_r+n)}{\Gamma(\gamma_1+n)\,\Gamma(\gamma_2+n)\ldots\Gamma(\gamma_s+n)} \frac{x^n}{n!},
\end{aligned} \tag{67}
$$

where we use a well-known notation due to Pochhammer; henceforth the parameters α and γ may be called the numerator and denominator elements respectively. When $r = 2$, $s = 1$, the ordinary hypergeometric series comes out, in which case the suffixes r and s are usually omitted so as to have

$$
\begin{aligned}
F(\alpha, \beta; \gamma; x) = 1 + \frac{\alpha\beta}{\gamma}\frac{x}{1!} + \frac{\alpha(\alpha+1)\beta(\beta+1)}{\gamma(\gamma+1)}\frac{x^2}{2!} \\
+ \frac{\alpha(\alpha+1)(\alpha+2)\beta(\beta+1)(\beta+2)}{\gamma(\gamma+1)(\gamma+2)}\frac{x^3}{3!} + \ldots.
\end{aligned} \tag{68}
$$

A very simple example is provided by the logarithmic function, namely,

$$\log(1+x) = xF(1, 1; 2; -x).$$

As to the convergence of (67), there are three different cases to be distinguished, viz.

(1) $r-s<1$; the series is convergent for all values of x. Examples of entire functions that can be defined in this manner are:

$$e^x = {}_0F_0(;\,;x), \tag{69}$$

$$L_n(x) = n!\,{}_1F_1(-n; 1; x), \tag{70}$$

$$I_\nu(2\sqrt{x}) = \frac{x^{\frac{1}{2}\nu}}{\Pi(\nu)}\,{}_0F_1(;\nu+1; x), \tag{71}$$

$$M_{k,m}(x) = x^{m+\frac{1}{2}}\,e^{-\frac{1}{2}x}\,{}_1F_1(m-k+\tfrac{1}{2}; 2m+1; x). \tag{72}$$

The last of them is known by the name of *Whittaker function*, of which, amongst others, the Bessel function is the specialization obtained when $k = 0$.

(2) $r-s = 1$; the series (67) is convergent only for $|x| < 1$, or perhaps on the unit circle too.

An example is the ordinary hypergeometric series (68); others are

$$(1+x)^\nu = {}_1F_0(-\nu;\,;-x), \tag{73}$$

$$P_\nu^\mu(x) = \frac{1}{\Pi(-\mu)}\left(\frac{x+1}{x-1}\right)^{\frac{1}{2}\mu} {}_2F_1\left(\nu+1, -\nu; 1-\mu; \frac{1-x}{2}\right). \tag{74}$$

The last function is the associated Legendre function of the first kind, being an extension of $P_\nu(x) = P_\nu^0(x)$.

(3) $r-s>1$; outside the origin the series (67) is divergent throughout. A series like this is only a *formal* solution of (66) in such a sense that, after termwise differentiation and substitution in the left-hand member of (66), and arranging according to powers of x, the coefficients vanish. In the sense of asymptotic series these formal series determine certain special solutions of the corresponding equations (66); whence their importance with respect to the operational treatment. Just a few examples:

$$x\,e^{-x}\,\mathrm{Ei}\,(x) \approx {}_2F_0\left(1, 1;\,;\,\frac{1}{x}\right) \quad (x \to \infty), \tag{75}$$

$$K_\nu(x) \approx \sqrt{\left(\frac{\pi}{2x}\right)}\,e^{-x}\,{}_2F_0\left(\tfrac{1}{2}+\nu, \tfrac{1}{2}-\nu;\,;\,-\frac{1}{2x}\right) \quad (x \to \infty), \tag{76}$$

$$W_{k,m}(x) \approx x^k\,e^{-\frac{1}{2}x}\,{}_2F_0\left(m-k+\tfrac{1}{2}, -m-k+\tfrac{1}{2};\,;\,-\frac{1}{x}\right) \quad (x \to \infty). \tag{77}$$

The first two of them were given in (VII, 24) and (32); the function corresponding to the series (77) is a linear combination of the Whittaker functions $M_{k,m}$ and $M_{k,-m}$. The relation (77) only means that the function

$$y(x) = x^{-k}\,e^{\frac{1}{2}x}\,W_{k,m}(x)$$

satisfies the hypergeometric differential equation

$$\left[\left\{-\frac{1}{x}\frac{d}{d(-1/x)}+m-k+\tfrac{1}{2}\right\}\left\{-\frac{1}{x}\frac{d}{d(-1/x)}-m-k+\tfrac{1}{2}\right\}-\frac{d}{d(-1/x)}\right]y = 0,$$

and possesses the following asymptotic development:

$$y \sim 1-(m-k+\tfrac{1}{2})(-m-k+\tfrac{1}{2})\frac{1}{x}$$
$$+(m-k+\tfrac{1}{2})(m-k+\tfrac{3}{2})(-m-k+\tfrac{1}{2})(-m-k+\tfrac{3}{2})\frac{1}{2x^2}+\ldots \quad (x \to \infty).$$

After this brief account we are able to survey most easily the fundamental importance of the hypergeometric differential equation from the point of view of the operational calculus. First of all, it is simpler to transpose the hypergeometric series than the original differential equation. In so doing we identify in (67) x with $\pm 1/p$, and then multiply by $p^{1-\nu}$. According to the theory of VII, § 8, term-by-term transposition is legitimate if the radius of convergence does not vanish, i.e. for $r \leqslant s+1$. Thus we are led to a new series for the original, which is easily recognized as a hypergeometric series too; it is thus found that

$$\frac{\Gamma(\nu)}{p^{\nu-1}}\,{}_rF_s\left(\alpha;\gamma;\pm\frac{1}{p}\right) \doteqdot t^{\nu-1}\,{}_rF_{s+1}(\alpha;\gamma,\nu;\pm t)\,U(t),$$
$$0<\mathrm{Re}\,p<\infty \ \ (r \leqslant s), \ \ 1<\mathrm{Re}\,p<\infty \ \ (r=s+1) \ \ (\mathrm{Re}\,\nu>0), \tag{78}$$

in which α is an abbreviation for the whole set of the elements $\alpha_1, \ldots, \alpha_r$, and γ for the set $\gamma_1, \ldots, \gamma_s$. Therefore, when starting from a hypergeometric series for the image, we find the original to be another hypergeometric series having the same number of numerator elements α, but with an extra denominator element, namely, the parameter ν. On the other hand, it is also possible to derive from (78) a relation between hypergeometric functions containing the same number of denominator elements, while the number of numerator elements at the right is one less than that at the left. To this end we only add the extra parameter ν to the α-elements of (78), whereby in the *right* member the element ν occurs in both numerator and denominator, and consequently, according to (67), can be omitted since it cancels out. In the manner described we then obtain a second operational relation, namely,

$$\frac{\Gamma(\nu)}{p^{\nu-1}}\,{}_{r+1}F_s\left(\alpha,\nu;\gamma;\pm\frac{1}{p}\right) \doteqdot t^{\nu-1}\,{}_rF_s(\alpha;\gamma;\pm t)\,U(t),$$
$$0<\operatorname{Re}p<\infty\ (r<s),\ 1<\operatorname{Re}p<\infty\ (r=s)\ (\operatorname{Re}\nu>0). \quad (79)$$

Either of the operational relations (78) and (79) may be valid if $r>s+1$ and $r>s$, respectively. For, even when starting with an *asymptotic* series for the original, the series obtained by operational transposition will be asymptotic too (we are dealing with power series, see theorem VI of chapter VII). As a consequence, the corresponding p-function, for which the series mentioned is an asymptotic expression for $p\to\infty$, is recognized as a solution of the corresponding hypergeometric differential equation (66), though in general this solution cannot be fixed uniquely in this way.

As to the convergence of the hypergeometric series of the originals and images of (78) and (79), we may distinguish four different cases which will now be discussed. At the same time we shall illustrate with examples how many special relations are included in the two mentioned.

(1) Each of the series for original and image converges everywhere.

An example is provided by taking no numerator element and only one denominator element in (79); thus

$$t^{\nu-1}\,{}_0F_1(;\gamma;t)\,U(t) \doteqdot \frac{\Gamma(\nu)}{p^{\nu-1}}\,{}_1F_1\left(\nu;\gamma;\frac{1}{p}\right) \quad (\operatorname{Re}p>0;\ \operatorname{Re}\nu>0).$$

When interpreting this result with the help of (71) and (72) we obtain the following relation between the Bessel function and the Whittaker function:

$$p^{k+1}\,e^{1/2p}\,M_{k,m}\left(\frac{1}{p}\right) \doteqdot \frac{\Pi(2m)}{\Pi(m-k-\frac{1}{2})}\,\frac{I_{2m}(2\sqrt{t})}{t^{k+\frac{1}{2}}}\,U(t),$$
$$0<\operatorname{Re}p<\infty\ (m-k>-\tfrac{1}{2}). \quad (80)$$

(2) The series for the original converges everywhere, that for the image only in a part of the p-plane.

As an example, in (78) we take no denominator element and only one numerator element. Upon applying (72) and (73) to the result we arrive at a simple image for the Whittaker function, namely,

$$t^{m-\frac{1}{2}} M_{k,m}(t)\, U(t) \doteqdot \Pi(2m)\, p \frac{(p-\frac{1}{2})^{k-m-\frac{1}{2}}}{(p+\frac{1}{2})^{k+m+\frac{1}{2}}}, \quad \tfrac{1}{2} < \operatorname{Re} p < \infty \ (\operatorname{Re} m > -\tfrac{1}{2}). \tag{81}$$

Another example is provided by the following relation, existing between Whittaker functions and the restricted hypergeometric functions:

$$p^{1-\alpha}\, {}_2F_1\!\left(\alpha, \beta; \gamma; \frac{1}{p}\right) \doteqdot \frac{e^{\frac{1}{2}t}\, t^{\alpha-\frac{1}{2}\gamma-1}}{\Gamma(\alpha)} M_{\frac{1}{2}\gamma-\beta, \frac{1}{2}(\gamma-1)}(t)\, U(t),$$
$$1 < \operatorname{Re} p < \infty \ (\operatorname{Re} \alpha > 0), \tag{82}$$

which can be proved by use of (72), and taking in (79) one element α and one element γ.

Moreover, either as a special case of (82), or directly from

$$\frac{n!}{p^n}\, {}_2F_1\!\left(n+1, -n; 1; \frac{1}{p}\right) \doteqdot t^n\, {}_1F_1(-n; 1; t)\, U(t) \quad (\operatorname{Re} p > 0),$$

one finds a relation between Legendre polynomials and Laguerre polynomials, viz.

$$\frac{P_n\!\left(1-\dfrac{2}{p}\right)}{p^n} \doteqdot \frac{t^n L_n(t)}{(n!)^2} U(t), \quad 0 < \operatorname{Re} p < \infty \ (n \geqslant 0). \tag{83}$$

(3) The series for the original converges for certain values of t, that for the image does not converge at all (i.e. it is only asymptotic).

As an example we mention

$$\frac{U(t)}{t^\mu}\, {}_2F_1(\nu+1, -\nu; 1-\mu; -t) \doteqdot \Pi(-\mu)\, p^\mu\, {}_2F_0\!\left(\nu+1, -\nu; ; -\frac{1}{p}\right)$$
$$(\operatorname{Re} p > 0;\ \operatorname{Re} \mu < 1).$$

Upon working it out with the aid of (74) and (76), it is found that

$$\frac{P_\nu^\mu(t)}{(t^2-1)^{\frac{1}{2}\mu}} U(t-1) \doteqdot \sqrt{\left(\frac{2}{\pi}\right)}\, p^{\mu+\frac{1}{2}} K_{\nu+\frac{1}{2}}(p), \quad 0 < \operatorname{Re} p < \infty \ (\operatorname{Re} \mu < 1), \tag{84}$$

which is an extension of (57).

(4) The series for both original and image diverge throughout (are asymptotic).

As this case is of little importance, it may suffice to quote just a single example, namely,

$${}_2F_0(1, 1; ; -t)\, U(t) \doteqdot {}_3F_0\!\left(1, 1, 1; ; -\frac{1}{p}\right) \quad (\operatorname{Re} p > 0).$$

In concluding this section, and the present chapter, we would give a few applications of the type of operational relation which has presented itself in the course of our discussion of the hypergeometric differential equation.

Example 1. *Relations between hypergeometric series.* So far as the integrals concerned are convergent, we at once obtain from the definition integrals of (78) and (79) the following formulae:

$$ {}_rF_s\left(\alpha;\gamma;\frac{1}{p}\right) = \frac{p^\nu}{\Gamma(\nu)}\int_0^\infty e^{-pt}t^{\nu-1}\,{}_rF_{s+1}(\alpha;\gamma,\nu;t)\,dt, $$

$$ \text{Re}\,p>0\ (r\leqslant s),\quad \text{Re}\,p>1\ (r=s+1)\quad (\text{Re}\,\nu>0), \qquad (85) $$

$$ {}_{r+1}F_s\left(\alpha,\nu;\gamma;\frac{1}{p}\right) = \frac{p^\nu}{\Gamma(\nu)}\int_0^\infty e^{-pt}t^{\nu-1}\,{}_rF_s(\alpha;\gamma;t)\,dt, $$

$$ \text{Re}\,p>0\ (r<s),\quad \text{Re}\,p>1\ (r=s)\quad (\text{Re}\,\nu>0). \qquad (86) $$

By repeatedly applying these reduction formulae, it is possible to express hypergeometric functions of arbitrary elements α, γ by means of multiple integrals over hypergeometric functions of continually increasing simplicity. In addition to this, they are also useful for the operational deduction of other general identities. For instance, when (86) and (85) are applied in succession, a repeated integral is obtained in which, after reversing the order of the integration, the inner integral is easily recognized as the definition integral corresponding to (III, 31). In this manner one can find the following formula:

$$ {}_{r+1}F_s\left(\alpha,\nu;\gamma;\frac{1}{p}\right) = \frac{2p^{\frac{1}{2}(\mu+\nu)}}{\Gamma(\mu)\,\Gamma(\nu)}\int_0^\infty u^{\frac{1}{2}(\mu+\nu)-1}K_{\mu-\nu}\{2\sqrt{(pu)}\}\,{}_rF_{s+1}(\alpha;\gamma,\mu;u)\,du, $$

$$ \text{Re}\,p>0\ (r<s),\quad \text{Re}\,p>1\ (r=s)\quad (\text{Re}\,\mu>0;\ \text{Re}\,\nu>0). $$

As a special case of this we may mention a formula in which the restricted hypergeometric function is expressed in the form of an integral over the product of two Bessel functions:

$$ F\left(\alpha,\beta;\gamma;\frac{1}{p}\right) = 2\frac{\Gamma(\gamma)}{\Gamma(\alpha)\,\Gamma(\beta)}p^{\frac{1}{2}(\alpha+\beta)}\int_0^\infty K_{\alpha-\beta}(s\sqrt{p})\,I_{\gamma-1}(s)\left(\frac{s}{2}\right)^{\alpha+\beta-\gamma}ds $$

$$ (\text{Re}\,p>1;\ \text{Re}\,\alpha>0;\ \text{Re}\,\beta>0). \qquad (87) $$

Relations like this will be investigated more fully in XI, § 5.

Example 2. *Properties of Whittaker functions.* These are easily derived from (81), since the latter relation reduces them to simpler functions. By multiplying (81) by the operators

$$ \frac{d}{dt}\pm\tfrac{1}{2} \doteq p\pm\tfrac{1}{2}, $$

and then determining the original of the new image, we find the following differential properties:

$$ \left(\frac{d}{dt}\pm\frac{1}{2}\right)\{t^{m-\frac{1}{2}}M_{k,m}(t)\} = 2mt^{m-1}M_{k\mp\frac{1}{2},\,m-\frac{1}{2}}(t). $$

In another application a recurrence relation between three Whittaker functions, all of the same parameter m, but of consecutive values of k, is obtained. For this purpose (81) is multiplied by t, then added to two other equations obtained from (81) by replacing k by $k\pm 1$, and finally multiplying by $k\pm m+\frac{1}{2}$. The resulting image can then easily be transformed into the t-language; the final identity is found to be

$$ (k+m+\tfrac{1}{2})\,M_{k+1,m}(t)+(k-m-\tfrac{1}{2})\,M_{k-1,m}(t) = (2k-t)\,M_{k,m}(t). $$

Still another property of these functions is obtained when applying the composition-product rule to the following factorization:

$$ \frac{1}{p}\times\Pi(2m_1)\,p\frac{(p-\frac{1}{2})^{k_1-m_1-\frac{1}{2}}}{(p+\frac{1}{2})^{k_1+m_1+\frac{1}{2}}}\times\Pi(2m_2)\,p\frac{(p-\frac{1}{2})^{k_2-m_2-\frac{1}{2}}}{(p+\frac{1}{2})^{k_2+m_2+\frac{1}{2}}} $$

$$ = \Pi(2m_1)\,\Pi(2m_2)\,p\frac{(p-\frac{1}{2})^{k_1+k_2-m_1-m_2-1}}{(p+\frac{1}{2})^{k_1+k_2+m_1+m_2+1}} $$

and using (81). The resulting integral property is as follows:

$$\int_0^t \tau^{m_1-\frac{1}{2}}(t-\tau)^{m_2-\frac{1}{2}} M_{k_1,m_1}(\tau) M_{k_2,m_2}(t-\tau)\,d\tau$$
$$= \frac{\Pi(2m_1)\,\Pi(2m_2)}{\Pi(2m_1+2m_2+1)} t^{m_1+m_2} M_{k_1+k_2,\,m_1+m_2+\frac{1}{2}}(t). \quad (88)$$

Though the derivation of (88) by operational methods is quite simple, it is of great importance, as can be seen from its particular cases. For instance, using the known expression for the Bessel function in terms of Whittaker functions, viz.

$$I_\nu(x) = \frac{1}{2^{2\nu+\frac{1}{2}}\Pi(\nu)} \frac{M_{0,\nu}(2x)}{\sqrt{x}},$$

we readily deduce from (88) that

$$\int_0^t I_{\nu_1}(\tau)\, I_{\nu_2}(t-\tau)\, \tau^{\nu_1}(t-\tau)^{\nu_2} d\tau = \frac{\Pi(\nu_1-\frac{1}{2})\,\Pi(\nu_2-\frac{1}{2})}{\sqrt{(2\pi)}\,\Pi(\nu_1+\nu_2)} t^{\nu_1+\nu_2+\frac{1}{2}} I_{\nu_1+\nu_2+\frac{1}{2}}(t)$$
$$(\operatorname{Re}\nu_1 > -\tfrac{1}{2};\ \operatorname{Re}\nu_2 > -\tfrac{1}{2}).$$

In like manner many other results could have been derived. Our main aim, however, was to illustrate the fact that, whenever the image of some function or other is simple, the operational method very quickly produces relations (especially when use is made of the composition-product rule) that are either unknown or at best unwieldy to tackle otherwise.

CHAPTER XI

OPERATIONAL RULES OF MORE COMPLICATED CHARACTER

1. Introduction

This chapter deals with operational rules that are not as elementary as those discussed in chapter IV. The simplest of them are obtained either by

(*a*) replacing p in $f(p)$ by some function of p, and then investigating the corresponding transformation of the original (§ 2), or by

(*b*) replacing t in $h(t)$ by some function of t, and investigating the corresponding transformation of the image (§ 3).

The transformation from t to e^t (§ 4) proves to be of particular importance. It allows the interrelation of a function and the coefficients of the corresponding power series to be treated operationally (§ 5). Moreover, in § 6, numerous rules concerning series are worked out in some detail. The last two sections, §§ 7 and 8, are devoted to identities derivable from given operational relations.

2. Rules obtained when p is replaced by a function of p

To make the general method clear, we shall give a simple example, confining ourselves to one-sided functions. If we have to express the original of $f(1/p)$ in terms of that of $f(p)$, we first recall that a given operational relation,

$$f(p) \doteq h(t)\,U(t) \quad (\operatorname{Re} p > 0),$$

leads, by means of its definition integral, to

$$f\left(\frac{1}{p}\right) = \frac{1}{p}\int_0^\infty e^{-s/p}\,h(s)\,ds \quad (\operatorname{Re} p > 0).$$

Let us next suppose that the original of the right-hand side can be obtained by transposition of the integrand (this will be legitimate under some mild restrictions); then, upon applying (x, 20) for $\nu = 1$, we find that

$$f\left(\frac{1}{p}\right) \doteq U(t)\int_0^\infty \sqrt{\left(\frac{t}{s}\right)}\,J_1\{2\sqrt{(ts)}\}\,h(s)\,ds \quad (\operatorname{Re} p > 0). \tag{1}$$

In like manner, by using (x, 20) for arbitrary ν, the general result is found to be

$$\frac{f(1/p)}{p^{\nu-1}} \doteq U(t)\int_0^\infty \left(\frac{t}{s}\right)^{\frac{1}{2}\nu} J_\nu\{2\sqrt{(ts)}\}\,h(s)\,ds \quad (\operatorname{Re} p > 0;\ \operatorname{Re}\nu > -1), \tag{2}$$

of which the particular case $\nu = 0$ may be written as follows:

$$pf\left(\frac{1}{p}\right) \doteq U(t)\int_0^\infty J_0\{2\sqrt{(ts)}\}\,h(s)\,ds \quad (\operatorname{Re} p > 0). \tag{3}$$

The operational rule (1) is of primary importance in network theory with respect to a comparison of the properties of low-pass and high-pass filters (these are filters letting through only frequencies below/above a certain frequency ω_0 without any appreciable attenuation). For instance, from the homogeneous low-pass filter described in the first example of IX, § 7, a corresponding high-pass filter can be derived by interchanging the self-inductances and capacitances in the circuit of fig. 61, or, in other words, by interchanging the series impedance $Q(p) = Lp$ and the shunt impedance $Z(p) = 1/Cp$. As a consequence, amongst others, the admittance of the high-pass filter in question is obtainable from that of the low-pass filter (IX, 37) by replacing p in the latter by $1/CLp = \omega_0^2/4p$†. Therefore, designating the operational relation corresponding to the current $i_\delta(t)$ through the low-pass filter (see (IX, 38)) by $h(t) \risingdotseq f(p)$, the image of the current $i_\Gamma(t)$ for the low-pass filter becomes $f(p)/p$ and for the high-pass filter $\frac{4p}{\omega_0^2} f\left(\frac{\omega_0^2}{4p}\right)$. Thus, with rule (3), we have for the current through the high-pass filter in response to a unit-function voltage across its terminals

$$U(t)\int_0^\infty J_0\{\omega_0 \sqrt{(ts)}\} h(s)\, ds.$$

This may be worked out for the special form of $h(s)$ according to (IX, 38), and leads to the result that

$$i_\Gamma(t) = U(t)\sqrt{\left(\frac{C}{L}\right)}\int_0^\infty J_0\{\sqrt{(\omega_0 ts)}\} J_{2n}(s)\, ds.$$

Again considering the way that led us to (1) and (2) we clearly see how in other cases, after substituting $\phi(p)$ for p, we may possibly express the new original in terms of that of the initial operational relation. For that purpose it is only necessary that the integrand of the definition integral of the given relation, when $\phi(p)$ replaces p, should be operationally transposable by some known relation.

Of most importance for practical applications are those rules which are derived with the aid of one-sided relations $f(p) \risingdotseq h(t)\, U(t)$ valid for $\text{Re}\, p > 0$. In this way it is easy to prove the following rules, using some suitable relation as basis:

(1) When in (VII, 40) p is replaced by s^2p, and afterwards the result obtained is differentiated with respect to s, then it follows that

$$\sqrt{p}\, e^{-s\sqrt{p}} \risingdotseq U(t)\frac{e^{-s^2/4t}}{\sqrt{(\pi t)}}, \quad 0 < \text{Re}\, p < \infty. \tag{4}$$

† See Balth. van der Pol, 'A theorem on electrical networks with an application to filters', *Physica*, 's-Grav., I, 521, 1934.

Proceeding in the same manner as indicated above we then come to

$$f(\sqrt{p}) \doteqdot \frac{U(t)}{\sqrt{(\pi t)}} \int_0^\infty e^{-s^2/4t} h(s)\, ds \quad (\operatorname{Re} p > 0). \tag{5}$$

(2) Let us apply the shift rule to (x, 20) for $\nu = 0$, then

$$\exp\left\{-s\left(p+\frac{1}{p}\right)\right\} \doteqdot J_0(2\sqrt{\{s(t-s)\}})U(t-s) \quad (\operatorname{Re} p > 0),$$

whence it follows that

$$\frac{f\left(p+\frac{1}{p}\right)}{p+\frac{1}{p}} \doteqdot U(t)\int_0^t J_0(2\sqrt{\{s(t-s)\}})\, h(s)\, ds \quad (\operatorname{Re} p > 0). \tag{6}$$

This rule is particularly useful for computing the transient phenomena† of a band-pass filter (a filter letting through only frequencies in a finite interval), once the analogues for a low-pass filter are known; a similar situation was seen to hold for (3) with respect to the transition from a low-pass to a high-pass filter.

(3) From the operational relation (x, 48) for $\nu = 0$ we have

$$J_0\{\sqrt{(t^2-s^2)}\}\, U(t-s) \doteqdot \frac{p}{\sqrt{(p^2+1)}} e^{-s\sqrt{(p^2+1)}} \quad (\operatorname{Re} p > 0),$$

and consequently the new rule

$$\frac{p}{p^2+1} f\{\sqrt{(p^2+1)}\} \doteqdot U(t)\int_0^t J_0\{\sqrt{(t^2-s^2)}\}\, h(s)\, ds \quad (\operatorname{Re} p > 0). \tag{7}$$

Numerous rules of similar character can likewise be deduced‡. Yet, other simple transformations do not necessarily lead to simple rules at all. For instance, the transformation $p \to p^2$ leads to e^{-sp^2} which has no original whatever. We would also remark that, in addition to the *two* operational relations always involved in rules such as those above, there sometimes exists a *third* relation of equal simplicity, especially if the rule under consideration contains an integral of the exponential type. In the case of (5), for instance, when denoting by $h_1(t)$ the original of $f(\sqrt{p})$, we have first the identity

$$h_1(t) = \frac{1}{\sqrt{(\pi t)}} \int_0^\infty e^{-s^2/4t} h(s)\, ds \quad (t > 0),$$

and then, putting $p = 1/4t$, $s = \sqrt{u}$,

$$\sqrt{(\pi p)}\, h_1\left(\frac{1}{4p}\right) = p\int_0^\infty e^{-pu} \frac{h(\sqrt{u})}{\sqrt{u}}\, du \quad (p > 0),$$

† See S. Ekelöf, *Elekt. Nachr.-Tech.* XII, 100, 1935.

‡ See, for example, K. F. Niessen, *Phil. Mag.* XX, 977, 1935.

which, clearly, is nothing but the definition integral of a new relation. Consequently, an alternative way of writing the rule (5) is in the form of the following three coexistent operational relations:

$$\left.\begin{aligned} &h(t)\,U(t) \doteq f(p), \\ &f(\sqrt{p}) \doteq \frac{U(t)}{\sqrt{(\pi t)}} \int_0^\infty e^{-s^2/4t}\, h(s)\, ds \equiv h_1(t)\, U(t), \\ &\sqrt{(\pi p)}\, h_1\!\left(\frac{1}{4p}\right) \doteq \frac{h(\sqrt{t})}{\sqrt{t}}\, U(t). \end{aligned}\right\} \quad (\operatorname{Re} p > 0). \qquad (8)$$

Let us now discuss some applications of the rules obtained above.

Example 1. *Integrals involving products of Bessel functions.* Since in rules like (3) a Bessel function occurs, it is quite easy to evaluate certain integrals involving a product of Bessel functions, by simply taking another Bessel function for $h(s)$. For instance, when (2) is applied to

$$e^{-\lambda t}\left(\frac{t}{a}\right)^{\frac{1}{2}\nu} J_\nu\{2\sqrt{(at)}\}\, U(t) \doteq \frac{p}{(p+\lambda)^{\nu+1}}\, e^{-a/(p+\lambda)} \quad (\operatorname{Re} p > -\lambda;\ \operatorname{Re}\nu > -1),$$

which may be found from (x, 20) with the aid of the attenuation rule, it follows that

$$\frac{p}{(1+\lambda p)^{\nu+1}} \exp\left\{\frac{a}{\lambda(1+\lambda p)}\right\} \doteq U(t)\, e^{a/\lambda}\left(\frac{t}{a}\right)^{\frac{1}{2}\nu} \int_0^\infty e^{-\lambda s} J_\nu\{2\sqrt{(as)}\}\, J_\nu\{2\sqrt{(ts)}\}\, ds$$
$$(\operatorname{Re} p > 0;\ \lambda > 0;\ \operatorname{Re}\nu > -1).$$

Moreover, by means of (x, 26) and the attenuation rule, the original corresponding to the left-hand member is readily determined, whence the identity

$$\exp\left\{-\frac{(t+a)}{\lambda}\right\} I_\nu\left(\frac{2}{\lambda}\sqrt{(at)}\right) = \lambda \int_0^\infty e^{-\lambda s} J_\nu\{2\sqrt{(as)}\}\, J_\nu\{2\sqrt{(ts)}\}\, ds \quad (\lambda > 0;\ \operatorname{Re}\nu > -1).$$

Finally, when t is replaced by b, and λ is identified with p, we obtain the definition integral corresponding to the following new relation:

$$e^{-(a+b)/p} I_\nu\left(\frac{2\sqrt{(ab)}}{p}\right) \doteq J_\nu\{2\sqrt{(at)}\}\, J_\nu\{2\sqrt{(bt)}\}\, U(t), \quad 0 < \operatorname{Re} p < \infty \quad (\operatorname{Re}\nu > -1). \qquad (9)$$

Example 2. After some slight changes of notation, the relation (9) can also be written as follows:

$$e^{-(a^2+b^2)/4p} I_\nu\left(\frac{ab}{2p}\right) \doteq J_\nu(a\sqrt{t})\, J_\nu(b\sqrt{t})\, U(t) \quad (\operatorname{Re} p > 0;\ \operatorname{Re}\nu > -1),$$

which, as a matter of fact, is just the third relation of the triplet (8), provided we put

$$h(t) = tJ_\nu(at)\, J_\nu(bt), \quad h_1(x) = 2\sqrt{\left(\frac{x}{\pi}\right)}\, e^{-(a^2+b^2)x} I_\nu(2abx).$$

Furthermore, the image of $h_1(t)\,U(t)$ is easily found with the help of (x, 64); the result is

$$h_1(t)\,U(t) \doteq -\frac{p}{\pi(ab)^{\frac{3}{2}}}\, Q'_{\nu-\frac{1}{2}}\left(\frac{p+a^2+b^2}{2ab}\right), \quad \operatorname{Re} p > -(a-b)^2 \ (\operatorname{Re}\nu > -\tfrac{3}{2}).$$

In its turn the last relation may be considered as the second of the system (8), if the image is put equal to $f(\sqrt{p})$. The image $f(p)$ of the first relation of (8) then becomes known also. This first relation then leads by way of the rule (IV, 38) to the final result:

$$J_\nu(at)\, J_\nu(bt)\, U(t) \doteq \frac{p}{\pi\sqrt{(ab)}}\, Q_{\nu-\frac{1}{2}}\left(\frac{p^2+a^2+b^2}{2ab}\right), \quad 0<\operatorname{Re} p<\infty \quad (\operatorname{Re}\nu > -\tfrac{1}{2}), \tag{10}$$

which, once more, demonstrates the intimate connexion existing between Legendre functions and Bessel functions.

Example 3. *Images for θ-functions.* Theta functions are generally characterized by infinite series of exponential terms, the exponents containing the variable of summation, n, to the second degree. A special θ-function was encountered in (VI, 39); for arbitrary arguments they are defined below:

$$\left.\begin{aligned}
\theta_0(v,t) &= \sum_{n=-\infty}^{\infty} e^{-\pi n^2 t + 2\pi i n(v+\frac{1}{2})} = \frac{1}{\sqrt{t}} \sum_{n=-\infty}^{\infty} e^{-\pi(n+v+\frac{1}{2})^2/t},\\
\theta_1(v,t) &= \sum_{n=-\infty}^{\infty} e^{-\pi(n-\frac{1}{2})^2 t + 2\pi i(n-\frac{1}{2})(v+\frac{1}{2})} = \frac{1}{\sqrt{t}} \sum_{n=-\infty}^{\infty} (-1)^n e^{-\pi(n+v+\frac{1}{2})^2/t},\\
\theta_2(v,t) &= \sum_{n=-\infty}^{\infty} e^{-\pi(n-\frac{1}{2})^2 t + 2\pi i(n-\frac{1}{2})v} = \frac{1}{\sqrt{t}} \sum_{n=-\infty}^{\infty} (-1)^n e^{-\pi(n+v)^2/t},\\
\theta_3(v,t) &= \sum_{n=-\infty}^{\infty} e^{-\pi n^2 t + 2\pi i n v} = \frac{1}{\sqrt{t}} \sum_{n=-\infty}^{\infty} e^{-\pi(n+v)^2/t}.
\end{aligned}\right\} \tag{11}$$

The pairs of corresponding series are equal in virtue of Poisson's sum formula (VI, 34). Let us now deduce their respective images, by using (5); the results being quite simple, we shall treat only $\theta_3(v,t)$ in any detail. To this end the second series of θ_3 is put into the form

$$\theta_3(v,t)\, U(t) = \frac{U(t)}{\sqrt{(\pi t)}} \int_0^\infty e^{-s^2/4t} \sum_{n=0}^{\infty} \left[\delta\left\{\frac{s}{\sqrt{\pi}} - 2(n+v)\right\} + \delta\left\{\frac{s}{\sqrt{\pi}} - 2(n-v+1)\right\}\right] ds, \tag{12}$$

which, however, only holds when the value of s at the singularities of the separate terms is positive throughout, that is, when $0<v<1$. In this case the right-hand member of (12) is easily recognized as the original of (5), provided $h(s)$ is taken equal to the sum of the impulse functions and accordingly we put

$$f(p) = \sqrt{\pi}\, p \left\{ \sum_{n=0}^{\infty} e^{-2\sqrt{\pi}\, p(n+v)} + \sum_{n=0}^{\infty} e^{-2\sqrt{\pi}\, p(n-v+1)} \right\}.$$

When $\operatorname{Re} p>0$ the summation above is easily carried out; it leads to the following expression for the image:

$$f(p) = \sqrt{\pi}\, p\, \frac{\cosh\{(2v-1)\sqrt{\pi}\, p\}}{\sinh(\sqrt{\pi}\, p)}.$$

According to (5) the image of $\theta_3(v,t)\, U(t)$ is obtained by substituting $p \to \sqrt{p}$, so as to have

$$\theta_3(v,t)\, U(t) \doteq \sqrt{(\pi p)}\, \frac{\cosh\{(2v-1)\sqrt{(\pi p)}\}}{\sinh\{\sqrt{(\pi p)}\}}, \quad 0<\operatorname{Re} p<\infty \quad (0 \leqslant v \leqslant 1). \tag{13}$$

The images of the remaining θ-functions defined in (11) are dealt with in like manner; for results the reader may be referred to the dictionary at the end of the book.

3. Rules obtained when t is replaced by a function of t

An example may illustrate how this type of rule can actually be established. As before, the starting-point is a given one-sided operational relation convergent to the right of the imaginary axis; thus

$$f(p) \doteqdot h(t)\,U(t) \quad (\operatorname{Re} p > 0),$$

and this time the question is whether the image of $h(1/t)\,U(t)$ can be expressed in terms of that of $h(t)$. The image to be investigated can be written with the help of the definition integral of the given relation in the following form:

$$p\int_0^\infty e^{-ps}\,h\left(\frac{1}{s}\right)ds = \int_0^\infty \frac{e^{-p/w}}{w/p}\,h(w)\,\frac{dw}{w}.$$

Now, the first factor of the integrand of the right-hand integral is expressible as a definite integral, namely,

$$\frac{e^{-p/w}}{w/p} = \frac{w}{p}\int_0^\infty e^{-ws/p}\,\sqrt{s}\,J_1(2\sqrt{s})\,ds,$$

as is easily verified by means of the definition integral corresponding to (x, 20) for $\nu = 1$. Consequently, the image in question is written in the form of a repeated integral; if then the order of integration is reversed the integral over w is easily recognized as the definition integral of the initial relation. Finally, replacing s by ps we obtain

$$h\left(\frac{1}{t}\right)U(t) \doteqdot \int_0^\infty J_1\{2\sqrt{(ps)}\}\sqrt{\left(\frac{p}{s}\right)}\,f(s)\,ds. \tag{14}$$

This rule has sense only in so far as the integral involved does converge, which, however, will be the case under suitable conditions (the image of $h(1/t)\,U(t)$ has to exist of course).

As an example of (14) we may study its application to the relation (III, 3), with the result that

$$\frac{t^{-\nu}}{\Pi(\nu)}\,U(t) \doteqdot \int_0^\infty J_1(u)\left(\frac{4p}{u^2}\right)^\nu du \quad (\operatorname{Re} p > 0;\ -\tfrac{1}{4} < \operatorname{Re}\nu < 1).$$

Since another image is known for the original at the left, we simply have to identify the two in order to get the formula

$$\int_0^\infty \frac{J_1(u)}{u^{2\nu}}\,du = \frac{\Pi(-\nu)}{4^\nu\,\Pi(\nu)} \quad (-\tfrac{1}{4} < \operatorname{Re}\nu < 1).$$

In like manner to (14), other rules can be established when the original is of the type $h\{\phi(t)\}\,U(t)$. To this end we substitute $\phi(s) = w$ in the corresponding definition integral, usually leading to a new integral that can be transformed into a repeated integral with the aid of some known operational relation. Upon reversing the order of integration, the inner integral is often easily recognized as the definition integral of the relation initially given. Some further examples of this procedure are quoted below.

(1) Using the definition integral corresponding to (4) it is found that

$$h(t^2)\,U(t) \doteqdot \frac{1}{\sqrt{\pi}}\int_0^\infty e^{-s} f\left(\frac{p^2}{4s}\right)\frac{ds}{\sqrt{s}} \quad (\operatorname{Re} p > 0). \tag{15}$$

For example, when (x, 20) for the Bessel function $t^{\frac{1}{2}\nu}J_\nu(2\sqrt{t})\,U(t)$ is assumed to be known, then the relation (x, 18) concerning $t^\nu J_\nu(t)\,U(t)$ is obtained at once.

(2) From the definition integral of (x, 12) it follows that

$$h(\lambda \sinh t)\,U(t) \doteqdot p\int_0^\infty J_p(\lambda s)\frac{f(s)}{s}\,ds \quad (\operatorname{Re} p > 0). \tag{16}$$

(3) Next, with the help of the definition integral of

$$\frac{p}{(p+\lambda)^{\alpha+1}} \doteqdot e^{-\lambda t}\frac{t^\alpha}{\Pi(\alpha)}\,U(t) \quad (\operatorname{Re} p > -\lambda),$$

we obtain

$$h\{\lambda(e^t-1)\}\,U(t) \doteqdot \frac{1}{\Gamma(p)}\int_0^\infty e^{-u}u^{p-1} f\left(\frac{u}{\lambda}\right)du \quad (\operatorname{Re} p > 0). \tag{17}$$

Numerous rules of similar character can thus be deduced, while, on the other hand, certain transformations, of equal simplicity, often do not lead to any rule at all; e.g. the case of $t \to \sqrt{t}$, where e^{-p^2} does not have an original. Again, whenever the rule contains an integral of the exponential type, a third rule can be added. For instance, when the image of (15) is considered as some new definition integral, the complete set of coexistent relations is as follows:

$$\left.\begin{aligned} f(p) &\doteqdot h(t)\,U(t), \\ h(t^2)\,U(t) &\doteqdot \frac{1}{\sqrt{\pi}}\int_0^\infty e^{-s} f\left(\frac{p^2}{4s}\right)\frac{ds}{\sqrt{s}} \equiv f_1(p), \\ \sqrt{(\pi p)}\,f_1(\sqrt{p}) &\doteqdot \frac{f(1/4t)}{\sqrt{t}}. \end{aligned}\right\} \quad (\operatorname{Re} p > 0) \tag{18}$$

This system is closely related to that given before in (8).

As to possible applications of the rules discussed so far, the transformation $t \to e^t$ proves very useful; it is therefore thought worth while to treat this transformation in some detail in the following section.

4. The exponential transformation: $t \to e^t$

Whereas all rules thus far discussed in the present chapter deal with one-sided originals, whether in the given relation or in the new relation obtained upon transformation of either p or t, it proves simplest to take the new relation two-sided when studying the special transformation $t \to e^t$. This amounts to asking for the image of $h(e^t)$ when starting with

$$h(t)\,U(t) \doteqdot f(p) \quad (\operatorname{Re} p > 0).$$

Before going on we may emphasize that for the transformation $t = e^\tau$ the initial original has to be one-sided, necessarily, since negative values of t would otherwise correspond to non-real values of τ.

Proceeding then in the same manner as in the foregoing section, we first write the definition integral of the image in question as follows:

$$h(e^t) \doteq p \int_{-\infty}^{\infty} e^{-ps} h(e^s)\, ds.$$

After putting $e^s = w$ we obtain next

$$h(e^t) \doteq p \int_0^{\infty} \frac{h(w)}{w^{p+1}}\, dw,$$

of which the right-hand side can also be written in the form of a repeated integral, thus:

$$\frac{1}{\Gamma(p)} \int_0^{\infty} dw\, h(w) \int_0^{\infty} ds\, e^{-ws} s^p \quad (\mathrm{Re}\, p > -1).$$

Upon reversing the order of integration, the integral with respect to w is easily recognized as the definition integral of the operational relation originally given; we have consequently

$$h(e^t) \doteq \frac{1}{\Gamma(p)} \int_0^{\infty} s^{p-1} f(s)\, ds. \tag{19}$$

Again, this rule has sense only if the image of $h(e^t)$ exists and the integral involved converges. In virtue of the simple character of the integrand in (19), the exponential substitution $s = e^{-\tau}$ allows the image to be considered as the definition integral corresponding to a new third relation. As in (8) and (18), a triplet of corresponding operational relations consequently is obtained, viz.

$$\left.\begin{aligned} &h(t)\, U(t) \doteq f(p), \\ &h(e^t) \doteq \frac{1}{\Gamma(p)} \int_0^{\infty} s^{p-1} f(s)\, ds \equiv f_1(p), \\ &f(e^{-t}) \doteq \Pi\,(p) f_1(p). \end{aligned}\right\} \tag{20}$$

We shall give now a number of operational relations that are most rapidly deduced by means of the exponential transformation under consideration.

(1) When (19) is applied to

$$P(\nu+1, t)\, U(t) = U(t) \int_0^t e^{-s} s^\nu\, ds \doteq \frac{\Pi(\nu)}{(p+1)^{\nu+1}} \quad (\mathrm{Re}\, p > 0;\ \mathrm{Re}\, \nu > -1),$$

which is a consequence of (III, 3), it is found that

$$P(\nu+1, e^t) \doteq \frac{\Pi(\nu)}{\Gamma(p)} \int_0^{\infty} \frac{s^{p-1}}{(s+1)^{\nu+1}}\, ds, \quad 0 < \mathrm{Re}\, p < \mathrm{Re}\, \nu + 1.$$

Now, the image can be evaluated by using Euler's first integral formula, if the substitution $s = 1/u - 1$ is made as well. The result is found to be

$$P(\nu+1, e^t) \doteq \Gamma(\nu+1-p), \quad 0 < \operatorname{Re} p < \operatorname{Re} \nu + 1, \tag{21}$$

in which it is striking that both a complete and an incomplete Γ-function are present. In addition, the third relation of (20) in this case becomes

$$\frac{1}{(1+e^{-t})^{\nu+1}} \doteq \frac{\Pi(p)\,\Pi(\nu-p)}{\Pi(\nu)}, \quad 0 < \operatorname{Re} p < \operatorname{Re} \nu + 1. \tag{22}$$

(2) Let us apply (19) to the relation (x, 20) involving the Bessel function. We then arrive at

$$e^{\frac{1}{2}\nu t} J_\nu(2e^{\frac{1}{2}t}) \doteq \frac{1}{\Gamma(p)} \int_0^\infty e^{-1/s} s^{p-\nu-1}\, ds \quad (\tfrac{1}{2}\operatorname{Re}\nu - \tfrac{3}{4} < \operatorname{Re} p < \operatorname{Re}\nu).$$

By the substitution $s = 1/w$ the integral is readily transformed into that of Euler for the Γ-function. Consequently

$$J_\nu(2e^{\frac{1}{2}t}) \doteq p \frac{\Gamma(\frac{1}{2}\nu - p)}{\Gamma(\frac{1}{2}\nu + p + 1)}, \quad -\tfrac{3}{4} < \operatorname{Re} p < \tfrac{1}{2}\operatorname{Re}\nu \ (\operatorname{Re}\nu > -\tfrac{3}{2}). \tag{23}$$

As to the determination of the left-hand boundary of the strip of convergence in the case of (23), this is not so simple; a short account may suffice. Its value specified above follows from the asymptotic series for J_ν corresponding to (x, 32), giving that the integrand of the definition integral belonging to (23) behaves, for $t \to +\infty$, like that of the definition integral for $\sin(e^{-t})$ for $t \to -\infty$, when in the latter the variable p is replaced by $-\frac{1}{2} - 2p$. And consequently, since according to (vi, 24) $\sin(e^{-t})$ has for its right-hand boundary $\operatorname{Re} p = 1$, the left-hand boundary of (23) is determined by $\operatorname{Re}(-\frac{1}{2} - 2p) = 1$, that is, $\operatorname{Re} p = -\frac{3}{4}$.

Furthermore, for the interval $-\frac{3}{2} < \operatorname{Re}\nu < \frac{3}{2}$, ν can be replaced by $-\nu$ in (23), and thus, by simply combining the two relations and using (x, 14), an image of the Hankel function of exponential argument is found. After some simplification, using a few elementary rules together with the Γ-function property expressed by (iv, 17), the simplest form of the operational relation under discussion is found to be as follows:

$$H_\nu^{(1)(2)}(2e^{-\frac{1}{2}t}) \doteq \mp \frac{i}{\pi} p\, e^{\pm i\pi(p-\frac{1}{2}\nu)}\, \Gamma(p+\tfrac{1}{2}\nu)\, \Gamma(p-\tfrac{1}{2}\nu),$$
$$\tfrac{1}{2}|\operatorname{Re}\nu| < \operatorname{Re} p < \tfrac{3}{4} \ (-\tfrac{3}{2} < \operatorname{Re}\nu < \tfrac{3}{2}). \tag{24}$$

Finally, by a suitable complex shift, namely, $t \to t - \pi i$, an image is obtained for the function K_ν (see (x, 29)), viz.

$$K_\nu(2e^{-\frac{1}{2}t}) \doteq \tfrac{1}{2} p\, \Gamma(p+\tfrac{1}{2}\nu)\, \Gamma(p-\tfrac{1}{2}\nu), \quad \tfrac{1}{2}|\operatorname{Re}\nu| < \operatorname{Re} p < \infty. \tag{25}$$

(3) Operational relations involving Legendre functions of exponential arguments, as obtained by applying (20) to (x, 84). We have, simultaneously,

$$\frac{P_\nu^\mu(e^t)}{(e^{2t}-1)^{\frac{1}{2}\mu}}U(t) \doteqdot \sqrt{\left(\frac{2}{\pi}\right)}\frac{1}{\Gamma(p)}\int_0^\infty s^{p+\mu-\frac{1}{2}}K_{\nu+\frac{1}{2}}(s)\,ds = f_1(p),$$

$$\sqrt{\left(\frac{2}{\pi}\right)}e^{-(\mu+\frac{1}{2})t}K_{\nu+\frac{1}{2}}(e^{-t}) \doteqdot \Pi(p)f_1(p),$$

in which the function $f_1(p)$ can be determined by transposing the original of the second relation with the help of (25) just proved. Accordingly,

$$\frac{P_\nu^\mu(e^t)}{(e^{2t}-1)^{\frac{1}{2}\mu}}U(t) \doteqdot \frac{2^{p+\mu-1}}{\sqrt{\pi}}\frac{\Gamma\left(\dfrac{p+\mu-\nu}{2}\right)\Gamma\left(\dfrac{p+\mu+\nu+1}{2}\right)}{\Gamma(p)},$$

$$|\operatorname{Re}\nu+\tfrac{1}{2}|-\operatorname{Re}\mu-\tfrac{1}{2}<\operatorname{Re}p<\infty \quad (\operatorname{Re}\mu<1). \qquad (26)$$

In concluding this section, some examples are given showing how the exponential relations above can lead to many properties of the functions treated in the preceding chapter; they are completely different from those obtained before with the help of originals not containing exponential arguments.

Example 1. *Properties of Bessel functions.* First of all, from the definition integral of

$$e^{-\mu t}J_\nu(2e^{\frac{1}{2}t}) \doteqdot p\frac{\Gamma(\frac{1}{2}\nu-\mu-p)}{\Gamma(\frac{1}{2}\nu+\mu+p+1)}, \quad -\tfrac{3}{4}-\operatorname{Re}\mu<\operatorname{Re}p<\tfrac{1}{2}\operatorname{Re}\nu-\operatorname{Re}\mu,$$

which itself follows from (23), and upon performing the substitutions $s = 2e^{\frac{1}{2}t}$ and $p = \frac{1}{2}(\lambda-1-2\mu)$, the following integral of Weber is obtained:

$$\int_0^\infty \frac{J_\nu(s)}{s^\lambda}ds = \frac{\Gamma\left(\dfrac{\nu-\lambda+1}{2}\right)}{2^\lambda\Gamma\left(\dfrac{\nu+\lambda+1}{2}\right)} \quad (-\tfrac{1}{2}<\operatorname{Re}\lambda<\operatorname{Re}\nu+1). \qquad (27)$$

Secondly, we are able to express the differential quotient of $J_\nu(x)$ with respect to the order ν in a manner quite different from that already given in the third example of x, § 5. Thus, by differentiation of (23) with respect to ν, the image is obtained as the product of a function of p, whose original again follows from (23), and another function of p consisting of the difference of two ψ-functions, whose operational image (in the sense of principal values) is known from (VI, 2). Therefore, the original of the complete image can be obtained by applying the composition-product rule; putting $x = 2e^{\frac{1}{2}t}$, the following result comes out:

$$\frac{\partial}{\partial\nu}J_\nu(x) = \frac{1}{2}\int_{-\infty}^\infty \frac{e^{-\frac{1}{2}\nu|\tau|}}{e^\tau-1}J_\nu(xe^{-\frac{1}{2}\tau})\,d\tau \quad (\operatorname{Re}\nu>-1).$$

Example 2. *Properties of Legendre functions.* In this case the relation (26) proves useful. Particularly simple results are obtained for the Legendre polynomials, where $\nu = n > 0$ and $\mu = 0$, since then the arguments of the Γ-functions in the numerator, except for the number $\frac{1}{2}p$, are equal to an integral multiple of $\frac{1}{2}$, and therefore the image can be written as a simple rational function, by using the duplication formula of the Γ-function (IV, 18)

$$\left.\begin{aligned} P_{2n}(\mathrm{e}^t)\,U(t) &\doteq \frac{(p+1)(p+3)(p+5)\dots(p+2n-1)}{(p-2)(p-4)(p-6)\dots(p-2n)}, \quad 2n<\operatorname{Re}p<\infty, \\ P_{2n+1}(\mathrm{e}^t)\,U(t) &\doteq \frac{p(p+2)(p+4)\dots(p+2n)}{(p-1)(p-3)(p-5)\dots(p-2n-1)}, \quad 2n+1<\operatorname{Re}p<\infty. \end{aligned}\right\} \tag{28}$$

The importance of these simple images is self-evident; when in the corresponding definition integral a change of variable is made, viz. $t = \log s$, then we obtain the formulae

$$\int_1^\infty \frac{P_{2n}(s)}{s^\alpha}\,ds = \frac{\alpha(\alpha+2)(\alpha+4)\dots(\alpha+2n-2)}{(\alpha-1)(\alpha-3)(\alpha-5)\dots(\alpha-2n-1)} \quad (\alpha>2n+1),$$

$$\int_1^\infty \frac{P_{2n+1}(s)}{s^\alpha}\,ds = \frac{(\alpha+1)(\alpha+3)(\alpha+5)\dots(\alpha+2n-1)}{(\alpha-2)(\alpha-4)(\alpha-6)\dots(\alpha-2n-2)} \quad (\alpha>2n+2).$$

The rational images of (28) cannot represent either driving point impedances or driving point admittances, since all their poles are to the right of the imaginary axis. However, it appears that these functions can represent transfer quantities, as e.g. Z_{ik}^{-1} or A_{ik}^{-1} (see IX, § 3). In this connexion it is remarked that in the first example of IX, § 8 we discussed a circuit allowing the impedance representation of a function closely related to (28).

Yet another application of (26) is possible, which is analogous to the previous example of differentiating the Bessel function with respect to its order, and yields an expression for the derivative of the Legendre function, $\frac{\partial}{\partial\nu}P_\nu(x)$. Proceeding in the same manner, we shall take, instead of (VI, 2), the relation

$$p\{\psi(p+a)-\psi(p+b)\} \doteq U(t)\,\frac{\mathrm{e}^{-bt}-\mathrm{e}^{-at}}{\mathrm{e}^t-1} \quad (-1-\min(a,b)<\operatorname{Re}p<\infty), \tag{29}$$

which is closely related to (III, 15), and may also be derived with the aid of (III, 14). The final result for the Legendre function thereupon becomes, for $\mu = 0$,

$$\frac{\partial}{\partial\nu}P_\nu(x) = \int_{1/x}^1 P_\nu(xu)\,\frac{u^\nu-u^{-\nu-1}}{u^2-1}\,du \quad (x>1).$$

For integral values of ν, when the functions become elementary,

$$\left\{\frac{\partial}{\partial\nu}P_\nu(x)\right\}_{\nu=0} = \log\left(\frac{1+x}{2}\right),$$

$$\left\{\frac{\partial}{\partial\nu}P_\nu(x)\right\}_{\nu=1} = x-1+x\log\left(\frac{1+x}{2}\right),$$

etc.

Example 3. *Hypergeometric functions.* In like manner it would be possible to derive exponential relations for hypergeometric functions by applying (19) to (x, 78) or (x, 79) for example, but, as will be demonstrated in the case of the restricted hypergeometric series (x, 68), these results are better derived in a straightforward manner.

Closely related to the relation (22) is the following:

$$(1-\mathrm{e}^{-t})^\mu\,U(t) \doteq \Gamma(\mu+1)\,\frac{\Gamma(p+1)}{\Gamma(p+\mu+1)} \quad (\operatorname{Re}p>0)\ (\operatorname{Re}\mu>-1), \tag{30}$$

already derived in (IV, 25) by reducing the corresponding definition integral to one of Euler's integrals. Further, we have

$$(1-e^{-t})^{\gamma-1} F(\alpha,\beta;\gamma;1-e^{-t})\,U(t) \doteq \Gamma(\gamma)\frac{\Gamma(p+1)}{\Gamma(p+\gamma)}F(\alpha,\beta;p+\gamma;1)$$

$$(\mathrm{Re}\,p>\max\{0,\ \mathrm{Re}\,(\alpha+\beta-\gamma)\};\ \mathrm{Re}\,\gamma>0), \quad (31)$$

as can be verified by expanding the original in powers of $(1-e^{-t})$ and applying (30) to the individual terms, provided the series for the image converges. Moreover, the right-hand member of (31) can be evaluated explicitly in terms of Γ-functions. To see this, we first remark that the series defined in (X, 68) is also expressible in the form of a definite integral, viz.

$$F(\alpha,\beta;\gamma;x) = \frac{\Gamma(\gamma)}{\Gamma(\beta)\,\Gamma(\gamma-\beta)}\int_0^1 u^{\beta-1}(1-u)^{\gamma-\beta-1}(1-xu)^{-\alpha}\,du \quad (\mathrm{Re}\,\gamma>\mathrm{Re}\,\beta>0). \quad (32)$$

Then, as $x\to 1$, we obtain the well-known result

$$F(\alpha,\beta;\gamma;1) = \frac{\Gamma(\gamma)\,\Gamma(\gamma-\alpha-\beta)}{\Gamma(\gamma-\alpha)\,\Gamma(\gamma-\beta)}, \quad (33)$$

by which (31) is simplified to

$$(1-e^{-t})^{\gamma-1} F(\alpha,\beta;\gamma;1-e^{-t})\,U(t) \doteq \Gamma(\gamma)\frac{\Gamma(p+1)\,\Gamma(p-\alpha-\beta+\gamma)}{\Gamma(p-\alpha+\gamma)\,\Gamma(p-\beta+\gamma)},$$

$$\max\{0,\ \mathrm{Re}\,(\alpha+\beta-\gamma)\}<\mathrm{Re}\,p<\infty \quad (\mathrm{Re}\,\gamma>0). \quad (34)$$

Incidentally, we mention the corresponding inversion integral which can be written as follows:

$$\frac{x^{\gamma-1}}{\Gamma(\gamma)}F(\alpha,\beta;\gamma;x)\,U(x) = \frac{1}{2\pi i}\int_{c-i\infty}^{c+i\infty}\frac{\Gamma(p)\,\Gamma(p-\alpha-\beta+\gamma)}{\Gamma(p-\alpha+\gamma)\,\Gamma(p-\beta+\gamma)}\frac{dp}{(1-x)^p}$$

$$(c>0;\ x<1;\ \mathrm{Re}\,\gamma>0).$$

This integral is automatically zero whenever x is negative, unlike the related Barnes integral to be investigated in the next section.

Let us once more discuss the Legendre function, this time in connexion with (34). For the particular values $\alpha=\nu+1$, $\beta=-\nu$, $\gamma=1$, it is found with the help of (X, 74) that

$$P_\nu(2e^{-t}-1)\,U(t) \doteq \frac{\Gamma(p)\,\Gamma(p+1)}{\Gamma(p+\nu+1)\,\Gamma(p-\nu)}, \quad 0<\mathrm{Re}\,p<\infty, \quad (35)$$

which, again, is particularly simple for Legendre polynomials ($\nu = n$), namely,

$$P_n(2e^{-t}-1)\,U(t) \doteq \frac{(p-1)(p-2)(p-3)\dots(p-n)}{(p+1)(p+2)(p+3)\dots(p+n)}, \quad 0<\mathrm{Re}\,p<\infty. \quad (36)$$

In view of the position of the poles and zeros, the rational function of (36) can represent a transfer frequency admittance (see the first example of IX, § 8), or a transfer frequency impedance.

5. Exponential operational relations for power series, in particular hypergeometric series

We now proceed to the discussion of another class of operational relations of the exponential type, which concern power series.

Assuming the power series

$$h(x) = \sum_{n=0}^{\infty} c_n x^n \quad (|x|<\rho),$$

then we may determine the coefficients c_n by Cauchy's formula

$$c_n = \frac{1}{2\pi i}\oint\frac{h(u)}{u^{n+1}}\,du, \quad (37)$$

in which the path of integration encircles the origin of the u-plane counter-clockwise, while lying inside the circle of convergence of radius ρ. Now, the integral (37), within wide limits, can be interpreted as a definition integral as follows. Let us assume that $h(u)$ is analytic in the neighbourhood of the negative axis; then the path of integration in (37) may be deformed into the contour L of fig. 12, which has already been discussed in connexion with Hankel's expression for the Γ-function, and starts and ends at $u = -\infty$, encircling the origin once counter-clockwise. Of course, it is to be supposed that the function $h(u)$ tends to zero rapidly enough to make the integral convergent. Whereas initially the function c_n is defined by the modified expression (37) only for integral, non-negative, values of n, it may be possible to extend this definition, so as to obtain a function $c(n)$ defined for general real values of the variable n,

$$c(n) = \frac{1}{2\pi i}\int_L \frac{h(u)}{u^{n+1}}\,du, \tag{38}$$

in which the many-valued function u^{-n-1} in the integrand is specified so as to assume real values for positive u. In order that (38) be valid even for large negative values of n, it is obvious that $h(u)$ should be subjected to quite severe conditions, which, however, will be taken for granted in the following.

For negative values of n the path of integration L may be contracted to the negative u-axis, this being the cut of the function u^{-n-1}. If, moreover, the substitution $u = -s$ is performed, then the expression (38) will lead to

$$c(n) = -\frac{\sin(\pi n)}{\pi}\int_0^\infty \frac{h(-s)}{s^{n+1}}\,ds \quad (n<0). \tag{39}$$

Let us now replace n by $-p$, and substitute $s = \mathrm{e}^{-t}$; then the expression (39) becomes simply the definition integral corresponding to the following operational relation:

$$h(-\mathrm{e}^{-t}) \risingdotseq \frac{\pi p}{\sin(\pi p)}c(-p) \quad (\operatorname{Re} p>0), \tag{40}$$

which will be valid provided that the function at the left does have an image. The definition integral corresponding to (40) has also been investigated by Ramanujan†.

An interesting result is obtained when, first, to the general relation (40) the complex shifts $t \to t \pm \pi i$ are applied, and then the resulting relations are subtracted from each other; the factor $\sin(\pi p)$ being eliminated, we are led to

$$h(\mathrm{e}^{-t}+i0) - h(\mathrm{e}^{-t}-i0) = 2i\pi p\, c(-p). \tag{41}$$

This relation illustrates that only if $h(x)$ has a cut along at least a finite part of the positive x-axis is it possible that $h(\mathrm{e}^{-t})$ and $h(-\mathrm{e}^{-t})$ have images

† See G. H. Hardy, *Ramanujan*, Cambridge, 1940, p. 186.

simultaneously. For, if such a cut were absent, the original of (41) would be equal to zero, which is not the case.

As to possible applications of (40), which has not been proved so much as made plausible heuristically, it is usually necessary to investigate how the function $c(n)$, given for integral non-negative values of n, has to be continued for negative non-integral values of n, since the rigorous formula (39) often proves to be of little value. In certain cases, however, e.g. when the coefficient c_n contains only factors of the form $(n+k)!$, the continuation is performed by taking, quite naturally, $\Pi(n+k)$ for *all* n.

Numerous applications present themselves in the case of the general hypergeometric functions, defined by

$$ {}_rF_s(\alpha;\gamma;x) = \sum_{n=0}^{\infty} c_n x^n, $$

in which, according to (x, 67), c_n has the value

$$ c(n) = \frac{\Gamma(\gamma_1)\ldots\Gamma(\gamma_s)}{\Gamma(\alpha_1)\ldots\Gamma(\alpha_r)} \frac{\Gamma(\alpha_1+n)\ldots\Gamma(\alpha_r+n)}{\Gamma(\gamma_1+n)\ldots\Gamma(\gamma_s+n)} \frac{1}{\Pi(n)}. $$

Whenever the formula (39) leads to the same result for arbitrary, not necessarily integral non-negative values of n, then the relation (40) reduces to

$$ {}_rf_s(\alpha;\gamma;-e^{-t}) \equiv \frac{\Gamma(\alpha_1)\ldots\Gamma(\alpha_r)}{\Gamma(\gamma_1)\ldots\Gamma(\gamma_s)}\, {}_rF_s(\alpha;\gamma;-e^{-t}) $$

$$ \fallingdotseq \frac{\Gamma(p+1)\,\Gamma(\alpha_1-p)\ldots\Gamma(\alpha_r-p)}{\Gamma(\gamma_1-p)\ldots(\Gamma(\gamma_s-p)}, \quad 0<\operatorname{Re}p<\min(\operatorname{Re}\alpha_1,\ldots,\operatorname{Re}\alpha_r), \tag{42} $$

whose strip of convergence (if existing at all) will lie between the imaginary axis and the pole of the image farthest to the left, provided this pole itself is situated to the right of the said axis. Here we have introduced a new notation, ${}_rf_s$, in order to simplify subsequent relations and formulae in which the combination of ${}_rF_s$ and its Γ-functions $\Gamma(\alpha)$ and $\Gamma(\gamma)$ frequently occurs.

Before proceeding to the discussion of (42) we shall first generalize it. For this purpose we start with the following identity:

$$ {}_rf_s(\alpha;\gamma;-\lambda+x) = \sum_{n=0}^{\infty} {}_rf_s(\alpha+n;\gamma+n;-\lambda)\frac{x^n}{n!}, \tag{43} $$

which is not difficult to verify if the convergence of the right-hand member has been assumed. Then (40) at once leads to the generalization in question, viz.

$$ {}_rf_s(\alpha;\gamma;-\lambda-e^{-t}) \fallingdotseq \Gamma(p+1)\,{}_rf_s(\alpha-p;\gamma-p;-\lambda), $$
$$ 0<\operatorname{Re}p<\min(\operatorname{Re}\alpha_1,\ldots,\operatorname{Re}\alpha_r). \tag{44} $$

In passing we would stress the heuristic value of the rule (40), which becomes obvious from the possibility of deducing a relation like (44) so rapidly.

Instead of proving that the replacement of n by $-p$ in the coefficient of x^n occurring in (43) is actually in agreement with formula (39), it is far simpler to establish (42) and (44) in the following straightforward manner. In the first place, the convergence of the inversion integral corresponding to (42) is investigated, which is done more easily than investigating the convergence of the definition integral, since in the former case we only need to use Stirling's approximation (see (VII, 28)) for the individual Γ-functions occurring in the image. In so doing we are led to distinguish three different cases if $\lambda = 0$, namely:

(1) $r \geqslant s$. The inversion integral of (42) converges absolutely;

(2) $r = s-1$. It converges absolutely provided

$$c < \tfrac{1}{2}\operatorname{Re}(\gamma_1 + \ldots + \gamma_s - \alpha_1 - \ldots - \alpha_r - 1);$$

(3) $r < s-1$. The inversion integral diverges absolutely.

From this survey we learn that there is no question of (42) when $r < s-1$. Further, since $f(p)/p$ has first-order poles at the points contained in the decreasing sequence of numbers $p = 0, -1, -2, \ldots$ and the r increasing sequences $\alpha_m, \alpha_m+1, \alpha_m+2, \ldots$ $(m = 1, 2, \ldots, r)$, one and the same image can only exist within a strip lying between two poles, consecutive with respect to their real parts. We shall confine ourselves to the case where $\alpha_1, \alpha_2, \ldots, \alpha_r$ have positive real parts (see fig. 77), the two systems of poles mentioned above then being completely separated from each other by the strip of convergence specified in (42). As a consequence, the path of integration of the inversion integral also separates the two sets of poles.

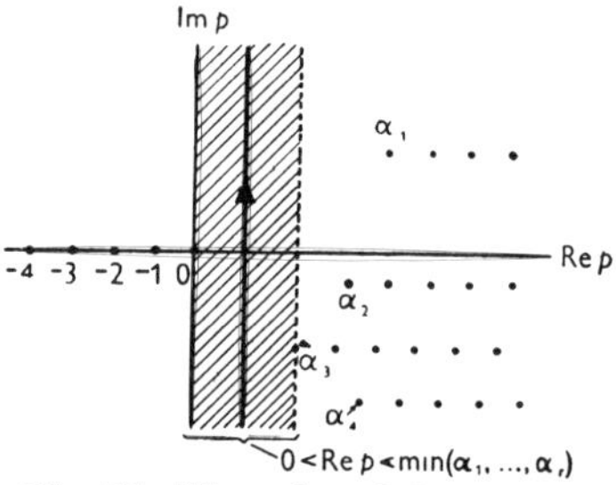

Fig. 77. The poles of the image of a hypergeometric function with exponential argument.

In the important cases $r = s-1$ (if $\operatorname{Re}(\gamma_1 + \ldots + \gamma_s - \alpha_1 - \ldots - \alpha_r) > 1$), $r = s$ and $r = s+1$ the validity of (42) can now be proved as follows. First, with the help of Stirling's formula, it is shown that the path of integration of the inversion integral may be closed towards the left at infinity (compare VI, § 11) for any t when $r = s$, and for $t > 0$ when $r = s+1$. Then the sum of the residues of all the poles thereby enclosed actually yields the original of (42) for those special values of t. Because of the analytic character of the inversion integral, the said original holds just as well for $t < 0$ when $r = s+1$, so that (42) is proved for $r = s$ and $r = s \pm 1$. Moreover, if $r = s+1$, the left-hand side of (42) represents an *asymptotic* expression $(t \to +\infty)$ for the inversion integral. This can be proved by working out the definition (VII, 23) after a transformation of the path of integration of the inversion integral into a curve passing between two special poles

$p = -N$ and $p = -N-1$. Next the validity of (42) follows from the fact that the corresponding hypergeometric differential equation (satisfied by the inversion integral) will have only *one* solution showing the asymptotic behaviour under consideration.

Most strikingly, for the cases $r > s+1$ (t arbitrary) and $r = s+1$ ($t < 0$) in which the path of integration cannot be closed at infinity towards the left, it may just be closed towards the right. In these cases the sum of all residues reduces to a sum of r *convergent* hypergeometric series. Upon comparing the expression thus obtained with the hypergeometric function in the left-hand member of (42), we get the identity

$$\frac{\Gamma(\alpha_1)\ldots\Gamma(\alpha_r)}{\Gamma(\gamma_1)\ldots\Gamma(\gamma_s)}\,{}_rF_s(\alpha;\gamma;-x) = \Gamma(\alpha_1)\frac{\Gamma(\alpha_2-\alpha_1)\ldots\Gamma(\alpha_r-\alpha_1)}{\Gamma(\gamma_1-\alpha_1)\ldots\Gamma(\gamma_s-\alpha_1)}$$

$$\times\frac{{}_{s+1}F_{r-1}\left(\alpha_1,\alpha_1-\gamma_1+1,\ldots,\alpha_1-\gamma_s+1;\,\alpha_1-\alpha_2+1,\ldots,\alpha_1-\alpha_r+1;\,\dfrac{(-1)^{r+s}}{x}\right)}{x^{\alpha_1}}$$

$$+\text{cycl.}\quad (x>0)\;(r \geqslant s+1), \qquad (45)$$

in which 'cycl.' denotes that for the complete expression those terms should be added which are obtained by replacing α_1 in the term fully written out by $\alpha_2, \alpha_3, \ldots, \alpha_r$ successively.

The theory discussed above is usually given for the restricted hypergeometric series, for which $r = 2$ and $s = 1$. The inversion integral of (42), as an analytic representation of this series (even when diverging), was so introduced by Barnes. Moreover, a similar integral representation applies even when the numbers $\alpha_1, \alpha_2, \ldots, \alpha_r$ do not lie on the right-hand side of the imaginary axis. In this case, however, the straight path of integration has to be changed into a curved path, L, from $-i\infty$ to $+i\infty$, again separating the two systems of poles, $-n$ and $\alpha_1+n, \ldots, \alpha_r+n$ (see fig. 78); as a consequence, the Barnes integral can no longer be interpreted as the inversion integral of some operational relation.

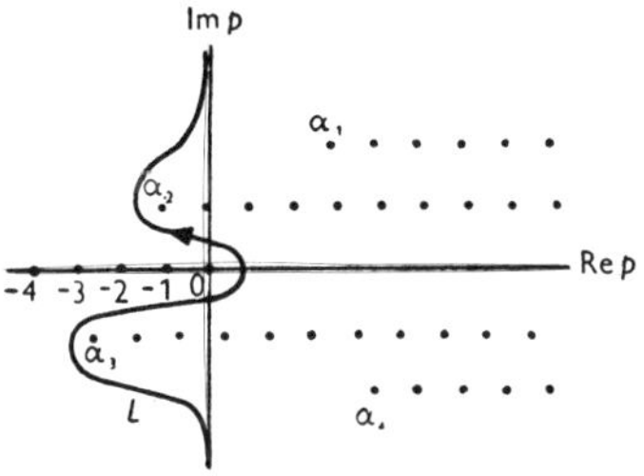

Fig. 78. The curved path of integration of the Barnes integral.

The significance of (42) for $r > s+1$ may be illustrated by taking $r = 2$ and $s = 0$ and using (x, 76):

$$p\Gamma(p+\nu)\,\Gamma(p-\nu)\,\Gamma(\tfrac{1}{2}-p) \doteqdot \frac{\sqrt{\pi}}{\cos(\nu\pi)}\,e^{\frac{1}{2}e^{-t}}K_\nu(\tfrac{1}{2}e^{-t}), \quad |\operatorname{Re}\nu| < \operatorname{Re}p < \tfrac{1}{2}. \quad (46)$$

Concerning other applications of (42), this general operational relation enables us to establish interesting properties of numerous functions that are included by hypergeometric functions. The usefulness of this two-sided relation is mainly due to its power of reducing even the most intricate hypergeometric function to simple Γ-functions. In contradistinction with this, the exponential transformation according to (19), when applied to relations such as (x, 78) or (x, 79), leads to one-sided relations that at most give rise to a relationship between two *different* hypergeometric functions.

Finally, we remark that (44), being an extension of (42), can be derived by transposing term by term the hypergeometric series in the image of (44), whereupon a summation is possible with the aid of the identity (43).

Example 1. *Integral representation of the restricted hypergeometric function.* In this special case the relation (42) reduces to

$$\frac{\Gamma(\alpha)\,\Gamma(\beta)}{\Gamma(\gamma)}\,F(\alpha,\beta;\gamma;-\mathrm{e}^{-t}) \doteqdot \frac{\Gamma(p+1)\,\Gamma(\alpha-p)\,\Gamma(\beta-p)}{\Gamma(\gamma-p)},$$
$$0<\operatorname{Re}p<\min(\operatorname{Re}\alpha,\operatorname{Re}\beta). \quad (47)$$

Now, there are several possibilities of splitting up the image into two factors, and, upon applying the rule of the composition product, we can obtain as many integral representations of the hypergeometric function under discussion. Let us quote some examples that are not too artificial:

$$\frac{\Gamma(p+1)\,\Gamma(\alpha-p)\,\Gamma(\beta-p)}{\Gamma(\gamma-p)} = \begin{cases} \dfrac{1}{p}\times\dfrac{\Gamma(p+1)}{\Gamma(\gamma-p)}\times p\Gamma(\alpha-p)\,\Gamma(\beta-p), \\[2ex] \dfrac{1}{p}\times\dfrac{\Gamma(p+1)\,\Gamma(\alpha-p)}{\Gamma(p+\alpha)}\times\dfrac{p\Gamma(p+\alpha)\,\Gamma(\beta-p)}{\Gamma(\gamma-p)}, \\[2ex] \dfrac{1}{p}\times\Gamma(p+1)\,\Gamma(\alpha-p)\times\dfrac{p\Gamma(\beta-p)}{\Gamma(\gamma-p)}, \\[2ex] \dfrac{1}{p}\times\dfrac{\Gamma(p+1)\,\Gamma(\alpha-p)}{\Gamma(\gamma-p)}\times p\Gamma(\beta-p). \end{cases}$$

The originals of the separate factors are either known from relations given previously or are easy to derive from similar relations. Upon performing the necessary transformations of the composition-product rule, and replacing e^{-t} by x, we find the following identities for $x>0$, the first of which (for other values of x) is already known from (x, 87):

$$\frac{\Gamma(\alpha)\,\Gamma(\beta)}{\Gamma(\gamma)}\,F(\alpha,\beta;\gamma;-x)$$
$$= \begin{cases} \dfrac{4}{x^{\frac{1}{2}(\alpha+\beta)}}\displaystyle\int_0^\infty K_{\alpha-\beta}\left(\frac{2s}{\sqrt{x}}\right) J_{\gamma-1}(2s)\,s^{\alpha+\beta-\gamma}\,ds \quad (\operatorname{Re}\alpha>0;\ \operatorname{Re}\beta>0), \\[2ex] \dfrac{\Gamma(\alpha+\beta)}{\Gamma(\alpha+\gamma)}\sqrt{\left(\dfrac{\pi}{x}\right)}\displaystyle\int_0^\infty \exp\left\{-\frac{s}{2}\left(1+\frac{1}{x}\right)\right\} I_{\alpha-\frac{1}{2}}\left(\frac{s}{2x}\right) M_{\beta+\frac{1}{2}(\alpha-\gamma),\frac{1}{2}(\alpha+\gamma-1)}(s)\,s^{\frac{1}{2}(\alpha-\gamma-1)}\,ds \\ \hfill (\operatorname{Re}\alpha>0;\ \operatorname{Re}\beta>0), \\[2ex] \dfrac{\Gamma(\alpha)}{\Gamma(\gamma-\beta)}\displaystyle\int_1^\infty \frac{s^{\alpha-\gamma}(s-1)^{\gamma-\beta-1}}{(s+x)^\alpha}\,ds \quad (\operatorname{Re}\gamma>\operatorname{Re}\beta>0), \\[2ex] \dfrac{\Gamma(\alpha)}{\Gamma(\gamma)}\dfrac{1}{x^\beta}\displaystyle\int_0^\infty \exp\left\{-\left(\frac{1}{x}+\frac{1}{2}\right)s\right\} M_{\alpha-\frac{1}{2}\gamma,\frac{1}{2}(\gamma-1)}(s)\,s^{\beta-\frac{1}{2}\gamma-1}\,ds \quad (\operatorname{Re}\beta>0). \end{cases}$$

To the list above there could, of course, also be added those expressions which are obtained by interchanging α and β, since then the left-hand side remains unaltered.

Example 2. *An integral of Barnes extended.* Let us apply (44) to the restricted hypergeometric series for $\lambda = -1$, thereby simplifying the image with the help of (33). The resulting operational relation is as follows:

$$\Gamma(\gamma-\alpha)\,\Gamma(\gamma-\beta)\,f(\alpha,\beta;\,\gamma;\,1-e^{-t}) \doteq \Gamma(p+1)\,\Gamma(\alpha-p)\,\Gamma(\beta-p)\,\Gamma(p+\gamma-\alpha-\beta),$$
$$\max\{0,\,\mathrm{Re}\,(\alpha+\beta-\gamma)\} < \mathrm{Re}\,p < \min(\mathrm{Re}\,\alpha, \mathrm{Re}\,\beta), \quad (48)$$

in which use is made of the f-notation introduced at (42); the image is simply the product of four Γ-functions. Again, the image allows of many different factorizations with a view to applying the rule of the composition product. Moreover, after some slight modification the inversion integral of (48) is easily shown to be equivalent to the following integral:

$$\Gamma(b+c)\,\Gamma(b+d)\,e^{-at}f(a+c, a+d;\, a+b+c+d;\, 1-e^{-t})$$
$$= \frac{1}{2\pi i}\int_{c_0-i\infty}^{c_0+i\infty} e^{pt}\,\Gamma(p+a)\,\Gamma(p+b)\,\Gamma(c-p)\,\Gamma(d-p)\,dp,$$
$$-\mathrm{Re}\,a < c_0 < \min(\mathrm{Re}\,c, \mathrm{Re}\,d), \quad (49)$$

which is an extension of Barnes's integral obtained when $t = 0$. The integral remains valid in the case of a curved path of integration, whenever this path separates the systems of poles (of the integrand) extending to the left ($-a,\ -a-1, \ldots$ and $-b,\ -b-1, \ldots$) from those extending to the right ($c, c+1, \ldots$ and $d, d+1, \ldots$).

Example 3. *Integral relations existing between hypergeometric functions.* Properties of Γ-functions serve a useful purpose for deducing many integral properties of hypergeometric functions by simple operational reasoning. As an exercise we start with the following factorization:

$$\frac{\pi}{4^{\alpha+\beta+\gamma-1}}\,\frac{\Gamma(2p+1)\,\Gamma(2\alpha-2p-1)\,\Gamma(2\beta-2p-1)}{\Gamma(2\gamma-2p-1)}$$
$$= \frac{1}{p}\times\frac{\Gamma(p+1)\,\Gamma(\alpha-p)\,\Gamma(\beta-p)}{\Gamma(\gamma-p)}\times\frac{p}{p+\frac{1}{2}}\,\frac{\Gamma(p+\frac{3}{2})\,\Gamma(\alpha-p-\frac{1}{2})\,\Gamma(\beta-p-\frac{1}{2})}{\Gamma(\gamma-p-\frac{1}{2})},$$

which is true on account of the duplication formula of the Γ-function. Using (47) and applying the rule of the composition product, we arrive at

$$\int_0^\infty f(\alpha,\beta;\,\gamma;\,-xs)\,f\left(\alpha,\beta;\,\gamma;\,-\frac{x}{s}\right)\frac{ds}{\sqrt{s}} = \frac{\pi}{4^{\alpha+\beta-\gamma-1}}\,\frac{f(2\alpha-1, 2\beta-1;\, 2\gamma-1;\, -x)}{\sqrt{x}},$$
$$\min(\mathrm{Re}\,\alpha, \mathrm{Re}\,\beta) > \tfrac{1}{2};\ x > 0.$$

Another very general integral relation concerning hypergeometric functions follows immediately from the definition integral of (44) when making the changes $u = e^{-t}$ and $p = x$. We thus find

$$\Gamma(x)\,{}_rf_s(\alpha-x;\,\gamma-x;\,-\lambda) = \int_0^\infty {}_rf_s(\alpha;\,\gamma;\,-\lambda-u)\,u^{x-1}\,du.$$

As will be evident from the foregoing, the methods described above are capable of extension in various directions, thus showing how extremely useful the operational calculus is in leading to a wealth of relations between hypergeometric functions that might otherwise only be obtained in an elaborate manner.

6. Rules concerning series expansions

In the present section we will indicate how various operational relations are to be found from known series expansions. In so doing we shall not concern ourselves about questions of validity and rigorous proofs. We rather refer to our previous statements that the operational calculus may simply be considered as a means for rapidly approaching results that, if necessary, can afterwards be verified rigorously.

With this in mind, we now proceed to give a short account of a number of rules that are obtainable from corresponding series expansions.

(1) Given the one-sided relation

$$f(p) \doteqdot h(t)\,U(t) \quad (\mathrm{Re}\,p > \alpha),$$

then it follows that

$$p\int_0^\infty \frac{h(s)}{\mathrm{e}^{ps}-1}\,ds \doteqdot U(t)\sum_{n=1}^{\infty}\frac{h(t/n)}{n} = H(t)\,U(t), \quad \mathrm{Re}\,p > \max(\alpha, 0), \tag{50}$$

while, moreover, we have (see example 4 of IV, § 3)

$$h(t) = \sum_{n=1}^{\infty}\frac{\mu(n)}{n}H\left(\frac{t}{n}\right).$$

In this $\mu(n)$ is the function of Möbius (see (IV, 10)). The rule (50) is obtained when expanding the integrand in the following manner:

$$\frac{h(s)}{\mathrm{e}^{ps}-1} = h(s)\sum_{n=1}^{\infty}\mathrm{e}^{-nps} \quad (\mathrm{Re}\,p > 0),$$

which leads to a sum of definition integrals.

(2) Let us assume $\lambda(x)$ to be expanded in a power series

$$\lambda(x) = \sum_{n=0}^{\infty} c_n x^n \quad (|x| < \rho), \tag{51}$$

then, provided the radius of convergence is not less than 1, there follow numerous relations of which either the image or the original is expressed in the form of some type of series. For example:

(*a*) A series of Bessel functions, viz.

$$\frac{\lambda(\mathrm{e}^{-1/p})}{p^\nu} \doteqdot U(t)\,t^{\frac{1}{2}\nu}\sum_{n=0}^{\infty}\frac{c_n}{n^{\frac{1}{2}\nu}}J_\nu\{2\sqrt{(nt)}\} \quad (\mathrm{Re}\,p > 0;\ \mathrm{Re}\,\nu > -1). \tag{52}$$

In particular, when $\nu = \frac{1}{2}$,

$$\frac{\lambda(\mathrm{e}^{-1/p})}{\sqrt{p}} \doteqdot U(t)\sum_{n=0}^{\infty}\frac{c_n}{\sqrt{(\pi n)}}\sin\{2\sqrt{(nt)}\} \quad (\mathrm{Re}\,p > 0). \tag{53}$$

(*b*) A series of Laguerre polynomials, viz.

$$\lambda\left(1-\frac{1}{p}\right) \doteqdot U(t)\sum_{n=0}^{\infty}\frac{c_n}{n!}L_n(t) \quad \left(\mathrm{Re}\,p > \frac{1}{\rho+1}\right). \tag{54}$$

(*c*) The following extension of a theta series:

$$\sqrt{p}\,\lambda(\mathrm{e}^{-\sqrt{p}}) \doteqdot \frac{U(t)}{\sqrt{(\pi t)}}\sum_{n=0}^{\infty}c_n\,\mathrm{e}^{-n^2/4t} \quad (\mathrm{Re}\,p > 0). \tag{55}$$

(*d*) The Dirichlet series:

$$\lambda(\mathrm{e}^{-\mathrm{e}^{-t}})-\lambda(0) \doteqdot \Pi(p)\sum_{n=1}^{\infty}\frac{c_n}{n^p} = \Pi(p)f(p), \quad \mathrm{Re}\,p>\max(\alpha,0), \tag{56}$$

in which α denotes the abscissa of convergence of the Dirichlet series. Closely related is the following relation, containing a series of Lambert:

$$\Pi(p)\,\zeta(p)f(p) \doteqdot \sum_{n=1}^{\infty}\frac{c_n}{\mathrm{e}^{n\,\mathrm{e}^{-t}}-1} = \sum_{n=1}^{\infty}\{\lambda(\mathrm{e}^{-n\,\mathrm{e}^{-t}})-\lambda(0)\}, \quad \mathrm{Re}\,p>\max(\alpha,1). \tag{57}$$

(*e*) The factorial series:

$$\lambda(1-\mathrm{e}^{-t})\,U(t) \doteqdot \sum_{n=0}^{\infty}\frac{n!\,c_n}{(p+1)(p+2)\dots(p+n)}, \quad \mathrm{Re}\,p>\max(\alpha,0), \tag{58}$$

in which α is the same number as in (56).

(*f*) The following series:

$$\log p\,\lambda\left(\frac{1}{p}\right) \doteqdot -U(t)\sum_{n=0}^{\infty}c_n\frac{d}{dn}\left\{\frac{t^n}{\Pi(n)}\right\}, \quad \mathrm{Re}\,p>1/\rho. \tag{59}$$

It may be observed that, by taking $\lambda(x) = 1/(1-x)$ in (59), and using (IV, 46) we arrive at the series

$$\mathrm{e}^t\,\mathrm{Ei}\,(-t) = \sum_{n=0}^{\infty}\frac{d}{dn}\left\{\frac{t^n}{\Pi(n)}\right\} \quad (t>0),$$

which is the counterpart of (VI, 4*a*).

Any of the rules above results from a term-by-term transposition of the corresponding series, in which use is made of already known formulae. Still other systems of rules can be obtained by starting with a type of series different from the power series, e.g. the Dirichlet series.

(3) Given some generating function of the form

$$\lambda(x)\,\mathrm{e}^{xt} = \sum_{n=0}^{\infty}\phi_n(t)\,x^n, \quad |x|<\rho(t), \tag{60}$$

then it follows

$$\phi_n(t)\,U(t) \doteqdot \left[\left[\frac{\lambda(p)}{p^n}\right]\right], \quad \mathrm{Re}\,p>0, \tag{61}$$

in which, as before, [[ ]] indicates that in the Laurent expansion one should omit the positive powers of p. For a proof of (61) we transpose

$$\phi_n(t)\,U(t) = \frac{U(t)}{n!}\frac{\partial^n}{\partial x^n}\{\lambda(x)\,\mathrm{e}^{xt}\}_{x=0},$$

which is a consequence of (60), leading first to

$$\phi_n(t)\,U(t) \doteqdot \frac{1}{n!}\frac{\partial^n}{\partial x^n}\left\{\frac{\lambda(x)}{1-x/p}\right\}_{x=0} \quad (\mathrm{Re}\,p>0).$$

Then the power-series expansion of { } is used, and (61) is obtained.

(4) When it is given that the relation $f(p) \doteqdot h(t)$ is valid at least for $\operatorname{Re} p = 0$, then we have the following equality:

$$\sum_{n=-\infty}^{\infty} \frac{f(2\pi i n)}{2\pi i n} = \sum_{n=-\infty}^{\infty} h(n). \tag{62}$$

This rule is the operational equivalent of the Poisson series given in (VI, 34).

Example 1. Let us consider the function

$$\operatorname{Lm}(z) = \sum_{n=1}^{\infty} \frac{z^n}{1-z^n} = \sum_{n=1}^{\infty} d(n) z^n \quad (|z|<1), \tag{63}$$

in which $d(n)$ denotes the number of divisors of n (1 and n included). The series (63) is a so-called Lambert series, for which an operational relation may be found by taking $\lambda(x) = 1/(1-x)$; then $c_n = 1$, and $f(p)$ in (56) becomes equal to $\zeta(p)$. Consequently, from (57),

$$\Pi(p)\,\zeta^2(p) \doteqdot \operatorname{Lm}(e^{-e^{-t}}), \quad 1<\operatorname{Re} p<\infty. \tag{64}$$

This relation is useful in studying the function $\operatorname{Lm}(z)$. For instance, its numerical behaviour for real values of z tending to 1. To that end we first rewrite the relation in the equivalent form

$$p\Pi(p)\,\zeta^2(p+1) \doteqdot e^{-t}\operatorname{Lm}(e^{-e^{-t}}) \quad (\operatorname{Re} p>0). \tag{65}$$

Then we can apply the Maclaurin expansion of the entire function $\{\zeta(p+1)-1/p\}^2$, which, according to (VII, 7), will lead to

$$\zeta^2(p+1) = \frac{1}{p^2} + \frac{2C}{p} + \sum_{n=0}^{\infty} c_n p^n.$$

After substitution of this series in the image of (65), the term-by-term transposition with the aid of (III, 16) yields

$$e^{-t}\operatorname{Lm}(e^{-e^{-t}}) = -\operatorname{Ei}(-e^{-t}) + 2C\,e^{-e^{-t}} + \sum_{n=0}^{\infty} c_n \frac{d^n}{dt^n}\{e^{-t}e^{-e^{-t}}\},$$

so that the manner of approaching infinity of the Lambert series in question as $x \to 1-0$ is determined by

$$\operatorname{Lm}(x) = \frac{\operatorname{Ei}(\log x)}{\log x} - 2C\frac{x}{\log x} + \ldots.$$

Moreover, when again taking $c_n = 1$, we have at once from (58) the following simple factorial series:

$$\frac{1}{x-1} = \sum_{n=0}^{\infty} \frac{n!}{x(x+1)(x+2)\ldots(x+n)} \quad (\operatorname{Re} x>1).$$

Example 2. In order to derive properties of the Möbius function (IV, 10), we take in (51) $c_n = \mu(n)$ by which $f(p)$ in (56) becomes equal to $1/\zeta(p)$; thus from (57)

$$\Pi(p) \doteqdot \sum_{n=1}^{\infty} \frac{\mu(n)}{e^{n\,e^{-t}}-1} \quad (\operatorname{Re} p>1).$$

Now, an alternative expression for the original of $\Pi(p)$ is known from (III, 16). After identifying the two expressions and putting $e^{-e^{-t}} = a$ we get

$$\sum_{n=1}^{\infty} \mu(n)\frac{a^n}{1-a^n} = a \quad (0<a<1).$$

Moreover, by putting $a = e^{-p}$ and translating the last identity into the t-language with the help of (V, 8), another property of the Möbius function is found, viz.

$$\sum_{n=1}^{\infty} \mu(n)\left[\frac{t}{n}\right] = U(t-1).$$

Example 3. An example showing how certain operational relations are obtainable from generating functions is provided by the Bernoulli polynomials, which are defined by

$$\frac{e^{xt}}{\int_0^1 e^{xt}\,dt} = \frac{x\,e^{xt}}{e^x - 1} = \sum_{n=0}^{\infty} \frac{B_n(t)}{n!} x^n \quad (|x| < 2\pi). \tag{66}$$

In this case the operational relation (61) becomes

$$\frac{B_n(t)}{n!} U(t) \doteqdot \left[\left[\frac{1}{p^{n-1}(e^p - 1)}\right]\right], \quad 0 < \operatorname{Re} p < \infty. \tag{67}$$

One of the elementary properties of the Bernoulli polynomials follows, for instance, from the fact that the image of (67) can be written alternatively

$$\sum_{k=0}^{n} \frac{B_k}{k!} \frac{1}{p^{n-k}},$$

in virtue of (66) and because the Bernoulli numbers B_m are equal to $B_m(0)$. Upon taking the original of this series term by term, we arrive at the following well-known explicit expression for the polynomials under consideration:

$$B_n(x) = \sum_{k=0}^{n} \binom{n}{k} B_k x^{n-k}.$$

7. Equalities in connexion with a single operational relation

As already remarked in VII, § 6, from any operational rule one can derive an identity by applying the theorem of Abel. These identities, involving simultaneously the functions of the original and the image, but no longer the mapping itself, are often most easily derived in a straightforward manner by making use of the definition integral or of the inversion integral. Examples of such identities, originating from a single operational relation $f(p) \doteqdot h(t)$ with strip of convergence $\alpha < \operatorname{Re} p < \beta$, are the following:

(1)
$$\int_a^b \frac{f(s)}{s}\,ds = \int_{-\infty}^{\infty} \frac{e^{-as} - e^{-bs}}{s} h(s)\,ds \quad (a,b) \subset (\alpha,\beta). \tag{68}$$

To verify (68), the function $f(s)$ at the left is replaced by the corresponding definition integral and then the order of integration is reversed. Of course, the side condition $\alpha \leqslant a \leqslant b \leqslant \beta$ should be satisfied, amongst others. This is indicated by $(a,b) \subset (\alpha,\beta)$, denoting that the interval $a < x < b$ lies inside that of $\alpha < x < \beta$.

(2)
$$\frac{1}{\pi i} \int_{-\infty}^{\infty} f(i\omega) \sin(a\omega) \frac{d\omega}{\omega^2} = \int_{-a}^{a} h(s)\,ds \quad (\alpha \leqslant 0 \leqslant \beta), \tag{69}$$

which may be justified along similar lines; use is to be made of Dirichlet's integral.

(3)
$$\frac{1}{\Pi(\nu)} \int_0^{\infty} f(s)\, s^{\nu-1}\,ds = \int_0^{\infty} \frac{h(s)}{s^{\nu+1}}\,ds \quad (\alpha \leqslant 0;\ \operatorname{Re}\nu > -1). \tag{70}$$

It is important to note that (70) is valid only for one-sided relations; the proof is analogous to those above.

Example. The identity (70) at once leads to an interesting property of electrical networks, namely,

$$\frac{1}{\Pi(\nu)}\int_0^\infty A(p)\,p^{\nu-1}\,dp = \int_0^\infty \frac{i_\Gamma(t)}{t^{\nu+1}}\,dt,$$

since the time admittance, $i_\Gamma(t)$, is just the original of the frequency admittance, $A(p)$.

8. Equalities in connexion with a pair of simultaneous relations (exchange identity)

Of course, identities containing the images and originals of more than one operational relation are equally possible. For instance, the following, which is of utmost importance:

$$\int_0^\infty h_1(\alpha s)\frac{f_2(s)}{s}\,ds = \int_0^\infty h_2(\alpha s)\frac{f_1(s)}{s}\,ds \quad (\alpha_1 \leqslant 0;\ \alpha_2 \leqslant 0;\ \alpha > 0), \tag{71}$$

and which involves two different given one-sided operational relations, viz.

$$f_1(p) \doteqdot h_1(t)\,U(t) \quad (\operatorname{Re} p > \alpha_1),$$

$$f_2(p) \doteqdot h_2(t)\,U(t) \quad (\operatorname{Re} p > \alpha_2).$$

Because in this identity the given data occur in such a way that a mere exchange of the two relations transforms either side into the other, we shall refer to it as an *exchange identity*†.

The general lines of the proof are similar to those indicated in the preceding section. To verify (71), the functions f_1 and f_2 are replaced by the corresponding definition integrals. The two resulting repeated integrals prove to be equal to each other, after a reversal of the order of integration in one of them, no matter which.

An analogue of this exchange identity exists for two-sided relations, in which case the limits of integration become $-\infty$ and $+\infty$. Then, however, it is necessary that the strip of convergence of each given relation coincides with the whole plane of p. Therefore the result is particularly important in the case of two originals that are cut off somewhere at the left as well as at the right. Thus, given the operational relations

$$f_1(p) \doteqdot h_1(t)\{U(t-a_1) - U(t-b_1)\}, \quad -\infty < \operatorname{Re} p < \infty,$$

$$f_2(p) \doteqdot h_2(t)\{U(t-a_2) - U(t-b_2)\}, \quad -\infty < \operatorname{Re} p < \infty,$$

the exchange identity, in its simplest form, can be written as follows:

$$\int_{a_1}^{b_1} h_1(s)\frac{f_2(s)}{s}\,ds = \int_{a_2}^{b_2} h_2(s)\frac{f_1(s)}{s}\,ds. \tag{72}$$

† It is dealt with by S. Goldstein, *Proc. Lond. Math. Soc.* (2) xxxiv, 103, 1932.

For a rigorous proof of the identities above it would be necessary to show that reversing the order of integration in certain integrals is legitimate. Anyhow, this is guaranteed whenever the integral of either the left side or the right side is absolutely convergent.

Another identity, also containing as data two given relations, though not of the exchange type, is the following:

$$\int_{-\infty}^{\infty} h_1(s)\, h_2(-s)\, ds = \frac{1}{2\pi i}\int_{c-i\infty}^{c+i\infty} \frac{f_1(s)\, f_2(s)}{s^2}\, ds, \tag{73}$$

in which the path of integration, $\operatorname{Re} s = c$, should be situated inside the common strip of convergence of the individual relations. The formula (73) is obtained when t is taken zero in the inversion integral corresponding to the composition product.

As to possible applications of rules involving two operational relations as data, we would remark that the exchange identity leads particularly to many results in quite an easy manner; for instance, it often reveals at once the symmetry, if any, between two parameters as contained in certain integrals.

Example 1. Let us illustrate the usefulness of the exchange identity by some exercises:

(*a*)

$$\int_0^{\infty} J_\mu(s)\, e^{-\nu \operatorname{arcsinh} s} \frac{ds}{\sqrt{(s^2+1)}} = \int_0^{\infty} J_\nu(s)\, e^{-\mu \operatorname{arcsinh} s} \frac{ds}{\sqrt{(s^2+1)}} \quad (\operatorname{Re}\mu > -1;\ \operatorname{Re}\nu > -1).$$

This follows from (71) when we start with the two relations obtained from the fundamental Bessel-function relation (X, 12) for two different orders, μ and ν. The substitution $s = \sinh u$ simplifies the identity to

$$\int_0^{\infty} J_\mu(\sinh u)\, e^{-\nu u}\, du = \int_0^{\infty} J_\nu(\sinh u)\, e^{-\mu u}\, du \quad (\operatorname{Re}\mu > -1;\ \operatorname{Re}\nu > -1).$$

(*b*) Starting from the relations

$$U(t)\left\{1 - \frac{t^2}{2!} + \frac{t^4}{4!} - \dots + \frac{t^{4n}}{(4n)!} - \cos t\right\} \doteqdot \frac{1}{p^{4n}(p^2+1)} \quad (\operatorname{Re} p > 0),$$

$$\frac{t^{4n+1}}{(4n+1)!} U(t) \doteqdot \frac{1}{p^{4n+1}} \quad (\operatorname{Re} p > 0),$$

we obtain by means of (71) that

$$\int_0^{\infty} \left\{1 - \frac{s^2}{2!} + \frac{s^4}{4!} - \dots + \frac{s^{4n}}{(4n)!} - \cos s\right\} \frac{ds}{s^{4n+2}} = \frac{\pi}{2(4n+1)!},$$

in which the integrand contains the 'remainder' of the power-series expansion of $\cos s$.

(*c*) The two-sided exchange identity may be applied to two relations obtained when in (III, 30), for the K-function, the argument takes on two different values, a and b. Then we find the following equality:

$$\int_{-\infty}^{\infty} K_s(a)\, e^{-b \cosh s}\, ds = \int_{-\infty}^{\infty} K_s(b)\, e^{-a \cosh s}\, ds \quad (\operatorname{Re} a > 0;\ \operatorname{Re} b > 0),$$

in which the integrations extend over the *order* of the Bessel function.

(*d*) The exchange identity (72) for cut-off functions, when applied to the relation (x, 60) for two Legendre polynomials of orders n and m, leads to the result

$$\int_{-1}^{1} P_n(s)\, I_{m+\frac{1}{2}}(s) \frac{ds}{\sqrt{s}} = (-1)^{m+n} \int_{-1}^{1} P_m(s)\, I_{n+\frac{1}{2}}(s) \frac{ds}{\sqrt{s}}.$$

The integrands are even or odd functions of s, depending on whether $n+m$ is even or odd. Only in the former case is the result non-trivial, viz.

$$\int_0^1 P_n(s)\, I_{m+\frac{1}{2}}(s) \frac{ds}{\sqrt{s}} = \int_0^1 P_m(s)\, I_{n+\frac{1}{2}}(s) \frac{ds}{\sqrt{s}} \quad (m+n \text{ even}).$$

Example 2. A final example may refer to the identity (73) which is based upon the rule of the composition product. For this purpose we start with

$$2pK_{p+\alpha}(x) \doteqdot \mathrm{e}^{-x\cosh t - \alpha t}, \quad -\infty < \operatorname{Re} p < \infty \quad (\operatorname{Re} x > 0),$$

$$2pK_{p-\beta}(y) \doteqdot \mathrm{e}^{-y\cosh t + \beta t}, \quad -\infty < \operatorname{Re} p < \infty \quad (\operatorname{Re} y > 0),$$

which follow from (III, 30). Upon applying (73) an identity is found of which one member is reducible to a K-function with the help of (III, 29). The final result contains an integral in which the integration is carried out over complex values of the order of the Bessel function, viz.

$$\int_{c-i\infty}^{c+i\infty} K_{s+\alpha}(x)\, K_{s-\beta}(y)\, ds = \pi i K_{\alpha+\beta}(x+y),$$

or alternatively, when $c = 0$,

$$\frac{1}{\pi} \int_{-\infty}^{\infty} K_{is+\alpha}(x)\, K_{is-\beta}(y)\, ds = K_{\alpha+\beta}(x+y) \quad (\operatorname{Re} x > 0;\ \operatorname{Re} y > 0).$$

CHAPTER XII

STEP FUNCTIONS AND OTHER DISCONTINUOUS FUNCTIONS

1. Introduction

A step function, as discussed in the present chapter, is defined as a real-valued function that is sectionally constant. That is to say, it suddenly jumps when the independent variable passes certain definite values, while it remains wholly constant between any two consecutive points of discontinuity. Obviously, the simplest possible is the unit function $U(t)$ jumping only at $t = 0$, which we have often dealt with in the course of the preceding investigations. The remaining step functions, many of which have already been encountered, are conveniently arranged in increasing order of complexity as follows:

(1) Step functions whose points of discontinuity are equidistant, the jumps being of equal magnitude.

An example is provided by the function $[t]$ of (v, 8), with unit jumps at integral values of t.

(2) Step functions having equidistant points of discontinuity, where the absolute values of the jumps are all the same, but the function may jump either down or up.

An example is given by the square-sine function, $\mathfrak{Sin}\ t = \sin t/|\sin t|$ of (iv, 7), jumping (on the positive side) at $0, \pi, 2\pi, 3\pi, \ldots$ by amounts of $2, -2, 2, -2, \ldots$, respectively.

(3) Step functions of equidistant discontinuities, jumping with arbitrary amplitudes.

Such are the originals $\sum\limits_{0}^{[t]} a_n$ of the relations (vi, 14), with jumps a_n at the positive integral points $t = n$.

(4) Step functions jumping at non-equidistant intervals, though all jumps are equal in magnitude.

For example, the original $[e^t]$ of (vi, 11) which is discontinuous at $t = \log 1, \log 2, \log 3, \ldots$, all jumps being of unit magnitude.

(5) Most generally, step functions with non-equidistant discontinuities, and arbitrary jumps.

To this type belong the originals $\sum\limits_{1}^{[e^t]} a_n$ of the relation (vi, 10) for Dirichlet series; in this example the function jumps at $t = \log 1, \log 2, \log 3, \ldots$, the corresponding amounts being $a_1, a_2, a_3, \ldots$.

In this chapter the step functions will be considered from a broad point of view. To begin with, step functions with equidistant discontinuities at integral points are discussed in § 2. In virtue of its close relationship to step functions, the so-called saw-tooth function will be treated in § 3. Further, the importance of step functions with a view to possible operational applications to arithmetic problems will become clear from §§ 4 and 5, where the said problems are studied in their relation to θ-functions and Dirichlet series. Next, in § 6, it is shown how any arbitrary series may lead to a step function, by considering the separate terms as elements of a function of the summation variable n. Finally, the composition-product rule when especially applied to step functions is treated in § 7.

First of all, some general remarks. The success of the operational method in case of these discontinuous step functions is due to their images being analytic functions throughout. Moreover, it is of little importance what values are actually ascribed to step functions at the discontinuities themselves, since any change therein merely corresponds to the addition of some null function to the original, which does not influence the image at all. On the other hand, it should be borne in mind that the inversion integral always produces the mean value at any point of discontinuity. It may finally be noted that, even in the case of quite simple images, the originals may be step functions that are more easily surveyed by means of a diagram than by a mathematical formula.

2. Operational relations involving step functions jumping at integral values of *t*

From any function $h(t)$ a step function can be derived having equidistant points of discontinuity, by replacing $h(t)$ inside the interval $(n, n+1)$ by $h(n)$, this being the value assumed at the left end-point of the interval; here n is an integer (see fig. 79). It will be evident that this new function is represented by $h([t])$, and its image is to be found as follows:

$$p\int_{-\infty}^{\infty} e^{-pt} h([t])\, dt = p \sum_{n=-\infty}^{\infty} \int_{n}^{n+1} e^{-pt} h([t])\, dt = p \sum_{n=-\infty}^{\infty} h(n) \int_{n}^{n+1} e^{-pt} dt.$$

The remaining integrals are readily evaluated; we then find the following operational relation:

$$h([t]) \doteqdot (1-e^{-p}) \sum_{n=-\infty}^{\infty} h(n)\, e^{-np}. \quad (1)$$

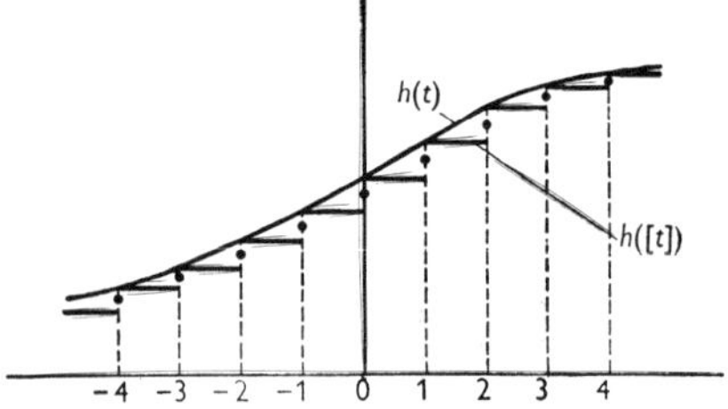

Fig. 79. The step function $h[t]$.

In particular, for one-sided step functions the relation (1) leads to images that are expanded in a series of powers of e^{-p}; the radius of convergence, ρ, is related to the strip of convergence, the latter being $\operatorname{Re} p > -\log \rho$. Conversely, to any convergent power series there corresponds an operational relation involving a step function: we only have to construct, from the given coefficients $h(n)$, a function that is equal to $h(n)$ inside $(n, n+1)$ for all integral values of n. For instance, let us consider the restricted hypergeometric function (x, 68) first rewritten as follows:

$$\frac{\Gamma(\alpha)\,\Gamma(\beta)}{\Gamma(\gamma)} F(\alpha,\beta;\gamma;\mathrm{e}^{-p}) = \sum_{n=0}^{\infty} \frac{\Gamma(\alpha+n)\,\Gamma(\beta+n)}{\Gamma(\gamma+n)} \frac{\mathrm{e}^{-pn}}{n!} \quad (\mathrm{Re}\, p > 0).$$

Then, we at once obtain the operational relation

$$\frac{\Gamma(\alpha+[t])\,\Gamma(\beta+[t])}{\Gamma(\gamma+[t])\,[t]!} \doteq \frac{\Gamma(\alpha)\,\Gamma(\beta)}{\Gamma(\gamma)} (1-\mathrm{e}^{-p})\, F(\alpha,\beta;\gamma;\mathrm{e}^{-p}), \quad 0 < \mathrm{Re}\, p < \infty. \tag{2}$$

In like manner, from the geometric series, we have that

$$\lambda^{[t]} U(t) \doteq \frac{1-\mathrm{e}^{-p}}{1-\lambda\,\mathrm{e}^{-p}}, \quad \log\lambda < \mathrm{Re}\, p < \infty, \tag{3}$$

and, from the exponential series, furthermore

$$\frac{\lambda^{[t]}}{[t]!} \doteq (1-\mathrm{e}^{-p})\, \mathrm{e}^{\lambda \mathrm{e}^{-p}}, \quad -\infty < \mathrm{Re}\, p < \infty. \tag{4}$$

In the case of (4) the strip of convergence covers the whole p-plane, since the corresponding power series converges everywhere (the function represented being an entire function). Moreover, as in (2) the factor $U(t)$ could be dropped here, because even without $U(t)$ the original vanishes automatically when $t < 0$, because $\Pi(-n) = \infty$ $(n = 1, 2, 3, \ldots)$. Thus it is seen that any of these cases leads to an *image* of exponential argument, whilst in the preceding chapter we encountered many *originals* of exponential argument.

It will be obvious that the definition integrals of the relations above cannot deliver new properties, since they are only a different way of writing the corresponding power series. But the inversion integral can do so; it often leads to a non-trivial formula, which, in particular, represents the mean values at any discontinuity of the step function under consideration. For example, when in the inversion integral of (4) we take $c = 0$ and $p = i\omega$, then the following (real) expression is finally obtained:

$$\frac{\lambda^{[t]}}{[t]!} = \frac{2}{\pi} \int_0^\infty \mathrm{e}^{\lambda \cos\omega} \sin \tfrac{1}{2}\omega \cos\{(t-\tfrac{1}{2})\,\omega - \lambda \sin\omega\} \frac{d\omega}{\omega}.$$

As will have been observed, the images above all contain the factor $1-\mathrm{e}^{-p}$. This factor can be eliminated by first multiplying the image by

$$(1-\mathrm{e}^{-p})^{-1} = \sum_{k=0}^{\infty} \mathrm{e}^{-kp} \quad (\mathrm{Re}\, p > 0),$$

and then determining the new original with the help of the shift rule. In so doing, the previous relation (VI, 14) is obtained anew for one-sided functions $h(t)$. Moreover, the operational relations discussed above bear some connexion to those existing for Dirichlet series (see (VI, 10)). In the latter case, too, there is the question of a relationship between a series and a step function, but the latter jumps at $t = \log 1, \log 2, \log 3, \ldots$.

Totally different relations involving step functions, however, are obtained when the discontinuities are not determined by the independent variable, but by the numerical values of the function $h(t)$ itself. An example of this is provided by

$$[h(t)]\, U\{h(t)\} \doteqdot \sum_{n=1}^{[h(\infty)]} e^{-ph^{-1}(n)} \quad (\mathrm{Re}\, p > 0), \tag{5}$$

which, too, is very general since $t = h^{-1}(s)$ is the inverse function of some arbitrary function $s = h(t)$ supposed to be monotonically increasing. This rule is easily proved by subdividing the range of integration in the definition integral into separate intervals whose end-points now coincide with those points where $h(t)$ passes through integral values. The operational relation (VI, 39) involving the θ_3-function is derivable from (5) by putting $h(t) = \sqrt{(t/\pi)}$. More generally, the function $h(t) = t^{1/n}$ will lead to

$$[\sqrt[n]{t}]\, U(t) \doteqdot \sum_{k=1}^{\infty} e^{-pk^n}, \quad 0 < \mathrm{Re}\, p < \infty,$$

of which the image is an extended θ-function.

3. The saw-tooth function

Let us consider the periodic function (of period 1) that decreases linearly from $+1$ to 0 inside any interval $n < t < n+1$ (n = integer). When plotted in a diagram, it shows the character of the teeth of a saw. Therefore, a convenient notation for this *saw-tooth function* is $\mathrm{Sa}(t)$, Sa being an abbreviation of the Latin 'serra' meaning saw. In the operational treatment below we shall be concerned mainly with the corresponding one-sided function, which is drawn in fig. 80 and is representable by the following formula:

$$\mathrm{Sa}(t)\, U(t) = \{[t] - t + \tfrac{1}{2}\}\, U(t). \tag{6}$$

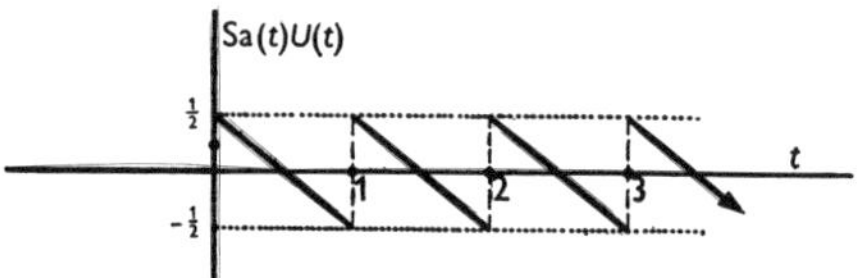

Fig. 80. The one-sided saw-tooth function.

In passing it may be noted that this function is not only important in mathematics, but also in electrical engineering (e.g. in television).

With a view to operational application the two following relations are of fundamental importance:

$$(1) \quad \mathrm{Sa}(t)\, U(t) \doteqdot \frac{1}{e^p - 1} - \frac{1}{p} + \frac{1}{2} = \frac{d}{dp} \log\left(\frac{\sinh \frac{1}{2}p}{\frac{1}{2}p}\right), \quad 0 < \mathrm{Re}\, p < \infty, \tag{7}$$

which can easily be inferred from (6) and (V, 8).

(2) The operational relation (VI, 43), which now may be written in the alternative form

$$\mathrm{Sa}(e^t) \doteqdot \zeta(p), \quad -1 < \mathrm{Re}\, p < 0. \tag{8}$$

The relation (7) enables us to deduce two important series representing the saw-tooth function. In the first place, let us differentiate the infinite product

$$\frac{\sinh p}{p} = \prod_{n=1}^{\infty}\left(1+\frac{p^2}{\pi^2 n^2}\right) \tag{9}$$

logarithmically, and transpose the resulting series term by term for $\mathrm{Re}\, p > 0$. We then get the following Fourier series:

$$\mathrm{Sa}(t) = \sum_{n=1}^{\infty} \frac{\sin(2\pi n t)}{\pi n}. \tag{10}$$

Secondly, an expansion in square-sine functions can be given. For that purpose, take the logarithmic derivative of the following infinite product:

$$\frac{\sinh \frac{1}{2}p}{\frac{1}{2}p} = \cosh\frac{p}{2^2}\cosh\frac{p}{2^3}\cosh\frac{p}{2^4}\ldots, \tag{11}$$

which was first given by Euler, and can be proved by using the property

$$A(\alpha) = \cosh \tfrac{1}{2}\alpha\, A(\tfrac{1}{2}\alpha),$$

valid for the function

$$A(\alpha) = \frac{\sinh \alpha}{\alpha}.$$

Upon transposing this logarithmic derivative (occurring in (7)) term by term, making use of (IV, 7) for the individual terms, we are led to the following result:

$$\mathrm{Sa}(t) = \sum_{n=1}^{\infty} \frac{\mathfrak{S}\mathrm{in}(2^n \pi t)}{2^{n+1}}. \tag{12}$$

The series (12) is simpler than (10), inasmuch as it does not show Gibbs's phenomenon. That is to say, the partial sum tends everywhere to the value of the function, even at the discontinuities, whilst in the case of (10) this applies only when the series is taken as a first-order Cesàro limit. On the other hand, the expansion (12) is an expansion in orthogonal functions that do not form a complete set; more precisely, there exist odd, non-identically vanishing, functions that are orthogonal to all square-sine functions occurring in (12).

The example below shows how many properties of the saw-tooth function may be derived by the operational method.

Example. In two different ways, it is possible to deduce by means of the exchange identity integrals that contain the saw-tooth function:

(1) Let us apply (XI, 71) to the relation (7) together with some arbitrary one-sided relation $h(t)\, U(t) \doteqdot f(p)$. It is found that

$$\int_0^{\infty} \mathrm{Sa}(s)\frac{f(s)}{s}\,ds = \int_0^{\infty}\left(\frac{1}{e^s-1}-\frac{1}{s}+\frac{1}{2}\right)\frac{h(s)}{s}\,ds. \tag{13}$$

In particular, taking $h(t) = e^{-at}t^{\nu}$, we can express the right-hand member of (13) in terms of the ζ-function; the first term of the integrand leads to this function on account of (III, 37) whilst the remaining integrals are elementary. The result obtained reads as follows:

$$\zeta(\nu, a) - \frac{1}{(\nu-1)a^{\nu-1}} - \frac{1}{2a^{\nu}} = \nu \int_0^{\infty} \frac{\mathrm{Sa}(s)}{(s+a)^{\nu+1}}\, ds \quad (\mathrm{Re}\,\nu > 1;\ \mathrm{Re}\,a > 0).$$

Furthermore, when $h(t) = e^{-at}U(t)$, whence $f(p) = p/(p+a)$, then the right-hand member of (13) equals the definition integral of the relation (VII, 31) for $p\mu(p)$, provided p is replaced by the parameter a. The following result is then easily obtained:

$$\mu(a) = \int_0^{\infty} \frac{\mathrm{Sa}(s)}{s+a}\, ds \quad (\mathrm{Re}\,a > 0), \tag{14}$$

which expresses the logarithm of the correction factor of Stirling's formula in terms of the saw-tooth function. Further, since (14) has the form of Stieltjes's moment integrals, according to example 4 of V, § 8 the function $\mu(z)$ has a cut along the negative part of the real axis. At this cut the real part of the function is continuous (except at the poles at $z = -1, -2, -3, \ldots$, of course), while the imaginary part is given by

$$\mathrm{Im}\,\mu(-x \pm i0) = \mp \pi\, \mathrm{Sa}(x) \quad (x > 0). \tag{15}$$

Closely related to (13) is

$$\int_0^{\infty} \mu'(s)\, h(s)\, ds = -\int_0^{\infty} \left(\frac{1}{e^s - 1} - \frac{1}{s} + \frac{1}{2}\right) \frac{f(s)}{s}\, ds,$$

which can be derived by applying (XI, 71) to a pair of one-sided operational relations, one of which is wholly arbitrary, the other being

$$-p\mu'(p) \doteq \left(\frac{1}{e^t - 1} - \frac{1}{t} + \frac{1}{2}\right) U(t), \quad 0 < \mathrm{Re}\,p < \infty, \tag{16}$$

which, in turn, may be found from (VII, 31) by application of the rule concerning multiplication of the original by t.

(2) When (XI, 71) is applied to both derivatives of (7) and $f(p) \doteq h(t)\, U(t)$, we obtain the formula

$$\int_0^{\infty} f(s)\, d\{\mathrm{Sa}(s)\, U(s)\} = \lim_{N\to\infty} \left\{\tfrac{1}{2}f(0) + f(1) + f(2) + \ldots + f(N-1) + \tfrac{1}{2}f(N) - \int_0^N f(s)\, ds\right\}$$
$$= \int_0^{\infty} \left(\frac{1}{e^s - 1} - \frac{1}{s} + \frac{1}{2}\right) h'(s)\, ds, \tag{17}$$

the proof of which requires the 'derivative' of $\mathrm{Sa}(t)\, U(t)$ to be determined from (6) by means of a series of impulse functions. In particular, for $h(t) = \dfrac{1}{a}(1 - e^{-at})$ the right-hand member becomes identical with the definition integral of (16); consequently the following limiting relation holds:

$$\lim_{N\to\infty} \left\{\frac{1}{2a} + \frac{1}{a+1} + \frac{1}{a+2} + \ldots + \frac{1}{a+N-1} + \frac{1}{2(a+N)} - \int_0^N \frac{ds}{a+s}\right\}$$
$$= -\mu'(a) = \log a + \frac{1}{2a} - \psi(a). \tag{18}$$

This formula allows of a simple physical interpretation, viz. when positive point charges of unit magnitude are situated equidistantly at $-1, -2, -3, \ldots$, and an extra charge of amount $\frac{1}{2}$ is present at the point 0, and, moreover, the negative axis is uniformly covered with negative charges such that the net charge between any two integral points is zero, then the potential due to this charge distribution, as a function of the position a, and on the line of charges itself, is given by (18), that is, by the derivative of Stirling's μ-function with respect to the co-ordinate a.

4. Arithmetic functions in connexion with θ-functions

As will become clear from this section and the next, the operational calculus is particularly useful in the study of numerous arithmetic functions, since in this way they are transformed into analytic p-functions.

As the first example of a non-elementary arithmetic problem we shall consider the determination of the number of lattice points (these being points of integral co-ordinates: $x = m$, $y = n$) inside the circle $x^2+y^2 = R^2$ having as centre the lattice point $m = n = 0$ (see fig. 81) and of radius R. To make the formulae as simple as possible, it is convenient to introduce the variable t by $t = R^2$ rather than by $t = R$. Accordingly, $A_2(t)$ is defined as the one-sided function that represents the number of lattice points inside the circle of radius $R = \sqrt{t}$. Now, this function is operationally related to the θ_3-function, as can be seen at once.

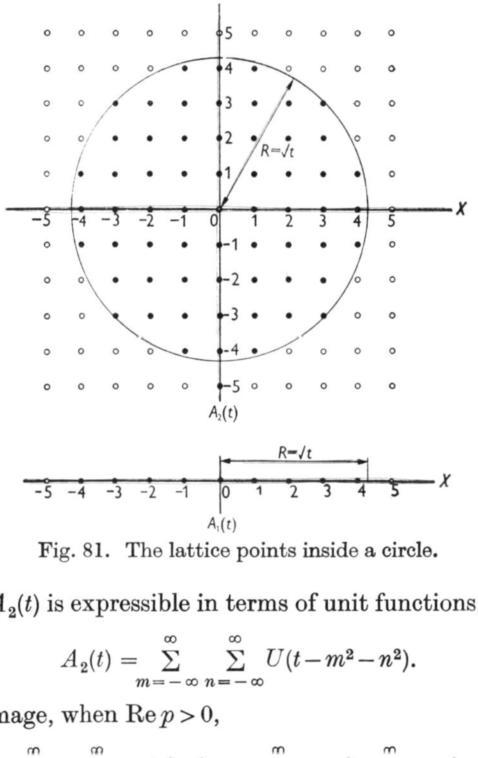

Fig. 81. The lattice points inside a circle.

First of all, $A_2(t)$ is expressible in terms of unit functions, as follows:

$$A_2(t) = \sum_{m=-\infty}^{\infty} \sum_{n=-\infty}^{\infty} U(t-m^2-n^2).$$

Thus, for the image, when $\operatorname{Re} p > 0$,

$$\sum_{m=-\infty}^{\infty} \sum_{n=-\infty}^{\infty} e^{-(m^2+n^2)p} = \sum_{m=-\infty}^{\infty} e^{-m^2p} \sum_{n=-\infty}^{\infty} e^{-n^2p},$$

of which either factor is equal to $\theta_3(0, p/\pi)$ on account of (VI, 39). Consequently, we have the very simple relation

$$A_2(t) \doteqdot \left\{\theta_3\left(0, \frac{p}{\pi}\right)\right\}^2, \quad 0 < \operatorname{Re} p < \infty.$$

It is not difficult to generalize this result. Let $A_n(t)$ denote the number of lattice points inside the n-dimensional sphere

$$x_1^2+x_2^2+\ldots+x_n^2 = t = R^2,$$

then, obviously, $$A_n(t) \doteqdot \left\{\theta_3\left(0, \frac{p}{\pi}\right)\right\}^n, \quad 0<\operatorname{Re} p<\infty. \tag{19}$$

In this connexion we remark that for $n = 1$ the relation (19) becomes equivalent to (VI, 39); the number of lattice points between $\sqrt{t}$ and $-\sqrt{t}$ is equal to $2[\sqrt{t}]+1$, thus

$$A_1(t) = \{2[\sqrt{t}]+1\}\,U(t) \doteqdot \theta_3\left(0, \frac{p}{\pi}\right) \quad (\operatorname{Re} p>0).$$

Therefore, by way of the operational calculus the problem of lattice points inside the n-sphere is reduced to considerations concerning the θ-function. For example, an asymptotic expression for $A_n(t)$ $(t\to\infty)$ is easily found, namely, by transformation of p into $1/p$ with the help of the last line of (XI, 11). On account of this relation, (19) may also be written in the following manner:

$$A_n(t) \doteqdot \left(\frac{\pi}{p}\right)^{\frac{1}{2}n} \left\{\theta_3\left(0, \frac{\pi}{p}\right)\right\}^n \quad (\operatorname{Re} p>0), \tag{19a}$$

or, alternatively, when expressed in terms of the familiar series:

$$A_n(t) \doteqdot \left(\frac{\pi}{p}\right)^{\frac{1}{2}n} (1+2\,\mathrm{e}^{-\pi^2/p}+2\,\mathrm{e}^{-4\pi^2/p}+2\,\mathrm{e}^{-9\pi^2/p}+\ldots)^n \quad (\operatorname{Re} p>0). \tag{20}$$

Since $A_n(t)$ is a monotonically increasing function of t, the Tauber theorem (VII, 15) for $\nu = \frac{1}{2}n$ leads to the asymptotic formula

$$A_n(t) \sim \frac{(\pi t)^{\frac{1}{2}n}}{\Pi(\frac{1}{2}n)} \quad (t\to\infty). \tag{21}$$

This result merely expresses the fact that the number of lattice points inside the n-dimensional sphere is, to a first approximation, equal to the volume of the sphere, which, of course, was to be expected from geometrical considerations (see also (IV, 24)). In addition, the corrections on this first approximation can be found by working out (20) so as to obtain the series:

$$A_n(t) \doteqdot \left(\frac{\pi}{p}\right)^{\frac{1}{2}n} \left\{1+\sum_{m=1}^{\infty} r_n(m)\,\mathrm{e}^{-\pi^2 m/p}\right\},$$

in which $r_n(m)$ denotes the number of possible splittings $m = m_1^2+\ldots+m_n^2$, when each m_j is any integer (positive, negative or zero). A transposition of this series with the aid of (X, 20) yields

$$A_n(t) = \frac{(\pi t)^{\frac{1}{2}n}}{\Pi(\frac{1}{2}n)} + \sum_{m=1}^{\infty} r_n(m) \left(\frac{t}{m}\right)^{\frac{1}{2}n} J_{\frac{1}{2}n}\{2\pi\sqrt{(mt)}\},$$

whence it follows that the corrections are representable by Bessel functions. A still simpler, closed expression for $A_2(t)$ is obtainable, when a well-known identity of the θ-function is used, viz.

$$\frac{1}{4}\left\{\theta_3\left(0,\frac{p}{\pi}\right)\right\}^2-\frac{1}{4}=\frac{1}{e^p-1}-\frac{1}{e^{3p}-1}+\frac{1}{e^{5p}-1}-\frac{1}{e^{7p}-1}+\dots$$

For, from (19) and (v, 8), we may infer that

$$\tfrac{1}{4}\{A_2(t)-1\}=\left[\frac{t}{1}\right]-\left[\frac{t}{3}\right]+\left[\frac{t}{5}\right]-\left[\frac{t}{7}\right]+\dots \quad (22)$$

This series terminates, of course, in a number of terms depending on t. An alternative way of interpreting (22) is to consider it as a relationship between the functions A_2 and A_1.

Before proceeding to another application of the operational calculus to the function $A_n(t)$, it may be noted that the theory above is easily extended so as to include the problem of lattice points inside multi-dimensional ellipsoids. The image equivalent to that of (19) consists of a product of θ-functions of mutually different second arguments. Further, when the centre of the sphere does not coincide with a lattice point, then the image involves the general θ-function, $\theta_3(v,p/\pi)$, of non-vanishing first argument, v.

Example. We shall now derive an integral equation for the lattice-point function $A_n(t)$. First, we have the following operational relation:

$$\left\{A_n(t)-\frac{(\pi t)^{\frac{1}{2}n}}{\Pi(\frac{1}{2}n)}\right\}U(t)\doteq\left(\frac{\pi}{p}\right)^{\frac{1}{2}n}(\{\theta_3(0,\pi/p)\}^n-1)\quad(\mathrm{Re}\,p>0),$$

which may be obtained from (19*a*) and the relation corresponding to the asymptotic expression (21) of $A_n(t)$, by mere subtraction. Then, using (xi, 2) applied for $\nu=1+\frac{1}{2}n$, we get another relation whose image contains the function $\theta_3^n(0,\pi p)$. Once more transposing the image with the aid of (19), we finally obtain

$$A_n(x)-1=\pi\int_0^\infty\left(\frac{x}{s}\right)^{\frac{1}{4}n+\frac{1}{2}}J_{\frac{1}{2}n+1}\{2\pi\sqrt{(xs)}\}\left\{A_n(s)-\frac{(\pi s)^{\frac{1}{2}n}}{\Pi(\frac{1}{2}n)}\right\}ds, \quad (23)$$

which may be considered as an *inhomogeneous* integral equation for A_n. On the other hand, the equation (23) is equivalent to a *homogeneous* equation for the function

$$\lambda_n(x)=\frac{A_n(x)-\dfrac{(\pi x)^{\frac{1}{2}n}}{\Pi(\frac{1}{2}n)}-1}{x^{\frac{1}{4}n+\frac{1}{2}}},$$

the numerator of which is just the correction corresponding to the approximate representation of $A_n(x)$ by the volume of the n-sphere of radius $\sqrt{x}$, whilst the term -1 indicates that the centre of the sphere should not be counted as a lattice point.

To derive this homogeneous integral equation we simply use

$$\pi\int_0^\infty\left(\frac{x}{s}\right)^{\frac{1}{4}n+\frac{1}{2}}J_{\frac{1}{2}n+1}\{2\pi\sqrt{(xs)}\}\,ds=\frac{(\pi x)^{\frac{1}{2}n}}{\Pi(\frac{1}{2}n)},$$

which, itself, follows from (xi, 27). The identity (23) leads to

$$\lambda_n(x)=\pi\int_0^\infty J_{\frac{1}{2}n+1}\{2\pi\sqrt{(xs)}\}\,\lambda_n(s)\,ds. \quad (24)$$

In particular, for the number of lattice points inside a circle, we find for $n = 2$ the following homogeneous integral equation:

$$\frac{A_2(x)-\pi x-1}{x} = \pi \int_0^\infty J_2\{2\pi\sqrt{(xs)}\}\frac{A_2(s)-\pi s-1}{s}\,ds.$$

5. Arithmetic functions in connexion with Dirichlet series

That many arithmetic functions are intimately connected with Riemann's ζ-function, $\zeta(p)$, is a consequence of the possibility of deriving from its series (see (III, 36))

$$\zeta(p) = \frac{1}{1^p}+\frac{1}{2^p}+\frac{1}{3^p}+\dots \quad (\operatorname{Re} p > 1)$$

by quite simple transformations, other Dirichlet series involving number-theoretic functions. For instance,

$$\zeta^2(p) = \sum_{n=1}^{\infty}\frac{d(n)}{n^p} \quad (\operatorname{Re} p > 1),$$

which was discussed in (VII, 16). In like manner, a great many Dirichlet series having number-theoretic coefficients are derivable from the infinite-product development of the ζ-function:

$$\zeta(p) = \Pi\left(1-\frac{1}{N^p}\right)^{-1} \quad (\operatorname{Re} p > 1), \tag{25}$$

in which N runs through the set of prime numbers: 2, 3, 5, 7, 11, ...†

Examples of series originating from this infinite product, after suitable operations, are the following:

(1) $$\frac{1}{\zeta(p)} = \Pi\left(1-\frac{1}{N^p}\right) = \sum_{n=1}^{\infty}\frac{\mu(n)}{n^p}$$
$$= 1-\frac{1}{2^p}-\frac{1}{3^p}-\frac{1}{5^p}+\frac{1}{6^p}-\frac{1}{7^p}+\frac{1}{10^p}-\dots \quad (\operatorname{Re} p > 1), \tag{26}$$

in which $\mu(n)$ denotes the Möbius function (see (IV, 10));

(2) $$\frac{\zeta(p)}{\zeta(2p)} = \Pi\left(1+\frac{1}{N^p}\right) = \sum_{n=1}^{\infty}\frac{\mu^2(n)}{n^p}$$
$$= 1+\frac{1}{2^p}+\frac{1}{3^p}+\frac{1}{5^p}+\frac{1}{6^p}+\frac{1}{7^p}+\frac{1}{10^p}+\dots \quad (\operatorname{Re} p > 1). \tag{27}$$

This series is obtainable from that of $\zeta(p)$ by deleting those terms n^{-p} for which n possesses as factor a square > 1;

(3) $$\frac{\zeta(2p)}{\zeta(p)} = \Pi\left(1+\frac{1}{N^p}\right)^{-1} = \sum_{n=1}^{\infty}\frac{(-1)^{\Omega(n)}}{n^p}$$
$$= 1-\frac{1}{2^p}-\frac{1}{3^p}+\frac{1}{4^p}-\frac{1}{5^p}+\frac{1}{6^p}-\frac{1}{7^p}-\frac{1}{8^p}+\frac{1}{9^p}+\frac{1}{10^p}-\dots \quad (\operatorname{Re} p > 1), \tag{28}$$

† Instead of the conventional p for prime number, the notation N is used, in order to avoid confusion with the variable p of the operational calculus.

where $\Omega(n)$ denotes the total number of prime factors of n, that is,

$$\Omega(n) = \alpha_1 + \alpha_2 + \ldots + \alpha_k,$$

when the canonic factorization of n is given by

$$n = N_1^{\alpha_1} N_2^{\alpha_2} \ldots N_k^{\alpha_k}.$$

(4) $$\frac{\zeta^2(p)}{\zeta(2p)} = \Pi \left(\frac{1+\dfrac{1}{N^p}}{1-\dfrac{1}{N^p}} \right) = \sum_{n=1}^{\infty} \frac{2^{\nu(n)}}{n^p}$$

$$= 1 + \frac{2}{2^p} + \frac{2}{3^p} + \frac{2}{4^p} + \frac{2}{5^p} + \frac{4}{6^p} + \frac{2}{7^p} + \frac{2}{8^p} + \frac{2}{9^p} + \frac{4}{10^p} + \ldots \quad (\operatorname{Re} p > 1), \qquad (29)$$

in which $\nu(n)$ is the number of *different* prime factors contained in n; $\nu(1) = 0$.

The verification of any of the formulae above is easily established, by working out the infinite product and investigating how frequent a certain term, n^{-p}, occurs. The frequency of this term becomes the coefficient of the Dirichlet series in question.

The originals corresponding to the foregoing series can be obtained with the aid of (VI, 10). Then it is found that

$$\frac{1}{\zeta(p)} \doteq \sum_{n=1}^{[e^t]} \mu(n), \quad 1 < \operatorname{Re} p < \infty, \qquad (26a)$$

$$\frac{\zeta(p)}{\zeta(2p)} \doteq \sum_{n=1}^{[e^t]} \mu^2(n) = Q([e^t]), \quad 1 < \operatorname{Re} p < \infty, \qquad (27a)$$

$$\frac{\zeta(2p)}{\zeta(p)} \doteq \sum_{n=1}^{[e^t]} (-1)^{\Omega(n)}, \quad 1 < \operatorname{Re} p < \infty, \qquad (28a)$$

$$\frac{\zeta^2(p)}{\zeta(2p)} \doteq \sum_{n=1}^{[e^t]} 2^{\nu(n)}, \quad 1 < \operatorname{Re} p < \infty, \qquad (29a)$$

where the function $Q(x)$ occurring in (27*a*) may be interpreted as the total number of square-free numbers not exceeding x.

At the end of the present section we shall illustrate by some examples how numerous properties of the arithmetic functions are deducible, operationally, from the relations listed above. In addition, we may remark that these functions, though initially defined only for integers, become, as operational originals, step functions which thus have definite values at the intermediate points. It is just this continuation (interpolation) which makes it possible to study the arithmetic function by means of the analytic image.

Moreover, it appears that still different arithmetic functions may lead to images that, unlike those above, do not contain the ζ-function algebraically only. This holds, in particular, with respect to functions occurring in the famous distribution law of prime numbers which will now be investigated in some detail.

For that purpose, the starting-point is the series for the logarithmic derivative of the ζ-function of (25):

$$\frac{\zeta'(p)}{\zeta(p)} = -\Sigma \frac{\log N}{N^p - 1} = -\Sigma \frac{\log N}{e^{p \log N} - 1}.$$

In virtue of (v, 8) the corresponding operational relation reads as follows:

$$\frac{\zeta'(p)}{\zeta(p)} \doteqdot -\sum_{N<e^t} \log N \left[\frac{t}{\log N}\right] = -\psi_T(e^t) \quad (\operatorname{Re} p > 1), \tag{30}$$

where we have introduced Tchebycheff's well-known function

$$\psi_T(x) \equiv \sum_{N<x} \log N \left[\frac{\log x}{\log N}\right]. \tag{31}$$

As is shown with not too much difficulty, this function can be written equivalently in the following two ways:

(1) $$\psi_T(x) = \sum_{n=1}^{[x]} \Lambda(n),$$

where $\Lambda(n) = \log n$ or $= 0$, depending on whether or not n is a power of some prime number;

(2) $$\psi_T(n) = \log\{1, 2, 3, 4, \ldots, n\}, \tag{32}$$

in which $\{1, 2, 3, 4, \ldots, n\}$ denotes the least common multiple of the numbers from 1 to n inclusive.

The distribution law of prime numbers in question follows at once when the complex Tauber theorem V (see VII, § 5) is applied to

$$-\frac{p}{p+1} \frac{\zeta'(p+1)}{\zeta(p+1)} \doteqdot e^{-t} \psi_T(e^t) \quad (\operatorname{Re} p > 0),$$

which relation easily follows from (30). For, on the one hand, the original of this relation multiplied by $-e^t$, i.e. the function $\psi_T(e^t)$, is monotonic, and on the other hand, we know that the image above divided by p, as well as $\zeta(p+1)$ itself, is singular on the boundary $\operatorname{Re} p = 0$ only at the first-order pole at $p = 0$. Therefore, the limit of the image being 1 for $p \to 0$, the Tauber theorem leads at once to the following asymptotic property:

$$\psi_T(e^t) \sim e^t \quad (t \to \infty),$$

or alternatively $$\psi_T(x) \sim x \quad (x \to \infty). \tag{33}$$

Further, it is easily shown† that the property (33) is equivalent to the distribution law

$$\pi(x) \sim \frac{x}{\log x}, \tag{34}$$

† See, for example, D. V. Widder, *The Laplace Transform*, Princeton, 1941, p. 226.

in which the one-sided function $\pi(x)$ denotes the number of primes below x. This may be made plausible in the following manner. Inasmuch as $[x]$ may be approximated by x, the brackets in (31) can be omitted, which leads to

$$\psi_T(x) \sim \sum_{N<x} \log x = \pi(x) \log x,$$

and then (33) at once yields the prime-number theorem (34).

In retrospect, the prime-number theorem appears to be based substantially on the known behaviour of $\zeta(p)$ for $\operatorname{Re} p \geqslant 1$. For, the simple fact that in this domain the function exhibits only a singularity at the pole $p = 1$ enables us to apply the complex Tauber theorem. It is further obvious that a sharper result concerning the behaviour of the difference $\pi(x) - x/\log x$ will be obtainable from further properties of the ζ-function relating to its behaviour to the left of the line $\operatorname{Re} p = 1$; in this way the zeros of $\zeta(p)$ inside the strip $0 < \operatorname{Re} p < 1$ prove important, but will not be investigated here. For the rest, the derivation of the distribution law was reduced to the property (33) of the function $\psi_T(x)$, and, in connexion with (32), this property can also be stated as follows:

$$\log\{1, 2, 3, \ldots, n\} \sim n \quad (n \to \infty),$$

or, alternatively,

$$\lim_{n\to\infty} \sqrt[n]{\{1, 2, 3, \ldots, n\}} = e.$$

On the other hand, it is still possible to deduce the properties of $\pi(x)$ more directly, without using the function $\psi_T(x)$. This will be seen in the third of the examples below, where other asymptotic properties of the prime numbers will also be put forward.

Example 1. *Distribution law of squares.* Information concerning the density of squares in the sequence of natural numbers is obtained when the following operational relation is studied:

$$\frac{p}{p+1} \frac{\zeta(p+1)}{\zeta(2p+2)} \doteqdot e^{-t} Q([e^t]) \quad (\operatorname{Re} p > 0).$$

This relation can be derived from (27 a) with the help of the attenuation rule, $Q(x)$ denoting the number of square-free numbers below x. Since $Q(x)$ is monotonic, and the image divided by p is singular on the boundary of convergence $\operatorname{Re} p = 0$ only at the first-order pole $p = 0$, the complex Tauber theorem is again applicable, now leading to

$$\lim_{t\to\infty} e^{-t} Q([e^t]) = \lim_{p\to 0} \frac{p}{p+1} \frac{\zeta(p+1)}{\zeta(2p+2)}.$$

Further, since the residue of $\zeta(p)$ at $p = 1$ is equal to 1, the value of the limit at the right is equal to $1/\zeta(2) = 6/\pi^2$; hence, asymptotically,

$$e^{-t} Q([e^t]) \sim \frac{6}{\pi^2} \quad (t \to \infty),$$

that is,

$$Q(n) \sim \frac{6}{\pi^2} n \quad (n \to \infty).$$

Consequently, $6/\pi^2$ of all integers, that is, approximately 60 % are square-free.

Example 2. *Connexion between the number of prime factors of n and the function of Möbius.* A property concerning the function $\nu(n)$, denoting the number of distinct prime factors of n, is found when the relation (27 *a*) is multiplied by the series

$$\zeta(p) = \sum_{n=1}^{\infty} e^{-p\log n} \quad (\operatorname{Re} p > 1).$$

A term-by-term transposition then leads to

$$\frac{\zeta^2(p)}{\zeta(2p)} \doteqdot Q\left(\left[\frac{e^t}{n}\right]\right) \quad (\operatorname{Re} p > 1). \tag{35}$$

On the other hand, the original of the left-hand side is known from (29 *a*). Consequently, upon putting $e^t = x$ the following identity is obtained:

$$\sum_{n=1}^{[x]} 2^{\nu(n)} = \sum_{n=1}^{\infty} Q\left(\left[\frac{x}{n}\right]\right) \quad (x > 1).$$

In like manner, by multiplying by $\zeta(2p)$ instead of by $\zeta(p)$ we get from (27 *a*) the property

$$\sum_{n=1}^{\infty} Q\left(\left[\frac{x}{n^2}\right]\right) = [x] \quad (x > 0).$$

Example 3. *Further asymptotic properties of the prime numbers.* In order to find these we start not with the logarithmic derivative of the infinite product (25) but with the series for the logarithm itself, namely,

$$\log\zeta(p) = -\Sigma\log\left(1-\frac{1}{N^p}\right) = \Sigma\frac{1}{N^p} + \tfrac{1}{2}\Sigma\frac{1}{N^{2p}} + \tfrac{1}{3}\Sigma\frac{1}{N^{3p}} + \dots \quad (\operatorname{Re} p > 1).$$

The first term of the right-hand member, where the summation has to be extended over all primes $N > 1$, is determinable from Möbius's inversion formula (IV, 11) leading to:

$$\Sigma\frac{1}{N^p} = \sum_{n=1}^{\infty}\frac{\mu(n)}{n}\log\zeta(np) \quad (\operatorname{Re} p > 1). \tag{36}$$

Upon transforming the left-hand side we arrive at

$$\pi(e^t) \doteqdot \sum_{n=1}^{\infty}\frac{\mu(n)}{n}\log\zeta(np), \quad 1 < \operatorname{Re} p < \infty, \tag{37}$$

by means of which the distribution function $\pi(x)$ of the prime numbers is directly associated with Riemann's ζ-function.

From this relation we shall derive consecutively three asymptotic properties of prime numbers†.

(1) To approximate the following sum of inverse primes,

$$\sum_{N<x}\frac{1}{N},$$

we first derive from (37) with the aid of elementary rules the new relation

$$\int_{-\infty}^{t} e^{-\tau}\pi(e^{\tau})\,d\tau \doteqdot \frac{1}{p+1}\sum_{n=1}^{\infty}\frac{\mu(n)}{n}\log\zeta(np+n) \quad (\operatorname{Re} p > 0).$$

† These are treated by Balth. van der Pol, 'On the application of the symbolic calculus to the theory of prime numbers', *Phil. Mag.* XXVI, 921, 1938.

Since on the boundary only the term $n = 1$ of the image becomes singular, and then only at $p = 0$, where it behaves as $-\log p$, we subtract (III, 26) from the relation above. Assuming that the Tauber theorem $h(\infty) = f(0)$ is actually applicable in this case, we obtain the following result:

$$\operatorname*{Lim}_{t\to\infty}\left\{\int_{-\infty}^{t} e^{-\tau}\pi(e^{\tau})\,d\tau-(\log t+C)\right\} = \sum_{n=2}^{\infty}\frac{\mu(n)}{n}\log\zeta(n) = -0{\cdot}3157\ldots.$$

The integral at the left can be worked out by partial integration, using the identity

$$\int_{-\infty}^{t} e^{-\tau}\,d\pi(e^{\tau}) = \int_{-\infty}^{t} \Sigma\,\delta(e^{\tau}-N)\,d\tau = \sum_{N<e^t}\frac{1}{N}.$$

When, moreover, the result (34) is used, we are finally led to

$$\operatorname*{Lim}_{x\to\infty}\left(\sum_{N<x}\frac{1}{N}-\log\log x\right) = C+\sum_{n=2}^{\infty}\frac{\mu(n)}{n}\log\zeta(n) = 0{\cdot}2615\ldots.$$

This approximation, which has been proved rigorously by Kluyver†, is worth while comparing with the well-known limiting relation

$$\operatorname*{Lim}_{x\to\infty}\left(\sum_{n<x}\frac{1}{n}-\log x\right) = C = 0{\cdot}5772\ldots,$$

where the summation is extended over *all* natural numbers inferior to x.

(2) To investigate the sum

$$\sum_{N<x}\log N$$

we first differentiate the identity (36) with respect to p; the relation (VI, 10) for the resulting Dirichlet series then becomes

$$\sum_{N<e^t}\log N \doteqdot -\sum_{n=1}^{\infty}\mu(n)\frac{\zeta'(np)}{\zeta(np)} \quad (\operatorname{Re}p>1).$$

Upon multiplying the original by e^{-t} we obtain

$$e^{-t}\sum_{N<e^t}\log N \doteqdot -\frac{p}{p+1}\sum_{n=1}^{\infty}\mu(n)\frac{\zeta'(np+n)}{\zeta(np+n)} \quad (\operatorname{Re}p>0), \tag{38}$$

the image of which possesses the limiting value 1 as $p\to 0$. This limit is fixed by the term with $n = 1$, since the other terms do not contribute at all, in virtue of the series

$$\sum_{n=2}^{\infty}\mu(n)\frac{\zeta'(n)}{\zeta(n)} = 0{\cdot}74\ldots$$

having a finite sum. Application of the complex Tauber theorem to (38) is allowed (upon multiplication by e^t the original becomes a monotonic function). Therefore,

$$1 = \operatorname*{Lim}_{t\to\infty}\left\{e^{-t}\sum_{N<e^t}\log N\right\},$$

thus

$$\sum_{N<x}\log N\sim x \quad (x\to\infty).$$

In other words, the nth root of the product of the primes below n tends to e as $n\to\infty$.

(3) To obtain an approximation for the sum

$$\sum_{N<x}\frac{\log N}{N},$$

† *Verslagen Akad. van Wetenschap. Amsterdam*, VIII, 679, 1900.

we first differentiate (36) with respect to p, then replace p by $p+1$, and next form the relation (VI, 10). The result becomes

$$\sum_{N<e^t} \frac{\log N}{N} \doteqdot -\sum_{n=1}^{\infty} \mu(n)\frac{\zeta'(np+n)}{\zeta(np+n)} \quad (\operatorname{Re} p > 0).$$

Upon subtracting the operational relation $tU(t) \doteqdot 1/p$ we finally obtain, again assuming the applicability of the complex Tauber theorem,

$$\operatorname*{Lim}_{x\to\infty} \left\{ \sum_{N<x} \frac{\log N}{N} - \log x \right\} = -\sum_{n=2}^{\infty} \mu(n)\frac{\zeta'(n)}{\zeta(n)} = -0{\cdot}74\ldots.$$

6. Step functions of argument equal to the summation variable in a series

An operational relation containing a step function of this type can be derived from any generating function of the form

$$\chi(x,\alpha) = \sum_{n=-\infty}^{\infty} \phi_n(\alpha)\,x^n \quad (a<|x|<b), \tag{39}$$

where α is some parameter. Thus, upon identifying x with e^{-p} the Laurent series (39) may be transposed term by term for $\operatorname{Re} p > 0$, thus leading to

$$\chi(e^{-p},\alpha) \doteqdot \sum_{n=-\infty}^{[t]} \phi_n(\alpha), \quad -\log b < \operatorname{Re} p < -\log a,$$

where the *annular* domain of convergence for x transforms into a *strip* of convergence for p.

A still simpler original is obtained when the image is multiplied by $1-e^{-p}$ and the shift rule is applied, so as to spot a single term of the series, thus:

$$(1-e^{-p})\,\chi(e^{-p},\alpha) \doteqdot \phi_{[t]}(\alpha), \quad -\log b < \operatorname{Re} p < -\log a. \tag{40}$$

Therefore we have an example in which the summation variable (of the generating function), indeed, occurs as the argument of a step function. Again, it is of little importance how the step function is defined at the points of discontinuity, though the inversion integral corresponding to (40), if convergent, will produce the mean value, that is, $\frac{1}{2}\{\phi(n-1)+\phi(n)\}$.

As an explicit example, involving a two-sided infinite Laurent series, we mention the generating function of the Bessel coefficients

$$e^{\frac{1}{2}\alpha(x-1/x)} = \sum_{n=-\infty}^{\infty} J_n(\alpha)\,x^n.$$

In this case the relation (40) becomes

$$(1-e^{-p})\,e^{-\alpha\sinh p} \doteqdot J_{[t]}(\alpha), \quad -\infty < \operatorname{Re} p < \infty. \tag{41}$$

The general relation (40) can be further applied to any given power series:

$$\lambda(x) = \sum_{n=0}^{\infty} \phi_n x^n,$$

since $\lambda(x)$ may here be considered as generating function of ϕ_n. The operational relations so obtained are automatically one-sided. Accordingly, the power-series expansions

$$\sum_{n=0}^{\infty} \sin(n\vartheta)\, x^n = \frac{x \sin\vartheta}{1-2x\cos\vartheta+x^2} \quad (|x|<1), \tag{42}$$

$$\sum_{n=0}^{\infty} P_n(\cos\vartheta)\, x^n = \frac{1}{\sqrt{(1-2x\cos\vartheta+x^2)}} \quad (|x|<1)\ (\vartheta \text{ real}), \tag{43}$$

$$\sum_{n=0}^{\infty} \frac{L_n(\alpha)}{n!} x^n = \frac{e^{-\alpha x/(1-x)}}{1-x} \quad (|x|<1) \tag{44}$$

lead in turn to the following relations:

$$\frac{\sin\vartheta(1-e^{-p})}{2(\cosh p-\cos\vartheta)} \doteq \sin(\vartheta[t])\, U(t), \quad 0<\operatorname{Re} p<\infty, \tag{42 a}$$

$$\frac{\sqrt{2}\sinh\frac{1}{2}p}{\sqrt{(\cosh p-\cos\vartheta)}} \doteq P_{[t]}(\cos\vartheta)\, U(t), \quad 0<\operatorname{Re} p<\infty, \tag{43 a}$$

$$e^{\alpha/(1-e^p)} \doteq \frac{L_{[t]}(\alpha)}{[t]!}, \quad 0<\operatorname{Re} p<\infty. \tag{44 a}$$

To illustrate the fundamental importance of (40) for power series, it is interesting to see how (40) includes as a special case an extension of Widder's inversion formula (VII, 58). For this purpose the Taylor series is chosen as generating function:

$$\chi(\alpha-\alpha x) = \sum_{n=0}^{\infty} \frac{(-\alpha)^n}{n!} \chi^{(n)}(\alpha)\, x^n, \quad |x| < \frac{\rho(\alpha)}{|\alpha|}.$$

Applying (40) we first have

$$(1-e^{-p})\, \chi(\alpha-\alpha e^{-p}) \doteq \frac{(-\alpha)^{[t]}}{[t]!} \frac{d^{[t]}}{d\alpha^{[t]}} \chi(\alpha), \quad -\log\frac{\rho}{|\alpha|} < \operatorname{Re} p < \infty,$$

and then, upon replacing $\chi(p)$ by $f(p)/p$ and p by p/α,

$$f(\alpha-\alpha e^{-p/\alpha}) \doteq \frac{\alpha^{[\alpha t]+1}}{[\alpha t]!}\left(-\frac{d}{d\alpha}\right)^{[\alpha t]}\left\{\frac{f(\alpha)}{\alpha}\right\}, \quad -\alpha\log\frac{\rho}{\alpha} < \operatorname{Re} p < \infty \ (\alpha>0).$$

Let us now pass to the limit $\alpha\to\infty$. Then the image reduces to $f(p)$. Moreover, if it is assumed that the limit of the original is the original of the limit, then the right-hand member transforms into an expression for $h(t)$. In this way, and after putting $\alpha t = n$, we obtain

$$h(t) = \operatorname*{Lim}_{n\to\infty}\left\{\frac{\alpha^{[n]+1}}{[n]!}\left(-\frac{d}{d\alpha}\right)^{[n]}\frac{f(\alpha)}{\alpha}\right\}_{\alpha=n/t},$$

in which, in contrast with the previous formula, n may tend to infinity along non-integral values.

As to possible applications of (40) to special functions, we observe that this operational relation is particularly useful in the study of series of functions of integral order, as may appear from the following two examples.

Example 1. An approximation for Laguerre polynomials of high orders is obtained when the image of (44*a*) is expanded, apart from a factor $e^{-\alpha/p}$, in a power series of p. A transposition of this series leads to

$$\frac{L_{[t]}(\alpha)}{[t]!} = e^{\frac{1}{2}\alpha}\left(1-\frac{\alpha}{12}\frac{d}{dt}+\frac{\alpha^2}{288}\frac{d^2}{dt^2}+\dots\right)(J_0\{2\sqrt{(\alpha t)}\}U(t)).$$

In particular we derive from the first term:

$$L_n(\alpha)\sim e^{\frac{1}{2}\alpha}n!\,J_0\{2\sqrt{(\alpha n)}\} \quad (n \gg 1).$$

Example 2. The inversion integral corresponding to (40) is generally a complex integral representation for discontinuous functions. Sometimes it may be reduced to a real integral, for instance, if the path of integration can be taken along the imaginary axis. An example of this is provided by the inversion integral of the two-sided relation (41) for the Bessel function. After substituting $p = i\omega$ and reducing the interval for $\omega<0$ to that for $\omega>0$, we finally obtain the real integral representation

$$J_{[\nu]}(\alpha) = \frac{2}{\pi}\int_0^\infty \frac{\sin\frac{1}{2}\omega}{\omega}\cos\{(\nu-\tfrac{1}{2})\omega-\alpha\sin\omega\}\,d\omega.$$

It is almost unnecessary to emphasize that for integral values of ν ($= n$) the left-hand side should be replaced by $\frac{1}{2}\{J_n(\alpha)+J_{n-1}(\alpha)\}$, in agreement with the fact that the inversion integral always leads to the mean value at any point of discontinuity. This again is an example of a discontinuous function whose argument ν occurs as a continuously variable parameter under the sign of integration of the integral representing it.

7. Contragrade series

We conclude our considerations regarding step functions with the investigation of the composition product of two such functions. In so doing we shall confine ourselves to step functions with equidistant discontinuities occurring at the integral points. Therefore, let us assume that

$$h_1([t]) \doteqdot f_1(p), \quad \alpha_1<\operatorname{Re}p<\beta_1,$$

$$h_2([t]) \doteqdot f_2(p), \quad \alpha_2<\operatorname{Re}p<\beta_2.$$

Instead of working out the composition product of these functions directly, we follow a more rapid way. From the derivation of (1) given at the beginning of this chapter it follows that

$$f_1(p) = (1-e^{-p})\sum_{n=-\infty}^{\infty} h_1(n)\,e^{-np},$$

and hence that

$$\phi(t) \equiv \sum_{n=-\infty}^{\infty} h_1(n)\,h_2([t]-n) \doteqdot \sum_{n=-\infty}^{\infty} h_1(n)\,e^{-np}f_2(p) = \frac{f_1(p)\,f_2(p)}{1-e^{-p}}. \quad (45)$$

From this the composition product in question can be obtained by multiplying the image by $(1-e^{-p})/p$; that is to say, (45) has to be multiplied through by the operator

$$\mathrm{Lin}(\ldots) \equiv \left\{\int_{-\infty}^{t} d\tau - \int_{-\infty}^{t-1} d\tau\right\}(\ldots) = \int_{t-1}^{t} d\tau(\ldots) \fallingdotseq \frac{1}{p}(1-e^{-p})(\ldots), \quad \mathrm{Re}\, p > 0. \tag{46}$$

In other words, to obtain the composition product in question we have to apply the operator 'Lin' to $\phi(t)$, which means that $\phi(t)$ should be replaced throughout by its mean value in the interval $t-1<\tau<t$. Since $\phi(t)$ itself is a step function, this operation, in graphical terms, simply comes down to replacing $\phi(t)$ (i) at the jump $t = n$ by the value $\phi(n-0)$, and (ii) at the intermediate points by the straight-line segments connecting these values (see fig. 82). Therefore the function $\mathrm{Lin}\,\phi(t)$ originates from $\phi(t)$ by linear interpolation between the integral points.

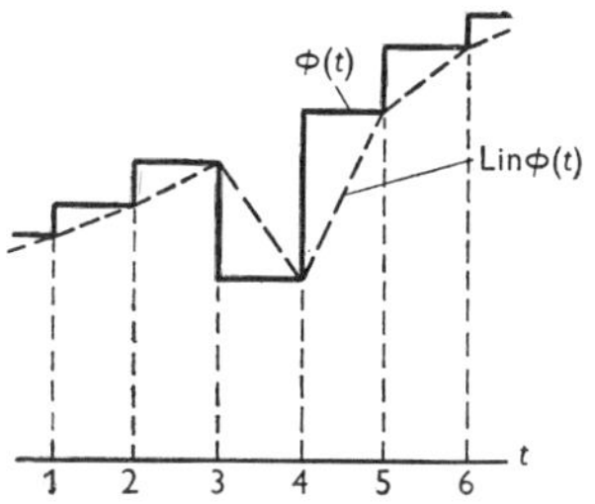

Fig. 82. Illustrating the application of the operator Lin to a step function.

Using the notation introduced above, we may write for the composition product of two step functions jumping at integral points

$$\frac{f_1(p)\,f_2(p)}{p} \fallingdotseq \mathrm{Lin} \sum_{n=-\infty}^{\infty} h_1(n)\,h_2([t]-n). \tag{47}$$

In general, the series (47) possesses an infinite number of terms. However, when the given operational relations are one-sided the series terminates automatically. In this case (47) becomes

$$\frac{f_1(p)\,f_2(p)}{p} \fallingdotseq U(t)\,\mathrm{Lin} \sum_{n=0}^{[t]} h_1(n)\,h_2([t]-n). \tag{48}$$

Summarizing, we may say that by the process above we are always led to series that involve the values of the initially given originals at the integral points, in such a sense that the arguments of the two originals run in opposite directions. It is for this that we introduce the term *contragrade series*, of which an explicit example will now be given.

Example. Of course, to obtain contragrade series we may equally well start with only one given relation and consider the (composition) product of a step function by itself. With this in mind we get from (43*a*) for Legendre functions, by applying (48), the following relation:

$$\frac{2\sinh^2 \frac{1}{2}p}{p(\cosh p - \cos\vartheta)} \fallingdotseq U(t)\,\mathrm{Lin} \sum_{k=0}^{[t]} P_k(\cos\vartheta)\,P_{[t]-k}(\cos\vartheta) \quad (\mathrm{Re}\, p > 0).$$

Since the left-hand member can be written as

$$\frac{1}{\sin\vartheta} \times \frac{1-e^{-p}}{p} \times e^{p} \frac{\sin\vartheta(1-e^{-p})}{2(\cosh p - \cos\vartheta)},$$

the original is also equal to

$$\frac{1}{\sin\vartheta}\operatorname{Lin}\{\sin(\vartheta[t+1])\,U(t+1)\} = \frac{U(t)}{\sin\vartheta}\operatorname{Lin}\sin\{\vartheta([t]+1)\},$$

as follows from (42a) and (46). We thus find the identity, by cancelling the operator Lin,

$$\frac{\sin\{\vartheta([t]+1)\}}{\sin\vartheta} = \sum_{k=0}^{[t]} P_k(\cos\vartheta)\,P_{[t]-k}(\cos\vartheta) \quad (t>0),$$

which for $t = n+0$ leads to the following result:

$$\frac{\sin\{(n+1)\vartheta\}}{\sin\vartheta} = \sum_{k=0}^{n} P_k(\cos\vartheta)\,P_{n-k}(\cos\vartheta).$$

We finally remark that contragrade series are also obtainable without the operational calculus, by squaring the generating series (in the example above, the series (43)).

CHAPTER XIII

DIFFERENCE EQUATIONS

1. Introduction

In addition to linear *differential* equations, linear *difference* equations are just as readily accessible to an operational treatment, as will be demonstrated in the present chapter by elaboration of a number of special problems†. In order to write difference equations as simply as possible, it is convenient to introduce symbols of abbreviation for the difference of two functional values occurring therein, similar to Δ_n used in IX, §§ 7, 8,

$$\Delta_n \phi(n) = \phi(n+\tfrac{1}{2}) - \phi(n-\tfrac{1}{2}). \tag{1}$$

If n is identified with the operational variable t, then, in virtue of the shift rule, the operation (1) reduces to multiplying the image by $2\sinh\frac{1}{2}p$. This property will be designated by

$$\Delta_t(\ldots) \doteqdot 2\sinh\tfrac{1}{2}p\,(\ldots), \tag{2}$$

which should be compared with the analogue for differentiation

$$\frac{d}{dt}(\ldots) \doteqdot p\,(\ldots).$$

Instead of (1) one usually considers the difference of the functions at points not symmetrically situated with respect to n, by generally defining

$$\underset{\omega}{\Delta_n} f(n) = \frac{f(n+\omega)-f(n)}{\omega}, \tag{3}$$

which difference quotient reduces to the differential quotient $f'(n)$ as $\omega \to 0$. In the body of this chapter we shall apply either the symmetric difference quotient (1) or the asymmetric (3) for $\omega = 1$, that is, the difference

$$\underset{1}{\Delta_n} f(n) = f(n+1) - f(n). \tag{4}$$

We shall deal especially with *linear* difference equations. The simplest of them, that of the sum function, will be treated in § 2, whilst in §§ 3 and 4 linear difference equations with constant and variable coefficients respectively will be discussed. Next, in § 5, the connexion between difference and differential equations is investigated. Finally, in § 6, the operational construction of difference equations for known functions is indicated.

† For a general survey of the theory of difference equations the reader may be referred to: N. E. Nörlund, *Vorlesungen über Differenzenrechnung*, Berlin, 1924; L. M. Milne-Thomson, *The Calculus of Finite Differences*, London, 1933.

2. Difference equation for the 'sum'

The difference equation for the 'sum' $g(x)$ of a function $\varphi(x)$ is defined as the following first-order difference equation:

$$\underset{1}{\Delta} g(x) = g(x+1)-g(x) = \varphi(x), \tag{5}$$

which is the difference analogue of the differential equation $g'(x) = \varphi(x)$. Equation (5) has infinitely many solutions, but knowledge of one of them is sufficient, inasmuch as any other solution can be found by adding an arbitrary periodic function, $\pi(x)$, of period 1†.

Particular solutions can be obtained operationally by identifying x with either of the operational variables t and p, though in many cases there is no room for choice; for instance, when $\varphi(x)$ does have an original but no image, then x has to be taken equal to p. In the following, however, we shall confine ourselves to identifying x with t.

Therefore, let us assume that the two operational relations

$$g(t) \risingdotseq G(p), \quad \varphi(t) \risingdotseq \phi(p),$$

for the unknown and the given function respectively, have a common strip of convergence. Inside the latter the operational transform of (5) becomes

$$(e^p-1)\,G(p) = \phi(p),$$

so that a mere division at once leads to the p-form of the unknown function

$$G(p) = \frac{\phi(p)}{e^p-1}. \tag{6}$$

Under suitable conditions, two different solutions of (5) may be derived from (6), as appears from the two expansions of (6) for $\operatorname{Re} p > 0$ and $\operatorname{Re} p < 0$ respectively:

$$G(p) = \sum_{n=1}^{\infty} e^{-np}\,\phi(p), \quad G(p) = -\sum_{n=0}^{\infty} e^{np}\,\phi(p).$$

For the corresponding originals, in the same order, are

$$g_1(t) = \sum_{n=1}^{\infty} \varphi(t-n), \tag{7a}$$

$$g_2(t) = -\sum_{n=0}^{\infty} \varphi(t+n). \tag{7b}$$

In general, these solutions have only a formal meaning, since the series above will often diverge, as in the analogous case of the differential equation $g'(x) = \phi(x)$ where we have, formally at least,

$$g_1(x) = \int_0^\infty \varphi(x-n)\,dn, \quad g_2(x) = -\int_0^\infty \varphi(x+n)\,dn.$$

† The symbol $\pi(x)$ should not be confused with that introduced in the preceding chapter for the distribution function of primes.

Fortunately, the series ($7\,a, b$) can often be made convergent by adjunction of suitable convergence factors satisfying the permanence condition (cf. VI, § 8). Then the expressions (7) may be considered as a shorthand notation for the series so made convergent. In the way indicated there exists in many cases the following solution†:

$$g_2(x) = \lim_{\eta \to 0}\left\{\int_0^\infty \varphi(s)\, e^{-\eta\lambda(s)}\, ds - \sum_{n=0}^{\infty} \varphi(t+n)\, e^{-\eta\lambda(t+n)}\right\},$$

which has the form of a Cauchy limit and in which the auxiliary function $\lambda(s)$ should be taken equal to

$$\lambda(s) = s^\alpha (\log s)^\beta \quad (\alpha \geqslant 1;\ \beta \geqslant 0).$$

This expression originates from ($7\,b$) after addition of the constant

$$\int_0^\infty \varphi(s)\, ds.$$

This solution is commonly called the *principal solution* of (5), or the 'sum' of the function $\varphi(x)$. Since it is obtained by addition of a constant to the operationally derived series ($7\,b$) it is clear that any solution derived from the case $\operatorname{Re} p < 0$ by the operational calculus will differ from the principal solution by at most an additive constant. In particular, the principal solution itself is found when $\int_0^\infty \varphi(s)\, ds = 0$. Moreover, the series $7\,a$ ($7\,b$) converges automatically for functions $\varphi(x)$ that are cut off at the left (right).

We now return to the image of (6), which can lead to a solution even without the use of the series expansion producing (7); any such solution, then, independently whether its strip of convergence is in $\operatorname{Re} p > 0$ or $\operatorname{Re} p < 0$, differs from the principal one by an additive constant at most. This follows from the fact that two solutions of (5) can only differ by a periodic function. A solution that has been derived as an operational original thus consists of the sum of the principal solution and a periodic function, the latter having no strip of convergence whatever (the condition $h(t+\Delta) = h(t)$ leads to $e^{\Delta p} f(p) = f(p)$ or $f(p) = 0$ ignoring δ-functions). Therefore the strip of any solution derived operationally has to be situated inside the strip of the principal solution, whereas the periodic part necessarily degenerates to a constant.

In addition to that given by the inversion integral, other expressions for the principal solution can be derived operationally by expanding the factor $1/(e^p - 1)$ in different ways. Amongst others we may mention the following possibilities:

(1) The *power series* with corresponding remainder:

$$\frac{1}{e^p - 1} = \sum_{n=0}^{N} \frac{B_n}{n!} p^{n-1} - \frac{p^N}{N!}\int_0^\infty B_N(s-[s])\, e^{-ps}\, ds \ (N \geqslant 1); \tag{8}$$

† See Nörlund, loc. cit. p. 42.

(2) The expansion in *partial fractions*:

$$\frac{1}{e^p-1} = \frac{1}{p} - \frac{1}{2} + \frac{2}{p}\sum_{n=1}^{\infty} \frac{p^2}{p^2+4\pi^2 n^2}; \tag{9}$$

(3) *Legendre's expansion:*

$$\frac{1}{e^p-1} = \frac{1}{p} - \frac{1}{2} + 2\int_0^{\infty} \frac{\sin(ps)}{e^{2\pi s}-1}\, ds. \tag{10}$$

As to an operational derivation of (10), the integrand, apart from $\sin ps$, is expanded into powers of $e^{-2\pi s}$; then the individual terms are readily recognized as definition integrals belonging to the operational relation for $\sin tU(t)$, whose sum proves to be equal to the series occurring in (9).

Now, by multiplying any of the three expansions (8), (9), (10) by $\phi(p)$, and performing the corresponding transposition for $\operatorname{Re} p > 0$ (provided that the strip of convergence of $\phi(p)$ is, partly at least, situated inside the right-hand half of the p-plane), we obtain consecutively the following representations of the solution g_1:

$$g_1(t) = \int_{-\infty}^{t} \varphi(s)\, ds + \sum_{n=1}^{N} \frac{B_n}{n!} \varphi^{(n-1)}(t) - \frac{1}{N!}\int_0^{\infty} B_N(s-[s])\, \varphi^{(N)}(t-s)\, ds (N \geqslant 1), \tag{11}$$

$$g_1(t) = \int_{-\infty}^{t} \varphi(s)\, ds - \tfrac{1}{2}\varphi(t) + 2\sum_{n=1}^{\infty} \int_{-\infty}^{t} \cos\{2\pi n(t-s)\}\, \varphi(s)\, ds, \tag{12}$$

$$g_1(t) = \int_{-\infty}^{t} \varphi(s)\, ds - \tfrac{1}{2}\varphi(t) - i\int_0^{\infty} \frac{\varphi(t+is)-\varphi(t-is)}{e^{2\pi s}-1}\, ds. \tag{13}$$

At the same time it is to be remarked that, in general, the series (11), taken as an infinite series without regard for the remainder, will diverge. However, without this remainder, we may eventually find an asymptotic expression of $g_1(t)$ for $t \to \infty$. The equation obtained by putting (13) equal to $(7a)$ is equivalent with a well-known formula of Plana.

Further, the derivation of (13) is based upon complex shifts and is significant only if $\varphi(t)$ can be continued analytically outside the real axis.

Several other expansions of $1/(e^p-1)$ might be used as starting-points; such as those of the associated function $p/\sinh p$, which will be treated in XIV, § 2.

Example. A difference equation that is readily accessible to an operational treatment is the following:

$$g(t+1) - g(t) = \frac{1}{\sqrt{\alpha}} e^{-\pi t^2/\alpha}.$$

In this case the right-hand side possesses the image $p\, e^{\alpha p^2/4\pi}$. Consequently the image (6) of the operational solutions becomes

$$g(t) \doteqdot \frac{p\, e^{\alpha p^2/4\pi}}{e^p - 1}.$$

The solution g_1, being the original for $\operatorname{Re} p > 0$ is, according to (7a), found to be the following series:

$$\frac{1}{\sqrt{\alpha}} \sum_{n=1}^{\infty} \mathrm{e}^{-\pi(n-t)^2/\alpha}.$$

Therefore it is one-half of the following two-sided series for θ_3 according to (XI, 11)

$$\theta_3(-t, \alpha) = \frac{1}{\sqrt{\alpha}} \sum_{n=-\infty}^{\infty} \mathrm{e}^{-\pi(n-t)^2/\alpha}.$$

By means of the preceding theory it is possible to obtain various expressions for this solution, and formulae for the θ-function itself may also be obtained. In this way (11) leads to

$$\sum_{n=1}^{\infty} \mathrm{e}^{-\pi(n-t)^2/\alpha} = \tfrac{1}{2}\sqrt{\alpha}\,\operatorname{erfc}\left(-t\sqrt{\frac{\pi}{\alpha}}\right)$$
$$+\mathrm{e}^{-\pi t^2/\alpha} \sum_{n=1}^{N} \frac{B_n}{n!} \left(\frac{\pi}{\alpha}\right)^{\frac{1}{2}(n-1)} \mathrm{He}_{n-1}\left(-t\sqrt{\frac{\pi}{\alpha}}\right)$$
$$-\frac{(\pi/\alpha)^{\frac{1}{2}N}}{N!} \int_0^{\infty} \mathrm{e}^{-\pi(s-t)^2/\alpha}\, \mathrm{He}_N\left\{(s-t)\sqrt{\frac{\pi}{\alpha}}\right\} B_N(s-[s])\, ds,$$

when use is made of the formula defining the Hermite polynomial

$$\mathrm{He}_n(-x) = \mathrm{e}^{x^2} \frac{d^n}{dx^n} (\mathrm{e}^{-x^2}).$$

It is striking that, when the result obtained with t replaced by $-t$, is added to the identity above, then an expression is found for the complete θ-function, viz.

$$\theta_3(t, \alpha) = 1 - \frac{(\pi/\alpha)^{\frac{1}{2}N}}{\sqrt{\alpha}\, N!} \int_{-\infty}^{\infty} \mathrm{e}^{-\pi(s-t)^2/\alpha}\, \mathrm{He}_N\left\{(s-t)\sqrt{\frac{\pi}{\alpha}}\right\} B_N(s-[s])\, ds \quad (N \geqslant 1),$$

in which the terms involving the Bernoulli numbers have disappeared, whereas the value of the right-hand side is at the same time independent of the positive integer N.

3. General linear difference equations with constant coefficients

The extension of the first-order difference equation with constant coefficients treated above to the corresponding difference equation of the order n has the general form

$$\sum_{n=0}^{N} a_n \Delta^n g(x) = \varphi(x). \tag{14}$$

In this equation the difference quotients of higher order, $\Delta^n g(x)$, can be defined as corresponding extensions of either the symmetric (1) or the asymmetric (4) first-order difference quotients. For instance, in the latter case the second-order quotient becomes

$$\underset{1}{\Delta^2} g(x) = \underset{1}{\Delta}\{\underset{1}{\Delta} g(x)\} = \underset{1}{\Delta}\{g(x+1) - g(x)\} = g(x+2) - 2g(x+1) + g(x).$$

Proceeding in the same manner to the difference quotients of higher order, the asymmetric one of order n is found to be

$$\underset{1}{\Delta^n} g(x) = \sum_{k=0}^{n} (-1)^{n-k} \binom{n}{k} g(x+k). \tag{15}$$

Consequently, by suitable rearrangement of the terms involved, the equation (14) may be transformed into the standard form

$$\sum_{n=0}^{N} c_n g(x+n) = \varphi(x). \tag{16}$$

As before, we are able to derive operationally quite simple particular solutions, by identifying x with t. For, by again assuming

$$g(t) \fallingdotseq G(p), \quad \varphi(t) \fallingdotseq \phi(p),$$

inside any possible common strip of convergence, and transposing (16), we are led to the relation

$$\sum_{n=0}^{N} c_n \, e^{np} \, G(p) = \phi(p).$$

Hence, the solutions in the language of p are given by

$$G(p) = \frac{\phi(p)}{\sum_{n=0}^{N} c_n \, e^{np}}, \tag{17}$$

from which by means of the inversion integral, for instance, a solution of the initial equation (16) is available.

It is quite natural to expand the function that occurs as factor of $\phi(p)$ into partial fractions with respect to the variable e^p, which in fact corresponds to applications of the second expansion theorem of Heaviside to differential equations (see VII, § 10). We therefore put

$$\frac{1}{\sum_{n=0}^{N} c_n \, e^{np}} = \sum_{j=1}^{N} \frac{\alpha_j}{e^p - x_j},$$

in which the numbers x_j are the roots of the following equation of degree n:

$$\sum_{n=0}^{N} c_n x^n = 0.$$

Then we may write (17) in the form

$$G(p) = \sum_{j=1}^{N} \alpha_j \frac{\phi(p)}{e^p - x_j}. \tag{18}$$

Suppose now that the roots x_j, which we shall assume to be mutually different, are ordered according to decreasing absolute value; then (18) will have different originals in the following strips of convergence:

$$\log |x_1| < \operatorname{Re} p < \infty,$$

$$\log |x_2| < \operatorname{Re} p < \log |x_1|,$$

$$\vdots$$

$$\log |x_N| < \operatorname{Re} p < \log |x_{N-1}|,$$

$$-\infty < \operatorname{Re} p < \log |x_N|,$$

inasmuch as $\varphi(t)$ does have an image there at all. As in the analogous considerations given before in the case of differential equations, we shall confine ourselves here to the strip $\operatorname{Re} p > \log |x_1|$, which is clearly the strip farthest to the right. Inside it the following expansion in partial fractions is valid for all j:

$$\frac{1}{e^p - x_j} = \sum_{n=0}^{\infty} x_j^n e^{-(n+1)p},$$

by means of which (18) transforms into

$$G(p) = \phi(p) \sum_{j=1}^{N} \alpha_j \sum_{n=0}^{\infty} x_j^n e^{-(n+1)p},$$

whose original is

$$g_0(t) = \sum_{j=1}^{N} \alpha_j \sum_{n=0}^{\infty} x_j^n \varphi(t-n-1). \tag{19}$$

The general form of this solution of (16) can be reduced so that the roots x_j no longer occur explicitly. To do this we use the following identity:

$$\frac{1}{\sum\limits_{n=0}^{N} \frac{c_n}{x^n}} = \sum_{j=1}^{N} \frac{\alpha_j}{\frac{1}{x} - x_j} = \sum_{j=1}^{N} \alpha_j \sum_{n=0}^{\infty} x_j^n x^{n+1} \quad \left(|x| < \frac{1}{|x_0|}\right),$$

whence we conclude that

$$\sum_{=1} \alpha_j x_j^n = \frac{1}{(n+1)!} \frac{d^{n+1}}{dx^{n+1}} \left\{ \frac{1}{\sum\limits_{k=0}^{N} \frac{c_k}{x^k}} \right\}_{x=0} .$$

Consequently (19) may be written in the alternative form

$$g_0(t) = \sum_{n=1}^{\infty} \frac{\varphi(t-n)}{n!} \frac{d^n}{dx^n} \left\{ \frac{1}{\sum\limits_{k=0}^{N} \frac{c_k}{x^k}} \right\}_{x=0} . \tag{20}$$

We can deal similarly with the strip of convergence $\operatorname{Re} p < -\log |x_N|$, which is farthest to the left; from the original of (18) we then arrive at the following solution:

$$g_N(t) = \sum_{n=0}^{\infty} \frac{\varphi(t+n)}{n!} \frac{d^n}{dx^n} \left\{ \frac{1}{\sum\limits_{k=0}^{N} c_k x^k} \right\}_{x=0} , \tag{21}$$

in which the arguments of $\varphi(t)$ increase rather than decrease.

The originals corresponding to the intermediate strips of convergence are less simple. Moreover, whenever one has succeeded in constructing some particular solution from any of the different strips, a more general solution of (16) is obtained by addition of the function

$$\sum_{j=1}^{N} \pi_j(t) x_j^t,$$

in which the N functions $\pi_j(t)$ are arbitrary periodic functions of period 1.

Example 1. Let us consider the difference equation

$$(\Delta^2+k^2)\,g(x) = \varphi(x), \tag{22}$$

which is the analogue of the differential equation

$$y''+k^2y = \varphi(x). \tag{23}$$

When Δ is understood to represent the symmetric difference quotient defined in (1) then Δ^2 has the following meaning:

$$\Delta^2 g(x) = g(x+1)-2g(x)+g(x-1),$$

and, consequently, (22) may be written as

$$g(x+1)+(k^2-2)\,g(x)+g(x-1) = \varphi(x).$$

Upon identifying x with t we see that the operational solutions of the difference equation above possess the image

$$G(p) = \frac{e^p}{e^{2p}+(k^2-2)\,e^p+1}\,\phi(p).$$

After having determined the two roots of the denominator, we find ($k^2<4$) for the original in the strip farthest to the right according to (19):

$$g_0(t) = \frac{1}{2ik\sqrt{(1-\frac{1}{4}k^2)}}\sum_{n=1}^{\infty}[\{1-\tfrac{1}{2}k^2+ik\sqrt{(1-\tfrac{1}{4}k^2)}\}^{n-1} \\ -\{1-\tfrac{1}{2}k^2-ik\sqrt{(1-\tfrac{1}{4}k^2)}\}^{n-1}]\,\varphi(t-n).$$

This solution (if convergent), when first simplified to

$$g_0(t) = \frac{1}{k\sqrt{(1-\frac{1}{4}k^2)}}\sum_{n=1}^{\infty}\sin\{2(n-1)\arcsin\tfrac{1}{2}k\}\,\varphi(t-n),$$

is obviously the analogue of the particular solution

$$y(x) = \frac{1}{k}\int_0^{\infty}\sin(kn)\,\varphi(x-n)\,dn$$

of the differential equation (23).

Example 2. An example showing that the foregoing theory is useful even in the case of difference equations of infinite order N is provided by the following equation:

$$\sum_{n=0}^{\infty}\frac{g(x+n)}{(n+1)!} = \Gamma(x-1). \tag{24}$$

Since in this example the series

$$\sum_{k=0}^{\infty}c_k x^k = \sum_{k=0}^{\infty}\frac{x^k}{(k+1)!} = \frac{e^x-1}{x}$$

is convergent, (21) leads to the solution

$$g_{\infty}(t) = \sum_{n=0}^{\infty}\frac{\Gamma(t+n-1)}{n!}\,\frac{d^n}{dx^n}\left(\frac{x}{e^x-1}\right)_{x=0} = \sum_{n=0}^{\infty}\frac{B_n}{n!}\Gamma(t+n-1).$$

This series is genuinely divergent; but it is an asymptotic representation of

$$\Gamma(t)\{\zeta(t)-1\} \quad \text{as} \quad t\to 1,$$

as can be shown by transforming the partial sum of order $N+1$, successively, with the help of Euler's integral for Γ, of (8) and of (III, 37), into

$$\sum_{n=0}^{N}\frac{B_n}{n!}\Gamma(t+n-1) = \Gamma(t)\{\zeta(t)-1\}+\frac{\Gamma(t+N)}{N!}\int_0^{\infty}\frac{B_N(u-[u])}{(u+1)^{t+N}}\,du.$$

On applying the definition (VII, 23), the asymptotic character will then be apparent. The function $\Gamma(t)\{\zeta(t)-1\}$, thus obtained from a divergent series, is a solution of (24) for all values of x outside the poles at $x = 1, 0, -1, -2, \ldots$. This statement yields the identity

$$\sum_{n=0}^{\infty} \frac{\Gamma(x+n)}{(n+1)!}\{\zeta(x+n)-1\} = \Gamma(x-1),$$

which may be verified operationally for $x>1$ by first identifying x with p (after multiplication by x), then transposing the individual terms with the aid of (III, 40) and (III, 16), and finally performing the summation over n in the original, which can be easily accomplished.

4. Linear difference equations with variable coefficients

Just as in the case of the previous differential equations we shall confine ourselves to difference equations with rational coefficients. For the standard form we may take

$$\sum_m \sum_n c_{m,n}\, x^m \Delta^n g(x) = \varphi(x), \tag{25}$$

in which Δ will denote the asymmetric operator (4).

Upon identifying x with t or with p the process of transposition leads to a differential equation, which can often be solved more easily than the original difference equation. In the following we shall restrict ourselves to indicating operationally the connexion between these equations and the factorial series, a problem extensively studied by Nörlund. For this purpose it is adequate to identify x with p, and to assume the existence of a common strip of convergence in the case of the two relations

$$pg(p) \risingdotseq G(t)\,U(t), \quad p\varphi(p) \risingdotseq \phi(t),$$

supposing the original of $pg(p)$ to be one-sided.

Then, by transposition after multiplication by p, equation (25) transforms into

$$\sum_m \sum_n c_{m,n} \frac{d^m}{dt^m}\{(e^{-t}-1)^n\, G(t)\,U(t)\} = \phi(t),$$

which, in terms of the variable $\tau = 1-e^{-t}$, becomes

$$\sum_m \sum_n c_{m,n}\left\{(1-\tau)\frac{d}{d\tau}\right\}^m \left\{(-\tau)^n\, G\left(\log\frac{1}{1-\tau}\right) U(\tau)\right\} = \phi\left(\log\frac{1}{1-\tau}\right).$$

Now let us suppose that we know a solution of the last differential equation in the form of a power series

$$G\left(\log\frac{1}{1-\tau}\right) = \sum_{n=0}^{\infty} c_n\, \tau^{\mu+n},$$

that is to say, we assume in terms of the original variable t that

$$G(t)\,U(t) = U(t)\sum_{n=0}^{\infty} c_n(1-e^{-t})^{\mu+n},$$

the image of which, according to (IV, 25), for $\mu > -1$ leads to

$$g(p) = \Gamma(p) \sum_{n=0}^{\infty} c_n \frac{\Pi(\mu+n)}{\Pi(p+\mu+n)} \quad (\operatorname{Re} p > 0),$$

or, alternatively,

$$g(p) = \frac{\Gamma(p)}{\Gamma(p+\mu)} \sum_{n=0}^{\infty} \frac{c_n \Pi(\mu+n)}{(p+\mu)(p+\mu+1)\dots(p+\mu+n)}. \tag{26}$$

This makes it clear that those solutions $g(x)$ of (25) for which $pg(p)$ has a one-sided original are, in general, always expressible as a factorial series. This holds, in particular, for the homogeneous equations (25), while the homogeneous difference equations with constant coefficients in general could not be treated operationally in such a simple manner.

Example. Let us consider the following difference equation,

$$\frac{g(x+h)+g(x-h)-2g(x)}{h^2} + i\alpha x g(x) = 0, \tag{27}$$

in which the increment of x is taken equal to the number h, in order that this second-order equation as $h \to 0$ reduces to the differential equation

$$\frac{d^2 g}{dx^2} + i\alpha x g(x) = 0. \tag{28}$$

Upon identifying x with p (after multiplication by p), and putting

$$pg(p) \risingdotseq G(t),$$

we obtain by transposition of (27) the following first-order differential equation:

$$(e^{ht} + e^{-ht} - 2)\, G(t) + i\alpha h^2 G'(t) = 0,$$

whose general solution is given by

$$G(t) = G(0) \exp\left\{\frac{2i}{\alpha h^3} \sinh(ht) - \frac{2it}{\alpha h^2}\right\}. \tag{29}$$

Therefore a solution of (27) is obtained from the definition integral corresponding to (29); this integral, however, only converges for $\operatorname{Re} p = 0$, and by means of the substitution $t = \dfrac{\tau}{h} + \dfrac{\pi i}{2h}$ it is reducible to the integral (III, 29) for K_ν or the corresponding Hankel function (compare (X, 29)). From the image of (29) we thus find the following solution of (27):

$$g(p) = \frac{\pi i}{h} G(0)\, H^{(1)}_{p/h + 2i/\alpha h^3}\left(\frac{2i}{\alpha h^3}\right) \quad (\operatorname{Re} \alpha > 0). \tag{30}$$

This function should reduce as $h \to 0$ to a solution of the differential equation (28). This limiting process can best be accomplished with respect to the original (29), which then becomes

$$G(0)\, e^{it^3/3\alpha}.$$

The image of the last function (divided by p), according to the well-known rainbow integral of Airy, becomes for $\operatorname{Re} p = 0$

$$\lim_{h \to 0} g(p) = \pi \sqrt{(\tfrac{1}{3}\alpha)}\, G(0)\, e^{i\frac{11}{12}\pi} \sqrt{p}\, H^{(1)}_{\frac{1}{3}}(\tfrac{2}{3}\sqrt{\alpha}\, p^{\frac{3}{2}}\, e^{\frac{5}{4}i\pi}) \quad (\operatorname{Re} \alpha > 0). \tag{31}$$

It is therefore to be expected that this solution of the *differential* equation (28), for small values of h, is an approximation to the solution (30) of the associated *difference* equation (27). In this manner, upon setting

$$\nu = \frac{p}{h} + \frac{2i}{\alpha h^3}, \quad x = \frac{2i}{\alpha h^3},$$

we arrive at the following asymptotic expression:

$$H_\nu^{(1)}(x) \sim e^{\frac{2}{3}i\pi} \sqrt{\left\{\frac{2}{3}\left(\frac{\nu}{x}-1\right)\right\}} H_{\frac{1}{3}}^{(1)}\left\{\frac{2^{\frac{3}{2}}}{3} x \left(\frac{\nu}{x}-1\right)^{\frac{3}{2}} e^{\frac{3}{2}i\pi}\right\}$$
$$= e^{\frac{1}{6}i\pi} \sqrt{\left\{\frac{2}{3}\left(1-\frac{\nu}{x}\right)\right\}} H_{\frac{1}{3}}^{(1)}\left\{\frac{2^{\frac{3}{2}}}{3} x \left(1-\frac{\nu}{x}\right)^{\frac{3}{2}}\right\} \quad \left(\nu\to\infty;\ x\to\infty;\ \frac{\nu}{x}\to 1\right).$$

Therefore, by means of transition from a difference equation to a differential equation, it has proved possible to link the Hankel functions of approximately equal order and argument with the special Hankel function of order $\frac{1}{3}$.

5. Connexion between differential and difference equations

An example showing the intimate connexion between the two types of equations referred to in the subtitle was encountered in the preceding section, in that a linear difference equation with rational coefficients was there reduced to a similar differential equation; the latter can often be solved more simply. Conversely, any differential equation whose solution is attempted in terms of a power series or a Laurent series automatically leads to a difference equation. For, upon substituting

$$g(x) = \sum_{n=-\infty}^{\infty} c_n x^n$$

in the equation

$$\sum_r \sum_s a_{rs} x^r \frac{d^s}{dx^s} g(x) = \varphi(x), \tag{32}$$

while also

$$\varphi(x) = \sum_{n=-\infty}^{\infty} \lambda_n x^n,$$

one obtains, from the condition that the coefficients of x^n have to be equal on right and left, the following result:

$$\sum_r \sum_s a_{rs}(n-r+1)(n-r+2)\dots(n-r+s)\, c_{n+s-r} = \lambda_n. \tag{33}$$

We are thus forced to study a difference equation; its solution is simpler than that of the differential equation, because one is only interested in the values of the function c_n at the integral points. In so far as we consider the solutions of (32) that are expansible in Laurent series, (32) is thus completely equivalent to (33).

We further note that this equation (33) can also be obtained operationally by putting

$$g(e^t) \doteqdot pG(p), \quad \varphi(e^t) \doteqdot p\phi(p),$$

whereupon the image of (32) becomes identical with equation (33), when n is replaced by p, c_n by $G(n)$, and λ_n by $\phi(n)$. An alternative way of expressing this is that the function $c(p)$ for integral values of p satisfies the same difference equation as the image of $g(e^t)$ divided by p, when $g(x)$ is a solution of (32) expansible into a Laurent series. For the rest, such a function $c(p)$ was encountered in the operational relation (XI, 40) for power series (for which $c(n)$ is one-sided), where it was defined in a certain definite manner for non-integral values of p.

Summarizing we may say that very often the solution of a *differential* equation is equivalent to that of a certain *difference* equation. Moreover, besides the exponential transformations mentioned, there exist other transformations that allow of the reduction of differential equations to more tractable difference equations. An example of this is given below, where a differential equation is linked up with a non-linear difference equation.

Example. Concerning the investigations relating to the Legendre polynomials $P_n(x) = P_n(\cos\vartheta)$ in chapter X, we always identified $x = \cos\vartheta$ with t. We shall now take the angle ϑ itself as the operational variable, which point of view leads to a difference equation.

We start with the differential equation of $P_n(\cos t)$, that is,

$$\left\{\frac{d^2}{dt^2}+\cot t\frac{d}{dt}+n(n+1)\right\} P_n(\cos t) = 0. \tag{34}$$

It is easily shown that the same homogeneous equation is satisfied by the cut-off function $P_n(\cos t)\,U(t)$. Now, let us assume that

$$P_n(\cos t)\,U(t) \doteqdot pf_n(p). \tag{35}$$

Further, multiply (34) by $\sin t$, and transpose the result; then, by applying some complex shifts with respect to the image, such as

$$e^{it}P_n(\cos t)\,U(t) \doteqdot pf_n(p-i),$$

the following difference equation is obtained:

$$(p+in)\{p-i(n+1)\}f_n(p-i) = (p-in)\{p+i(n+1)\}f_n(p+i). \tag{36}$$

A solution of this equation can be found after first multiplying it by itself for n replaced by $n-1$. The result of this transformation is a non-linear equation, viz.

$$\{(p-i)^2+n^2\}f_n(p-i)f_{n-1}(p-i) = \{(p+i)^2+n^2\}f_n(p+i)f_{n-1}(p+i).$$

The right-hand member is obtained from the left-hand one, for all values of p, by replacing $p-i$ by $p+i$; therefore the equation is satisfied whenever the following condition holds:

$$(p^2+n^2)f_n(p)f_{n-1}(p) = \lambda(n),$$

in which λ is independent of p.

Let us now apply the last relation n times in succession. Restricting n to even values for the present, we get

$$f_n(p) = \frac{\{p^2+(n-1)^2\}\{p^2+(n-3)^2\}\dots(p^2+1^2)}{(p^2+n^2)\{p^2+(n-2)^2\}\dots(p^2+2^2)}f_0(p)\,G(n),$$

in which $G(n)$ is some product of functions depending only on n. If this type of expression for the image of $P_n(\cos t)\,U(t)$ does exist, then $pf_0(p)$ must be, at least for $\operatorname{Re} p > 0$, the image 1 of $P_0(\cos t)\,U(t) = U(t)$. Hence only the function $G(n)$ is not yet determined.

It can be shown that $G(n) = 1$; this follows from $P_n(\cos t) = 1$ for $t = 0$ and the Abel theorem, $h(0) = f(\infty)$, applied to (35). We thus find for n even

$$P_n(\cos t)\,U(t) \doteqdot \frac{(p^2+1^2)(p^2+3^2)(p^2+5^2)\dots\{p^2+(n-1)^2\}}{(p^2+2^2)(p^2+4^2)(p^2+6^2)\dots(p^2+n^2)}, \quad 0<\operatorname{Re}p<\infty. \quad (37a)$$

Similarly for odd values of n:

$$P_n(\cos t)\,U(t) \doteqdot \frac{(p^2+0^2)(p^2+2^2)(p^2+4^2)\dots\{p^2+(n-1)^2\}}{(p^2+1^2)(p^2+3^2)(p^2+5^2)\dots(p^2+n^2)}, \quad 0<\operatorname{Re}p<\infty. \quad (37b)$$

These operational relations are thus obtained from a particular solution of the difference equation (36). The question whether, indeed, we have got the genuine solution is to be answered in the affirmative, as can afterwards be verified; for instance, by induction†.

We note finally that new properties of Legendre functions are derivable from (37) that are not so trivial from the point of view of the relations given earlier. For example, from the identity

$$\frac{1}{p}\phi_n(p)\,\phi_{n-1}(p) = \frac{p}{p^2+n^2},$$

which easily follows from (37), and which involves the image $\phi_n(p) = pf_n(p)$ of $P_n(\cos t)\,U(t)$, we obtain the following composition product‡:

$$\int_0^t P_n(\cos\tau)\,P_{n-1}\{\cos(t-\tau)\}\,d\tau = \frac{\sin(nt)}{n}.$$

6. Operational construction of difference equations

In addition to the fact that the operational calculus is very useful in solving difference equations, the converse problem of deriving such equations very often lends itself to an operational approach, which may lead to a better understanding of the functions involved. This procedure will be illustrated by some examples.

(1) *Bessel functions.* First we deduce from (XII, 41) that

$$(1-e^{-\frac{1}{2}p})\,e^{-\alpha\sinh\frac{1}{2}p} \doteqdot J_{[2t]}(\alpha), \quad -\infty<\operatorname{Re}p<\infty.$$

Now, the image satisfies the following equation:

$$\left(e^{\frac{1}{2}p}-e^{-\frac{1}{2}p}+2\frac{\partial}{\partial\alpha}\right)f(p) = 0.$$

Consequently, we have for the analogous property of the original,

$$h(t+\tfrac{1}{2})-h(t-\tfrac{1}{2})+2\frac{\partial h}{\partial\alpha} = 0,$$

† Extensions of these relations, concerning the functions P_n^m, are to be found in the list at the end of the book.

‡ Cf. Balth. van der Pol, 'On the operational solution of linear differential equations and an investigation of the properties of the solutions', *Phil. Mag.* LIII, 861, 1929.

a result which can be written as follows:

$$\left(\Delta_t + 2\frac{\partial}{\partial\alpha}\right) J_{[2t]}(\alpha) = 0,$$

or in particular for integral values of n

$$\left(\Delta_n + \frac{\partial}{\partial t}\right) J_{2n}(2t) = 0,$$

when Δ_n is understood to be the symmetric difference quotient (1). Applying once more the same operator, but replacing t by $-t$, we are led to

$$\left(\Delta_n^2 - \frac{\partial^2}{\partial t^2}\right) J_{2n}(2t) = 0, \tag{38}$$

which we discussed in (IX, 40), and which clearly shows the wave character of the Bessel function in virtue of its resemblance to the partial differential equation

$$\left(\frac{\partial^2}{\partial n^2} - \frac{\partial^2}{\partial t^2}\right) \phi(n, t) = 0. \tag{39}$$

Accordingly, whereas the solutions of (38) occur in the study of the homogeneous low-pass filter, the solutions of (39) play a part in problems concerning the homogeneous cable, to which the low-pass filter reduces when the number of cells in a cable of fixed length grows indefinitely.

(2) *Hypergeometric functions.* Some well-known difference relations for the ordinary hypergeometric functions immediately follow from the operational relation (XI, 47), viz.

$$h(t) = F(\alpha, \beta; \gamma; -e^{-t}) \doteqdot \frac{\Gamma(\gamma)}{\Gamma(\alpha)\,\Gamma(\beta)} \frac{\Gamma(p+1)\,\Gamma(\alpha-p)\,\Gamma(\beta-p)}{\Gamma(\gamma-p)},$$
$$0 < \operatorname{Re} p < \min(\operatorname{Re}\alpha, \operatorname{Re}\beta).$$

By means of elementary properties of the Γ-function it is quite easy to prove the following identities concerning the image:

$$\alpha\Delta_\alpha f = \beta\Delta_\beta f = (p-\gamma)(\Delta_\gamma f) = -pf,$$

where Δ here denotes the asymmetric difference quotient given in (4). A transposition of these formulae leads to

$$\alpha\Delta_\alpha h = \beta\Delta_\beta h = \left(\frac{d}{dt} - \gamma\right)(\Delta_\gamma h) = -\frac{dh}{dt},$$

which, when $-e^{-t}$ is replaced by x, transforms into recurrence relations for the function $F(\alpha, \beta; \gamma; x)$, viz.

$$\alpha\Delta_\alpha F = \beta\Delta_\beta F = -\left(\gamma + x\frac{\partial}{\partial x}\right)(\Delta_\gamma F) = x\frac{\partial F}{\partial x}.$$

(3) *Legendre polynomials.* From the operational relation (XII, 43*a*), that is,

$$\frac{\sqrt{2}\sinh\frac{1}{2}p}{\sqrt{(\cosh p-\cos\vartheta)}} \fallingdotseq P_{[t]}(\cos\vartheta)\,U(t) \quad (\operatorname{Re} p>0),$$

we obtain the following result for the image by differentiating with respect to ϑ:

$$2(\cosh p-\cos\vartheta)\frac{\partial f}{\partial\vartheta}+\sin\vartheta f=0.$$

The original of this equation yields as difference-differential equation for the Legendre polynomials (Δ is again the asymmetric difference quotient)

$$\left\{(\Delta_n^2+4\sin^2\tfrac{1}{2}\vartheta)\frac{d}{d\vartheta}+\sin\vartheta\right\}P_n(\cos\vartheta)=0.$$

CHAPTER XIV

INTEGRAL EQUATIONS

1. Introduction

Integral equations also, at least when they are linear, lend themselves to an operational treatment. The two following types are most common:

$$\phi(x) = \int_{-\infty}^{\infty} K(x, \xi) g(\xi) \, d\xi, \tag{1}$$

$$\phi(x) = g(x) - \int_{-\infty}^{\infty} K(x, \xi) g(\xi) \, d\xi. \tag{2}$$

They are the so-called integral equations of the first and second kinds respectively, in which the unknown function $g(x)$ is to be determined; the given function $K(x, \xi)$ is called the *kernel* of the integral equation.

The two kinds of integral equations above can be reduced to one another by means of impulse functions. For, when in (2) a new kernel is introduced, given by

$$K_1(x, \xi) = \delta(x-\xi) - K(x, \xi),$$

then (1) is obtained; conversely, (1) can be considered as a special case of (2), as is obvious from the following alternative way of writing:

$$\lambda\phi(x) = g(x) - \int_{-\infty}^{\infty} \{\delta(x-\xi) - \lambda K(x, \xi)\} g(\xi) \, d\xi,$$

in which the parameter λ is still arbitrary.

Moreover, the integral equations commonly called after Volterra are those in which the upper limit of integration is x, rather than ∞. For instance, the Volterra equation of the first kind is

$$\phi(x) = \int_{-\infty}^{x} K(x, \xi) g(\xi) \, d\xi, \tag{3}$$

which is a specialization of (1), the kernel now being $K(x, \xi)\, U(x-\xi)$.

We thus see that the impulse and unit functions, which are so frequently used in the present book, provide us a means of considering any of the particular integral equations above as some special case of one and the same general type of equation.

It can further be stated that, broadly speaking, for all integral equations that have explicitly been solved in the past the kernel either depends only on the difference $x-\xi$ or is reducible to such a form by means of some transformation or other of the variables involved. In the first case we have $K(x, \xi) = K(x-\xi)$, which we shall call a *difference kernel*. Of such integral equations the term with the integral has the form of a composition product, wherefore this particular type of equation is immediately accessible to an operational approach.

Accordingly, after first discussing in § 2 the problem of the 'moving average', which may provide a typical example of integral equations originating from practice, we shall deal in § 3 with the simplest possible case, that is, the integral equations of the first kind involving a difference kernel. Thereupon, in § 4, we shall study integral equations of the first kind that possess a kernel reducible to a difference kernel. Next, in § 5, the theory of equations of the second kind with a difference kernel is given. Finally, §§ 6 and 7 contain some general remarks on homogeneous integral equations and the process of finding integral relations of known functions operationally.

2. The integral equation for the moving average

The 'moving average' or 'sliding mean' of a function $g(x)$ with respect to the interval a is a new function, $g^*(x)$, which for any value of x is the mean value of $g(x)$ over the interval between $x-\frac{1}{2}a$ and $x+\frac{1}{2}a$ of length a; thus

$$g^*(x) = \frac{1}{a}\int_{x-\frac{1}{2}a}^{x+\frac{1}{2}a} g(\xi)\,d\xi = \frac{1}{a}\int_{-\frac{1}{2}a}^{\frac{1}{2}a} g(x+s)\,ds. \tag{4}$$

The function $g^*(x)$, which thus comes into existence by taking the mean value over an interval of constant length, is generally smoother than the initial function $g(x)$. Therefore this concept is very useful in sciences such as economics; for instance, in the case of the whimsical daily course of prices and currency values, whose irregularities may be smoothed away by taking monthly averages.

So far g^* has to be calculated from the given function g; in other cases, however, the smoother function g^* is given and the question is to determine the original function g. Then (4) presents an integral equation of the first kind for the unknown function g. As an example we may mention the photo-electric reproduction of sound; the film containing the sound impressions passes a narrow slit through which light falls on a photocell which transforms the variations in light intensity into corresponding variations of electric currents. Owing to the non-zero width of the slit, this reproduction procedure is slightly inaccurate in that the function $g(x)$, represented by the varying degree of transparency of the film (x being measured in the longitudinal direction of the film), is smoothed by the slit, so that the photocell can give at most a moving average of $g(x)$. A removal of the effect of the finite slit, if possible, comes down to a process that enables us to determine the function g from the moving average g^*.

The procedure of smoothing is easy to treat by the operational calculus. For this purpose, that is, to determine g^* from g, we identify x with t and accordingly put

$$g(t) \risingdotseq f(p), \quad g^*(t) \risingdotseq f^*(p),$$

whereupon transposition of (4) (*in casu* the right-hand way of writing) leads to

$$f^*(p) = \frac{1}{a}\int_{-\frac{1}{2}a}^{\frac{1}{2}a} e^{ps} f(p)\, ds = \frac{\sinh(\frac{1}{2}ap)}{\frac{1}{2}ap} f(p). \tag{5}$$

As far as the image is concerned the introduction of the moving average reduces to a multiplication by $(\sinh \frac{1}{2}ap)/\frac{1}{2}ap$. The operator 'Lin', which was introduced in (XII, 46) in connexion with the composition product of step functions and which amounts in the language of p to multiplying by $1/p(1-e^{-p})$, can be considered as forming a certain moving average, the independent variable t of which lies at the boundary and not in the centre of the interval of integration (this means that the function under consideration is merely shifted).

In order to get the initial function from its moving average we first determine $f(p)$ from (5) by a simple division:

$$f(p) = \frac{\frac{1}{2}ap}{\sinh(\frac{1}{2}ap)} f^*(p), \tag{6}$$

and then calculate the corresponding original, $g(t)$. This can be achieved, for instance, by means of the inversion integral. However, to obtain expressions for $g(t)$ that are more useful for numerical calculations, we proceed as in XIII, § 2, with the difference equation of the 'sum'. Just as we used there certain expansions of the factor $1/(e^p - 1)$ occurring in the image (XIII, 6), we now develop the factor $\frac{1}{2}ap/\sinh \frac{1}{2}ap$.

The simplest expansions (taking $a = 1$) are

$$\frac{\frac{1}{2}p}{\sinh \frac{1}{2}p} = \begin{cases} p(e^{-\frac{1}{2}p} + e^{-\frac{3}{2}p} + e^{-\frac{5}{2}p} + \ldots) & (\operatorname{Re} p > 0), \\ -p(e^{\frac{1}{2}p} + e^{\frac{3}{2}p} + e^{\frac{5}{2}p} + \ldots) & (\operatorname{Re} p < 0). \end{cases}$$

Upon multiplying these series by $f^*(p)$ and transposing the results term by term we obtain

$$g_1(t) = \frac{d}{dt}\{g^*(t-\tfrac{1}{2}) + g^*(t-\tfrac{3}{2}) + g^*(t-\tfrac{5}{2}) + \ldots\}, \tag{7a}$$

$$g_2(t) = -\frac{d}{dt}\{g^*(t+\tfrac{1}{2}) + g^*(t+\tfrac{3}{2}) + g^*(t+\tfrac{5}{2}) + \ldots\}. \tag{7b}$$

These series terminate whenever $g(t)$ is cut off at both sides; they should be compared with those given in (XIII, 7) as solutions of the difference equation for the sum function. An important difference is that the series (XIII, 7), if convergent, are not equal (their difference being given by $\sum_{-\infty}^{\infty} \varphi(t+n)$), whereas in the case under discussion the series (7 a) and (7 b) are identical. For, upon subtraction it is found that the difference is

$$\frac{d}{dt}\sum_{n=-\infty}^{\infty} g^*(t+n+\tfrac{1}{2}),$$

in which, according to (4), the sum is equal to the constant $\int_{-\infty}^{\infty} g(\xi)\,d\xi$. For the rest, the connexion between our integral equation for the moving average and the difference equation for the sum function is obvious, since by differentiation of (4) with respect to x we immediately arrive at such a difference equation, namely,

$$a\frac{d}{dx}g^*(x) = g(x+\tfrac{1}{2}a)-g(x-\tfrac{1}{2}a).$$

To illustrate the simple meaning of a series such as (7 a), let us consider once more the example of the sound film. Besides the original film, with transparency corresponding to the function $g(x)$, we assume a second with transparency $g^*(x)$ obtained by catching on a slit of finite width $a = 1$ the light that has passed the first film. According to (7 a) it would then be possible to obtain anew the original film by moving the second film along an infinite number of extremely narrow slits a distance $a = 1$ apart, and by using the total amount of light passed through the system of slits in the construction of a third film, provided the transparency of the last film be determined by the differential coefficient of the quantity of light received (which can be achieved by means of common amplifier designs).

There are many other representations for the function $g(x)$ in terms of its moving average $g^*(x)$, corresponding to different expansions of the factor $\frac{1}{2}p/\sinh\frac{1}{2}p$ (we again take $a = 1$). Since in practice we are usually concerned with moving averages that are not so much given in analytic form as in graphical terms, we need expansions differing from those applied previously for the related factor $1/(e^p-1)$. In this connexion we would mention the following series expansions:

$$\frac{\frac{1}{2}p}{\sinh\frac{1}{2}p} = e^{\frac{1}{2}p}\sum_{n=0}^{\infty}\frac{B_n}{n!}p^n \qquad (|p|<2\pi), \tag{8}$$

$$\frac{\frac{1}{2}p}{\sinh\frac{1}{2}p} = 1-\tfrac{1}{24}(2\sinh\tfrac{1}{2}p)^2+\tfrac{3}{640}(2\sinh\tfrac{1}{2}p)^4-\tfrac{5}{7168}(2\sinh\tfrac{1}{2}p)^6+\dots$$
$$(|\sinh\tfrac{1}{2}p|<1), \tag{9}$$

$$\frac{\frac{1}{2}p}{\sinh\frac{1}{2}p} = \sum_{n=1}^{\infty}p^{n-1}\frac{e^{-\frac{1}{2}np}}{n}\left(\frac{\sinh\frac{1}{2}p}{\frac{1}{2}p}\right)^{n-1} \qquad (|1-e^{-p}|<1), \tag{10}$$

$$\frac{\frac{1}{2}p}{\sinh\frac{1}{2}p} = \sum_{n=0}^{\infty}\left(1-\frac{\sinh\frac{1}{2}p}{\frac{1}{2}p}\right)^n = \sum_{n=0}^{\infty}\sum_{k=0}^{n}(-1)^k\binom{n}{k}\left(\frac{\sinh\frac{1}{2}p}{\frac{1}{2}p}\right)^k$$
$$(|1-(\sinh\tfrac{1}{2}p)/\tfrac{1}{2}p|<1), \tag{11}$$

in which the first expansion corresponds to (XIII, 8).

It cannot be expected, however, that a term-by-term transposition, after multiplication of these series by $f^*(p)$, will generally lead to convergent expressions for $g(t)$, since the domains of convergence of the series above do not extend to infinity. On the other hand, these non-convergent series can represent $g(t)$ asymptotically as $t \to \infty$; this is plausible since the series under consideration, when cut off before transposition, are approximations to the image near $p = 0$ (cf. VII, § 2). A closer inspection will reveal that for one-sided functions in particular these series are useful for $t \gg 1$; that is, for time functions that present themselves at the moment $t = 0$, when a considerable number of averaging periods (of lengths 1) have elapsed.

To survey as simply as possible the results following from the series (8)–(11) inclusive, we first remark that

(1) The factors $(2\sinh\frac{1}{2}p)^n$ occurring in (9), when multiplied by $f^*(p)$, are transposed into the powers Δ_t^n of the symmetric difference operator Δ_t of (XIII, 2).

(2) In the case of the powers
$$\left(\frac{\sinh\frac{1}{2}p}{\frac{1}{2}p}\right)^n$$
occurring in (10) and (11), applying the factor $(\sinh\frac{1}{2}p)/\frac{1}{2}p$ once, according to (5) means that we should take one moving average with unit interval of integration.

Starting with
$$\frac{\sinh\frac{1}{2}p}{\frac{1}{2}p}f(p) \doteqdot g^*(t) = \int_{t-\frac{1}{2}}^{t+\frac{1}{2}} g(\xi)\,d\xi,$$
the result of the square of the factor under consideration becomes
$$\left(\frac{\sinh\frac{1}{2}p}{\frac{1}{2}p}\right)^2 f(p) \doteqdot \{g^*(t)\}^* = \int_{t-\frac{1}{2}}^{t+\frac{1}{2}} d\xi\, g^*(\xi) = \int_{t-\frac{1}{2}}^{t+\frac{1}{2}} d\xi \int_{\xi-\frac{1}{2}}^{\xi+\frac{1}{2}} d\eta\, g(\eta). \quad (12)$$

We now have the moving average of the second order, which for shortness may be designated by $g^{**}(t)$. Similarly, moving averages of arbitrary order are conveniently defined by
$$\left(\frac{\sinh\frac{1}{2}p}{\frac{1}{2}p}\right)^n f(p) \doteqdot g^{\overbrace{**\cdots*}^{n}}(t), \quad (13)$$
in which there are n asterisks to the g-function at the right.

After these preliminary considerations we are able to write down the expansions of $g(t)$ that correspond to the series (8)–(11), viz.
$$g(t) \sim \sum_{n=0}^{\infty} \frac{B_n}{n!}\frac{d^n}{dt^n} g^*(t+\tfrac{1}{2}), \quad (14)$$
$$g(t) \sim g^*(t) - \tfrac{1}{24}\Delta^2 g^*(t) + \tfrac{3}{640}\Delta^4 g^*(t) - \tfrac{5}{7168}\Delta^6 g^*(t) + \ldots = \frac{2}{\Delta}\operatorname{arc\,sinh}(\tfrac{1}{2}\Delta)g^*(t), \quad (15)$$
$$g(t) \sim g^*(t-\tfrac{1}{2}) + \frac{1}{2}\frac{d}{dt}g^{**}(t-1) + \frac{1}{3}\frac{d^2}{dt^2}g^{***}(t-\tfrac{3}{2}) + \ldots, \quad (16)$$
$$g(t) \sim g^*(t) + \{g^*(t) - g^{**}(t)\} + \{g^*(t) - 2g^{**}(t) + g^{***}(t)\} + \ldots. \quad (17)$$

Concerning numerical calculation of g when g^* is given in graphical form, the second series (15) is the most simple in that differentiations and integrations are not necessary. However, when in some way or other the higher-order moving averages are easily determinable (17) also is tractable. By breaking off the series of (17) at a finite number of terms, and arranging together the averages of equal order, the following approximations are successively obtained:

$$\left.\begin{array}{l} g^*(t), \\ 2g^*(t)-g^{**}(t), \\ 3g^*(t)-3g^{**}(t)+g^{***}(t), \\ \text{etc.} \end{array}\right\} \qquad (18)$$

After these considerations concerning the resolution of the integral equation for the moving average the following general remarks may be made:

(1) Comparison of (5) with the theory of (VIII, 17) shows that under electrical circumstances the transformation of g^* to g could be obtained by applying an electromotive force $g^*(t)$ at the input terminals of an electric system of transfer admittance

$$\frac{\frac{1}{2}ap}{\sinh(\frac{1}{2}ap)},$$

since then the output current would become equal to $g(t)$. The required admittance could be realized by means of an amplifier and a continuous cable (cf. XV, § 4).

(2) Another consequence of formula (5) is that in the frequency spectrum of g^* there are no frequencies that are multiples of $1/a$. This follows from the inversion integral of $g^*(t)$ for $\operatorname{Re} p = 0$, since the amplitudes of the individual angular frequencies $\omega = p/i$ are proportional to

$$\frac{1}{p}f^*(p) = \frac{\sinh(\frac{1}{2}ap)}{\frac{1}{2}ap}\frac{f(p)}{p},$$

and therefore vanish for p equal to any integral multiple of $2\pi i/a$. The result is that any component occurring in $g(t)$ with period equal to a multiple of $1/a$, cannot influence g^* and, consequently cannot be retrieved from it. This indeterminateness is due to the fact that the periodic components mentioned above are solutions of the homogeneous integral equation corresponding with (4). On the other hand, any cut-off function that is periodic where it does not vanish identically does contribute to the moving average. Thus, for the one-sided periodic function

$$g(t) = \sin\left(\frac{2\pi}{a}t\right)U(t)$$

the moving average with respect to the interval a is given by

$$g^*(t) = \frac{a}{\pi}\cos^2\left(\frac{\pi t}{a}\right)\{U(t+\tfrac{1}{2}a)-U(t-\tfrac{1}{2}a)\}.$$

(3) The operational relation (13) for the moving average of order n can be written as follows:

$$\overbrace{g^{**\dots*}}^{n}(t) \doteqdot \frac{1}{p}\frac{(2\sinh\frac{1}{2}p)^n}{p^{n-1}} f(p). \tag{19}$$

We now introduce the function $D_n(t)$ with image given by

$$D_n(t) \doteqdot \frac{(2\sinh\frac{1}{2}p)^n}{p^{n-1}}, \quad -\infty < \operatorname{Re} p < \infty. \tag{20}$$

This function is seen to vanish outside the interval $-\frac{1}{2}n < t < \frac{1}{2}n$. Accordingly the composition product following from (19) can be written as

$$\overbrace{g^{**\dots*}}^{n}(t) = \int_{t-\frac{1}{2}n}^{t+\frac{1}{2}n} D_n(t-\tau)\, g(\tau)\, d\tau.$$

Returning once more to the example of the sound film, we notice that the last formula signifies that applying the process of the moving average n times in succession is substantially equivalent to applying a single slit of n-fold width, provided the transparency of slit is not uniform, but depends on $D_n(t)$. The function $D_n(t)$, being the original of the relation (20), is simply expressible as follows:

$$D_n(t) = \frac{1}{(n-1)!}\Delta_t^n\{t^{n-1}U(t)\}, \tag{21}$$

in which Δ_t is the symmetric difference operator of

$$\Delta_t \phi(t) = \phi(t+\tfrac{1}{2}) - \phi(t-\tfrac{1}{2}).$$

When n increases the function $D_n(t)$ becomes smoother and smoother while the area below the curve remains equal to 1, as can be inferred from the Tauber theorem $f(0) \Rightarrow h(\infty)$ when applied to the relation obtained from (20) by means of the integration rule.

Very strikingly, the function D_n tends to Gauss's error function when n grows indefinitely, as was observed by Laplace† who gave a clear interpretation of the law of errors by proceeding along these lines. This transition to the error function follows operationally at once from an alternative of (20), viz.

$$\sqrt{n}\, D_n(\sqrt{n}\, t) \doteqdot p\left\{\frac{\sinh\left(\dfrac{p}{2\sqrt{n}}\right)}{\dfrac{p}{2\sqrt{n}}}\right\}^n = p\left(1+\frac{p^2}{24n}+\dots\right)^n. \tag{22}$$

This leads, approximating the right-hand side by $p\, e^{p^2/24}$, to:

$$D_n(t) \sim \sqrt{\left(\frac{6}{\pi n}\right)}\, e^{-6t^2/n} \quad (n \to \infty).$$

(4) The extension of the above unsmoothing procedure to two dimensions can be interpreted by producing a sharp picture from a blurred photograph or television image.

† P. S. Laplace, *Théorie analytique des probabilités*, Paris, 1812.

It will be obvious that the theory of the higher-order moving averages may easily be generalized so as to include cases in which the length of the interval, a, is not constant in the course of performing the n successive processes of averaging.

Example. A set of functions for which the moving average is simpler than the initial function is provided by the Bernoulli polynomials. In this case we have the identity

$$\int_{x-\frac{1}{2}}^{x+\frac{1}{2}} B_n(s)\, ds = (x-\tfrac{1}{2})^n,$$

which in the notation introduced above can be written as

$$B_n^*(x) = (x-\tfrac{1}{2})^n. \tag{23}$$

The maxima and mimima outside the point $x = \frac{1}{2}$, as they are shown by $B_n(x)$, have thus disappeared from the moving average. Moreover, we remark that (23) may be considered as the special result obtained for $\nu = -n$ (n = integer) from

$$\int_{x-\frac{1}{2}}^{x+\frac{1}{2}} \zeta(\nu, s)\, ds = \frac{1}{(\nu-1)(x-\frac{1}{2})^{\nu-1}}, \tag{24}$$

which is of equal simplicity; $\zeta(\nu, s)$ denotes the generalized ζ-function, which includes the Bernoulli polynomials according to

$$B_n(x) = -n\,\zeta(1-n, x).$$

Therefore, corresponding to $g(x) = \zeta(\nu, x)$ we have also the following moving average:

$$\zeta^*(\nu, x) = \frac{1}{(\nu-1)(x-\frac{1}{2})^{\nu-1}}.$$

As a verification, we may substitute this function in the formulae giving g in terms of g^*; for instance, when $\operatorname{Re}\nu > 1$ formula (7b) leads to the definition series (III, 35). Moreover, (14) supplies us with the series

$$\zeta(\nu, x) \sim \frac{1}{(\nu-1)x^{\nu-1}} + \frac{1}{2x^\nu} + \sum_{n=2}^{\infty} \frac{B_n}{n!} \frac{\nu(\nu+1)\dots(\nu+n-2)}{x^{\nu+n-1}}, \tag{25}$$

which represents the ζ-function asymptotically for $x \to \infty$ if $\nu > 1$. This follows from theorem VI of chapter VII applied to the relation (III, 38). For the rest, (25) may equally be derived from a well-known formula of Hermite:

$$\zeta(\nu, x) = \frac{1}{(\nu-1)x^{\nu-1}} + \frac{1}{2x^\nu} - \frac{i}{x^\nu} \int_0^\infty \left\{\left(1 - i\frac{y}{x}\right)^{-\nu} - \left(1 + i\frac{y}{x}\right)^{-\nu}\right\} \frac{dy}{e^{2\pi y}-1}.$$

Formula (25) is obtained from this expression upon term-by-term integration when the integrand has been expanded into powers of $-y^2/x^2$, though this expansion is convergent only for $y < x$. Moreover, for $\nu = 1-n$ the right-hand member of (25) yields the rigorous expression for the Bernoulli polynomials, since then the series terminates.

We would remark finally that for a complete operational treatment of (24) as an integral equation it is simpler to identify x with p than with t, as has been done here.

3. Integral equations of the first kind with difference kernel

According to (1) the standard form of these integral equations, of which that of the moving average is a particular case, is the following:

$$\varphi(x) = \int_{-\infty}^{\infty} K(x-\xi)\, g(\xi)\, d\xi. \tag{26}$$

A more general physical example of this equation is encountered if, given some electrical network (for instance, a two-terminal network), one asks for the voltage $e(t)$ that is necessary to obtain the given current $i(t)$, when it is only known that $i_\delta(t)$ is the response of the system to the impulse voltage $\delta(t)$. For in this case $e(t)$ has to be determined according to (VIII, 37) from

$$i(t) = \int_{-\infty}^{\infty} i_\delta(t-\tau)\, e(\tau)\, d\tau.$$

The general equation (26) can be solved operationally at once, since it is written in the form of a composition product. To do this assume the following operational relations:

$$g(t) \risingdotseq G(p), \quad \varphi(t) \risingdotseq \phi(p), \quad K(t) \risingdotseq \mathscr{K}(p),$$

for the required solution $g(t)$ and the given functions $\varphi(t)$ and K, respectively. When they possess a common strip of convergence we have

$$\phi(p) = \frac{1}{p} \mathscr{K}(p)\, G(p),$$

from which the solution in the language of p is obtained by a simple division, thus

$$G(p) = \frac{p\phi(p)}{\mathscr{K}(p)}. \tag{27}$$

The required solution, $g(t)$, follows from (27) by determination of the corresponding original, again with the aid of the composition-product rule. The special form of the function $\mathscr{K}(p)$ in a given problem usually indicates how to factorize the right-hand member of (27) as favourably as possible, or how to expand the factor $p/\mathscr{K}(p)$ adequately in order to proceed along the same lines as was done for the function $p/\sinh p$ in the preceding section. On the other hand, if from the point of view of the general theory we are not willing to use any particular properties of the functions ϕ and $\mathscr{K}$, we may first write (27) as follows:

$$G(p) = \frac{1}{p} \times \frac{p^2}{\mathscr{K}(p)} \times \phi(p),$$

and then assume the original of $p^2/\mathscr{K}(p)$ to be known, thus

$$\frac{p^2}{\mathscr{K}(p)} \risingdotseq K^*(t), \tag{28}$$

whence the solution of (26) easily follows, viz.

$$g(t) = \int_{-\infty}^{\infty} K^*(t-\tau)\,\varphi(\tau)\,d\tau = \int_{-\infty}^{\infty} K^*(\tau)\,\varphi(t-\tau)\,d\tau. \tag{29}$$

Consequently, whereas in the original integral equation the known function is given as an integral involving the kernel and the unknown function, the solution of the integral equation comes out in quite a similar form where the integral contains the known function together with a new kernel, K^*. The last function is called the *reciprocal kernel* of K, while (29) is the reciprocal integral equation corresponding to (26).

In the foregoing considerations both the given kernel K and the reciprocal kernel K^* depend only on the difference of two variables. As a consequence, the moments of the given function, viz.

$$M_n = \int_{-\infty}^{\infty} \varphi(\tau)\,\tau^n\,d\tau,$$

are very important for solving this type of integral equation. For it follows from (29), provided $K^*(t-\tau)$ is an entire function of τ, that

$$g(t) = \sum_{n=0}^{\infty} \frac{(-1)^n}{n!} M_n \frac{d^n}{dt^n} K^*(t).$$

As to the integral equations of the first kind occurring in practice, it is remarkable that the kernels involved are often singular; for instance, they may become infinitely large somewhere. This might be anticipated, however, since integral equations of the first kind of regular kernel in general do not possess solutions that are continuous. In agreement with this, the kernel of Abel's classical integral equation is singular at $x = \xi$:

$$\varphi(x) = \int_0^x \frac{g(\xi)}{(x-\xi)^\alpha}\,d\xi \quad (x>0). \tag{30}$$

Let us indicate how this well-known equation may be solved operationally. When $\phi(p)$ and $G(p)$ denote the images of the one-sided functions $\varphi(t)\,U(t)$ and $g(t)\,U(t)$ respectively, the transposition of (30) yields

$$\phi(p) = \frac{1}{p}\Pi(-\alpha)\,p^\alpha G(p) \quad (\alpha<1),$$

whence it follows that

$$G(p) = \frac{p^{1-\alpha}}{\Pi(-\alpha)}\phi(p) = \frac{\sin(\pi\alpha)}{\pi}\frac{1}{p} \times \frac{\Pi(\alpha-1)}{p^{\alpha-1}} \times p\phi(p), \tag{31}$$

and consequently, if we confine ourselves to the interval $0<\alpha<1$ as is usual, the solution may be written down almost at once:

$$g(t) = \frac{\sin(\pi\alpha)}{\pi}\left\{\int_0^t \frac{\varphi'(t-\tau)}{\tau^{1-\alpha}}\,d\tau + \frac{\varphi(0)}{t^{1-\alpha}}\right\} = \frac{\sin(\pi\alpha)}{\pi}\frac{d}{dt}\int_0^t \frac{\varphi(\tau)}{(t-\tau)^{1-\alpha}}\,d\tau \quad (t>0). \tag{32}$$

The most familiar case met in practice is a mechanical one obtained for $\alpha = \frac{1}{2}$. The problem is to construct the path of a falling particle (in the earth's field of gravitation) under the condition that the time of its falling from P (at the variable height x) to the origin (at $x = 0$) shall be some prescribed function of x, $T(x)$ say (see fig. 83).

Let v be the velocity at some intermediate point at the height ξ, then

$$T(x) = \int_0^P \frac{ds}{v} = \frac{1}{\sqrt{(2g)}}\int_0^P \frac{ds}{\sqrt{(x-\xi)}} = \frac{1}{\sqrt{(2g)}}\int_0^x \frac{ds/d\xi}{\sqrt{(x-\xi)}}d\xi \quad (x>0), \quad (33)$$

in which $ds = (ds/d\xi)\,d\xi$ is the line element of the required curve (this curve may be approximated mechanically by a smooth tube). Expression (33) is an Abel integral equation (30) for the function $ds/d\xi$, and its appropriate solution is therefore given by

$$s'(x) = \frac{\sqrt{(2g)}}{\pi}\frac{d}{dx}\int_0^x \frac{T(\xi)}{\sqrt{(x-\xi)}}d\xi.$$

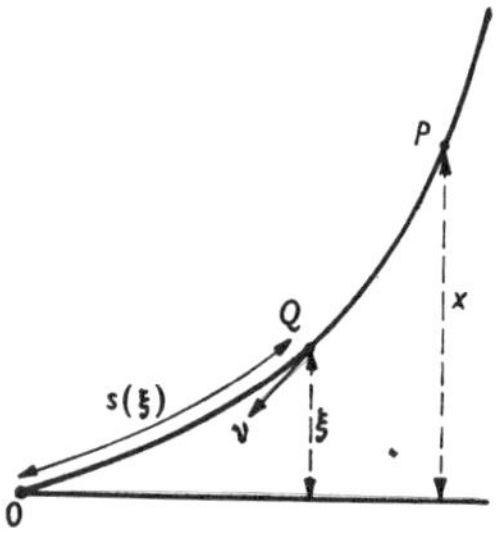

Fig. 83. Illustrating the theory concerning a particle falling along a prescribed curve.

Upon integration we obtain for the arc length s of the required tube, measuring from the origin $(x = 0)$,

$$s(x) = \frac{\sqrt{(2g)}}{\pi}\int_0^x \frac{T(\xi)}{\sqrt{(x-\xi)}}d\xi, \qquad (34)$$

by means of which it is easy to deduce the actual form of the path.

In the tautochrone problem of Huygens the time of falling is independent of the height (at least for some x-interval); we thus have $T(x) = T_0$ and

$$s(x) = \frac{2T_0}{\pi}\sqrt{(2gx)},$$

from which it readily follows that the path has to be a cycloid.

We would emphasize that our operational way of solving Abel's integral equation (30) is easily extended so as to include negative values of α, which are usually not considered. For this purpose we start by writing (31) as follows:

$$G(p) = \frac{p^{1-[\alpha]}}{\Pi(-\alpha)}\times\frac{1}{p}\times\frac{1}{p^{\alpha-[\alpha]-1}}\times\phi(p),$$

which leads to a solution that is valid for $\alpha < 1$, viz.

$$g(t) = \frac{1}{\Pi(-\alpha)\,\Pi(\alpha-[\alpha]-1)}\frac{d^{1-[\alpha]}}{dt^{1-[\alpha]}}\int_0^t \frac{\varphi(\tau)}{(t-\tau)^{[\alpha]-\alpha+1}}d\tau \quad (t>0).$$

This is obviously a generalization of (32).

Example 1. Here follows a list of selected kernels $K(t)$ for which the solution of the corresponding integral equation can be written down almost at once; the integral equation is understood to be

$$\varphi(t) = \int_{-\infty}^{\infty} K(\tau)\, g(t-\tau)\, d\tau.$$

(a) $K\ t) = \cos t\, U(t) \doteqdot \dfrac{p^2}{p^2+1} \quad (\operatorname{Re} p > 0),$

(b) $K(t) = L_n(t)\, U(t) \doteqdot n!\left(1-\dfrac{1}{p}\right)^n \quad (\operatorname{Re} p > 0),$

(c) $K(t) = t^{\frac{1}{2}\nu} J_\nu(2\sqrt{t})\, U(t) \doteqdot \dfrac{e^{-1/p}}{p^\nu} \quad (\operatorname{Re}\nu > -1;\ \operatorname{Re} p > 0),$

(d) $K(t) = P_n(\cos t)\, U(t) \doteqdot \dfrac{\{p^2+(n-1)^2\}\{p^2+(n-3)^2\}\ldots}{(p^2+n^2)\{p^2+(n-2)^2\}\ldots} \quad (\operatorname{Re} p > 0),$

(e) $K(t) = \theta_3(0,t)\, U(t) \doteqdot \sqrt{(\pi p)}\coth\{\sqrt{(\pi p)}\} \quad (\operatorname{Re} p > 0),$

(f) $K(t) = \log|\coth \tfrac{1}{2}t| \doteqdot \pi\tan(\tfrac{1}{2}\pi p) \quad (-1 < \operatorname{Re} p < 1).$

The corresponding images $G(p)$ of $g(t)$ are found to be successively, if $\phi(p)$ represents the image of the given function $\varphi(t)$,

(a′) $G(p) = \left(p+\dfrac{1}{p}\right)\phi(p),$

(b′) $G(p) = \dfrac{p}{n!\left(1-\dfrac{1}{p}\right)^n}\phi(p) = \dfrac{1}{n!}\dfrac{1}{p}\times p\left(1+\dfrac{1}{p-1}\right)^n \times p\phi(p),$

(c′) $G(p) = p^{\nu+1} e^{1/p}\phi(p) = \dfrac{1}{p}\times\dfrac{e^{1/p}}{p^{[\nu]-\nu}}\times p^{[\nu]+2}\phi(p),$

(d′) $$G(p) = p\,\frac{(p^2+n^2)\{p^2+(n-2)^2\}\ldots}{\{p^2+(n-1)^2\}\{p^2+(n-3)^2\}\ldots}\phi(p)$$
$$= \frac{1}{p}\times\frac{(p^2+n^2)\{p^2+(n-2)^2\}\ldots}{\{p^2+(n+1)^2\}\{p^2+(n-1)^2\}\ldots}\times\{p^2+(n+1)^2\}\phi(p),$$

(e′) $G(p) = \sqrt{\left(\dfrac{p}{\pi}\right)}\tanh\{\sqrt{(\pi p)}\}\,\phi(p) = \dfrac{1}{\pi}\dfrac{1}{p}\times\sqrt{(\pi p)}\tanh\{\sqrt{(\pi p)}\}\times p\phi(p),$

(f′) $G(p) = \dfrac{p}{\pi}\cot\left(\dfrac{\pi p}{2}\right)\phi(p) = \dfrac{1}{\pi^2}\dfrac{1}{p}\times\dfrac{\pi p}{p-1}\tan\{\tfrac{1}{2}\pi(p-1)\}\times(p-p^2)\phi(p).$

By operational transposition of these images, using the rule of the composition product in the sense of the factorization specified above, we obtain the following solutions with the aid of certain operational transforms discussed before. (The original integral equations are also written down.)

$$\text{(A)}\quad \begin{cases} \varphi(t) = \displaystyle\int_0^\infty \cos\tau\, g(t-\tau)\, d\tau, \\ g(t) = \varphi'(t) + \displaystyle\int_{-\infty}^{t} \varphi(\tau)\, d\tau; \end{cases}$$

$$\text{(B)}\quad \begin{cases} \varphi(t) = \displaystyle\int_0^\infty L_n(\tau)\, g(t-\tau)\, d\tau, \\ g(t) = \dfrac{\varphi'(t)}{n!} + \dfrac{1}{(n!)^2}\displaystyle\int_0^\infty e^{-\tau} L_n'(\tau)\,\varphi'(t+\tau)\, d\tau; \end{cases}$$

$$\text{(C)}\quad \begin{cases} \varphi(t) = \displaystyle\int_0^\infty \tau^{\frac{1}{2}\nu} J_\nu(2\sqrt{\tau})\, g(t-\tau)\, d\tau, \\ g(t) = \dfrac{d^{[\nu]+2}}{dt^{[\nu]+2}} \displaystyle\int_0^\infty \tau^{\frac{1}{2}\{[\nu]-\nu\}} I_{[\nu]-\nu}(2\sqrt{\tau})\, \varphi(t-\tau)\, d\tau \quad (\operatorname{Re}\nu > -1); \end{cases}$$

$$\text{(D)}\quad \begin{cases} \varphi(t) = \displaystyle\int_0^\infty P_n(\cos\tau)\, g(t-\tau)\, d\tau, \\ g(t) = \left\{\dfrac{d^2}{dt^2} + (n+1)^2\right\} \displaystyle\int_0^\infty P_{n+1}(\cos\tau)\, \varphi(t-\tau)\, d\tau; \end{cases}$$

$$\text{(E)}\quad \begin{cases} \varphi(t) = \displaystyle\int_0^\infty \theta_3(0,\tau)\, g(t-\tau)\, d\tau, \\ g(t) = \dfrac{1}{\pi} \displaystyle\int_0^\infty \theta_2(0,\tau)\, \varphi'(t-\tau)\, d\tau; \end{cases}$$

$$\text{(F)}\quad \begin{cases} \varphi(t) = \displaystyle\int_{-\infty}^\infty \log|\coth\tfrac{1}{2}\tau|\, g(t-\tau)\, d\tau, \\ g(t) = \dfrac{1}{\pi^2}\left(\dfrac{d}{dt} - \dfrac{d^2}{dt^2}\right) \displaystyle\int_{-\infty}^\infty e^\tau \log|\coth\tfrac{1}{2}\tau|\, \varphi(t-\tau)\, d\tau. \end{cases}$$

Of course, it depends on the function $\varphi(t)$ whether any specific inversion formula makes sense or not; the integrals involved should at least be convergent.

Example 2. Another type of integral equation is that in which the kernel is a step function. In this case the given as well as the reciprocal integral equation reduces to an infinite series. For instance, take the relation (VI, 11),

$$[e^t] \doteq \zeta(p) \quad (\operatorname{Re} p > 1),$$

whose original is the kernel of the following integral equation:

$$\varphi(t) = \sum_{n=1}^\infty n \int_{\log n}^{\log(n+1)} g(t-\tau)\, d\tau.$$

In terms of the variable p this integral equation is

$$\phi(p) = \frac{1}{p}\zeta(p)\, G(p),$$

with the solution

$$G(p) = \frac{1}{p} \times \frac{1}{\zeta(p)} \times p^2 \phi(p).$$

By application of the rule for the composition product, and using (XII, 26a), we obtain the solution $g(t)$ in the following form:

$$\begin{aligned} g(t) = \int_{-\infty}^\infty \sum_{k=1}^{[e^\tau]} \mu(k)\, \varphi''(t-\tau)\, d\tau &= \sum_{n=1}^\infty \int_{\log n}^{\log(n+1)} \sum_{k=1}^n \mu(k)\, \varphi''(t-\tau)\, d\tau \\ &= \sum_{n=1}^\infty \sum_{k=1}^n \mu(k) \left[\varphi'\{t-\log n\} - \varphi'\{t-\log(n+1)\}\right]. \end{aligned}$$

The last series can still be simplified somewhat; the final result may be given in the form of the reciprocal system:

$$\varphi(t) = \sum_{n=1}^\infty n \int_{\log n}^{\log(n+1)} g(t-\tau)\, d\tau, \qquad g(t) = \sum_{n=1}^\infty \mu(n)\, \varphi'(t-\log n).$$

A mere glance reveals that this system is quite similar to that of (IV, 11); as a matter of fact, the latter is obtained after substitution of

$$g(x) = h(e^x), \quad \varphi(x) = \int^x H(e^s)\,ds.$$

Example 3. In concluding this section we discuss the integral equation

$$\varphi(t) = \int_0^t (\log\tau + C)\,g(t-\tau)\,d\tau \quad (t>0)$$

with the purpose of showing that the operational method may lead to a solution even when the reciprocal kernel has no original at all (or is not known). To solve the logarithmic integral equation above, we put $g(t)\,U(t) \doteqdot G(p)$, $\varphi(t)\,U(t) \doteqdot \phi(p)$; the solution in the language of p is then given by

$$G(p) = -\frac{p}{\log p}\phi(p),$$

from which we cannot obtain the original directly by the composition-product rule, since thus far we have not met an image with a logarithmic denominator. However, when $G(p)$ is first rewritten as

$$G(p) = -\frac{1}{p} \times p^2\phi(p) \times \int_0^\infty \frac{ds}{p^s} \quad (|\,p\,|>1),$$

application of the composition-product rule does become possible, and the result is found to be

$$\begin{aligned} g(t) &= -\frac{d^2}{dt^2}\int_0^t d\tau\,\varphi(t-\tau)\int_0^\infty ds\,\frac{\tau^s}{\Pi(s)} \\ &= -\int_0^t d\tau\,\varphi''(t-\tau)\int_0^\infty ds\,\frac{\tau^s}{\Pi(s)} - \varphi'(0)\int_0^\infty ds\,\frac{t^s}{\Pi(s)} - \varphi(0)\int_{-1}^\infty ds\,\frac{t^s}{\Pi(s)} \quad (t>0). \end{aligned}$$

4. Integral equations of the first kind with kernel reducible to a difference kernel

Some typical examples (which are the most frequently occurring in physical problems) of this kind of integral equation are those for which the kernel is a function of either the product, $x\xi$, or the ratio, x/ξ, of the two variables x and ξ. For one-sided kernels these equations read

$$\varphi(x) = \int_0^\infty K(x\xi)\,g(\xi)\,d\xi, \quad \varphi(x) = \int_0^\infty K(x/\xi)\,g(\xi)\,d\xi.$$

Upon replacing x by e^t and ξ by either $e^{-\tau}$ or e^τ these equations are transformed into new ones, both with a kernel depending only on the difference $t-\tau$. These new equations therefore can be solved in the manner of the foregoing section; this will be illustrated by a few examples.

Example 1. An integral equation that was exhaustively investigated by Stieltjes is the following:

$$\varphi(x) = \int_0^\infty \frac{g(\xi)}{x+\xi}\,d\xi, \tag{35}$$

where the kernel is thus determined by

$$K(x,\xi) = \frac{U(\xi)}{x+\xi}.$$

Before proceeding to the solution of (35) the reader is reminded that in (v, 47) a solution was shown of a more general equation; there we took the lower limit of integration equal to $-\infty$, though, on the other hand, we had to assume that $\varphi(x)$ should be analytically continuable outside the real axis.

At the present moment we are able to solve (35) without the restriction on $\varphi(x)$ mentioned above. That is, upon performing the transformations $x = e^t$ and $\xi = e^\tau$, there is obtained an integral equation with a difference kernel:

$$\varphi(e^t) = \int_{-\infty}^{\infty} \frac{g(e^\tau)}{e^{t-\tau}+1}\, d\tau.$$

Proceeding in the usual manner, we put

$$\varphi(e^t) \risingdotseq \phi(p),$$

$$g(e^t) \risingdotseq G(p).$$

Then, on account of (III, 12), we get

$$\phi(p) = -\frac{\pi}{\sin(\pi p)} G(p) \quad (-1 < \operatorname{Re} p < 0),$$

so as to obtain for the solution in the language of p

$$G(p) = -\frac{\sin(\pi p)}{\pi}\phi(p). \tag{36}$$

From this formula several expressions can be deduced for the unknown function $g(x)$, among which the following may be mentioned particularly.

(1) Corresponding to the inversion integral we have

$$g(x) = \frac{i}{2\pi^2}\int_{c-i\infty}^{c+i\infty} dp\, x^p \sin(\pi p) \int_0^\infty du \frac{\varphi(u)}{u^{p+1}} \quad (x>0;\ -1<c<0).$$

(2) When in (36) we write $\sin(\pi p) = (e^{p\pi i} - e^{-p\pi i})/2i$ and apply the shift rule to the two exponential functions, then the solution g is found in the form

$$g(x) = \frac{i}{2\pi}\{\varphi(-x+i0) - \varphi(-x-i0)\} \quad (x>0).$$

This formula is similar to that of (v, 49); its validity again requires that $\varphi(x)$ be analytically continuable inside the complex plane, where the negative axis is a cut of $\varphi(x)$.

(3) Let the factor $\sin(\pi p)$ be replaced by its infinite-product representation

$$\sin(\pi p) = \pi p \prod_{n=-\infty}^{\infty}{}' \left(1+\frac{p}{n}\right),$$

then transpose operationally. As a result the following infinite product is obtained:

$$g(x) = -x\frac{d}{dx}\prod_{n=-\infty}^{\infty}{}' \left(1+\frac{x}{n}\frac{d}{dx}\right)\varphi(x) \quad (x>0),$$

in which the prime at the product sign means that the term $n = 0$ should be deleted.

Example 2. Consider the following integral equation:

$$\varphi(x) = \int_0^x F\left(\alpha,\beta;\gamma;\ 1-\frac{\xi}{x}\right)\left(1-\frac{\xi}{x}\right)^{\gamma-1} g(\xi)\, d\xi, \tag{37}$$

in which the kernel contains the hypergeometric function. Upon performing the substitutions $x = e^t$, $\xi = e^\tau$, we get

$$\varphi(e^t) = \int_{-\infty}^{t} F(\alpha,\beta;\gamma;\ 1-e^{\tau-t})(1-e^{\tau-t})^{\gamma-1} g(e^\tau)\, e^\tau\, d\tau.$$

Now, let us put $\varphi(e^t) \doteqdot \phi(p), \quad e^t g(e^t) \doteqdot G(p),$

then, with the help of (XI, 34), the operational transform of the integral equation under consideration is found to be

$$\phi(p) = \frac{1}{p}\Gamma(\gamma)\frac{\Gamma(p+1)\,\Gamma(p-\alpha-\beta+\gamma)}{\Gamma(p-\alpha+\gamma)\,\Gamma(p-\beta+\gamma)}G(p) \quad (\mathrm{Re}\,p>0;\ \mathrm{Re}\,\gamma>0).$$

It thus follows that

$$G(p) = \frac{1}{\Gamma(\gamma)}\frac{1}{p}\times\frac{p\Gamma(p-\alpha+\gamma)\,\Gamma(p-\beta+\gamma)}{\Gamma(p+1)\,\Gamma(p-\alpha-\beta+\gamma)}\times p\phi(p).$$

The next step is to apply the rule of the composition product and use (XI, 34) once more. If we return to the variable $x = e^t$, the solution of (37) proves to be

$$g(x) = \frac{\sin(\gamma\pi)}{\pi}x^{\alpha-1}\int_0^x F\left(-\alpha, 1+\beta-\gamma;\ 1-\gamma;\ 1-\frac{\xi}{x}\right)\frac{\xi^{\gamma-\alpha}}{(x-\xi)^\gamma}\varphi'(\xi)\,d\xi \quad (\mathrm{Re}\,\gamma<1).$$

Of many other equations involving hypergeometric kernels the reciprocal integral equations may be found similarly. The system (II, 7) concerning the Hankel identity is also of this type.

5. Integral equations of the second kind with a difference kernel

The canonical form of this type of integral equation is provided by

$$\varphi(x) = g(x) - \int_{-\infty}^{\infty} K(x-\xi)\,g(\xi)\,d\xi. \tag{38}$$

It, too, can be readily solved by operational means. Assuming

$$\varphi(t) \doteqdot \phi(p), \quad g(t) \doteqdot G(p), \quad K(t) \doteqdot \mathscr{K}(p),$$

we may write (38) in terms of p as follows:

$$\phi(p) = G(p) - \frac{\mathscr{K}(p)\,G(p)}{p},$$

from which we obtain for the image of the solution in question:

$$G(p) = \frac{\phi(p)}{1-\dfrac{\mathscr{K}(p)}{p}}. \tag{39}$$

The well-known expansion of Liouville and Neumann can now be derived in a simple manner, by expanding $1\Big/\left(1-\dfrac{\mathscr{K}(p)}{p}\right)$ into powers of $\mathscr{K}(p)/p$:

$$G(p) = \sum_{n=0}^{\infty}\left\{\frac{\mathscr{K}(p)}{p}\right\}^n\phi(p) = \sum_{n=0}^{\infty} G_n(p) \doteqdot \sum_{n=0}^{\infty} g_n(t).$$

The function G_n clearly fulfils the following recurrence relation:

$$G_n(p) = \frac{1}{p}\mathscr{K}(p)\,G_{n-1}(p),$$

while $G_0(p) \equiv \phi(p)$. Therefore the corresponding originals, $g_n(t)$, are successively determinable from

$$g_0(t) = \varphi(t), \quad g_n(t) = \int_{-\infty}^{\infty} K(t-\tau)\, g_{n-1}(\tau)\, d\tau,$$

which yields

$$g_n(t) = \int_{-\infty}^{\infty} d\tau_1 K(t-\tau_1) \int_{-\infty}^{\infty} d\tau_2 K(\tau_1-\tau_2) \ldots \int_{-\infty}^{\infty} d\tau_n K(\tau_{n-1}-\tau_n)\, \varphi(\tau_n)$$
$$= \int_{-\infty}^{\infty} K_n(t-\tau)\, \varphi(\tau)\, d\tau,$$

in which we have introduced the *iterated kernels* $K_n(t)$, which may be defined by the operational relation

$$K_n(t) \doteq \frac{\{\mathscr{K}(p)\}^n}{p^{n-1}}.$$

It often happens in practical applications that the individual terms of the Liouville-Neumann expansion, $\sum_0^{\infty} g_n(t)$, have definite physical meanings, as will be illustrated by the example of fig. 84 which shows, schematically, a conventional feedback-amplifier device.

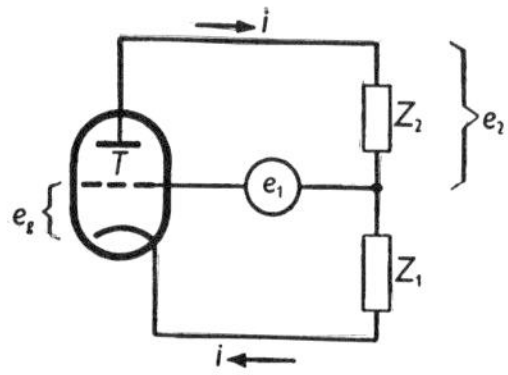

Fig. 84. Scheme of a feedback amplifier.

Let the input voltage $e_1(t)$ be amplified by some triode or pentode valve, so as to obtain the output voltage $e_2(t)$; further, let $Z_1(p)$ denote the frequency impedance of a feedback-circuit element that is part of the grid circuit as well as of the anode circuit of the valve T. Provided the internal resistance of the latter be infinitely large, the variation of the anode current (i) is approximately proportional to that of the grid voltage (e_g); thus $i(t) = Se_g(t)$, in which S is the mutual conductance (or slope) of the valve. If the grid current is neglected, and if we put

$$e_1(t) \doteq E_1(p), \quad e_2(t) \doteq E_2(p), \quad i(t) \doteq I(p),$$

then the grid voltage has the image

$$E_1(p) + Z_1(p)\, I(p) \doteq e_g(t).$$

If now $Z_2(p)$ denotes the frequency impedance at the output side (at which there is the amplified voltage, $e_2(t)$), we have $E_2(p) = Z_2(p)\, I(p)$; on the other hand, it follows that $I(p) = S\{E_1(p) + Z_1(p)\, I(p)\}$. Consequently, by eliminating $I(p)$, the following operational relationship between E_1 and E_2 is seen to exist:

$$E_2 - SZ_1E_2 = SZ_2E_1. \tag{40}$$

If, moreover, $\quad pZ_1(p) \doteq e_{1,\delta}(t), \quad pZ_2(p) \doteq e_{2,\delta}(t),$

then (40) is the image equation of the integral equation

$$e_2(t) - S\int_{-\infty}^{\infty} e_{1,\delta}(t-\tau)\, e_2(\tau)\, d\tau = S\int_{-\infty}^{\infty} e_{2,\delta}(t-\tau)\, e_1(\tau)\, d\tau,$$

from which the function $e_2(t)$, given $e_1(t)$, can be determined, provided the functions $e_{1,\delta}$ and $e_{2,\delta}$ are also supposed to be known; they are those voltages which, when applied to the terminals of the impedances Z_1 and Z_2 respectively, generate a current $\delta(t)$ therein.

The Liouville-Neumann series for the required function $e_2(t)$ in terms of p is then given by

$$E_2 = \frac{SZ_2E_1}{1-SZ_1} = \sum_{n=0}^{\infty} S^{n+1}Z_1^n Z_2 E_1, \tag{41}$$

which easily follows from (40). As to the physical meaning of the individual terms we remark that the original of the first term, that is, of Z_2SE_1, is the output voltage due to the current $I_0 = SE_1$ which would exist without feedback. However, if there is feedback, then the current $I_0(p)$ generates an additional input voltage which is given by $-Z_1I_0$. Its effect is an additional output current, $I_1 = -SZ_1I_0$, to which there corresponds an output voltage represented by the second term of (41), $Z_2I_1 = S^2Z_1Z_2E_1$; and so on and on. Therefore, the successive terms of the series (41) correspond to an analysis of the feedback mechanism, in which any term is the direct consequence of the preceding one. Generalizing we may say that in physical applications the Liouville-Neumann solution is interpretable as the building up of the final solution from results that are successively obtained by elementary processes.

Example. Consider the following integral equation:

$$\varphi(t) = g(t) + \lambda\int_0^{\infty} \frac{g(t-\tau)}{\sqrt{(\pi\tau)}}\, d\tau.$$

Its operational equivalent is easily found to be

$$\phi(p) = G(p) + \frac{\lambda}{\sqrt{p}}\, G(p),$$

whence for the image of the solution in question

$$G(p) = \frac{\phi(p)}{1+\dfrac{\lambda}{\sqrt{p}}} = \frac{1}{p} \times \frac{\sqrt{p}}{\sqrt{p}+\lambda} \times p\phi(p). \tag{42}$$

Now, we infer from (XI, 5) that

$$\frac{\sqrt{p}}{\sqrt{p}+\lambda} \doteqdot e^{\lambda^2 t}\operatorname{erfc}(\lambda\sqrt{t})\, U(t) \quad (\operatorname{Re} p > 0), \tag{43}$$

which leads to the solution:

$$g(t) = \int_0^{\infty} e^{\lambda^2\tau}\operatorname{erfc}(\lambda\sqrt{\tau})\, \varphi'(t-\tau)\, d\tau.$$

On the other hand, when (42) is expanded into powers of λ the solution is obtained in the form of the following Liouville-Neumann series:

$$g(t) = \varphi(t) + \sum_{n=1}^{\infty} \frac{(-\lambda)^n}{\Gamma(\frac{1}{2}n)} \int_0^{\infty} \tau^{\frac{1}{2}n-1} \varphi(t-\tau)\, d\tau.$$

6. Homogeneous integral equations

In general, homogeneous integral equations having a difference kernel do not lend themselves to a simple operational treatment. For instance, the equation of the first kind:

$$\int_{-\infty}^{\infty} K(x-\xi)\, g(\xi)\, d\xi = 0$$

has an image equation equal to

$$\frac{\mathscr{K}(p)}{p} G(p) = 0,$$

which, however, would lead to images of g that are not analytic in p, such as $G(p) = \delta\{\mathscr{K}(p)/p\}$. On the other hand, there does exist an operational method of solving those homogeneous equations whose kernel is not exactly a difference kernel but is reducible to such. This may be illustrated by the discussion of the following example:

$$g(x) = \int_0^{\infty} \sqrt{(x\xi)}\, J_\nu(x\xi)\, g(\xi)\, d\xi, \tag{44}$$

the solutions of which are called self-reciprocal with respect to the Bessel-function transformation. For instance, the lattice-point function λ_n of (XII, 24), or more precisely, the function $\sqrt{x}\,\lambda_n(x^2/2\pi)$, satisfies (44) for $\nu = \frac{1}{2}n+1$. Furthermore, a whole class of solutions of (44) is obtained by adding to any arbitrary function $\phi(x)$ the corresponding Bessel-function transform; the sum of these two functions is a solution of (44) in virtue of the Hankel identity. Titchmarsh† treats other less trivial methods in which (44) is reduced to a simple functional equation; these methods particularly lend themselves to an operational approach. This may be exemplified by substitution of $x = e^{-t}$, $\xi = e^{\tau}$ in (44) through which the equation is transformed into

$$g(e^{-t}) = \int_{-\infty}^{\infty} e^{-\frac{1}{2}(t-\tau)} J_\nu(e^{\tau-t})\, e^{\tau} g(e^{\tau})\, d\tau. \tag{45}$$

Though in this equation a difference kernel occurs, (45) is not of the types treated above, since it contains the unknown function differently at the left and right sides. Yet the operational method will lead to a solution in virtue of the fact that, as in the case of the integral equations treated before, the right-hand side is in the form of a composition product.

† E. C. Titchmarsh, *Introduction to the Theory of Fourier Integrals*, Oxford, 1937, p. 245.

Let us assume that the imaginary axis is situated inside the strip of convergence of the following operational relation:

$$g(\mathrm{e}^{-t}) \doteqdot pF(p), \quad \alpha < \operatorname{Re} p < \beta. \tag{46}$$

Then (46) has some strip in common with the analogous relation

$$g(\mathrm{e}^{t}) \doteqdot pF(-p), \quad -\beta < \operatorname{Re} p < -\alpha.$$

Moreover, using a formula that follows from (XI, 23), namely

$$\mathrm{e}^{-\frac{1}{2}t} J_\nu(\mathrm{e}^{-t}) \doteqdot p 2^{p-\frac{1}{2}} \frac{\Gamma\{\frac{1}{2}(\nu+p)+\frac{1}{4}\}}{\Gamma\{\frac{1}{2}(\nu-p)+\frac{3}{4}\}} \quad -\operatorname{Re}\nu - \tfrac{1}{2} < \operatorname{Re} p < 1 \ (\operatorname{Re}\nu > -\tfrac{3}{2}),$$

we obtain as the transposed equation of (45) if $\operatorname{Re}\nu > \max(\alpha-\frac{3}{2}, -\beta-\frac{1}{2})$, $\alpha < 0$ and $\beta > \frac{1}{2}$:

$$F(p) = 2^{p-\frac{1}{2}} \frac{\Gamma\{\frac{1}{2}(\nu+p)+\frac{1}{4}\}}{\Gamma\{\frac{1}{2}(\nu-p)+\frac{3}{4}\}} F(1-p).$$

We have thus found a functional equation for $F(p)$ indicating that the following function is even in p:

$$\frac{F(p+\frac{1}{2})}{2^{\frac{1}{2}p}\Gamma\{\frac{1}{2}(\nu+p+1)\}} = E(p). \tag{47}$$

Conversely, when starting with an even function $E(p)$ that is arbitrary in so far as the corresponding $pF(p)$ has an original for $0 > \alpha < \operatorname{Re} p < \beta > \frac{1}{2}$, ν being restricted to $\operatorname{Re}\nu > \max(\alpha-\frac{3}{2}, -\beta-\frac{1}{2})$, we may obtain a solution of (44). For instance, when $E(p) = 1$, we have

$$g(\mathrm{e}^{-t}) \doteqdot pF(p) = 2^{\frac{1}{2}p-\frac{1}{4}} p\Gamma\{\tfrac{1}{2}(\nu+p)+\tfrac{1}{4}\},$$

and from its original we see that the function $x^{\nu+\frac{1}{2}}\mathrm{e}^{-\frac{1}{2}x^2}$, which is proportional to $g(x)$, is self-reciprocal with respect to the Bessel-function transformation of order ν if $\operatorname{Re}\nu > -\frac{1}{2}$.

Further, a very general solution of (44) can be obtained from the inversion integral corresponding to (46), by taking $F(p)$ equal to that function which in virtue of (47) corresponds to the arbitrary even function $E(p)$; the result proves to be

$$g(x) = \frac{A}{\sqrt{x}} \int_{c-i\infty}^{c+i\infty} \frac{2^{\frac{1}{2}p}}{x^p} \Gamma\left(\frac{\nu+p+1}{2}\right) E(p)\, dp,$$

in which, of course, $E(p)$ and c should have been chosen so that the integral converges.

Whereas solutions have here been found by applying an exponential substitution to (44), still other transformations are possible. As an example we mention the substitutions $x = 2\sqrt{t}$, $\xi = \sqrt{s}$, which lead to the simple result that $g(x)$ solves the equation, whenever the image of

$$t^{\frac{1}{2}\nu-\frac{1}{4}} g(\sqrt{t})\, U(t) \doteqdot G(2p)$$

satisfies the functional equation

$$G\left(\frac{1}{p}\right) = p^{\nu-1} G(p).$$

7. The operational construction of integral equations

What we observed at the end of the preceding chapter in connexion with difference equations is also true for integral equations. The operational calculus can also be applied to the converse problem, that is, to derive new integral equations from known operational relations. It may suffice to discuss this question with two examples, which are from quite different fields.

Example 1. *Integral equation for the general hypergeometric function.* It follows from (XI, 42) that

$$\frac{p}{p-1} f(p-1) = \frac{1}{p} \times pA(p) \times f(p), \tag{48}$$

in which $f(p)$ is the image of ${}_rF_s(\alpha; \gamma; -e^{-t})$, and

$$A(p) = \frac{(\alpha_1 - p)(\alpha_2 - p) \ldots (\alpha_r - p)}{(p-1)(\gamma_1 - p)(\gamma_2 - p) \ldots (\gamma_s - p)}.$$

Assuming that $r - s \leqslant 1$, so as to secure the convergence of the power series for ${}_rF_s(\alpha; \gamma; x)$ for $|x| < 1$, we may interpret the image $A(p)$ in the following way:

$$pA(p) \doteqdot i_\delta(t), \quad \mathrm{Re}\, p > \max\,(1, \mathrm{Re}\,\gamma_1, \ldots, \mathrm{Re}\,\gamma_s).$$

In this relation, according to (VIII, 35), the function $i_\delta(t)$ represents the output current of a four-terminal network of frequency admittance equal to the meromorphic function $A(p)$, in response to the impulse voltage $\delta(t)$.

By transposition of (48), applying the rule of the composition product, the following homogeneous integral equation results:

$$e^t {}_rF_s(\alpha; \gamma; -e^{-t}) = \int_{-\infty}^{\infty} i_\delta(t-\tau)\, {}_rF_s(\alpha; \gamma; -e^{-\tau})\, d\tau,$$

which may be simplified somewhat, so as to yield

$$\frac{{}_rF_s(\alpha; \gamma; -x)}{x} = \int_0^\infty i_\delta\left(\log \frac{\xi}{x}\right) \frac{{}_rF_s(\alpha; \gamma; -\xi)}{\xi}\, d\xi \quad (x > 0).$$

In deriving this formula we must suppose that the operational relations for the three functions $f(p)$, $\dfrac{p}{p-1} f(p-1)$ and $pA(p)$ have a common domain of convergence, which amounts to

$$\max\,(1, \mathrm{Re}\,\gamma_1, \ldots, \mathrm{Re}\,\gamma_s) < \min\,(\mathrm{Re}\,\alpha_1, \ldots, \mathrm{Re}\,\alpha_r) > 1.$$

Example 2. *Non-linear integral relations satisfied by θ-functions.* Relations of this character are derivable from elementary properties of hyperbolic functions which represent θ-functions in terms of p. For instance, the image $f(p)$ of (VII, 51), thus

$$\sqrt{(\pi p)} \coth \{\sqrt{(\pi p)}\} \doteqdot \theta_3(0, t)\, U(t) \quad (\mathrm{Re}\, p > 0)$$

fulfils the equation $$-2p \frac{d}{dp}\left\{\frac{f(p)}{p}\right\} = -\pi + \frac{f(p)}{p} + \frac{f^2(p)}{p},$$

which by transposition according to elementary rules leads to Bernstein's relation†:

$$\int_0^x \theta_3(0, \xi)\{1 + \theta_3(0, x-\xi)\}\, d\xi = \pi + 2x\theta_3(0, x) \quad (x > 0). \tag{49}$$

† F. Bernstein, *S.B. Berl. Akad.* 1920, pp. 735–47.

Extensions of these type of relations are easily made; e.g. for functions $\theta(v, x)$ with non-vanishing first argument, v. As an example, consider the following set of operational relations:

$$\left.\begin{aligned}\theta_3(v,t)\,U(t) &\doteq \sqrt{(\pi p)}\frac{\cosh\{(1-2v)\sqrt{(\pi p)}\}}{\sinh\{\sqrt{(\pi p)}\}} = f_3(v,p),\\ \theta_2(v,t)\,U(t) &\doteq \sqrt{(\pi p)}\frac{\sinh\{(1-2v)\sqrt{(\pi p)}\}}{\cosh\{\sqrt{(\pi p)}\}} = f_2(v,p),\end{aligned}\right\}\quad (\operatorname{Re} p>0;\ 0\leqslant v\leqslant 1),$$

of which the first has been derived before (see (XI, 13)). It is then easily found that

$$-p\frac{d}{dp}\left\{\frac{f_3(v,p)}{p}\right\} = \frac{1}{2p}f_3(v,p)\{1+f_3(0,p)\}+(v-\tfrac{1}{2})\frac{1}{p}f_2(v,p)f_3(0,p),$$

whose original is simply

$$2x\theta_3(v,x) = \int_0^x \theta_3(v,\xi)\{1+\theta_3(0,x-\xi)\}\,d\xi+(2v-1)\int_0^x \theta_2(v,\xi)\,\theta_3(0,x-\xi)\,d\xi \quad (x>0).$$

The particular case $v = 0$ leads, in combination with (49), to

$$\int_0^x \theta_2(0,\xi)\,\theta_3(0,x-\xi)\,d\xi = \pi, \quad (x>0).$$

8. Note on the operational interpretation of the Wiener-Hopf technique

Many problems of mathematical physics depend on an integral equation of the first kind with a difference kernel, which is only valid for $x \geqslant 0$, viz.

$$\varphi(x) = \int_0^\infty \mathrm{K}(x-\xi)\,g(\xi)\,d\xi, \quad (x\geqslant 0). \tag{50}$$

The theory of section 3 can only be applied after introducing the functions

$$g^*(x) = g(x)\,U(x);\quad \varphi^*(x) = \varphi(x)\,U(x);\quad h^*(x) = U(-x)\int_0^\infty K(x-\xi)\,g(\xi)\,d\xi.$$

The new integral equation

$$\int_{-\infty}^\infty K(x-\xi)\,g^*(\xi)\,d\xi = \varphi^*(x)+h^*(x), \tag{51}$$

is then valid for *all* x, but depends on two unknown functions g^* and h^*. The method of the Wiener-Hopf technique† amounts to the introduction of the operational image

$$\mathscr{G}(p)\frac{\mathscr{K}(p)}{p} = \phi(p)+\mathscr{H}(p)$$

of (51), in which $\mathscr{G}$, ϕ and $\mathscr{H}$ are the images of the above introduced functions, and $\mathscr{K}(p)$ that of $K(x)$. The latter equation is solved by splitting $\mathscr{K}(p)/p$ into two factors $L_1(p)$ and $L_2(p)$, such that the first and second member of

$$\mathscr{G}(p)\,L_2(p)-P(p) = \frac{\phi(p)+\mathscr{H}(p)}{L_1(p)}-P(p) = 0$$

become regular in the overlapping strips $\operatorname{Re} p<\beta$ and $\operatorname{Re} p>\alpha$ respectively $(\alpha<\beta)$ for a properly chosen function P. The image $\mathscr{G}(p)$, and consequently $g(x)$, can then be determined afterwards.

† See, e.g., C. J. Bouwkamp, *Reports on Progress in Physics*, XVII, 1954, §7.

CHAPTER XV

PARTIAL DIFFERENTIAL EQUATIONS IN THE OPERATIONAL CALCULUS OF ONE VARIABLE

1. Introduction

Partial differential equations can often be solved operationally; either by transposition of a single variable or by transposition of all variables involved. Postponing a detailed investigation of the latter method to the last chapter, we shall treat here those cases for which the operational interpretation of just a single variable leads to the results aimed at. In so doing operational relations will be obtained for which, in general, the strip of convergence depends on the remaining variables, which have not been used in the actual transposition.

The theory of this and the last chapter will be illustrated by various examples, more specifically by some well-known differential equations of the second order with constant coefficients, which play an important part in mathematical physics. Usually the time variable, t, is most useful for transposition.

First of all we may summarize those equations which are the most important of those treated later on. Moreover, we would call attention to the corresponding Green functions, which are of considerable importance with a view to the operational calculus.

(1) *The one-dimensional wave equation*, which reads as follows:

$$\frac{\partial^2 u}{\partial x^2} - au - 2b\frac{\partial u}{\partial t} - \frac{1}{c^2}\frac{\partial^2 u}{\partial t^2} = -\varphi. \tag{1}$$

The second and third terms are introduced to account for dispersion, if necessary; that is, dependence of propagation velocity on frequency for sinusoidal vibrations is not excluded in advance. The character of the dispersion is recognized from the exponential solutions of the associated homogeneous equation, which are damped vibrations, viz.

$$e^{-bc^2t}\, e^{i\omega(t-x/v)},$$

the phase velocity being given by

$$v = \frac{c}{\sqrt{\left(1 - \dfrac{c^2(a-b^2c^2)}{\omega^2}\right)}}. \tag{2}$$

Thus the velocity v depends on the angular frequency ω, and only for very high frequencies is it equal to the velocity of propagation c existing without dispersion. Two different cases may be distinguished depending on whether

$a > b^2c^2$ or $a < b^2c^2$; the first is that of *normal dispersion*, where v is a decreasing function of ω (for instance, electromagnetic waves in the ionosphere); the second case is that of *anomalous dispersion* in which v increases with ω (e.g. vibrations travelling along a cable).

A further generalization of (1) is achieved by assuming the one-dimensional space, in which the propagation takes place, to be non-isotropic, such that the velocity of propagation (ignoring possible effects of dispersion) has different values for waves travelling to the right and those travelling to the left, c_1 and c_2 respectively. In this case, presenting itself for instance as propagation of sound in a windy atmosphere, the differential equation can be most simply written as follows:

$$\left(c_1\frac{\partial}{\partial x}+\frac{\partial}{\partial t}\right)\left(c_2\frac{\partial}{\partial x}-\frac{\partial}{\partial t}\right)u+\alpha u = -\varphi, \tag{3}$$

provided the dispersion term proportional to $\partial u/\partial t$ is absent.

The right-hand member $-\varphi$ of the wave equations (1) and (3) describes the vibrational sources that are present. As we saw before, in problems of a single variable x we can obtain a solution for an arbitrary right-hand member by a simple quadrature, involving the Green function $G(x; x_0)$ which itself is a particular solution of the same differential equation except that the right-hand member is equal to $-\delta(x-x_0)$. This particular solution represents, in physical applications, the effect of an external force impressed locally at $x = x_0$ (see v, § 8, example 2) when, in addition, certain boundary conditions are satisfied. In the same manner we can introduce the two-dimensional Green function $G(x, t; x_0, t_0)$ in the case of partial differential equations like (1) and (3), which for shortness may be written $Lu = -\varphi$. Accordingly, $G(x, t; x_0, t_0)$ satisfies the equation

$$Lu = -\delta(x-x_0)\,\delta(t-t_0)$$

and represents the action of a source localized at $x = x_0$ at the moment $t = t_0$ (possible boundary conditions are unimportant in this connexion).

Sufficient for solving the differential equations with constant coefficients is the knowledge of a *special* Green function, say $G_0(x, t)$, for which the right-hand member is simply equal to $-\delta(x)\,\delta(t)$. Indeed, using G_0 in the construction of the function

$$u(x, t) = \int_{-\infty}^{\infty} d\xi \int_{-\infty}^{\infty} d\tau\, G_0(x-\xi, t-\tau)\,\varphi(\xi, \tau), \tag{4}$$

we see that this u-function satisfies $Lu = -\varphi$, as follows from the linearity of the differential equation under investigation:

$$\begin{aligned} Lu(x, t) &= \int_{-\infty}^{\infty} d\xi \int_{-\infty}^{\infty} d\tau\, \varphi(\xi, \tau)\, LG_0(x-\xi, t-\tau) \\ &= -\int_{-\infty}^{\infty} d\xi \int_{-\infty}^{\infty} d\tau\, \varphi(\xi, \tau)\,\delta(x-\xi)\,\delta(t-\tau) = -\varphi(x, t). \end{aligned}$$

(2) *The general n-dimensional wave equation*

$$\Delta u + au - \frac{1}{c^2}\frac{\partial^2 u}{\partial t^2} = -\varphi, \tag{5}$$

in which Δ is the Laplace operator

$$\Delta \equiv \frac{\partial^2}{\partial x_1^2} + \frac{\partial^2}{\partial x_2^2} + \ldots + \frac{\partial^2}{\partial x_n^2}.$$

In the study of this equation of n variables the constant a determines the dispersion (after elimination of a possible term proportional to $\partial u/\partial t$ by means of a suitable transformation); moreover, the space is assumed to be isotropic. Again, the corresponding Green function, which represents a source situated at the origin of co-ordinates at time $t = 0$, will be the solution of

$$\Delta G + aG - \frac{1}{c^2}\frac{\partial^2 G}{\partial t^2} = -\delta(x_1)\,\delta(x_2)\ldots\delta(x_n)\,\delta(t) \tag{6}$$

under proper boundary conditions.

It is very important that, in virtue of the space being isotropic, the product of the first n impulse functions, that is, $\delta(x_1)\,\delta(x_2)\ldots\delta(x_n)$ may be replaced by a single one-dimensional impulse function depending only on the radial distance from the origin, $r = \sqrt{(x_1^2 + x_2^2 + \ldots + x_n^2)}$. This property follows from the identity

$$\delta(x_1)\,\delta(x_2)\ldots\delta(x_n) = \frac{\Gamma(\frac{1}{2}n)}{\pi^{\frac{1}{2}n}}\frac{\delta(r)}{r^{n-1}} = \frac{2\Pi(\frac{1}{2}n)}{\pi^{\frac{1}{2}n}n!}\delta^{(n-1)}(-r). \tag{7}$$

For a proof of (7) we observe that the left and central members are certainly equal when $r \neq 0$, both then being zero. They will have been shown completely identical when we prove that integration over the n-sphere with radius R and centre at the origin leads to the same result for all three members. Denoting by $\Omega_n(r)$ the area of the n-sphere of radius r, we have for the decisive integrals of the first two members:

$$\int dx_1 \int dx_2 \ldots \int dx_n\, \delta(x_1)\,\delta(x_2)\ldots\delta(x_n)$$

and

$$\frac{\Gamma(\frac{1}{2}n)}{\pi^{\frac{1}{2}n}}\int_0^R \frac{\delta(r)}{r^{n-1}}\Omega_n(r)\,dr$$

respectively. In fact, either integral is equal to 1; this is obvious for the first one, and for the second it is easily verified with the aid of

$$\Omega_n(r) = \frac{dV_n(r)}{dr} = \frac{\pi^{\frac{1}{2}n}}{\Pi(\frac{1}{2}n)}nr^{n-1}, \tag{8}$$

which follows from (IV, 24). This completes the proof of the validity of the left part of (7). The equality of the second and third terms in (7) can be inferred from (V, 66).

In passing we remark that the factorization (7) of multi-dimensional impulse functions with respect to a system of mutually orthogonal directions may have an analogue in the case of *approximate* impulse functions. For instance, the two-dimensional impulse function has the following approximation according to (v, 33 *b*):

$$\frac{\delta(r)}{\pi r} = \delta(x)\,\delta(y) = \operatorname*{Lim}_{\lambda\to\infty}\frac{\lambda}{\sqrt{\pi}}\,e^{-\lambda^2x^2}\operatorname*{Lim}_{\lambda\to\infty}\frac{\lambda}{\sqrt{\pi}}\,e^{-\lambda^2y^2} = \operatorname*{Lim}_{\lambda\to\infty}\frac{\lambda^2}{\pi}\,e^{-\lambda^2r^2}.$$

(3) *The n-dimensional equation of heat conduction, or diffusion equation.* This equation is given by

$$\Delta u - \frac{1}{a^2}\frac{\partial u}{\partial t} = -\varphi, \tag{9}$$

in which Δ is the same operator as in the case (2). Again, the corresponding Green functions are of primary importance. G may represent the distribution of temperature in space as due to a source of heat that at time $t = 0$ is concentrated at the origin of co-ordinates; at proper normalizing conditions it fulfils the following equation:

$$\Delta G - \frac{1}{a^2}\frac{\partial G}{\partial t} = -\delta(x_1)\,\delta(x_2)\ldots\delta(x_n)\,\delta(t). \tag{10}$$

Any of the partial differential equations so far considered has constant coefficients, whereby (1) and (5) are of the *hyperbolic* type and (9) of the *parabolic* type; that is to say, the conic obtained upon replacing the operators $\partial/\partial x_k$ by x_k is a hyperbola or a parabola. It is to be emphasized that the equations of the elliptic type, of which the potential equation is the classic example ($\Delta u = -\varphi$), are less manageable in the operational treatment.

The actual start of the present chapter concerns investigations of homogeneous partial differential equations with constant coefficients, to be found in § 2. In § 3 the usual boundary conditions are taken into account. Next, in § 4, the quasi-stationary theory of cables is investigated; in this connexion we ought to remark that the type of problem relating to it (transients, etc.) forced Heaviside to introduce his symbolic calculus, the ancestor of present-day operational calculus. Finally, it is demonstrated in § 5 by a simple example how *inhomogeneous* partial differential equations, too, are accessible to an operational attack.

2. Homogeneous linear partial differential equations (general solutions)

To elucidate the operational method for this type of equation we shall consider the homogeneous one-dimensional wave equation (1) when $b = 0$; thus

$$\frac{\partial^2 u}{\partial x^2} - au - \frac{1}{c^2}\frac{\partial^2 u}{\partial t^2} = 0. \tag{11}$$

By transposition with respect to the variable t, where it is assumed that

$$u(x, t) \doteqdot f(x, p),$$

the *partial* differential equation in the language of t transforms into the *ordinary* differential equation in the language of p:

$$\frac{d^2 f}{dx^2} - \left(a + \frac{p^2}{c^2}\right) f = 0.$$

The general solution of this equation is given by the formula

$$f(x, p) = \lambda_1(p) \exp\left\{x \sqrt{\left(\frac{p^2}{c^2} + a\right)}\right\} + \lambda_2(p) \exp\left\{-x \sqrt{\left(\frac{p^2}{c^2} + a\right)}\right\}, \tag{12}$$

in which the constants of integration, $\lambda_1(p)$ and $\lambda_2(p)$, are arbitrary functions of the parameter p. In determining the general solution of (11) from the original of (12) we shall assume first that a is positive (normal dispersion). Now, we can apply the operational relation (x, 48) for $\nu = 0$, that is,

$$J_0\{\sqrt{(t^2 - a^2)}\}\, U(t-a) \doteqdot \frac{p}{\sqrt{(p^2+1)}} e^{-a\sqrt{(p^2+1)}}, \quad 0 < \operatorname{Re} p < \infty, \tag{13}$$

which holds for $a > 0$ as well as for $a < 0$.

To obtain the simplest possible form of the solution of (11), we shall introduce two new constants of integration χ_1 and χ_2 instead of λ_1 and λ_2, such that (12) may be written equivalently as follows:

$$f(x,p) = \frac{1}{p} \times \chi_1(p) \times \frac{p}{\sqrt{(p^2 + ac^2)}} \exp\left\{x \sqrt{\left(\frac{p^2}{c^2} + a\right)}\right\}$$
$$+ \frac{1}{p} \times \chi_2(p) \times \frac{p}{\sqrt{(p^2 + ac^2)}} \exp\left\{-x \sqrt{\left(\frac{p^2}{c^2} + a\right)}\right\}.$$

Upon applying (13) together with the rule of the composition product, the original of $f(x, p)$ is seen to be

$$u(x, t) = \int_{-x/c}^{\infty} \varphi_1(t-\tau)\, J_0(\sqrt{\{a(c^2\tau^2 - x^2)\}})\, d\tau$$
$$+ \int_{x/c}^{\infty} \varphi_2(t-\tau)\, J_0(\sqrt{\{a(c^2\tau^2 - x^2)\}})\, d\tau. \tag{14}$$

In this expression φ_1 and φ_2 are arbitrary functions of t, since they are the originals of the arbitrary functions $\chi_1(p)$ and $\chi_2(p)$ respectively. Therefore, (14) is the most general solution of (11).

It should be borne in mind that when x is positive the argument of the Bessel function in the first integral becomes imaginary for values of τ inside the interval $-x/c < \tau < x/c$; it is thus convenient to replace J_0 by I_0. The corresponding holds with respect to the second integral when x is negative.

For values of a that are negative (anomalous dispersion) (11) can be solved analogously, when instead of (13) the following relation is fundamental:

$$I_0\{\sqrt{(t^2-a^2)}\}\,U(t-a) \doteqdot \frac{p}{\sqrt{(p^2-1)}}\,e^{-a\sqrt{(p^2-1)}}, \quad 1<\operatorname{Re}p<\infty, \tag{15}$$

which may be proved along the same lines as (13).

As we have seen, the general solution always contains the Bessel function of order zero, which is a consequence of the dispersion term introduced in (11). In this respect it is an extension of the general solution

$$u = \phi_1(x+ct)+\phi_2(x-ct)$$

of the elementary differential equation of the vibrating string

$$\frac{\partial^2 u}{\partial x^2}-\frac{1}{c^2}\frac{\partial^2 u}{\partial t^2} = 0.$$

The method described above is extremely useful in solving any hyperbolic differential equation of the second order (two variables) with constant coefficients; in addition, any such equation is reducible to (11) by some suitable transformation of the variables.

In the case of the analogous elliptic differential equations, such as

$$\frac{\partial^2 u}{\partial x^2}+au+\frac{1}{c^2}\frac{\partial^2 u}{\partial t^2} = 0,$$

the method above is not directly applicable, since it would lead to images connected with the function

$$\frac{p}{\sqrt{(p^2+1)}}\,e^{ia\sqrt{(p^2+1)}},$$

which has no proper original at all. Yet we are able to determine the solution operationally; namely, by identifying the variable t with the operational variable p, thus:

$$\frac{\partial^2 u}{\partial x^2}+au+\frac{1}{c^2}\frac{\partial^2 u}{\partial p^2} = 0.$$

In this case, however, we have to apply operational relations possessing a line of convergence; the result so obtained is similar to (14), though the lower limits of integration are complex numbers. Also later on, in the discussion of simultaneous operators (see XVI, § 5), we shall have an opportunity to show that differential equations of the elliptic type in general lead to operational relations with a degenerate strip of convergence.

The general theory so far given may best be illustrated by some examples, in order to indicate further possibilities as to the procedure in which one of the variables of the partial differential equation is transposed operationally; whether this particular variable is identified with t or p depends on the particular problem under discussion.

Example 1. *Differential equation of the parabolic type.* A solution of the homogeneous one-dimensional equation of heat condition (9), that is, of

$$\frac{\partial^2 u}{\partial x^2}-\frac{1}{a^2}\frac{\partial u}{\partial y}=0, \tag{16}$$

that contains an arbitrary function is easily found upon identifying x with t, since then one is led to an ordinary differential equation of the first order. We thus put

$$u(t,y) \doteqdot f(p,y),$$

whence it follows that the new equation becomes

$$p^2 f-\frac{1}{a^2}\frac{df}{dy}=0,$$

of which the general solution is provided by the function

$$f(p,y)=\lambda(p)\,\mathrm{e}^{a^2p^2y}.$$

Further, by using the relation (II, 24) and applying the composition-product rule, the following solution of (16) is obtained:

$$u(x,y)=\frac{1}{\sqrt{y}}\int_{-\infty}^{\infty}\varphi(x-\tau)\,\mathrm{e}^{-\tau^2/4a^2y}\,d\tau,$$

in which the function φ is wholly arbitrary. The most general solution is found by addition of the analogous expression obtained when x is replaced by $-x$ and φ by some other arbitrary function.

Example 2. *Connexion between the wave equation and the heat-conduction equation.* It often appears that a solution of one or another differential equation leads to a solution of a related differential equation. To make this clearer, we shall consider the three-dimensional equations of wave propagation and heat conduction simultaneously. With proper choice of units we thus consider the system

$$\Delta u_1-\frac{\partial^2 u_1}{\partial t^2}=0,\quad \Delta u_2-\frac{\partial u_2}{\partial t}=0.$$

Upon putting

$$u_1(x,y,z,t)\doteqdot f_1(x,y,z,p),\quad u_2(x,y,z,t)\doteqdot f_2(x,y,z,p),$$

we obtain for the transposed equations, in the same order,

$$\Delta f_1-p^2 f_1=0, \tag{17a}$$

$$\Delta f_2-p f_2=0. \tag{17b}$$

From this it will be evident that, when some image function $f_1(x,y,z,p)$ satisfies (17*a*), the analogous

$$f_2(x,y,z,p)=f_1(x,y,z,\sqrt{p}) \tag{18}$$

may be a solution of (17*b*). Then the original of the image (18) is, on account of (XI, 5), expressible in terms of u_1 for $t>0$, if we assume that the homogeneous wave equation is still valid for u_1 taken one-sided (i.e. if $u_1=\partial u_1/\partial t=0$ for $t=0$). Thus we get

$$u_2(x,y,z,t)=\frac{1}{\sqrt{(\pi t)}}\int_0^{\infty}\mathrm{e}^{-s^2/4t}u_1(x,y,z,s)\,ds=\frac{2}{\sqrt{\pi}}\int_0^{\infty}\mathrm{e}^{-\lambda^2}u_1(x,y,z,2\lambda\sqrt{t})\,d\lambda\quad(t>0),$$

where we have at once a solution of the heat-conduction equation from one for the wave equation.

A somewhat more general solution of the differential equation of heat conduction is obtained when in (18) f_1 is first multiplied by some arbitrary function of p, which does not affect its being a solution of (17 a). Upon applying the composition-product rule in this case, we are led to the following solution of the heat-conduction equation:

$$\int_0^t d\tau\, \chi(t-\tau) \int_0^\infty d\lambda\, e^{-\lambda^2} u_1(x, y, z, 2\lambda\sqrt{\tau}).$$

Example 3. *Legendre functions.* In order to derive new relations for the functions of Legendre P_n^m, including those with $m \neq 0$, we start with the following differential equation, in which ρ, ϕ and z denote cylindrical co-ordinates:

$$\frac{\partial^2 \chi}{\partial \rho^2} + \frac{1}{\rho}\frac{\partial \chi}{\partial \rho} + \frac{\partial^2 \chi}{\partial z^2} + \frac{1}{\rho^2}\frac{\partial^2 \chi}{\partial \phi^2} = 0, \tag{19}$$

of which, amongst others, solutions are

$$\chi_n^{(1)}(\rho, z, \phi) = \frac{P_n^m(\cos\vartheta)}{r^{n+1}} e^{im\phi} = \frac{P_n^m\left(\dfrac{z}{\sqrt{(\rho^2+z^2)}}\right)}{(\rho^2+z^2)^{\frac{1}{2}(n+1)}} e^{im\phi},$$

and $$\chi_n^{(2)}(\rho, z, \phi) = r^n P_n^m(\cos\vartheta)\, e^{im\phi} = (\rho^2+z^2)^{\frac{1}{2}n}\, P_n^m\left(\frac{z}{\sqrt{(\rho^2+z^2)}}\right) e^{im\phi}.$$

By subsequently identifying z in (19) with either of the operational variables p and t, while keeping ρ constant, we are able to deduce operational relations for the functions $\chi_n^{(1)}$ and $\chi_n^{(2)}$ respectively:

(a) When it is supposed that

$$p\chi_n^{(1)}(\rho, p, \phi) \doteqdot h(\rho, t)\, e^{im\phi}, \tag{20}$$

and when (19) is transposed term by term, after being multiplied by $p \equiv z$, then the following ordinary differential equation for h is obtained:

$$\frac{d^2 h}{d\rho^2} + \frac{1}{\rho}\frac{dh}{d\rho} + \left(t^2 - \frac{m^2}{\rho^2}\right) h = 0.$$

This equation is the differential equation (x, 11) of Bessel functions of argument ρt. Since $J_m(\rho t)$ is the only solution that remains finite as ρ tends to zero, it may be concluded that the operational relation (20) for $\chi_n^{(1)}$ must have the form

$$p\, \frac{P_n^m\left(\dfrac{p}{\sqrt{(\rho^2+p^2)}}\right)}{(\rho^2+p^2)^{\frac{1}{2}(n+1)}} \doteqdot \lambda(t)\, J_m(\rho t).$$

So far the function $\lambda(t)$ is unknown. Its value can be calculated by letting ρ tend to zero, after dividing both sides by ρ^m. When use is made of well-known approximations for $P_n^m(x)$ as $x \to 1$ and $J_m(x)$ as $x \to 0$, we readily find that

$$\frac{(n+m)!}{(n-m)!}\frac{1}{p^{n+m}} \doteqdot t^m \lambda(t).$$

Since the original of the left-hand side is known (let $\mathrm{Re}\, p$ be positive), we then conclude that we have to take

$$\lambda(t) = \frac{t^n}{(n-m)!} U(t).$$

The complete relation (20) thus becomes

$$p\, \frac{P_n^m\left(\dfrac{p}{\sqrt{(\rho^2+p^2)}}\right)}{(\rho^2+p^2)^{\frac{1}{2}(n+1)}} \doteqdot \frac{1}{(n-m)!} t^n J_m(\rho t)\, U(t), \quad 0 < \mathrm{Re}\, p < \infty. \tag{21}$$

(b) In order to derive an image containing as data the function $\chi_n^{(2)}(\rho, t, \phi)$, it is simplest to take the corresponding one-sided function. We therefore assume

$$h^*(t) = \chi_n^{(2)}(\rho, t, \phi)\, U(t) = (\rho^2+t^2)^{\frac{1}{2}n}\, P_n^m\left(\frac{t}{\sqrt{(\rho^2+t^2)}}\right) e^{im\phi}\, U(t) \risingdotseq f(\rho, p)\, e^{im\phi}. \quad (22)$$

First of all, we have to construct from (19) the differential equation for the one-sided function $h^*(t)$. This can be achieved in a manner already known. Moreover, we suppose $n+m$ to be an even number. The differential equation in question then proves to be the following:

$$\frac{\partial^2 h^*}{\partial\rho^2}+\frac{1}{\rho}\frac{\partial h^*}{\partial\rho}+\frac{\partial^2 h^*}{\partial t^2}-\frac{m^2}{\rho^2}h^* = \rho^n P_n^m(0)\,\delta'(t)\,e^{im\phi}.$$

We now proceed in the same way as we have done so often before. Let $h^*(t)$ correspond to $f(p)$; the transposed form of the equation above is

$$\frac{d^2f}{d\rho^2}+\frac{1}{\rho}\frac{df}{d\rho}+\left(p^2-\frac{m^2}{\rho^2}\right)f = \rho^n P_n^m(0)\,p^2.$$

The corresponding homogeneous equation is the differential equation of $J_m(p\rho)$; a particular solution of the inhomogeneous equation can be found by a method similar to that used for the analogous equation (x, 7) for the (one-sided) Hermite polynomials, where the right-hand member also contained only a term with p^2.

Again using the notation with double brackets introduced in (IV, 4a), we easily verify that the function

$$f(\rho, p) = \frac{P_n^m(0)}{c_n}\left[\left[\frac{J_m(\rho p)}{p^n}\right]\right]$$

is a solution of the differential equation considered if c_n is the coefficient of x^n in $J_m(x)$. The final result thus becomes

$$(\rho^2+t^2)^{\frac{1}{2}n}\, P_n^m\left(\frac{t}{\sqrt{(\rho^2+t^2)}}\right) U(t) \risingdotseq (n+m)!\left[\left[\frac{J_m(\rho p)}{p^n}\right]\right], \quad 0<\mathrm{Re}\,p<\infty, \quad (23)$$

which is easily seen to be valid also for $n+m$ odd.

An interesting application of (23) is provided by its definition integral for $\rho = 1$. Upon substituting in the integrand $t = \cot\tau$, we arrive at

$$\left[\left[\frac{J_m(x)}{x^n}\right]\right] = \frac{x}{(n+m)!}\int_0^{\frac{1}{2}\pi} e^{-x\cot\tau}\,\frac{P_n^m(\cos\tau)}{\sin^{n+2}\tau}\,d\tau \quad (\mathrm{Re}\,x>0).$$

3. Homogeneous partial differential equations with boundary conditions

For equations with boundary conditions the operational transposition of only one of the variables answers its purpose if either the location of the boundary is independent of that particular variable, or depends on it in a very simple manner. As in the case of ordinary differential equations, it is often possible to account for the boundary condition by limiting the variables determining the boundary, by means of unit functions. This leads to a new differential equation, involving some additional terms in the right-hand member (see VIII, § 4). This will be illustrated by two problems of heat conduction; in the first example the boundary concerns only the time, in the second example a space variable also.

Example 1. The problem is to determine the distribution of temperature inside an infinitely long, thin bar, when the temperature is constant over the cross-section, and for $t = 0$ is given by $T(x, 0) = \varphi(x)$, when $\varphi(x)$ is a prescribed function; x is measured in the axial direction of the bar.

We have thus to solve the equation†

$$\frac{\partial^2 T}{\partial x^2} - \frac{1}{a^2}\frac{\partial T}{\partial y} = 0 \tag{24}$$

under the boundary condition: $T = \varphi(x)$ when $y = 0$. Since the location of the boundary is independent of x, we put

$$T(x,y) \doteqdot f(p,y), \quad \varphi(x) \doteqdot \phi(p),$$

by means of which (24) transforms into

$$p^2 f - \frac{1}{a^2}\frac{df}{dy} = 0.$$

Now, in the general solution of this equation, that is,

$$f(p,y) = \lambda(p)\, e^{a^2 p^2 y},$$

the function λ has to be identical with ϕ on account of the boundary condition at $y = 0$, i.e. $f(p, 0) = \phi(p)$. We therefore obtain when $y > 0$

$$T(x,y) \doteqdot \frac{1}{p} \times p\, e^{a^2 p^2 y} \times \phi(p),$$

from which the solution in question follows with the aid of a composition product, viz.

$$T(x,y) = \frac{1}{2a\sqrt{(\pi y)}} \int_{-\infty}^{\infty} e^{-(x-\xi)^2/4a^2 y} \varphi(\xi)\, d\xi \quad (y > 0).$$

Example 2. The next problem is to calculate the error in the reading of a mercury thermometer registering the air temperature around an aircraft rapidly losing height, which is due to a finite time of relaxation of the meter. To make the problem as simple as possible, we assume that the aircraft loses its height at a constant rate, and, moreover, that the temperature decreases linearly with the height. Thus the actual temperature of the air at the aircraft is $T = T_0 + \alpha t$.

The problem under consideration was discussed by Bromwich‡. The reading $\bar{T}$ of the thermometer at any moment corresponds to the space average of the temperatures T in its reservoir, which we take to be a sphere of radius R. Since the radially symmetrical temperature at any point depends only on the time t and the distance r from the centre, we have

$$\bar{T}(t) = \frac{4\pi \int_0^R T(r,t)\, r^2\, dr}{\frac{4}{3}\pi R^3} = \frac{3}{R^3}\int_0^R T(r,t)\, r^2\, dr. \tag{25}$$

Inside the reservoir the three-dimensional heat-conduction equation holds, depending only on r and t, too:

$$\frac{\partial^2}{\partial r^2}(rT) - \frac{1}{a^2}\frac{\partial}{\partial t}(rT) = 0. \tag{26}$$

This equation has to be solved under the conditions $T = T_0 + \alpha t$ when $r = R$ and $T = T_0$ when $t = 0$. Since the location of the boundaries depends on t more simply than on r, we assume $T^*(r,t) = T(r,t)\, U(t) \doteqdot f(r,p)$, whence (26) becomes

$$\frac{\partial^2}{\partial r^2}(rT^*) - \frac{1}{a^2}\frac{\partial}{\partial t}(rT^*) = -\frac{r}{a^2} T_0 \delta(t).$$

† Here the time variable is denoted by y instead of t, since later on x rather than the time is identified with the operational t-variable.

‡ *Phil. Mag.* XXXVII, 407, 1919.

The image of this inhomogeneous equation takes on the form

$$\frac{d^2}{dr^2}(rf)-\frac{p}{a^2}rf = -\frac{r}{a^2}T_0 p,$$

having as general solution

$$rf(r,p) = rT_0+\lambda_1(p)\, e^{(r/a)\sqrt{p}}+\lambda_2(p)\, e^{-(r/a)\sqrt{p}}. \tag{27}$$

The unknown functions λ_1 and λ_2 are determined by two conditions:

(1) $$T(R,t) = T_0+\alpha t \quad (t>0),$$

whose transform tells us that $f(R,p) = T_0+\dfrac{\alpha}{p}$,

and

(2) T is finite when $r = 0$; thus

$$\lim_{r\to 0}\{rT^*(r,t)\} = 0;$$

hence $$\lim_{r\to 0}\{rf(r,p)\} = 0.$$

From this it follows that $\lambda_1+\lambda_2 = 0$.

Since the actual determination of the functions λ is easy, it may suffice to state only the final result:

$$f(r,p) = T_0+\frac{R\alpha}{rp}\frac{\sinh\left(\frac{r}{a}\sqrt{p}\right)}{\sinh\left(\frac{R}{a}\sqrt{p}\right)}, \tag{28}$$

which is the image function of the temperatures obtaining in the thermometer's reservoir.

Instead of determining first its corresponding original, $T(r,t)$, and then calculating the mean value according to (25), it is far more simple to perform the process of averaging with respect to the image, followed by a calculation of $\overline{T}$ as original corresponding to the averaged image. We then have

$$\overline{T}(t) \doteqdot \frac{3}{R^3}\int_0^R f(r,p)\, r^2\, dr,$$

which, upon substitution of (28), yields the following result:

$$\overline{T}(t) \doteqdot T_0+\frac{3\alpha a}{Rp^{\frac{3}{2}}}\left\{\coth\left(\frac{R}{a}\sqrt{p}\right)-\frac{a}{R\sqrt{p}}\right\}.$$

The original of this function can be obtained with the aid of (VII, 51) and some composition product; the complete answer to our question is then provided by

$$\overline{T}(t) = T_0+\frac{3\alpha a^2}{R^2}\left\{\int_0^t (t-\tau)\,\theta_3\left(0,\frac{\pi a^2}{R^2}\tau\right)d\tau-\frac{t^2}{2}\right\}$$
$$= T_0+\alpha\left(t-\frac{R^2}{15a^2}+\frac{6R^2}{\pi^4 a^2}\sum_{n=1}^{\infty} n^{-4}\, e^{-\pi^2 a^2 n^2 t/R^2}\right) \quad (t>0).$$

In this expansion the first two terms indicate the actual temperature of the air so that the error due to the retardated adjustment is given by the remaining terms. After some time has elapsed the following approximation holds:

$$\overline{T}(t) \sim T_0+\alpha(t-t_0),$$

in which the time of retardation, $t_0 = R^2/15a^2$, is a measure of the time during which a temperature disturbance propagates itself through the spherical thermometer.

4. Quasi-stationary theory of electric cables

Whereas the rigorous theory of the coaxial cable is based upon Maxwell's equations, it is often sufficient to treat the cable quasi-stationarily after Kirchhoff, whereby it is replaced by two conductors carrying currents of equal amounts, $i(x, t)$, but of opposite directions. In its general form the problem involves four physical parameters, namely, the self-inductance $\bar{L}$ and series resistance $\bar{R}_s$ per unit length, both in the longitudinal direction, and the capacitance $\bar{C}$ and shunt resistance $\bar{R}_q$ per unit length, both in the transverse direction. The cable can therefore be looked upon as a ladder network (see IX, § 7) of an infinite number of cells, the elements of which (per unit of length) are given by the series impedance

$$\bar{Z}_s(p) = \bar{L}p + \bar{R}_s, \tag{29}$$

and the shunt admittance
$$\bar{A}_q(p) = \bar{C}p + \frac{1}{\bar{R}_q}. \tag{30}$$

An equivalent scheme of this cable is shown in fig. 85. In the quasi-stationary treatment both the current i in the longitudinal direction and the voltage e in the transverse direction are considered, and we generally wish to determine i and e in terms of x and t so that certain requirements at the boundaries at the end-points $x = 0$ and $x = l$ are fulfilled. For an operational treatment involving the transposition of a single variable (the transposition of both variables is postponed until the next chapter, § 3, second example), we shall take this variable to be the time, since the location of the boundaries is independent of t. We accordingly put

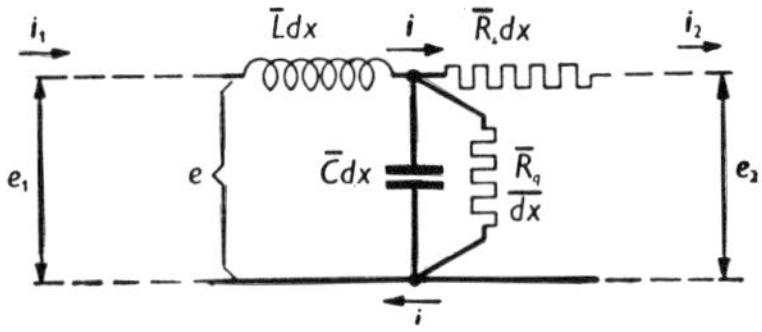

Fig. 85. Equivalent scheme for an electric cable.

$$i(x, t) \doteqdot I(x, p), \quad e(x, t) \doteqdot E(x, p).$$

The basic equations are then obtained from Ohm's law in its operational form (VIII, 17). On the one hand, when applied to a longitudinal element dx, across which there is an impedance $\bar{Z}_s dx$ and a potential drop

$$-\frac{\partial e}{\partial x}dx \doteqdot -\frac{\partial E}{\partial x}dx,$$

it gives the relation
$$\frac{\partial E(x, p)}{\partial x} = -\bar{Z}_s(p)\, I(x, p). \tag{31}$$

On the other hand, there is a transverse current equal to $-\frac{\partial i}{\partial x}dx$ flowing through the admittance $\bar{A}_q dx$; thus, moreover,

$$\frac{\partial I(x, p)}{\partial x} = -\bar{A}_q(p)\, E(x, p). \tag{32}$$

These two equations are valid for arbitrary inhomogeneous cables. In the following, however, we shall restrict ourselves to homogeneous cables in which $\bar{Z}_s$ and $\bar{A}_q$ are independent of x. It is then possible to eliminate either of the two variables E and I between (31) and (32), so as to obtain two second-order differential equations, viz.

$$\frac{\partial^2 I}{\partial x^2} - \lambda^2(p)\, I = 0, \tag{33}$$

$$\frac{\partial^2 E}{\partial x^2} - \lambda^2(p)\, E = 0, \tag{34}$$

where we have introduced the parameter λ by

$$\lambda(p) = \sqrt{\{\bar{Z}_s(p)\, \bar{A}_q(p)\}} = \sqrt{\left\{(\bar{L}p + \bar{R}_s)\left(\bar{C}p + \frac{1}{\bar{R}_q}\right)\right\}}. \tag{35}$$

First of all one may ask for the current i and the voltage e at an arbitrary point of the cable, when these quantities are prescribed functions of time, $i_1(t) \risingdotseq I_1(p)$ and $e_1(t) \risingdotseq E_1(p)$, at $x = 0$. For this purpose we start from the general solution of (33):

$$I(x, p) = \alpha(p)\, e^{\lambda(p)x} + \beta(p)\, e^{-\lambda(p)x},$$

then determine the corresponding voltage E (see (32)), and finally calculate the functions $\alpha(p)$ and $\beta(p)$ from the conditions for the boundary at $x = 0$. The result of these calculations determines the operational images of current and voltage as follows:

$$\left.\begin{aligned} I &= \cosh\{x\lambda(p)\}\, I_1(p) - A(p) \sinh\{x\lambda(p)\}\, E_1(p), \\ E &= -\frac{\sinh\{x\lambda(p)\}}{A(p)}\, I_1(p) + \cosh\{x\lambda(p)\}\, E_1(p), \end{aligned}\right\} \tag{36}$$

in which, in addition to the parameter (35), we have introduced

$$A(p) = \sqrt{\frac{\bar{A}_q(p)}{\bar{Z}_s(p)}} = \sqrt{\frac{\bar{C}p + 1/\bar{R}_q}{\bar{L}p + \bar{R}_s}}. \tag{37}$$

In particular, let us consider a cable of length l. By substitution of $x = l$ we then get a relationship between the current and voltage at the left hand and those at the right hand. Denoting the latter quantities by $i_2(t) \risingdotseq I_2(p)$ and $e_2(t) \risingdotseq E_2(p)$, we obtain for the currents at the ends, in terms of the voltages at those places:

$$\left.\begin{aligned} I_1 &= A(p)\left[\coth\{l\lambda(p)\}\, E_1 - \frac{E_2}{\sinh\{l\lambda(p)\}}\right], \\ I_2 &= A(p)\left[\frac{E_1}{\sinh\{l\lambda(p)\}} - \coth\{l\lambda(p)\}\, E_2\right]. \end{aligned}\right\} \tag{38}$$

From the structure of these formulae it is obvious that a cable or transmission line of length l is equivalent to a symmetric four-terminal network (see fig. 55), which, according to (IX, 13), possesses a transfer admittance equal to

$$A_{12}(p) = \frac{A(p)}{\sinh\{l\lambda(p)\}}, \tag{39}$$

and a driving-point admittance equal to

$$A_{11}(p) = A(p)\coth\{l\lambda(p)\}. \tag{40}$$

This clearly reveals the significance of the function $A(p)$ as driving-point admittance of an infinitely long transmission line. As follows from the above, all results of the theory of electrical networks treated in chapter IX may be applied to cables whenever they concern the special case of four-terminal networks. For instance, in addition to the frequency admittance (40), there exists a corresponding time admittance which is the image of the former when $\mathrm{Re}\, p > 0$; this time admittance represents the switch-on current at the left-hand side of the cable in response to a unit voltage $U(t)$ applied across it. Moreover, as in the discussion of filters (IX, § 7), the right-hand end of the cable may be short-circuited by an arbitrary impedance Z_2 (see fig. 86), so as to lead to $E_2(p) = Z_2(p)\, I_2(p)$. It is then possible by using (38) to express I_2 and I_1 in terms of E_1. If the result so obtained is substituted in (36), the following image for the current emerges:

$$I = A(p)\, E_1(p)\, \frac{e^{-x\lambda(p)} + R(p)\, e^{(x-2l)\lambda(p)}}{1 - R(p)\, e^{-2l\lambda(p)}}, \tag{41}$$

Fig. 86. Electric cable terminated by an impedance.

in which $R(p)$ is an abbreviation for

$$R(p) = \frac{1 - A(p)\, Z_2(p)}{1 + A(p)\, Z_2(p)}.$$

Herewith we have got the 'continuous' analogue of formula (IX, 33) for the current in a homogeneous filter. In consequence of this there are again two different ways of treating transient problems by operational methods, when it is assumed that $e_1(t)$ is known:

(1) *Method of D'Alembert.* Here (41) is expanded into a geometric series, viz.

$$I = A(p)\, E_1(p) \left\{ \sum_{n=0}^{\infty} R^n(p)\, e^{-(x+2nl)\lambda(p)} + \sum_{n=1}^{\infty} R^n(p)\, e^{(x-2nl)\lambda(p)} \right\},$$

the individual terms of which represent travelling waves, proceeding either to the right or to the left; $R(p)$ may be interpreted as 'reflexion coefficient' of the right-hand terminal.

(2) *Method of Bernoulli.* The image of (41) is expanded into partial fractions corresponding to the zeros p_k of the denominator, which determine the characteristic frequencies $p_k/2\pi i$ in their dependence on the

short-circuit impedance Z_2. The current itself is then expressible as an expansion into the cable's eigenfunctions $e^{p_k t}$.

The situation is the simplest possible if $Z_2(p)$ is equal to the so-called characteristic impedance $1/A(p)$. If so, then

$$Z_2(p) = \sqrt{\frac{\bar{Z}_s(p)}{\bar{A}_q(p)}} = \sqrt{\frac{\bar{L}p+\bar{R}_s}{\bar{C}p+1/\bar{R}_q}},$$

and the coefficient of reflexion, R, becomes zero. In this case the cable behaves as if it was infinitely long; according to (41) we then have

$$I = A(p)\,E_1(p)\,e^{-x\lambda(p)}. \tag{42}$$

It will be almost unnecessary to emphasize that by the theory given above it is possible to give a thorough discussion of all kinds of transient phenomena encountered in the technique of cables and transmission lines. Of these problems some typical examples will now be treated.

Example 1. *Transients on a cable matched by its characteristic impedance.* We are going to discuss the switch-on phenomena produced by an impulse voltage $\delta(t)$ which may be applied to the left end of the cable. According to (42) the current, i_δ, is determined by the relation

$$i_\delta(x,t) \doteq pA(p)\,e^{-x\lambda(p)} \quad (\mathrm{Re}\,p>0). \tag{43}$$

Let us substitute in the image the constants of the cable as given by (35) and (37). Upon introducing the new constants

$$c_0 = \frac{1}{\sqrt{(\bar{L}\bar{C})}}, \quad \mu_1 = \frac{\bar{R}_s}{2\bar{L}} - \frac{1}{2\bar{C}\bar{R}_q}, \quad \mu_2 = \frac{\bar{R}_s}{2\bar{L}} + \frac{1}{2\bar{C}\bar{R}_q}, \tag{44}$$

the operational relation above assumes the following simple form:

$$i_\delta(x,t) \doteq c_0 \frac{p(\bar{C}p+1/\bar{R}_q)}{\sqrt{\{(p+\mu_2)^2-\mu_1^2\}}} \exp\left\{-\frac{x}{c_0}\sqrt{\{(p+\mu_2)^2-\mu_1^2\}}\right\} \quad (\mathrm{Re}\,p>0). \tag{45}$$

Furthermore, upon applying (15) in combination with some elementary rules, we are led to

$$i_\delta(x,t) = \sqrt{\left(\frac{\bar{C}}{\bar{L}}\right)}\, e^{-\mu_2 t}\left(\frac{\partial}{\partial t}-\mu_1\right)\left[I_0\left\{\mu_1\sqrt{\left(t^2-\frac{x^2}{c_0^2}\right)}\right\}U\left(t-\frac{x}{c_0}\right)\right], \tag{46}$$

which, in order to interpret it physically, we may change into

$$i_\delta(x,t) = \sqrt{\left(\frac{\bar{C}}{\bar{L}}\right)}\, e^{-\mu_2 x/c_0}\,\delta\left(t-\frac{x}{c_0}\right) + \sqrt{\left(\frac{\bar{C}}{\bar{L}}\right)}\, e^{-\mu_2 t}\left(\frac{\partial}{\partial t}-\mu_1\right)\left[I_0\left\{\mu_1\sqrt{\left(t^2-\frac{x^2}{c_0^2}\right)}\right\}U\left(t-\frac{x}{c_0}\right)\right]. \tag{47}$$

Therefore, the switch-on current for an arbitrary cable consists of the sum of an impulse and an after-effect term. The first term represents a disturbance travelling at a velocity c_0 along the cable, its amplitude steadily decreasing on account of the attenuation factor $e^{-\mu_2 x/c_0}$. The after-effect term represents a disturbance that decreases exponentially with time at any fixed place x, at least if one waits long enough. The last follows from the exponential factor $e^{|x|}$ of $I_0(x)$ as $x \to \infty$, which tells us that the second term of (47) for $t \gg x/c_0$ decreases in proportion to the attenuation factor

$$e^{-(\mu_2-|\mu_1|)t},$$

that is

$$e^{-\bar{R}_s t/\bar{L}} \quad \text{if} \quad \mu_1<0, \qquad e^{-t/\bar{C}\bar{R}_q} \quad \text{if} \quad \mu_1>0.$$

It is very striking that the exponential decrement of the after-effect term depends either only on $\bar{R}_s$ and $\bar{L}$ or only on $\bar{C}$ and $\bar{R}_q$.

The general form (46) of the derivative of the indicial admittance is of exceptional importance in the theory and practice of cable technique, as may be illustrated by the following special cases:

(1) $\mu_1 = 0$ (*distortion-free cable*). We then have

$$\frac{\bar{R}_s}{\bar{L}} = \frac{1}{\bar{C}\bar{R}_q},$$

while at the same time the characteristic impedance becomes equal to the purely ohmic resistance $\sqrt{(\bar{L}/\bar{C})}$. Moreover, the after-effect is completely absent. As to the name of 'distortion-free cable' we may note that the expression for the transfer admittance at the point x,

$$A(p)\,\mathrm{e}^{-x\lambda(p)} = \sqrt{\left(\frac{\bar{C}}{\bar{L}}\right)}\,\mathrm{e}^{-x\bar{R}_s/c_0\bar{L}}\,\mathrm{e}^{-xp/c_0},$$

shows that undamped alternating voltages ($p = i\omega$), in their way along the cable, continually change in amplitude, but in a manner wholly independent of the frequency applied. Consequently, any undamped wave retains its spectral composition in travelling along the distortionless cable; it was this point of view from which this type of transmission line was first considered by Heaviside.

(2) $\bar{R}_s = 0$, $\bar{R}_q = \infty$ (*loss-free cable*). In this case $\mu_1 = \mu_2 = 0$. Since the cable does not contain ohmic resistances, it is the analogue of the low-pass filter discussed in example 1 of IX, § 7. The formula

$$i_\delta(x,t) = \sqrt{\left(\frac{\bar{C}}{\bar{L}}\right)}\,\delta\!\left(t-\frac{x}{c_0}\right),$$

determining the switch-on current due to an impulse voltage, was derived in (IX, 39), by a process in which the cable emerged as the limiting case of the filter. An arbitrary undamped wave, when travelling along the cable, not only retains its spectral composition but also its amplitude. However, the phases of the separate waves are changing continually according to the factor

$$\mathrm{e}^{-xp/c_0} = \mathrm{e}^{-ix\omega/c_0}.$$

(3) $\bar{L} = 0$ (*induction-free cable*). This particular case was first considered by Lord Kelvin. From (43), (37) and (35) we obtain with the aid of (XI, 4) the following elementary function for the switch-on current:

$$i_\delta(x,t) = U(t)\sqrt{\left(\frac{\bar{C}}{\pi\bar{R}_s}\right)}\left(\frac{1}{\bar{C}\bar{R}_q}+\frac{d}{dt}\right)\left\{\frac{\exp\left(-\dfrac{t}{\bar{C}\bar{R}_q}-\dfrac{\bar{R}_s\bar{C}}{4}\dfrac{x^2}{t}\right)}{\sqrt{t}}\right\}.$$

(4) $\bar{L} = 0$, $\bar{R}_q = \infty$ (*RC cable*). In this case the only elements of the cable are $\bar{R}_s$ and $\bar{C}$; the time admittance proves to be

$$i_{\sqcap}(x,t) = \sqrt{\left(\frac{\bar{C}}{\pi\bar{R}_s}\right)}\frac{\mathrm{e}^{-\bar{R}_s\bar{C}x^2/4t}}{\sqrt{t}}.$$

Furthermore, $$A(p) = \sqrt{\left(\frac{\bar{C}}{\bar{R}_s}p\right)}, \quad \lambda(p) = \sqrt{(\bar{R}_s\bar{C}p)}, \tag{48}$$

whence it follows that equation (33) reduces to

$$\frac{\partial^2 I}{\partial x^2} - \bar{R}_s\bar{C}pI = 0.$$

The original of the last equation expresses that for this cable the diffusion equation (9) holds; that is to say, in the form of

$$\frac{\partial^2 i}{\partial x^2} - \bar{R}_s\bar{C}\frac{\partial i}{\partial t} = 0.$$

Example 2. *Transient phenomena of RC cables.* In virtue of the special properties of the RC cable the transient phenomena remain very simple, even when the cable is not terminated by its characteristic impedance. For instance, when the right end is short-circuited (open), that is, when $Z_2 = 0$ (∞), then from (41) for the switch-on current in response to a unit voltage (thus $E_1 = 1$), with the aid of (48) we have

$$i_{\lrcorner}(x,t) \doteqdot \sqrt{\left(\frac{\bar{C}}{\bar{R}_s}p\right)} \frac{\cosh\{(l-x)\sqrt{(\bar{R}_s\bar{C}p)}\}}{\sinh\{l\sqrt{(\bar{R}_s\bar{C}p)}\}},$$

and

$$i_{\lrcorner}(x,t) \doteqdot \sqrt{\left(\frac{\bar{C}}{\bar{R}_s}p\right)} \frac{\sinh\{(l-x)\sqrt{(\bar{R}_s\bar{C}p)}\}}{\cosh\{l\sqrt{(\bar{R}_s\bar{C}p)}\}},$$

respectively. Furthermore, the operational relation (XI, 13) in combination with the corresponding relation for the function θ_2 leads to the following expressions for the current itself:

$$i_{\lrcorner}(x,t) = \frac{U(t)}{l\bar{R}_s}\theta_3\left(\frac{x}{2l}, \frac{\pi t}{l^2\bar{R}_s\bar{C}}\right),$$

and

$$i_{\lrcorner}(x,t) = \frac{U(t)}{l\bar{R}_s}\theta_2\left(\frac{x}{2l}, \frac{\pi t}{l^2\bar{R}_s\bar{C}}\right),$$

respectively. Finally, by setting x equal to zero it follows that the open RC cable has a driving-point admittance that can also be realized by the ladder network of (IX, 30).

5. Inhomogeneous partial differential equations

As an example of this type of equation, consider the inhomogeneous wave equation corresponding to (11):

$$\frac{\partial^2 u}{\partial x^2} - au - \frac{1}{c^2}\frac{\partial^2 u}{\partial t^2} = -\varphi.$$

As in § 2, let us transpose the variable t and assume

$$u(x,t) \doteqdot f(x,p), \quad \varphi(x,t) \doteqdot \phi(x,p).$$

Then the image is found to satisfy the ordinary differential equation

$$\frac{d^2 f}{dx^2} - \left(a + \frac{p^2}{c^2}\right) f = -\phi.$$

The general procedure for solving such an inhomogeneous ordinary differential equation is called the 'method of variation of constants', which, in the operational sense, reduces to introducing x as a second operational variable. One is thus quite naturally led to the methods of simultaneous operational calculus, which will be treated more fully in the following chapter. Only in simple problems does it make no sense to transpose both variables, of which an example may be given below.

Example. The Green function for the n-dimensional wave equation. Let us return to the wave equation with a dispersion term, as given in (5). Its corresponding Green function, with respect to a point source localized at the origin at $t = 0$, satisfies equation (6). On account of the radial symmetry of this Green function equation (6) may be transformed into

$$\left\{\frac{\partial^2}{\partial r^2}+\frac{n-1}{r}\frac{\partial}{\partial r}+a-\frac{1}{c^2}\frac{\partial^2}{\partial t^2}\right\}G_n = -\frac{\Gamma(\frac{1}{2}n)}{\pi^{\frac{1}{2}n}}\frac{\delta(r)}{r^{n-1}}\delta(t),$$

by using (7), and by expressing the n-dimensional Laplace operator in terms of spherical co-ordinates. If $G_n(r,t) \doteqdot f_n(r,p)$, then the equation for the image becomes

$$\left\{\frac{d^2}{dr^2}+\frac{n-1}{r}\frac{d}{dr}+\left(a-\frac{p^2}{c^2}\right)\right\}f_n = -\frac{\Gamma(\frac{1}{2}n)}{\pi^{\frac{1}{2}n}}p\frac{\delta(r)}{r^{n-1}}. \tag{49}$$

Outside the origin this differential equation is homogeneous; and it can then be reduced to (x, 11) valid for Bessel functions. Since in this case the argument of the Bessel function proves to be imaginary, only that solution which involves a K-function (see (x, 29)) does not increase exponentially as $r \to \infty$. Therefore the solution f_n required must be of the form $(r > 0)$:

$$f_n = \alpha_n(p)\frac{K_{\frac{1}{2}n-1}\left\{r\sqrt{\left(\frac{p^2}{c^2}-a\right)}\right\}}{r^{\frac{1}{2}n-1}}. \tag{50}$$

The unknown function $\alpha_n(p)$ has yet to be determined. In order to investigate its connexion with the right-hand member of (49), it is necessary to fix the function f_n on the interval $r < 0$ (the latter does not correspond to real space). That this can easily be done is due to the fact that the right-hand member of (49) is even or odd according as n is odd or even. Therefore it is simplest to continue f_n as an even or odd function in accordance. Keeping in mind that f_n as $r \to +0$ is proportional to r^{2-n}, as follows from (50), and applying to (49) the operator

$$\lim_{\epsilon\to 0}\int_{-\epsilon}^{\epsilon} dr\, r^{n-1}\ldots,$$

we further find that

$$\lim_{r\to+0}\left\{r^{n-1}\frac{df_n}{dr}\right\} = -\frac{p}{2}\frac{\Gamma(\frac{1}{2}n)}{\pi^{\frac{1}{2}n}}.$$

This relation fixes the value of α_n; if use is made of the explicit expression for the first term in the power series of the K-function, the following operational relation results:

$$G_n(r,t) \doteqdot \frac{p}{2\pi}\left(\frac{\sqrt{\left(\frac{p^2}{c^2}-a\right)}}{2\pi r}\right)^{\frac{1}{2}n-1} K_{\frac{1}{2}n-1}\left\{r\sqrt{\left(\frac{p^2}{c^2}-a\right)}\right\} \quad (r>0). \tag{51}$$

The Green function itself will now be determined; the method followed is different according as n is even or odd:

(*a*) *n is odd.* Since its order is an odd multiple of $\frac{1}{2}$, the K-function above is expressible in terms of elementary functions. Therefore (51) may be written equivalently as follows:

$$G_n(r,t) \doteqdot \frac{p}{4\pi}\left(-\frac{d}{2\pi r\,dr}\right)^{\frac{1}{2}(n-3)}\left[\frac{\exp\left\{-r\sqrt{\left(\frac{p^2}{c^2}-a\right)}\right\}}{r}\right].$$

Of this the original will be determined for $\operatorname{Re} p > c \operatorname{Re} \sqrt{a}$; as it turns out that this original involves a unit function, we thus obtain the Green function corresponding to a system that is initially at rest. The actual form of the original may be found with the aid of the operational relation that, in its turn, may be obtained by differentiation of (15) and (13) with respect to a ($a>0$ and <0, respectively). Independent of the algebraic sign of a, the final result then appears to be

$$G_n = \frac{c}{2}\left(-\frac{d}{2\pi r\,dr}\right)^{\frac{1}{2}(n-1)}\left(I_0(\sqrt{\{a(c^2t^2-r^2)\}})\,U(ct-r)\right) \quad (n \text{ odd}). \tag{52}$$

This expression indicates in a surveyable manner how a singular disturbance at $r=t=0$ propagates itself in odd-dimensional space. The simplest possible case corresponds to $a=0$ when there is no dispersion:

$$G_n = \frac{c}{2}\left(-\frac{d}{2\pi r\,dr}\right)^{\frac{1}{2}(n-1)}\{U(ct-r)\}. \tag{53}$$

For instance

$$G_1(x,t) = \frac{c}{2}\,U\left(t-\frac{|x|}{c}\right), \tag{54}$$

$$G_3(r,t) = \frac{\delta\left(t-\frac{r}{c}\right)}{4\pi r}, \tag{55}$$

$$G_5(r,t) = \frac{1}{8\pi^2}\left\{\frac{\delta\left(t-\frac{r}{c}\right)}{r^3}+\frac{\delta'\left(t-\frac{r}{c}\right)}{cr^2}\right\}.$$

The conclusion is that in a space with an odd number of dimensions the spherical wave travels with a sharp front; at any point it presents itself only during an infinitesimally short time (principle of Huygens), which property is lost in the presence of dispersion, however.

(*b*) *n is even.* Though the K-function is not expressible in elementary functions, the original can be calculated from (51) when this relation is first transformed by means of the following identity:

$$K_\nu\{\sqrt{(\alpha^2-\beta^2)}\} = \frac{1}{2}\left(\frac{\alpha-\beta}{\alpha+\beta}\right)^{\frac{1}{2}\nu}\int_{-\infty}^{\infty} e^{-\alpha\cosh s-\beta\sinh s-\nu s}\,ds \quad (\operatorname{Re}\alpha > |\operatorname{Re}\beta|),$$

which is an extension of formula (III, 29). Again restricting ourselves to $\operatorname{Re} p > c \operatorname{Re} \sqrt{a}$, we obtain an expression whose original leads to a sifting integral involving the $(\frac{1}{2}n-1)$th derivative of an impulse function; after some simplification, and confining ourselves to the case of no dispersion, we then get the following result:

$$G_n = \frac{c}{2\pi}\,U\left(t-\frac{r}{c}\right)\left(\frac{d}{2\pi cr\,dt}\right)^{\frac{1}{2}n-1}\left[\frac{\cosh\left\{(\frac{1}{2}n-1)\operatorname{arc\,cosh}\frac{ct}{r}\right\}}{\sqrt{(c^2t^2-r^2)}}\right] \quad (n \text{ even}). \tag{56}$$

In this way one has, for instance,

$$G_2(r,t) = \frac{1}{2\pi}\frac{U\left(t-\frac{r}{c}\right)}{\sqrt{\left(t^2-\frac{r^2}{c^2}\right)}}, \tag{57}$$

$$G_4(r,t) = -\frac{1}{4\pi^2c^2}\frac{U\left(t-\frac{r}{c}\right)}{\left(t^2-\frac{r^2}{c^2}\right)^{\frac{3}{2}}},$$

from which it may be concluded that, even if there is no dispersion at all, the radial propagation of waves in even-dimensional space is accompanied by an after-effect, occurring after the arrival of the first disturbance at the moment $t = r/c$. In this case the principle of Huygens is *not* valid.

In concluding this chapter we remark that the Green function for the equation of heat conduction in n dimensions can be determined similarly. For odd values of n it is found that

$$G_n^*(r, t) = a^2 \frac{e^{-r^2/4a^2t}}{(4\pi a^2 t)^{\frac{1}{2}n}} U(t).$$

This function is thus a solution of the differential equation (10).

CHAPTER XVI

SIMULTANEOUS OPERATIONAL CALCULUS

1. Introduction

The theory of simultaneous operators deals with the transformation of functions of more than one variable by means of multiple Laplace integrals. Not much emphasis has thus far been laid on this part of the operational calculus, yet some investigations trace back as far as Heaviside. A somewhat more systematic treatment was begun in 1931 by Van der Pol and Niessen†. The need for a simultaneous operational calculus presents itself quite naturally, when problems dependent on more than one variable are to be treated operationally in a manner similar to that developed in the preceding chapters with respect to problems involving a single variable. Therefore we shall state in § 2 some important features of the theory required for handling simultaneous operators. In §§ 3 and 4 it will be shown how for hyperbolic differential equations with constant coefficients the simultaneous calculus is helpful in determining certain particular solutions. One of the things then presenting itself is the well-known theory of characteristics of partial differential equations, which was not met in the foregoing chapter where we transposed only one of the variables of these same differential equations. In § 5 differential equations of the elliptic type are discussed; the operational calculus clearly indicates also to what extent this type of differential equation differs from the hyperbolic type. Finally, §§ 6 and 7 provide some examples showing how other topics, such as simultaneous partial differential equations and partial difference equations, are suited to simultaneous operational calculus.

2. General theory

In the present section the foundations of the simultaneous calculus will be discussed. However, in order to keep the discussion as simple as possible we shall only deal with two variables, since any extension to the case of more than two variables follows readily. As in the investigations involving one variable, the base is then formed by the definition integral, which maps the *original* $h(x, y)$ on the function

$$f(p,q) = pq\int_{-\infty}^{\infty} dx \int_{-\infty}^{\infty} dy\, e^{-px-qy}\, h(x,y), \tag{1}$$

which, again, is called the corresponding *image*. An appropriate notation for

† *Phil. Mag.* XI, 368, 1931.

the relationship existing between the original and the image functions is then provided by
$$h(x,y) \doteq f(p,q).$$

First of all, the multiple integral introduced above should have a value independent of the order of integration. If so, then the image can be found in two different ways as follows:

(*a*) In the first place the integration with respect to y is carried out; this leads to an operational relation involving the variable x as a parameter, the y-variable being transposed in terms of q:
$$q\int_{-\infty}^{\infty} dy\, e^{-qy}\, h(x,y) = h^*(x,q) \doteq h(x,y). \tag{2a}$$

The next step is performing the integration with respect to x, that is, transposing the variable x in terms of p. The final result of the twofold transposition is then found to be
$$p\int_{-\infty}^{\infty} dx\, e^{-px}\, h^*(x,q) = f_1(p,q) \doteq h(x,y). \tag{2b}$$

(*b*) Conversely, a first integration with respect to x leads to the image
$$p\int_{-\infty}^{\infty} dx\, e^{-px}\, h(x,y) = h^{**}(p,y) \doteq h(x,y), \tag{3a}$$
and the second integration yields
$$q\int_{-\infty}^{\infty} dy\, e^{-qy}\, h^{**}(p,y) = f_2(p,q) \doteq h(x,y). \tag{3b}$$
As already mentioned, the image has sense if, and only if, the reversal of the order of integration is immaterial; that is to say, if $f_1(p,q) \equiv f_2(p,q)$.

Conversely, when the image is given, the procedure to obtain the corresponding original consists of first transposing back one of the two variables, no matter which one, so as to construct an intermediate relation as (2*a*) or (3*a*), and then transposing back the remaining variable.

We shall now discuss some general properties which are extensions of properties already known from the calculus of a single variable; they can all be derived from the definition integral (1).

I. *The domain of convergence*

In general the domain of convergence of the definition integral will be determined by some inequality, involving simultaneously the four real quantities $\operatorname{Re} p$, $\operatorname{Re} q$, $\operatorname{Im} p$ and $\operatorname{Im} q$. Whereas in the case of one variable the domain of convergence depends only on $\operatorname{Re} p$, in general both the real as well as the imaginary parts of p and q are important here. Yet in many applications it is sufficient to consider only the real parts of p and q. It is then most convenient

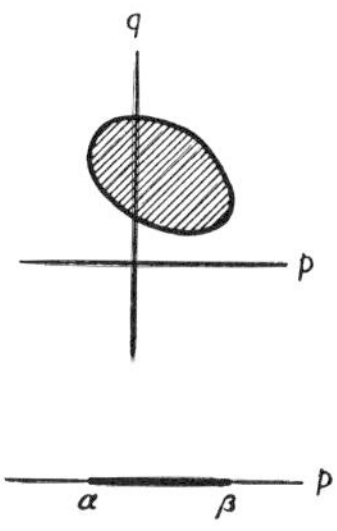

Fig. 87. The domain of convergence for real values of the operational variables.

to represent the domain of convergence graphically in a plane of rectangular Cartesian co-ordinates. The domain of convergence will then be represented by the inside of a closed contour, whether finite or extending to infinity in some directions. Such a region is clearly the two-dimensional analogue of the segment $\alpha < \mathrm{Re}\, p < \beta$ determining the strip of convergence in case of a single operational variable (see fig. 87).

Example. In order to construct the original corresponding to the function

$$f(p,q) = \frac{pq}{p+\operatorname{arc\,sinh} q},$$

we first transpose the variable p; it is then found that

$$f(p,q) \doteqdot q\, \mathrm{e}^{-x \operatorname{arc\,sinh} q}\, U(x), \qquad \mathrm{Re}\,(p+\operatorname{arc\,sinh} q) > 0.$$

Next the variable q is transposed with the aid of (x, 16); the result becomes

$$\frac{pq}{p+\operatorname{arc\,sinh} q} \doteqdot \frac{x}{y} J_x(y)\, U(x)\, U(y),$$
$$\mathrm{Re}\,(p+\operatorname{arc\,sinh} q) > 0;\ \mathrm{Re}\, q > 0. \quad (4)$$

For real values of p and q the domain of convergence of (4) is that shaded in fig. 88; it is partly bounded by the positive axis and partly by the transcendental curve $q = -\sinh p$.

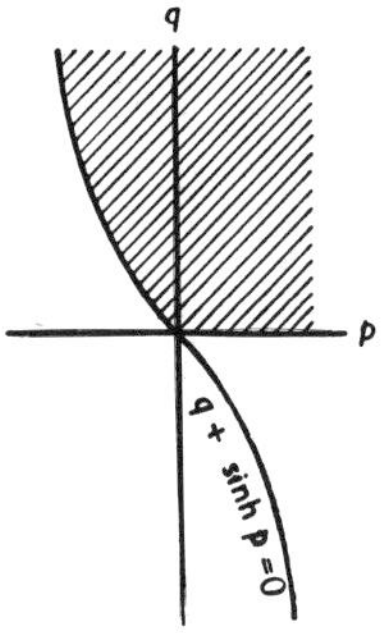

Fig. 88. The domain of convergence for the image of

$$\frac{x}{y} J_x(y)\, U(x)\, U(y).$$

II. *The inversion integral*

One can successively write down the inversion integrals (perhaps convergent only in the sense of a first-order Cesàro limit) corresponding to the two relations of either the set (2 a, b) or the set (3 a, b), i.e.

$$\left.\begin{array}{c} h(x,y) \doteqdot h^*(x,q) \doteqdot f(p,q), \\ \text{or} \\ h(x,y) \doteqdot h^{**}(p,y) \doteqdot f(p,q), \end{array}\right\} \quad (5)$$

respectively.

Then the following double integral for the dependence of the original on the image is found:

$$h(x,y) = \frac{1}{(2\pi i)^2} \int_{c_1-i\infty}^{c_1+i\infty} dp \int_{c_2-i\infty}^{c_2+i\infty} dq\, \mathrm{e}^{px+qy} \frac{f(p,q)}{pq}, \quad (6)$$

in which the constants c_1 and c_2 are such that the domain of integration, $\mathrm{Re}\, p = c_1$, $\mathrm{Re}\, q = c_2$, is wholly inside the domain of convergence.

III. *Operational rules in which the two variables are subjected to independent transformations*

To obtain the rules of the subtitle we have merely to extend those derived previously (such as in chapter IV), while bearing in mind that simultaneous operational relations are generated by the succession of two simple operational relations. For instance, the extension of the shift rule reads as follows:

$$h(x+\alpha, y+\beta) \fallingdotseq e^{\alpha p+\beta q} f(p,q);$$

that for the attenuation rule is transformed into

$$e^{-\alpha x-\beta y}\, h(x,y) \fallingdotseq \frac{pq}{(p+\alpha)(q+\beta)} f(p+\alpha, q+\beta).$$

IV. *The rotation rule*

In this rule both the co-ordinates p and q of the image as well as the co-ordinates x and y of the original are subjected to a linear transformation. Any of these transformations may be interpreted as a rotation of axes followed by a magnification, if necessary in combination with a reflexion. There exist two different formulae according as the given transformation of variables concerns the original or the image, namely,

$$h(\alpha x+\beta y, \gamma x+\delta y) \fallingdotseq \frac{pq}{|\Delta|}\left\{\frac{f(p',q')}{p'q'}\right\}_{p'=\frac{\delta p-\gamma q}{\Delta},\; q'=\frac{-\beta p+\alpha q}{\Delta}}, \tag{7a}$$

$$pq\left\{\frac{f(p',q')}{p'q'}\right\}_{p'=\alpha p+\beta q,\; q'=\gamma p+\delta q} \fallingdotseq \frac{1}{|\Delta|}\, h\left(\frac{\delta x-\gamma y}{\Delta}, \frac{-\beta x+\alpha y}{\Delta}\right), \tag{7b}$$

in which $\Delta = \alpha\delta-\beta\gamma$.

Of these rules, in which the two variables thus no longer transform independently, the first can be proved by simply introducing in the corresponding definition integral (1) a pair of new integration variables according to $x' = \alpha x + \beta y$ and $y' = \gamma x+\delta y$. Then it appears at the same time that the domain of convergence of the new operational relation is obtained by substituting p' and q' for p and q, respectively, in that of the initial relation. The structure of the corresponding rotations in the x, y-plane and the p, q-plane is determined by the requirement that the exponent of the definition integral be invariant, that is,

$$px+qy = p'x'+q'y'.$$

A further extension of the rotation rule, so as to include the case of n variables, is evident. The corresponding linear transformations, that of the variables x_i and that of the variables p_i, shall have to be such that

$$\sum_{i=1}^{n} p_i x_i = \sum_{i=1}^{n} p_i' x_i'.$$

Moreover, there exist many particular cases of the rotation rule, of which we may mention for two variables

$$h(x+y, y) \fallingdotseq q\frac{f(p, q-p)}{q-p},$$

and

$$p\frac{f(p+q)}{p+q} \fallingdotseq h(x)\, U(y-x) \qquad (\operatorname{Re} q > 0). \tag{8}$$

In the last relation it is understood that $f(p) \risingdotseq h(x)$ is a known operational relation in one variable. By adding the corresponding relation for x and y interchanged, we obtain the symmetric relation

$$f(p+q) \risingdotseq h\{\min(x, y)\} \qquad (\mathrm{Re}\, p > 0;\ \mathrm{Re}\, q > 0). \tag{9}$$

V. *Composition product*

The original of the product of two images in the presence of two operational variables p and q can be obtained by using the rule of the composition product for a single variable twice in succession; first the one and then the other variable is transposed with the aid of the intermediate relations (5) corresponding to the individual factors. In so doing it is found that

$$\frac{f_1(p,q)\, f_2(p,q)}{pq} \risingdotseq \int_{-\infty}^{\infty} d\xi \int_{-\infty}^{\infty} d\eta\, h_1(\xi, \eta)\, h_2(x-\xi, y-\eta), \tag{10}$$

in which it is assumed that $f_1(p,q) \risingdotseq h_1(x,y)$ and $f_2(p,q) \risingdotseq h_2(x,y)$. Moreover, (10) is valid in the overlap of the two domains of convergence of the initial relations. In this connexion it may be remarked that in virtue of (XV, 4) a solution of an inhomogeneous partial differential equation is obtainable from the composition product of its right member $\varphi(x,y)$ by its Green function $G_0(x,y)$.

VI. *Asymptotic expressions*

The theorems of Abel and of Tauber which were treated in chapter VII can be applied to either of the two operational variables, so as to reduce the original *simultaneous* relation to a pair of relations of a *single* variable. For instance, when the region ($\mathrm{Re}\, p > 0$, $\mathrm{Re}\, q > 0$) is wholly inside the domain of convergence of

$$h(x,y)\, U(x)\, U(y) \risingdotseq f(p,q),$$

then a formal application of Abel's theorem (VII, $2a$) leads to the following set of *simple* operational relations:

$$h(x, \infty)\, U(x) \risingdotseq f(p, 0), \tag{11a}$$

$$h(\infty, y)\, U(y) \risingdotseq f(0, q). \tag{11b}$$

The first of (11), for example, may be proved by first deriving the following equation by applying (VII, $2a$) to ($2a$):

$$h(x, \infty) = \lim_{q \to 0} q \int_0^{\infty} e^{-qy}\, h(x,y)\, dy. \tag{12}$$

Then the definition integral in one variable of $h(x,\infty)\, U(x)$ in general leads to the function $f(p, 0)$ as determined by the definition integral (1) in two variables. In order to make this true, however, a reversal of two limits is necessary, which is legitimate when, for instance, the integral of (12) tends to $h(x, \infty)$ uniformly in x for $x > 0$.

In addition, once again applying (VII, $2a$) to (11) leads us (under suitable conditions, of course) to

$$h(\infty, \infty) \Rightarrow f(0, 0), \tag{13}$$

which is the extended form of the formula $h(\infty) \Rightarrow f(0)$ valid in the case of one variable.

Before proceeding to a systematic discussion concerning possible applications of the simultaneous operational calculus, we would like to give two examples that undoubtedly show how useful the simultaneous calculus is for specific problems encountered on various occasions.

Example 1. *Integrals over products of Bessel functions.* An integral that can be evaluated (if convergent) by means of the operational calculus of three variables† is the following:

$$I = \int_0^\infty J_\nu(as)\, J_\nu(bs)\, J_\nu(cs)\, \frac{ds}{s^{\nu-1}}. \tag{14}$$

We put

$$a = 2\sqrt{x_1}, \quad b = 2\sqrt{x_2}, \quad c = 2\sqrt{x_3},$$

and then rewrite the integral as

$$(x_1 x_2 x_3)^{\frac{1}{2}\nu} I(x_1, x_2, x_3) = \int_0^\infty x_1^{\frac{1}{2}\nu} s^\nu J_\nu(2s\sqrt{x_1}) \times x_2^{\frac{1}{2}\nu} s^\nu J_\nu(2s\sqrt{x_2}) \times x_3^{\frac{1}{2}\nu} s^\nu J_\nu(2s\sqrt{x_3})\, \frac{ds}{s^{4\nu-1}}.$$

Now, the variables x_1, x_2 and x_3 occur separately and equivalently; therefore the integrand can be transposed immediately with the aid of (X, 20) if p_1, p_2 and p_3 are the respective image variables:

$$(x_1 x_2 x_3)^{\frac{1}{2}\nu} I(x_1, x_2, x_3) \overset{\cdots}{\underset{\cdots}{=}} \frac{1}{(p_1 p_2 p_3)^\nu} \int_0^\infty \exp\left\{-s^2\left(\frac{1}{p_1} + \frac{1}{p_2} + \frac{1}{p_3}\right)\right\} s^{2\nu+1}\, ds,$$

and consequently

$$(x_1 x_2 x_3)^{\frac{1}{2}\nu} I(x_1, x_2, x_3) \overset{\cdots}{\underset{\cdots}{=}} \frac{\Pi(\nu)}{2(p_1 p_2 p_3)^\nu \left(\dfrac{1}{p_1} + \dfrac{1}{p_2} + \dfrac{1}{p_3}\right)^{\nu+1}}$$

$$\operatorname{Re} p_{1,2,3} > 0 \quad (\operatorname{Re} \nu > -1). \tag{15}$$

From this result the value of the integral I can be calculated by a successive transposition of the variables p_1, p_2, p_3 at the right. The first transposition, with the aid of (III, 3) and the attenuation rule, then yields

$$(x_1 x_2 x_3)^{\frac{1}{2}\nu} I(x_1, x_2, x_3) \overset{\cdot\cdot}{\underset{\cdot\cdot}{=}} \frac{x_1^\nu}{2} \frac{p_2 p_3}{(p_2 + p_3)^{\nu+1}} \exp\left(-\frac{p_2 p_3}{p_2 + p_3} x_1\right) U(x_1) \quad (\operatorname{Re} p_{2,3} > 0).$$

Secondly, p_2 is transposed by use of (X, 26) together with the attenuation rule. The new result is found to be

$$x_3^{\frac{1}{2}\nu} I(x_1, x_2, x_3) \overset{\cdot}{\underset{\cdot}{=}} \frac{e^{-(x_1+x_2)p_3}}{2p_3^{\nu-1}} I_\nu\{2p_3\sqrt{(x_1 x_2)}\}\, U(x_1)\, U(x_2) \qquad (\operatorname{Re} p_3 > 0).$$

Finally, the relation (X, 28) enables us to determine the original with respect to p_3 if $\operatorname{Re} \nu > -\frac{1}{2}$; the integral in question proves to be

$$I = \frac{\{4x_1 x_2 - (x_1 + x_2 - x_3)^2\}^{\nu-\frac{1}{2}}}{4^{\nu+\frac{1}{2}} \sqrt{\pi}\, \Pi(\nu - \frac{1}{2})\, (x_1 x_2 x_3)^{\frac{1}{2}\nu}} U\{4x_1 x_2 - (x_1 + x_2 - x_3)^2\}\, U(x_1)\, U(x_2). \tag{16}$$

†See also O. Heaviside, *Electromagnetic Theory,* vol. III, 243.

This expression, which has to be symmetric in all three variables x_1, x_2, x_3, can be simplified considerably when the following function is introduced:

$$\Delta(a,b,c)\begin{cases} = \text{area of the triangle formed, if possible, with the sides } a,\, b,\, c; \text{ this area is understood to be positive.} \\ = 0 \text{ when the triangle degenerates into a line.} \\ = \text{negative in al remaining cases.}\end{cases}$$

This function satisfies

$$\Delta(a,b,c)\, U(\Delta) = \tfrac{1}{2}\sqrt{\{a^2b^2 - \tfrac{1}{4}(a^2+b^2-c^2)^2\}}\, U(a+b-c)\, U(b+c-a)\, U(c+a-b). \quad (17)$$

In terms of this function Δ the answer given by (16) is easily transformed into the simple result asked for:

$$\int_0^\infty J_\nu(as)\, J_\nu(bs)\, J_\nu(cs) \frac{ds}{s^{\nu-1}} = \frac{2^{\nu-1}}{\sqrt{\pi}\,\Pi(\nu-\frac{1}{2})} \frac{\{\Delta(a,b,c)\}^{2\nu-1}}{(abc)^\nu} U(\Delta) \qquad (\operatorname{Re}\nu > -\tfrac{1}{2}),$$

and at the same time the operational relation (15) becomes

$$\frac{\pi}{(p_1p_2p_3)^\nu \left(\dfrac{1}{p_1}+\dfrac{1}{p_2}+\dfrac{1}{p_3}\right)^{\nu+1}} \doteqdot \frac{\{\Delta(2\sqrt{x_1}, 2\sqrt{x_2}, 2\sqrt{x_3})\}^{2\nu-1}}{\Pi(2\nu)} U(\Delta), \quad \operatorname{Re} p_{1,2,3} > 0 \ (\operatorname{Re}\nu > -\tfrac{1}{2}). \quad (18)$$

Particularly interesting is the special case for $\nu = 0$, namely,

$$\frac{\pi}{\dfrac{1}{p_1}+\dfrac{1}{p_2}+\dfrac{1}{p_3}} \doteqdot \frac{U(\Delta)}{\Delta(2\sqrt{x_1}, 2\sqrt{x_2}, 2\sqrt{x_3})} \qquad \operatorname{Re} p_{1,2,3} > 0, \quad (19)$$

to which we may add a simple relation for the angles ϕ_1, ϕ_2, ϕ_3, of the triangle here considered; namely, from

$$\frac{\partial\phi_1}{\partial x_1} = \frac{1}{\Delta(2\sqrt{x_1}, 2\sqrt{x_2}, 2\sqrt{x_3})}, \quad (20)$$

together with the integration rule applied to (19) with respect to x_1, it follows that

$$\phi_1 = \arccos\left(\frac{x_2+x_3-x_1}{2\sqrt{(x_2x_3)}}\right) \doteqdot \frac{\pi}{p_1\left(\dfrac{1}{p_1}+\dfrac{1}{p_2}+\dfrac{1}{p_3}\right)}, \qquad \operatorname{Re} p_{1,2,3} > 0. \quad (21)$$

It is striking that this analytically complicated function, though elementary from the point of view of geometry, has such a simple operational image. Moreover, when (21) and its analogues for ϕ_2 and ϕ_3 are added together, then the resulting relation is equivalent to the property that the sum of the angles of the triangle is equal to π.

Example 2. *The problem of random flights.* Many problems, such as those occurring in Brownian movement, the spreading of plants and animals from a centre, the walk of a drunken man, can be idealized by the following problem of probabilities: To determine the probability $w_n(x,y)\,dx\,dy$ that some person shall reach by n steps the neighbourhood $(dx\,dy)$ of the point $P(x,y)$ when starting from a fixed point O; provisionally it is supposed that the n successive steps of lengths $a_1, a_2, \ldots, a_n$ are performed in a plane, while the direction of any step is completely at random, all directions being equally probable (see fig. 89).

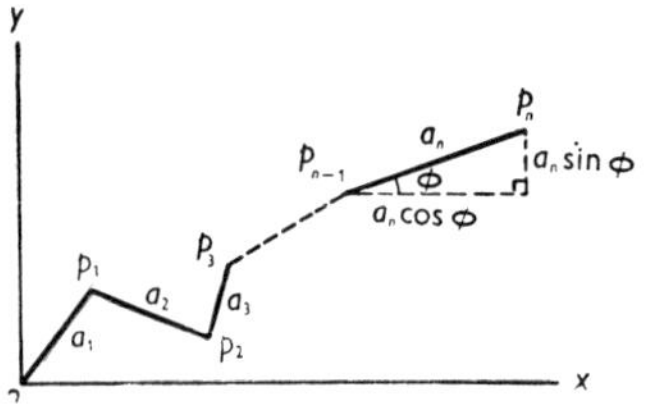

Fig. 89. Illustrating the theory of random flights.

The function in question satisfies the recurrence relation

$$w_n(x,y) = \frac{1}{2\pi}\int_0^{2\pi} w_{n-1}(x-\alpha_n\cos\phi, y-a_n\sin\phi)\,d\phi, \tag{22}$$

which expresses that, if the direction of the last step is at the angle ϕ with the x-axis, this last step has begun in the neighbourhood $dx\,dy$ of the point P_{n-1} with co-ordinates

$$x-a_n\cos\phi,\ y-a_n\sin\phi,$$

where all angles ϕ lying between 0 and 2π are equally probable.

Let us first determine the image, $f_n(p,q)$, of the function $w_n(x,y)$. For this purpose we transpose the integrand of (22) and apply the shift rule. Then the following recurrence relation for the image function is obtained:

$$f_n(p,q) = \frac{1}{2\pi}\int_0^{2\pi} f_{n-1}(p,q)\,e^{-a_n(p\cos\phi+q\sin\phi)}\,d\phi,$$

whence it further follows that

$$f_n(p,q) = I_0\{a_n\sqrt{(p^2+q^2)}\}\,f_{n-1}(p,q). \tag{23}$$

Obviously $w_0(x,y) = \delta(x)\,\delta(y)$, since after zero steps no other point than 0 itself can be reached, and it must be reached with certainty (normalization!). Therefore $f_0(p,q) = pq$. Upon applying (23) n times in succession we easily arrive at the complete image; thus

$$w_n(x,y) \doteqdot pq I_0\{a_1\sqrt{(p^2+q^2)}\}\,I_0\{a_2\sqrt{(p^2+q^2)}\}\ldots I_0\{a_n\sqrt{(p^2+q^2)}\}, \quad -\infty<\operatorname{Re}p,q<\infty. \tag{24}$$

The relation (24) converges for all values of p and q, as w_n is certainly zero for

$$x^2+y^2>(a_1+a_2+\ldots+a_n)^2,$$

since after n steps the distance $PO = \rho = \sqrt{(x^2+y^2)}$ can never be greater than $a_1+a_2+\ldots+a_n$.

In order to obtain an explicit expression for $w_n(x,y)$ itself, consider the inversion integral (6) corresponding to (24). By taking $c_1 = c_2 = 0$, and substituting $p = i\omega_1$ and $q = i\omega_2$, we obtain

$$w_n(x,y) = \frac{1}{4\pi^2}\int_{-\infty}^{\infty} d\omega_1 \int_{-\infty}^{\infty} d\omega_2\, e^{i(\omega_1 x+\omega_2 y)}\, J_0\{a_1\sqrt{(\omega_1^2+\omega_2^2)}\}\ldots J_0\{a_n\sqrt{(\omega_1^2+\omega_2^2)}\}.$$

This formula may be simplified by introduction of polar co-ordinates, both for the variables of integration ($\omega_1 = s\cos\psi$, $\omega_2 = s\sin\psi$) and the co-ordinates of P ($x = \rho\cos\phi$, $y = \rho\sin\phi$). Upon performing the integration with respect to ψ we then get the formula

$$w_n(x,y) = \frac{1}{2\pi}\int_0^{\infty} J_0(\rho s)\,J_0(a_1 s)\,J_0(a_2 s)\ldots J_0(a_n s)\,s\,ds, \tag{25}$$

which clearly shows the radial symmetry of the function w_n. In view of this symmetry it is natural to consider the probability function $G_n(R)$ describing the chance that, after n steps, a distance smaller than R is covered. This function is given by

$$G_n(R) = 2\pi\int_0^R w_n(\rho)\,\rho\,d\rho, \tag{26}$$

in which the integration with respect to ρ can be carried out with the aid of the well-known result

$$\int_0^x J_0(u)\,u\,du = xJ_1(x).$$

This leads to the following integral representation of $G_n(R)$:

$$G_n(R) = R\int_0^\infty J_0(a_1 s)\, J_0(a_2 s) \dots J_0(a_n s)\, J_1(Rs)\, ds.$$

When, in particular, all steps are of equal length: $a_1 = a_2 = \dots = a_n = a$, and R is taken equal to a, then we obtain

$$G_n(a) = a\int_0^\infty \{J_0(as)\}^n J_1(as)\, ds = -\frac{\{J_0(as)\}^{n+1}}{n+1}\bigg|_0^\infty = \frac{1}{n+1}, \tag{27}$$

which is a well-known result of Kluyver†, expressing that $1/(n+1)$ is the probability that, after n equal steps, a distance not greater than the length of one step is covered.

In considering once more the original probability function, it proves that the determination of the function w_n is most simple when from (25) a new relation is derived, containing instead of x and y the n operational variables

$$x_0 = \frac{\rho^2}{4}, \quad x_1 = \frac{a_1^2}{4}, \quad \dots, \quad x_n = \frac{a_n^2}{4}, \tag{28}$$

of which the corresponding image variables will be denoted by $p_0, p_1, \dots, p_n$. Then the new variables, x_j, occur separately in the integrand of (25). Thus a transposition with the aid of (x, 20) becomes possible, in a manner as indicated for the integral (14) of the preceding example. Proceeding in this way we obtain the following new relation:

$$4\pi w_n \;\risingdotseq\; \frac{1}{\dfrac{1}{p_0}+\dfrac{1}{p_1}+\dots+\dfrac{1}{p_n}} \qquad \mathrm{Re}\, p_{0,1,\dots,n} > 0, \tag{29}$$

from which, by successive transpositions of the p's, there follow expressions that have necessarily to be equal to (25); moreover, the inversion integral of (29) yields an expression for W_n free of Bessel functions. The lowest value of n not leading to some trivial result for (29) is $n = 2$; thus, from (29), (19) and (28),

$$w_2(\rho) = \frac{1}{4\pi^2}\frac{U(\Delta)}{\Delta(\rho, a_1, a_2)}. \tag{30}$$

It further follows from (26), in combination with (30) and (20), that

$$G_2(\rho) = \frac{U(\Delta)}{\pi}\phi_0(\rho, a_1, a_2), \tag{31}$$

this being the probability function describing the chance that a distance not greater than ρ is reached in two steps; ϕ_0 denotes the angle opposite the side ρ of the triangle. In contrast with this simple result for w_2, the function w_3 involves integrals of the elliptic type, not to mention the higher-order probability functions.

It is also possible to solve the problem of random flights in more than two, ν say, dimensions. For odd values of ν the results are expressible in elementary, though perhaps complicated, functions. Proceeding in a manner similar to that described above, we find for arbitrary values of ν,

$$w_n^{(\nu)} = \frac{2}{(4\pi)^{\frac{1}{2}\nu}}\{\Gamma(\tfrac{1}{2}\nu)\}^n \int_0^\infty \frac{J_{\frac{1}{2}\nu-1}(rs)}{(\frac{1}{2}rs)^{\frac{1}{2}\nu-1}} \prod_{k=1}^{n} \frac{J_{\frac{1}{2}\nu-1}(a_k s)}{(\frac{1}{2}a_k s)^{\frac{1}{2}\nu-1}}\, s^{\nu-1}\, ds,$$

which is obviously a generalization of (25). For odd values of ν the Bessel functions reduce to elementary functions. Further, in this case of odd ν, the integrand is even in s; thus one may replace

$$\int_0^\infty \quad \text{by} \quad \tfrac{1}{2}\int_{-\infty}^\infty,$$

† *Proc. K. Akad. Wet. Amst.* VIII, 341, 1906.

and then consider the new integral as the value at $t = 0$ of some inversion integral, when first s is put equal to ip. For instance, when $\nu = 3$ it is found that w_n is the value of the original (at $t = 0$) of the function

$$-\frac{p^3}{2\pi}\frac{\sinh(a_1 p)}{a_1 p}\cdots\frac{\sinh(a_n p)}{a_n p}\frac{\sinh(rp)}{rp}. \tag{32}$$

At the same time (32) clearly reveals the connexion of our problem with the problem of moving averages, as is seen from (XIV, 13) when we take $a_1 = a_2 = \ldots = a_n = r$.

The complete original of (32) can be obtained in two different ways. First, we transform the image into a sum of exponential terms each divided by p^{n-2}, and transpose term by term. Secondly, we start by transposing the factor $1/p^{n-2}$ and account for the successive hyperbolic-sine functions by applying the shift rule $(2n+1)$ times (twice for each factor). We thus obtain two distinct expressions for the original. If in both t is equated to zero the following strikingly simple solutions for the problem of random flights in three dimensions are found, respectively:

$$w_n^{(3)} = \frac{\sum\limits_{s}(\pm a_1 \pm a_2 \pm \ldots \pm a_n \pm r)^{n-2}(-1)^{\lambda(s)+1}\,U(\pm a_1 \pm a_2 \pm \ldots \pm a_n \pm r)}{2^{n+2}(n-2)!\,\pi a_1 a_2 \ldots a_n r},$$

$$w_n^{(3)} = -\frac{[\Delta_{a_1}\Delta_{a_2}\ldots\Delta_{a_n}\Delta_r\{t^{n-2}U(t)\}]_{t=0}}{16\pi(n-2)!}.$$

In the first expression the summation should be carried out over all possible combinations of signs (s), $\lambda(s)$ representing thereby the total number of minus signs in the separate terms. In the second expression the delta symbols are short notations for certain symmetric difference quotients; for instance, Δ_{a_2} means

$$\Delta_{a_2}\phi(t) = \frac{\phi(t+\frac{1}{2}a_2)-\phi(t-\frac{1}{2}a_2)}{a_2}.$$

To compare (30), for two steps in two dimensions, with the probability function $w_2^{(3)}$ for two steps in three dimensions, we take $n = 2$ in the formulae above so as to obtain

$$w_2^{(3)}(r) = \frac{1}{8\pi}\frac{1}{a_1 a_2 r},$$

provided a_1, a_2, r can be the lengths of the sides of a real triangle; otherwise the function is zero. Moreover, the analogue of (31) is the probability that after two steps in the three-dimensional space a point within the distance R is reached. This probability is different from zero only if

$$|a_1 - a_2| < R < a_1 + a_2;$$

its value is then given by

$$\frac{R^2 - |a_1 - a_2|^2}{4a_1 a_2}.$$

Finally, the analogue of (27) in three dimensions: that is, the probability that after n steps the drunken man is not farther from his starting-point than corresponds to a single step is

$$\frac{1}{\pi(n+1)}\int_{-\infty}^{\infty}\left(\frac{\sin u}{u}\right)^{n+1} du = \frac{1}{(n+1)!}\Delta^{n+1}\{t^n U(t)\}_{t=0} = \sum_{k=0}^{[\frac{1}{2}(n+1)]}\frac{(-1)^k}{k!}\frac{\{\frac{1}{2}(n+1)-k\}^n}{(n+1-k)!},$$

Δ being the symmetric difference operator (XIII, 1).

It is of particular importance to note that the theory above includes the special case of the theory of Brownian movement, where all the steps may be taken equal: $a_1 = a_2 = \ldots = a_n = a$. Then a represents the mean free path covered by some molecule between two successive collisions, and n is proportional to the time of observation. Accordingly, the probability for the position of a molecule with reference to some initial position $x = y = 0$ at $t = 0$ in two dimensions after n collisions follows from (24); thus

$$w_n(x, y) \doteqdot pq \left(I_0\{a\sqrt{(p^2+q^2)}\} \right)^n, \quad -\infty < \operatorname{Re} p,\ q < \infty.$$

Just as in the problem of moving averages (see XIV, end of § 2), involving the function D_n, an approximation for large values of n can be obtained. To this end we approximate the image by replacing I_0 by the first two terms of its power-series expansion. Then, wholly analogous to (XIV, 22), it follows that

$$n\,w_n(\sqrt{n}\,x, \sqrt{n}\,y) \doteqdot pq \left\{1 + \frac{a^2(p^2+q^2)}{4n} + \ldots \right\}^n.$$

Consequently, for large values of n the image may be approximated by

$$pq\,e^{\frac{1}{4}a^2(p^2+q^2)},$$

the original of which leads to an asymptotic expression for w_n itself, namely,

$$w_n(x, y) \sim \frac{e^{-(x^2+y^2)/a^2 n}}{\pi a^2 n} \quad (n \to \infty). \tag{33}$$

We have thus found a continuous approximation for the function w_n; it is to be borne in mind, however, that this approximation is not identically zero in the range $\rho = \sqrt{(x^2+y^2)} > na$, unlike the true function w_n. It is further to be remarked that (33) is a solution of the diffusion equation

$$\left(\frac{\partial^2}{\partial x^2} + \frac{\partial^2}{\partial y^2} - \frac{4}{a^2}\frac{\partial}{\partial n}\right) w_n = 0,$$

which, in this connexion, represents an approximation of the rigorous equation of the problem of random flights:

$$\left(I_0\left\{a\sqrt{\left(\frac{\partial^2}{\partial x^2} + \frac{\partial^2}{\partial y^2}\right)}\right\} - 1 - \Delta_n\right) w_n = 0,$$

if Δ_n is understood to be the asymmetric operator defined in (XIII, 4).

3. Second-order differential equations of the hyperbolic type with constant coefficients and two variables

The general form of these differential equations is given by

$$a_{11}\frac{\partial^2 u}{\partial x^2} + 2a_{12}\frac{\partial^2 u}{\partial x\,\partial y} + a_{22}\frac{\partial^2 u}{\partial y^2} + 2a_{13}\frac{\partial u}{\partial x} + 2a_{23}\frac{\partial u}{\partial y} + a_{33}u = -\varphi, \tag{34}$$

in which the coefficients satisfy an additional condition, viz.

$$a_{12}^2 - a_{11}a_{22} > 0.$$

The simplest standard form into which this equation may be transformed by a linear transformation of x and y is the following:

$$\frac{\partial^2 u}{\partial x\,\partial y} + \alpha u = -\varphi. \tag{35}$$

As an example showing the reduction to the standard form, let us consider the equation for the quasi-stationary coaxial cable, which, according to the original of (xv, 33), is given by

$$\frac{\partial^2 i}{\partial x^2}-\bar{L}\bar{C}\frac{\partial^2 i}{\partial t^2}-\left(\frac{\bar{L}}{\bar{R}_q}+\bar{C}\bar{R}_s\right)\frac{\partial i}{\partial t}-\frac{\bar{R}_s}{\bar{R}_q}i=0.$$

First we have, upon the substitution $i = e^{-\mu_2 t} i^*$, the equation

$$\frac{\partial^2 i^*}{\partial x^2}-\frac{1}{c_0^2}\left(\frac{\partial^2}{\partial t^2}-\mu_1^2\right)i^*=0,$$

in which c_0, μ_1 and μ_2 are as defined in (xv, 44). Then, by a second transformation, $\xi = x+c_0 t$, $\eta = x-c_0 t$, the resulting equation becomes

$$4\frac{\partial^2 i^*}{\partial\xi\,\partial\eta}+\frac{\mu_1^2}{c_0^2}i^*=0,$$

which is obviously of the required form (35).

As we already know from (xv, 4), a particular solution of (35) is expressible in terms of a Green function $G_0(x,y)$ which, under certain boundary conditions, satisfies the equation

$$\frac{\partial^2 G_0}{\partial x\,\partial y}+\alpha G_0=-\delta(x)\,\delta(y). \tag{36}$$

Therefore it will be sufficient to confine ourselves to the special equation (36); moreover, for the present we shall take α to be positive.

Now, when we put

$$G_0(x,y) \doteqdot f_0(p,q),$$

then (36) is transposed into

$$(pq+\alpha)\,f_0=-pq,$$

from which the image of the Green function follows at once, viz.

$$G_0(x,y) \doteqdot -\frac{pq}{pq+\alpha}. \tag{37}$$

To this image there correspond different originals G_0 depending on the particular domain of convergence chosen; once having obtained these different functions G_0 we may investigate afterwards what boundary conditions they fulfil. Let us first transpose the variable p in (37); then we get the two following operational relations with respect to q, y:

$$G_0(x,y) \doteq \begin{cases} -e^{-\alpha x/q}\,U(x), & \operatorname{Re}(p+\alpha/q)>0, \\ e^{-\alpha x/q}\,U(-x), & \operatorname{Re}(p+\alpha/q)<0. \end{cases} \tag{38}$$

The transposition of the remaining variable q can be achieved by using the relations (x, 20) and (x, 26) for $\nu = 0$, which for this purpose may be summarized as follows:

$$e^{-a/q} \doteqdot \begin{cases} J_0\{2\sqrt{(ay)}\}\, U(y), & \operatorname{Re} q > 0, \\ -J_0\{2\sqrt{(ay)}\}\, U(-y), & \operatorname{Re} q < 0. \end{cases} \tag{39}$$

In these formulae a may be positive as well as negative, whilst the function J_0 may be replaced by $I_0(2\sqrt{|ay|})$ when ay is negative. Since either of the operational relations (38) can be associated with both of (39), a total number of four different originals of (37) is found:

$$-\frac{pq}{pq+\alpha} \doteqdot \begin{cases} -J_0\{2\sqrt{(\alpha xy)}\}\, U(x)\, U(y), & \operatorname{Re}(p+\alpha/q) > 0;\ \operatorname{Re} q > 0, \\ J_0\{2\sqrt{(\alpha xy)}\}\, U(x)\, U(-y), & \operatorname{Re}(p+\alpha/q) > 0;\ \operatorname{Re} q < 0, \\ J_0\{2\sqrt{(\alpha xy)}\}\, U(-x)\, U(y), & \operatorname{Re}(p+\alpha/q) < 0;\ \operatorname{Re} q > 0, \\ -J_0\{2\sqrt{(\alpha xy)}\}\, U(-x)\, U(-y), & \operatorname{Re}(p+\alpha/q) < 0;\ \operatorname{Re} q < 0, \end{cases} \tag{40}$$

the domains of convergence of which are adjacent so as to include all values of p and q.

We have thus obtained four different solutions of our differential equation (36), from which we may derive as many solutions of the general equation (34) by linear transformation and by (xv, 4). Instead of applying the transformation mentioned, the general equation (34) may be treated operationally in a more direct manner, whereby the corresponding Green function is determined with the aid of the rotation rule (7) applied to (40). We would illustrate this method by the equation considered in (xv, 3) describing the propagation of waves in a one-dimensional, anisotropic space with dispersion; in this case the Green function $G_1(x,t)$, corresponding to a source at $x = t = 0$, satisfies

$$\left(c_1\frac{\partial}{\partial x}+\frac{\partial}{\partial t}\right)\left(c_2\frac{\partial}{\partial x}-\frac{\partial}{\partial t}\right)G_1+\alpha G_1 = -\delta(x)\,\delta(t), \tag{41}$$

in which the variable t plays the same role as the former y. By transposition of equation (41) the image of the Green function can be obtained. Accordingly,

$$G_1(x,t) \doteqdot -\frac{pq}{(c_1p+q)\,(c_2p-q)+\alpha}.$$

The original of the image function at the right can be determined by applying the transformation $p' = c_1p+q$, $q' = c_2p-q$ to (40) and using (7 *b*). First of all, however, to make the results as surveyable as possible, we shall introduce some auxiliary quantities. Let us put therefore

$$d_1 = \frac{c_1t-x}{\sqrt{(1+c_1^2)}},\quad d_2 = \frac{c_2t+x}{\sqrt{(1+c_2^2)}},\quad e = \frac{2\sqrt{\alpha}}{c_1+c_2}\sqrt[4]{\{(1+c_1^2)\,(1+c_2^2)\}}, \tag{42}$$

the meaning of which will be discussed in due course. Then, again, four different solutions are found which may be conveniently arranged as follows ($\alpha > 0$):

$$-\frac{(c_1+c_2)pq}{(c_1p+q)(c_2p-q)+\alpha} \doteqdot \begin{cases} -J_0\{e\sqrt{(-d_1d_2)}\}\,U(-d_1)\,U(d_2) = G_1^{(1)}(x,t), & \text{I} \\ I_0\{e\sqrt{(d_1d_2)}\}\,U(d_1)\,U(d_2) = G_1^{(2)}(x,t), & \text{II} \\ I_0\{e\sqrt{(d_1d_2)}\}\,U(-d_1)\,U(-d_2) = G_1^{(3)}(x,t), & \text{III} \\ -J_0\{e\sqrt{(-d_1d_2)}\}\,U(d_1)\,U(-d_2) = G_1^{(4)}(x,t). & \text{IV} \end{cases} \tag{43}$$

From elementary geometry it is obvious that d_1 and d_2 as defined in (42) are equal to the distances from the point $P(x,t)$ to the lines

$$A_1 \equiv x - c_1 t = 0, \quad A_2 \equiv x + c_2 t = 0,$$

respectively, all lying in the plane of x, t (see fig. 90). Either of these straight lines is perpendicular to one of the two asymptotes of the hyperbola with equation

$$(c_1 x + t)(c_2 x - t) + \alpha = 0.$$

Moreover, the parameter e is the linear eccentricity (distance from centre to focus) of the hyperbola. Further, when p and q are restricted to real numbers, the two different branches and the asymptotes of the hyperbola are just the boundaries of the four domains of convergence, I, II, III and IV, corresponding to the expressions (43), in such a way as may be seen from fig. 91.

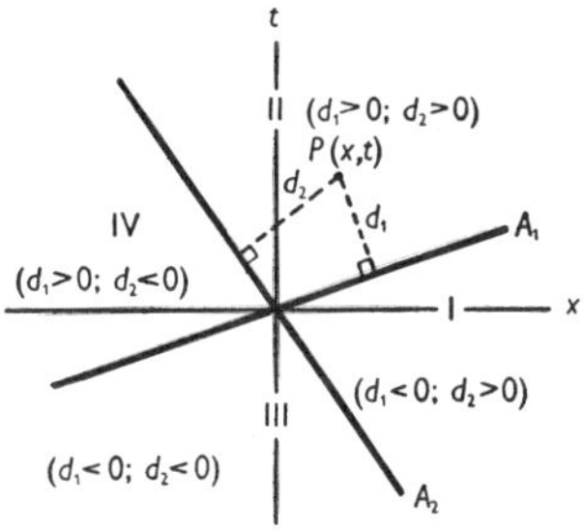

Fig. 90. Domains of non-vanishing values of the originals of (43).

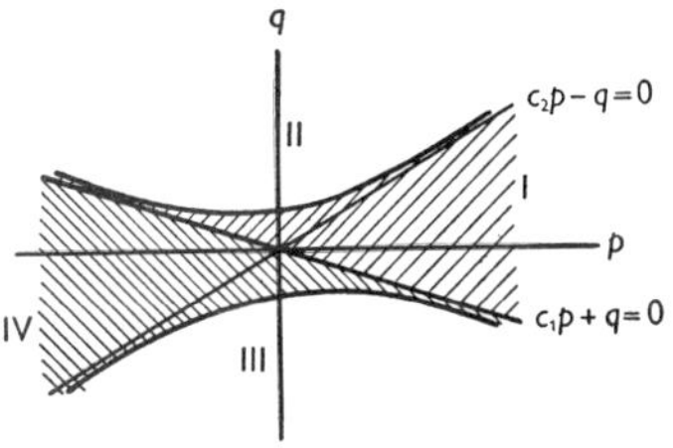

Fig. 91. Domains of convergence for the originals of (43) for real p and q.

Either of the four solutions of equation (41) that are obtained by dividing (43) by c_1+c_2 is characterized by certain boundary conditions. This is most clearly indicated by the unit functions in consequence of which $G_1^{(1)}$, e.g., is non-identically zero only inside the sector labelled I in fig. 90. As to physical applications, the function $G_1^{(2)}$ is the most important, since it belongs to the sector II for which we have $d_1 > 0$, $d_2 > 0$ or

$$-c_2 t < x < c_1 t.$$

These inequalities include those points of space which at the moment t could have been reached by a wave that, starting at $x = t = 0$, propagates itself in the positive ($x \to \infty$) direction at a velocity c_1 and in the negative direction at a velocity c_2. Consequently, the function $G_1^{(2)}$ describes a system that is initially ($t < 0$) at rest and suddenly excited at the moment $t = 0$ at the point $x = 0$.

On the other hand, $G_1^{(3)}$ is characteristic for a system that, while in action before $t = 0$, at $t = 0$ is suddenly hit by external forces restoring the situation of complete rest for $t > 0$. The functions $G_1^{(1)}$ and $G_1^{(4)}$ may be interpreted similarly; for instance, $G_1^{(1)}$ corresponds to the situation when the system in the past ($t < 0$) was at rest only for $x < 0$.

The lines A_1 and A_2 of fig. 90, along which any of the four Green functions shows discontinuities, are called the *characteristics* corresponding to $x = t = 0$, this being in accordance with the conventional terminology of the general theory of partial differential equations. Generally, in the theory of differential equations of order n, the characteristics are curves or surfaces along which a solution may possess discontinuous derivatives of orders up to n, the lower-order derivatives being continuous throughout. The importance of the conception of characteristics appears automatically when dealing with operational methods.

In deriving the four different Green functions (43) we tacitly supposed α to be positive (anomalous dispersion). In physical applications, however, α may be negative as well (normal dispersion). In the latter case e becomes imaginary and the functions J_0 and I_0 must replace each other, and, moreover, the domains of convergence shown in fig. 91 have to be altered. In order to understand the practical significance of the algebraic sign of α better, we shall now consider the function $G_1^{(2)}$ for a one-dimensional isotropic medium in the presence of dispersion. We start with the following equation (see (xv, 1)):

$$\frac{\partial^2 u}{\partial x^2} - au - 2b\frac{\partial u}{\partial t} - \frac{1}{c^2}\frac{\partial^2 u}{\partial t^2} = -\delta(x)\,\delta(t). \tag{44}$$

By the substitution $u = e^{-bc^2 t} G$ this equation is transformed into the standard type (41):

$$\frac{\partial^2 G}{\partial x^2} - \frac{1}{c^2}\frac{\partial^2 G}{\partial t^2} + \beta G = -\delta(x)\,\delta(t),$$

in which $\beta = b^2c^2 - a$. In determining $G^{(2)}$ according to (43) we have to distinguish between three cases; these are

(1) $\beta > 0$ (anomalous dispersion):

$$-\frac{pq}{p^2 - q^2/c^2 + \beta} \doteqdot \frac{c}{2} I_0(\sqrt{\{\beta(c^2t^2 - x^2)\}})\, U(ct - |x|)$$

$$\operatorname{Re} q > c\,|\operatorname{Re}\sqrt{(p^2+\beta)}|; \tag{45a}$$

(2) $\beta = 0$ (no dispersion at all):

$$-\frac{pq}{p^2-q^2/c^2} \fallingdotseq \frac{c}{2}\,U(ct-|x|), \qquad \operatorname{Re} q > c\,|\operatorname{Re} p|; \tag{45b}$$

(3) $\beta = -\beta' < 0$ (normal dispersion):

$$-\frac{pq}{p^2-q^2/c^2-\beta'} \fallingdotseq \frac{c}{2} J_0(\sqrt{\{\beta'(c^2t^2-x^2)\}})\,U(ct-|x|)$$
$$\operatorname{Re} q > c\,|\operatorname{Re}\sqrt{(p^2-\beta')}\,|. \tag{45c}$$

The domains of convergence belonging to these operational relations are most easily obtained when, starting with the original, one transposes first x and then t. It is seen from (45) that a disturbance located at $x = 0$ at the time $t = 0$ will become perceptible at the point x not earlier than at the moment $t = |x|/c$; after the occurrence of the discontinuity at the point under consideration, there remains the after-effect, which is constant only if dispersion is absent. The after-effect function $u^{(2)}$, consisting of the original of (45) multiplied by e^{-bc^2t}, is easily approximated when $t \gg |x|/c$ by using well-known asymptotic expressions for Bessel functions. The results are found to be:

$$\beta > 0: \qquad u^{(2)} \sim \frac{\sqrt{c}}{2\sqrt[4]{(b^2c^2-a)}}\,\frac{e^{-ct\{bc-\sqrt{(b^2c^2-a)}\}}}{\sqrt{(2\pi t)}},$$

$$\beta < 0: \qquad u^{(2)} \sim \frac{\sqrt{c}}{\sqrt[4]{(a-b^2c^2)}}\,\frac{e^{-bc^2t}\cos\{c\sqrt{(a-b^2c^2)}\,t-\frac{1}{4}\pi\}}{\sqrt{(2\pi t)}}, \tag{46}$$

showing that this Green function decreases aperiodically in the case of anomalous dispersion, and that it behaves as a damped oscillation in the case of normal dispersion (see fig. 92). When the functional values of $u^{(2)}$ in two different points, x_1 and x_2, are compared with each other, it is of course observed that the disturbance phenomenon presents itself first at x_1 and then at x_2, provided the latter is farther off the origin; but later on, after $t \gg |x_2|/c$, the values do not differ much. Therefore, in the case of normal dispersion, the zeros of $u^{(2)}(x_1, t)$ and $u^{(2)}(x_2, t)$ will finally coincide. On the other hand, it follows from the rigorous (45) that before then the two functions pass through zero an equal number of times. Consequently, the first few zeros of $u^{(2)}(x_2, t)$ are closer together

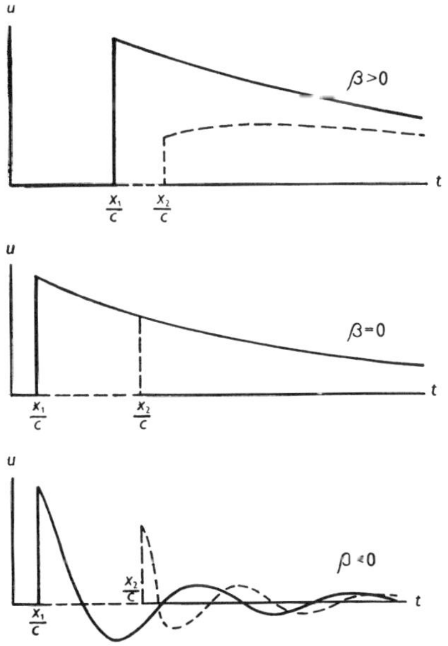

Fig. 92. After-effect function for different kinds of dispersion.

than those of $u^{(2)}(x_1, t)$. In the case of anomalous dispersion the function u proves to pass through a maximum if $|x| > 2bc/\beta$; an example of this is shown by the dashed curve of fig. 92.

Example 1. *Doppler effect for one-dimensional waves of sound in the presence of wind.* Let us imagine a moving point source of frequency $\omega/2\pi$ that starts at $t = 0$ with a constant velocity v to the right ($x = vt$). Let w be the wind velocity. The propagation of the sound waves emitted by the moving source is governed by the following differential equation:

$$\left\{(c+w)\frac{\partial}{\partial x}+\frac{\partial}{\partial t}\right\}\left\{(c-w)\frac{\partial}{\partial x}-\frac{\partial}{\partial t}\right\} u = -c^2\, e^{i\omega t}\delta(x-vt)\, U(t), \tag{47}$$

if the strength of the point source is suitably normalized. As before, we put

$$u(x,t) \mathrel{\ddot{=}} f(p,q).$$

In order to transpose (47) operationally we first determine the image of the right-hand member of (47). This is done in the order x, t as follows:

$$e^{i\omega t}\delta(x-vt)\, U(t) \doteq e^{i\omega t - vpt} p\, U(t) \mathrel{\ddot{=}} \frac{pq}{q+vp-i\omega} \qquad \mathrm{Re}\,(q+vp) > 0. \tag{48}$$

Likewise transposing the left-hand side of (47), we get for the image of the function in question, after a simple division,

$$u(x,t) \mathrel{\ddot{=}} -c^2 \frac{pq}{\{(c+w)\,p+q\}\{(c-w)\,p-q\}} \frac{1}{q+vp-i\omega} \qquad \mathrm{Re}\,(q+vp) > 0.$$

The original of the function at the right might be obtained as a composition product of two functions of two independent variables, but it is far more simple to use the expansion of the image in rational fractions. For instance, let us first transpose the variable q; the image is rewritten in the following way:

$$u(x,t) \mathrel{\ddot{=}} \frac{c}{2\{(c+w-v)\,p+i\omega\}} \frac{q}{q+(c+w)\,p} + \frac{c}{2\{(c+v-w)\,p-i\omega\}} \frac{q}{q-(c-w)\,p} + \frac{c^2 p}{\{(i\omega+wp-vp)^2-c^2p^2\}} \frac{q}{q+vp-i\omega}.$$

Performing the transposition of q for the strip of convergence extending to $q = +\infty$ (this corresponds to a solution for which the system is at rest before $t = 0$), we easily obtain

$$u \doteq \frac{c}{2} U(t) \left\{ \frac{e^{-(c+w)\,pt}}{(c+w-v)\,p+i\omega} + \frac{e^{(c-w)\,pt}}{(c+v-w)\,p-i\omega} + \frac{e^{i\omega t}}{i\omega}\left(\frac{p}{p+\dfrac{i\omega}{c+w-v}} - \frac{p}{p-\dfrac{i\omega}{c+v-w}} \right) e^{-vtp} \right\}.$$

A second transposition is still necessary, namely, that of p which is the operational variable corresponding to x. We again take the original corresponding to the strip extending towards $p = +\infty$, so as to get the solution that is zero at $x = +\infty$. We write the final expression only:

$$u = \frac{ic}{2\omega} U(t) \Bigg[U(x+ct-wt) - U(x-ct-wt) + \{U(x-ct-wt)-U(x-vt)\} \exp\left(-\frac{i\omega(x-ct-wt)}{c+w-v}\right) - \{U(x+ct-wt)-U(x-vt)\} \exp\left(\frac{i\omega(x+ct-wt)}{c+v-w}\right) \Bigg]. \tag{49}$$

Owing to our systematic introducing of unit functions we have thus obtained a closed expression that is valid for all different cases, though the character for $v<c+w$ and that for $v>c+w$ is quite distinct. Of course, the solution can also be given without using unit functions; for instance, when $x>0$, $t>0$ and $w<c$, the real part of (49) is

(1) when $v<c+w$:

$$u = \begin{cases} 0 & t<\dfrac{x}{c+w}, \\ -\dfrac{c}{2\omega}\sin\left\{\dfrac{\omega(x-ct-wt)}{c+w-v}\right\} & \dfrac{x}{c+w}<t<\dfrac{x}{v}, \\ \dfrac{c}{2\omega}\sin\left\{\dfrac{\omega(x+ct-wt)}{c+v-w}\right\} & t>\dfrac{x}{v}; \end{cases} \tag{50a}$$

(2) when $v>c+w$:

$$u = \begin{cases} 0 & t<\dfrac{x}{v}, \\ \dfrac{c}{2\omega}\left[\sin\left\{\dfrac{\omega(x-ct-wt)}{c+w-v}\right\}+\sin\left\{\dfrac{\omega(x+ct-wt)}{c+v-w}\right\}\right] & \dfrac{x}{v}<t<\dfrac{x}{c+w}, \\ \dfrac{c}{2\omega}\sin\left\{\dfrac{w(x+ct-wt)}{c+v-w}\right\} & t>\dfrac{x}{c+w}. \end{cases} \tag{50b}$$

In the first case (velocity of moving source smaller than the velocity of sound in the direction of moving of the source), the oscillations at the point of observation start at the time $t = c/(x+w)$, this being the moment at which the first disturbance (transmitted by the source at the moment it begins to move) arrives at x. Moreover, these oscillations show a discontinuity at $t = x/v$, while the frequency changes from the value

$$\omega_1 = \frac{c+w}{c+w-v}\,\omega$$

to the value

$$\omega_2 = \frac{c-w}{c+v-w}\,\omega,$$

in agreement with Doppler's principle.

In the remaining case, when $v>c+w$, the oscillations start not earlier than $t = x/v$, this being the moment at which the point source itself comes along. Thereupon oscillations of both frequencies, ω_1 and ω_2, are excited, the former of which, however, is suppressed at the moment $t = x/(c+w)$ mentioned above.

Example 2. *Quasi-stationary theory of coaxial cables.* Part of this theory was given in xv, § 4, where only one of the two independent variables, x and t, was transposed. We shall now apply the simultaneous operational calculus, and put

$$i(x,t) \doteqdot I(p,q), \quad e(x,t) \doteqdot E(p,q).$$

First of all, we have to replace the variable p occurring in (xv, 31 and 32) by q, since now q, and not p, is the operational variable corresponding to t. Also transposing the remaining variable x contained in these formulae, we get the following new system of basic equations:

$$pE(p,q)+\bar{Z}_s(q)\,I(p,q) = 0, \tag{51a}$$

$$\bar{A}_q(q)\,E(p,q)+p\,I(p,q) = 0. \tag{51b}$$

It is to be remarked that, as in chapter VIII (end of § 2), for equations in one independent variable, the eigenfunctions of the problem under discussion may be found by solving these homogeneous equations; this now amounts to equating to zero the determinant of the system (51).

Let us exemplify the actual method of calculation in this field of simultaneous operational calculus by a somewhat more concrete problem. Let us determine the distribution of currents in an infinitely long cable extending from $-\infty$ to $+\infty$, in response to a disturbing electromotive force localized at $x = t = 0$; this force may be due, for instance, to induction by a neighbouring high-current conductor. The external force may be accounted for by taking an extra term in equation (51a) in addition to the term $pE \doteqdot \partial e/\partial x$, which represents the voltage drop in the axial direction per unit of length. For reasons of simplicity let us take this additional term to be equal to $-pq \doteqdot -\delta(x)\,\delta(t)$, which implies, *inter alia*, that the external force is of a certain definite strength. Then the system of equations (51) changes into

$$pE(p,q) + \bar{Z}_s(q)\,I(p,q) = pq, \qquad \bar{A}_q(q)\,E(p,q) + p\,I(p,q) = 0,$$

from which the image of the current in question readily follows:

$$i(x,t) \doteqdot \frac{pq\,\bar{A}_q\,q)}{\lambda^2(q) - p^2},$$

λ being defined as in (xv, 35). Further, by using the notations introduced before, the image may be written as in

$$i(x,t) \doteqdot -\bar{C}\frac{pq(q+\mu_2-\mu_1)}{p^2 - \dfrac{(q+\mu_2)^2}{c_0^2} + \dfrac{\mu_1^2}{c_0^2}},$$

which leads after some reduction, for instance with the aid of (45a), to the required current

$$i = \frac{1}{2}\sqrt{\left(\frac{C}{L}\right)}\,\mathrm{e}^{-\mu_2 t}\left(\frac{\partial}{\partial t} - \mu_1\right)\left[I_0\left\{\mu_1\sqrt{\left(t^2 - \frac{x^2}{c_0^2}\right)}\right\} U\left(t - \frac{|x|}{c_0}\right)\right].$$

This formula is in agreement with the expression (xv, 46) describing the current flowing in a line that extends from $x = 0$ to $x = +\infty$ (this being equivalent to a line of finite length matched to its characteristic impedance), in response to an impulse voltage applied at the left-hand end, $x = 0$ (except for the factor $\frac{1}{2}$, of course). This factor indicates that, whenever a two-sided infinite line is excited by the impulse voltage, one-half of the original disturbance travels to the right, the other to the left.

4. Hyperbolic differential equations of the second order in more than two independent variables

Of this type of differential equation, so far as equations with constant coefficients are concerned, the multi-dimensional wave equation is the most important in physical applications. In this case, too, it is sufficient to know the Green function G referring to a source at the origin at $t = 0$ in order to obtain a particular solution of the inhomogeneous equation. Assuming the simplest possible form of dispersion in an isotropic, n-dimensional medium we have for this Green function the differential equation mentioned in (xv, 6):

$$\left(\frac{\partial^2}{\partial x_1^2} + \frac{\partial^2}{\partial x_2^2} + \dots + \frac{\partial^2}{\partial x_n^2} + a - \frac{1}{c^2}\frac{\partial^2}{\partial t^2}\right) G_n = -\delta(x_1)\,\delta(x_2)\dots\delta(x_n)\,\delta(t).$$

By now transposing *all* variables operationally, putting

$$G_n(x_1, x_2, \dots, x_n, t) \doteqdot f_n(p_1, p_2, \dots, p_n, q),$$

we at once obtain

$$G_n \doteqdot -\frac{p_1 p_2 \dots p_n q}{p_1^2 + p_2^2 + \dots + p_n^2 + a - q^2/c^2}. \tag{52}$$

We now proceed to indicate how, from this operational relation, the functions G_n may be determined successively for an increasing number of dimensions n. Let us begin with the operational relation for G_1, which was given in (45) for the different cases of dispersion and which we may summarize as follows:

$$-\frac{p_1 q}{p_1^2-q^2/c^2+a} \doteq \frac{c}{2} I_0(\sqrt{\{a(c^2t^2-x_1^2)\}})\, U(ct-|x_1|) \qquad \operatorname{Re} q > c\,|\operatorname{Re}\sqrt{(p_1^2+a)}\,|. \tag{53}$$

It is to be remarked that the original has been taken for the strip of convergence with respect to q that is farthest to the right. This is necessary in order to get a system that is at rest for $t<0$ and that corresponds to a one-sided original.

The two-dimensional Green function G_2, given by (52), follows from (53) when the image is first multiplied by p_2 and a is replaced by $a+p_2^2$, and then the original with respect to p_2 is constructed. For this purpose use is made of the relation

$$p_2 I_0\{\sqrt{(p_2^2+a)}\} \doteq \frac{U(1-x_2^2)}{\pi}\frac{\cosh\{a\sqrt{(1-x_2^2)}\}}{\sqrt{(1-x_2^2)}} \qquad -\infty<\operatorname{Re} p_2<\infty, \tag{54}$$

which is readily deduced by transposing the integrand of

$$p_2 I_0\{\sqrt{(p_2^2+a^2)}\} = \frac{p_2}{2\pi}\int_0^{2\pi} e^{-p_2\cos\phi-a\sin\phi}\,d\phi$$

with the help of an impulse function. The final result is then found to be†

$$-\frac{p_1p_2q}{p_1^2+p_2^2-q^2/c^2+a} \doteqdot \frac{c}{2\pi}\frac{\cosh[\sqrt{\{a(c^2t^2-x_1^2-x_2^2)\}}]}{\sqrt{(c^2t^2-x_1^2-x_2^2)}}U\{ct-\sqrt{(x_1^2+x_2^2)}\}$$
$$\operatorname{Re} q>c\,|\operatorname{Re}\sqrt{(p_1^2+p_2^2+a)}\,|. \tag{55}$$

This relation might have been derived more directly, that is, from the now threefold definition integral. The original G_2 represents the propagation in two-dimensional space of a disturbance in response to the excitation of the origin at the moment $t=0$.

If there is no dispersion, we have the simple relation

$$-\frac{p_1p_2q}{p_1^2+p_2^2-q^2/c^2} \doteqdot \frac{1}{2\pi}\frac{U(t-\rho/c)}{\sqrt{(t^2-\rho^2/c^2)}} \qquad \operatorname{Re} q>c\,|\operatorname{Re}\sqrt{(p_1^2+p_2^2)}\,|, \tag{56}$$

in which $\rho=\sqrt{(x_1^2+x_2^2)}$. This function is observed, for instance, if an infinite ideal membrane is suddenly excited at some point O. The oscillation at a point at distance ρ from the centre of disturbance starts at the moment $t=\rho/c$ with an infinitely large amplitude and decreases, as in the one-dimensional case, the more rapidly the greater the distance from the centre.

† Henceforward the co-ordinates in two and three dimensions will be denoted by x_1, x_2, x_3, instead of by x, y, z; so as to indicate that they correspond to the operational variables p_1, p_2, p_3.

One can proceed in the same way, and determine G_3 by multiplying the image of (55) by p_3, replacing a by $a+p_3^2$, and deducing the original with respect to the latter variable. It is equally possible, however, to determine the functions G_n by transposing only the variable t in the original differential equation, as has been done in xv, § 5. The general result is different according as n is odd or even, thus:

$$\left.\begin{aligned}
&-\frac{p_1 p_2 \cdots p_n q}{p_1^2+p_2^2+\ldots+p_n^2-q^2/c^2+a}\\
&\qquad \risingdotseq \frac{c}{2}\left(-\frac{d}{2\pi r\,dr}\right)^{\frac{1}{2}(n-1)}[I_0(\sqrt{\{a(c^2t^2-r^2)\}})\,U(ct-r)] \quad (n \text{ odd});\\
&-\frac{p_1 p_2 \cdots p_n q}{p_1^2+p_2^2+\ldots+p_n^2-q^2/c^2+a} \risingdotseq\\
&\frac{c}{2\pi}e^{ct\sqrt{a}}\,U\left(t-\frac{r}{c}\right)\left(\frac{d}{2\pi c r\,dt}\right)^{\frac{1}{2}n-1}\\
&\quad\times\left[\frac{e^{-ct\sqrt{a}}\cosh\left\{(\tfrac{1}{2}n-1)\operatorname{arc\,cosh}\frac{ct}{r}+\sqrt{\{a(c^2t^2-r^2)\}}\right\}}{\sqrt{(c^2t^2-r^2)}}\right] \quad (n \text{ even})\\
&\qquad\qquad \operatorname{Re} q > c\,|\operatorname{Re}\sqrt{(p_1^2+\ldots+p_n^2+a)}\,|,
\end{aligned}\right\} \quad (57)$$

in which $r = \sqrt{(x_1^2+x_2^2+\ldots+x_n^2)}$.

As already stated, the Green functions are helpful in deriving particular solutions of the n-dimensional wave equation with arbitrary right-hand member. As an example of the general formula (57) we may mention that, after some simplifying calculations, the function

$$G = \frac{1}{4\pi}\left[\frac{\delta(t-r/c)}{r} - \frac{\omega_0}{c}\frac{J_1\{\omega_0\sqrt{(t^2-r^2/c^2)}\}}{\sqrt{(t^2-r^2/c^2)}}\,U(t-r/c)\right]$$

is seen to be the Green function corresponding to the three-dimensional wave equation in the presence of normal dispersion as follows:

$$\left(\Delta - \frac{1}{c^2}\frac{\partial^2}{\partial t^2} - \frac{\omega_0^2}{c^2}\right) G = -\,\delta(x)\,\delta(y)\,\delta(z)\,\delta(t).$$

This function G may, for instance, describe the propagation of electromagnetic waves in the ionosphere, expanding from a point source. At some point a distance r off the source an impulse wave passes at the moment $t = r/c$, followed by the after-effect which is due to dispersion. The after-effect is described by an oscillating function whose individual oscillations (as in the one-dimensional case of fig. 92 when $\beta < 0$), in the beginning, follow in time the more rapidly the greater the distance from the point of observation to the centre of disturbance.

Concerning the general application to equations in more than two independent variables, it occasionally proves useful not to transpose *all* variables involved; for instance, the variable determining some boundary at which a certain condition has to be fulfilled may better not be transposed at all, as may be illustrated by the next example.

Example. To solve the two-dimensional equation of heat conduction,

$$\frac{\partial^2 T}{\partial x^2}+\frac{\partial^2 T}{\partial y^2}-\frac{1}{a^2}\frac{\partial T}{\partial t} = 0,$$

when the temperature at $t = 0$ is prescribed to be equal to $\lambda(x, y)$. By putting

$$h(x,y,t) = T(x,y,t)\, U(t),$$

and accounting for the boundary condition specified above, we are led to the following differential equation satisfied by h:

$$\frac{\partial^2 h}{\partial x^2}+\frac{\partial^2 h}{\partial y^2}-\frac{1}{a^2}\frac{\partial h}{\partial t} = -\frac{1}{a^2}\lambda(x,y)\,\delta(t).$$

Further, transposing the variables x and y under the assumption that

$$h(x,y,t) \mathrel{\ddot{=}} f(p_1, p_2, t), \quad \lambda(x,y) \mathrel{\ddot{=}} \Lambda(p_1, p_2),$$

we obtain an ordinary differential equation in terms of t, namely,

$$(p_1^2+p_2^2)\, f - \frac{1}{a^2}\frac{df}{dt} = -\frac{1}{a^2}\Lambda(p_1, p_2)\,\delta(t).$$

A particular solution of this equation is provided by

$$f = U(t)\exp\{a(p_1^2+p_2^2)\,t\}\,\Lambda(p_1, p_2),$$

the original of which can be determined from a composition product according to (10). It is then found that

$$T(x,y,t) = \frac{1}{4\pi a^2 t}\int_{-\infty}^{\infty} d\xi \int_{-\infty}^{\infty} d\eta \exp\left\{-\frac{(x-\xi)^2+(y-\eta)^2}{4a^2 t}\right\}\lambda(\xi,\eta),$$

which, indeed, satisfies the prescribed boundary condition at $t = 0$, as may be independently verified afterwards with the aid of the impulse function (v, 33a).

5. Elliptic differential equations

Whereas hyperbolic differential equations as considered in the preceding two sections are characteristic in dynamic problems, the equations of the elliptic type occur in static and stationary problems. A stationary problem of utmost importance is to determine the amplitude u, not depending on the time t, of vibrational motions

$$\psi = u\, e^{-i\omega t},$$

satisfying the familiar wave equation

$$(\Delta + a)\,\psi - \frac{1}{c^2}\frac{\partial^2 \psi}{\partial t^2} = 0.$$

The differential equation of the amplitude u, that is,

$$(\Delta + k^2)\, u = 0, \tag{58}$$

in which $k^2 = a+\omega^2/c^2$ will be called the *elliptic wave equation*. In what follows we confine ourselves to equations in two space variables. Again, to be able to derive a particular solution of the inhomogeneous equation

$$\frac{\partial^2 u}{\partial x_1^2}+\frac{\partial^2 u}{\partial x_2^2}+k^2 u = -\varphi, \tag{59}$$

involving an arbitrary function ϕ in its right-hand member, it is sufficient to know the Green function, L_2, satisfying

$$\left(\frac{\partial^2}{\partial x_1^2}+\frac{\partial^2}{\partial x_2^2}+k^2\right)L_2 = -\delta(x_1)\,\delta(x_2) = -\frac{\delta(\rho)}{\pi\rho} \quad (\rho = \sqrt{(x_1^2+x_2^2)}). \tag{60}$$

A transposition of this differential equation then immediately leads to the image of the Green function under discussion, viz.

$$L_2(x_1,x_2) \doteqdot -\frac{p_1 p_2}{p_1^2+p_2^2+k^2}. \tag{61}$$

However, when k is real the corresponding definition integral proves to be convergent only if

$$\operatorname{Re}\sqrt{(p_1^2+p_2^2)} = 0,$$

the domain of convergence thus being degenerate for real p_1 and p_2. Therefore, the situation is quite different in comparison with the two-dimensional equation of the hyperbolic type, in the discussion of which we found four distinct originals with domains of convergence covering all values of p_1 and p_2 (see (40)). Another feature, related to the difference pointed out above, is that elliptic differential equations do not possess Green functions that at infinity tend to zero only inside certain sectors of the complete x_1, x_2-plane. Accordingly, the Green function

$$-\frac{1}{2\pi}\log\rho$$

corresponding to the two-dimensional potential equation ($k = 0$) becomes infinitely large as $\rho\to\infty$, no matter in what direction.

Instead of determining the original of (61) directly, that is to say, by successively transposing the two variables p_1 and p_2, we now proceed in a different way, which also shows how one may often obtain solutions of elliptic differential equations by a certain process applied to operational relations for hyperbolic equations in one more independent variable. For this purpose we transpose with respect to t the right-hand member of the operational relation (56) which holds for the two-dimensional Green function of the hyperbolic wave equation when dispersion is absent. Upon using (x, 30) for $\nu = 0$ it is found that

$$-\frac{p_1 p_2 q}{p_1^2+p_2^2-q^2/c^2} \doteqdot \frac{q}{2\pi}K_0\left\{\frac{q}{c}\sqrt{(x_1^2+x_2^2)}\right\}, \qquad \operatorname{Re} q > c\,|\operatorname{Re}\sqrt{(p_1^2+p_2^2)}\,|,$$

in which q is now simply a parameter and may therefore be equated to $-ikc$. We thus obtain when $\operatorname{Im} k > 0$,

$$-\frac{p_1 p_2}{p_1^2+p_2^2+k^2} \doteqdot \frac{1}{2\pi} K_0\{-ik\sqrt{(x_1^2+x_2^2)}\} = \frac{i}{4} H_0^{(1)}\{k\sqrt{(x_1^2+x_2^2)}\},$$
$$|\operatorname{Re}\sqrt{(p_1^2+p_2^2)}| < \operatorname{Im} k. \quad (62)$$

In order that this relation really exists, it is necessary that k is chosen equal to that square root of k^2 for which $\operatorname{Im} k > 0$, though originally only k^2 is given. Nevertheless, it is possible to include the limiting case $\operatorname{Im} k = 0$. For, upon putting $k = \alpha + i\beta$, we obtain two different operational relations by taking both the real and imaginary parts of the relation (62) for real values of p_1 and p_2 (it follows from analytic continuation that both relations so obtained are equally valid for complex values of p). By letting α tend to zero, and putting $\beta = k$, we then find, respectively,

$$\frac{p_1 p_2}{p_1^2+p_2^2+k^2} \doteqdot \frac{1}{4} Y_0\{|k|\sqrt{(x_1^2+x_2^2)}\}, \qquad \operatorname{Re}\sqrt{(p_1^2+p_2^2)} = 0 \quad (k \text{ real}), \quad (63)$$

$$4\pi p_1 p_2 \delta(p_1^2+p_2^2+k^2) \doteqdot J_0\{k\sqrt{(x_1^2+x_2^2)}\}, \qquad \operatorname{Re}\sqrt{(p_1^2+p_2^2)} = 0 \quad (k \text{ real}). \quad (64)$$

These relations might also have been derived in a more straightforward manner. It is to be remarked that the function $Y_0 = \frac{1}{\pi}\mathbf{Y}_0 = -i(H_0^{(1)} - J_0)$ is Weber's solution of the second kind of Bessel's differential equation of order zero.

The foregoing paragraphs clearly indicate how the difference in character of the Green functions of hyperbolic compared to elliptic differential equations manifests itself. It is also striking that, both in (62) as well as in (63), the original does not tend to a finite limit as $k \to 0$. This property is closely related to the fact that for this limiting value of k the corresponding Green function is given by the logarithmic potential, $\log \rho$, which in virtue of its special behaviour at infinity, does not possess an image at all, whatever p_1 and p_2 may be. On the other hand, there certainly are operational relations for the potential function $\log \rho$ provided the latter is first cut off suitably. For instance, let us replace the function by zero in all quadrants of the $x_1 x_2$-plane except the first. Then the following relation can be derived:

$$\log\{\sqrt{(x_1^2+x_2^2)}\}\, U(x_1)\, U(x_2) \doteqdot \frac{\frac{1}{2}\pi p_1 p_2 - p_1^2 \log p_2 - p_2^2 \log p_1}{p_1^2+p_2^2} - C,$$
$$\operatorname{Re} p_1 > 0,\ \operatorname{Re} p_2 > 0. \quad (65)$$

Leaving the two-dimensional elliptic wave equation, we now proceed to the analogous equation in three dimensions. The well-known Green function of the three-dimensional equation

$$L_3 = \frac{e^{ikr}}{4\pi r} \quad (r = \sqrt{(x_1^2+x_2^2+x_3^2)})$$

is a solution of

$$\left(\frac{\partial^2}{\partial x_1^2}+\frac{\partial^2}{\partial x_2^2}+\frac{\partial^2}{\partial x_3^2}+k^2\right) L_3 = -\delta(x_1)\,\delta(x_2)\,\delta(x_3). \tag{66}$$

When $\operatorname{Im} k > 0$, this Green function certainly possesses an image in virtue of its exponential decay as $r \to \infty$. This does not hold for the function $(\cos kr)/4\pi r$ which satisfies the same differential equation. By transposition of (66) it readily follows that

$$\frac{\exp\{ik\sqrt{(x_1^2+x_2^2+x_3^2)}\}}{4\pi\sqrt{(x_1^2+x_2^2+x_3^2)}} \;\overset{\cdots}{\underset{\cdots}{=}}\; -\frac{p_1p_2p_3}{p_1^2+p_2^2+p_3^2+k^2}, \qquad |\operatorname{Re}\sqrt{(p_1^2+p_2^2+p_3^2)}| < \operatorname{Im} k. \tag{67}$$

This relation may also be verified with the aid of the definition integral, which now becomes a triple integral. The domain of convergence is easily seen in accordance with the specification given above. The limiting cases for $\operatorname{Im} k \to 0$ again apply; it is then found that

$$\frac{\cos\{k\sqrt{(x_1^2+x_2^2+x_3^2)}\}}{4\pi\sqrt{(x_1^2+x_2^2+x_3^2)}} \;\overset{\cdots}{\underset{\cdots}{=}}\; -\frac{p_1p_2p_3}{p_1^2+p_2^2+p_3^2+k^2},$$
$$\operatorname{Re}\sqrt{(p_1^2+p_2^2+p_3^2)} = 0 \quad (k \text{ real}), \tag{68}$$

$$\frac{\sin\{|k|\sqrt{(x_1^2+x_2^2+x_3^2)}\}}{4\pi\sqrt{(x_1^2+x_2^2+x_3^2)}} \;\overset{\cdots}{\underset{\cdots}{=}}\; \pi p_1p_2p_3\,\delta(p_1^2+p_2^2+p_3^2+k^2),$$
$$\operatorname{Re}\sqrt{(p_1^2+p_2^2+p_3^2)} = 0 \quad (k \text{ real}). \tag{69}$$

Note that in this case the value $k = 0$ does not raise difficulties, since for the three-dimensional, radially symmetric potential the following relation exists:

$$\frac{1}{4\pi\sqrt{(x_1^2+x_2^2+x_3^2)}} \;\overset{\cdots}{\underset{\cdots}{=}}\; -\frac{p_1p_2p_3}{p_1^2+p_2^2+p_3^2}, \qquad \operatorname{Re}\sqrt{(p_1^2+p_2^2+p_3^2)} = 0, \tag{70}$$

which, contrarily to (65), does not require the original to be limited to the first octant ($x_1 > 0$, $x_2 > 0$, $x_3 > 0$).

The corresponding operational relations may also be deduced for the elliptic wave equation in n dimensions, either in a direct manner, or by transposing (for $a = 0$) the variable t in the right-hand member of the relation (57) already obtained for the $(n+1)$-dimensional hyperbolic wave equation. In the particular case of the radially symmetric potential in n dimensions,

$$\frac{p_1p_2\cdots p_n}{p_1^2+p_2^2+\ldots+p_n^2} \;\overset{\cdots}{\underset{\cdots}{=}}\; -\frac{\Gamma(\frac{1}{2}n-1)}{4\pi^{\frac{1}{2}n}}\,\frac{1}{(x_1^2+x_2^2+\ldots+x_n^2)^{\frac{1}{2}n-1}},$$
$$\operatorname{Re}\sqrt{(p_1^2+p_2^2+\ldots+p_n^2)} = 0 \quad (n \geqslant 3). \tag{71}$$

Example 1. *Properties of three-dimensional sources of vibration.* The originals of (67) and (68) determine the amplitude of stationary vibrations (wave-length $2\pi/k$) excited by a source at the origin of co-ordinates. To investigate these vibrations one may use (67) for damped vibrations (k complex) and (68) for undamped vibrations

(k real). As examples of relations† that are most simply obtained by operational means we mention the following.

(*a*) *The inversion integral.* When k is real, (68) is valid, amongst others, for $\operatorname{Re} p_1 = \operatorname{Re} p_2 = \operatorname{Re} p_3 = 0$. Therefore, the substitutions $p_1 = is_1$, etc., in the corresponding threefold inversion integral, lead to

$$\frac{\cos(kr)}{r} = \frac{1}{2\pi^2}\int_{-\infty}^{\infty} ds_1 \int_{-\infty}^{\infty} ds_2 \int_{-\infty}^{\infty} ds_3 \frac{e^{i(s_1x_1+s_2x_2+s_3x_3)}}{s_1^2+s_2^2+s_3^2-k^2} \quad (k \text{ real}).$$

On account of the zero in the denominator of the integrand this formula has sense only if the integral is taken as a principal value. It can, moreover, be interpreted as a superposition of plane waves, the integration being extended over all space directions (ratios of s_1, s_2, s_3) as well as over all wave-lengths (wave number $2\pi/\lambda = \sqrt{(s_1^2+s_2^2+s_3^2)}$). The amplitudes of the individual wavelets are such that the result of the superposition is just a monochromatic spherical wave.

(*b*) *The composition product.* This, for the operational relation (67) by itself, leads to the new relation

$$\frac{p_1p_2p_3}{(p_1^2+p_2^2+p_3^2+k^2)^2} \doteqdot \frac{1}{16\pi^2}\int_{-\infty}^{\infty} d\xi_1 \int_{-\infty}^{\infty} d\xi_2 \int_{-\infty}^{\infty} d\xi_3$$
$$\times \frac{\exp\{ik\sqrt{(\xi_1^2+\xi_2^2+\xi_3^2)}\}}{\sqrt{(\xi_1^2+\xi_2^2+\xi_3^2)}} \frac{\exp\{ik\sqrt{[(x_1-\xi_1)^2+(x_2-\xi_2)^2+(x_3-\xi_3)^2]}\}}{\sqrt{[(x_1-\xi_1)^2+(x_2-\xi_2)^2+(x_3-\xi_3)^2]}},$$
$$|\operatorname{Re}\sqrt{(p_1^2+p_2^2+p_3^2)}| < \operatorname{Im} k.$$

However, the original of the left-hand function may also be obtained by applying to (67) the operator

$$\int_{-\infty}^{x_3} ds_3\, s_3(\ldots) \doteqdot -\frac{d}{dp_3}\left\{\frac{1}{p_3}(\ldots)\right\} \quad (\operatorname{Re} p_3 > 0).$$

Equating the two distinct expressions so obtained we arrive at an identity that may prove useful in problems of wave diffraction, viz.

$$\frac{1}{2\pi i}\iiint \frac{e^{ikOQ}}{OQ}\frac{e^{ikQP}}{QP}\, d\tau_Q = \frac{e^{ikOP}}{k} \quad (\operatorname{Im} k > 0),$$

in which the integration is over the complete three-dimensional space ($d\tau_Q = dx_1\,dx_2\,dx_3$) and the following distances have been introduced:

$$OQ = \sqrt{(\xi_1^2+\xi_2^2+\xi_3^2)},$$
$$QP = \sqrt{\{(x_1-\xi_1)^2+(x_2-\xi_2)^2+(x_3-\xi_3)^2\}},$$
$$OP = \sqrt{(x_1^2+x_2^2+x_3^2)}.$$

This integral represents the effect of secondary sources that are distributed through space with a density in proportion to the field of a primary source situated at the origin of co-ordinates.

(*c*) *A plane covered with sources of vibration.* Still other superpositions of vibrational sources can be tackled operationally. For instance, by writing down the definition integral of (68) with respect to p_2 and p_3, and then taking $p_2 = p_3 = 0$, we get a relation only involving p_1, viz.

$$-\frac{p_1}{p_1^2+k^2} \doteqdot \int_{-\infty}^{\infty} dx_2 \int_{-\infty}^{\infty} dx_3 \frac{\cos\{k\sqrt{(x_1^2+x_2^2+x_3^2)}\}}{4\pi\sqrt{(x_1^2+x_2^2+x_3^2)}}, \qquad \operatorname{Re} p_1 = 0 \quad (k \text{ real}).$$

† Similar relations are treated by, for instance, Lord Kelvin, *Papers on Electrostatics and Magnetism*, 1884, 2nd ed., p. 112.

Furthermore, the remaining variable p_1 may be transposed according to

$$\frac{p}{p^2+\alpha^2} \doteqdot \frac{\sin(\alpha\,|\,t\,|)}{2\alpha}, \quad \operatorname{Re} p = 0. \tag{72}$$

Then the final result is found to be

$$\int_{-\infty}^{\infty} dx_2 \int_{-\infty}^{\infty} dx_3 \frac{\cos\{k\sqrt{(x_1^2+x_2^2+x_3^2)}\}}{\sqrt{(x_1^2+x_2^2+x_3^2)}} = -2\pi \frac{\sin(k\,|\,x_1\,|)}{k} \quad (k \text{ real}). \tag{73}$$

Similarly, from (69) with the aid of (VI, 56),

$$\int_{-\infty}^{\infty} dx_2 \int_{-\infty}^{\infty} dx_3 \frac{\sin\{k\sqrt{(x_1^2+x_2^2+x_3^2)}\}}{\sqrt{(x_1^2+x_2^2+x_3^2)}} = 2\pi \frac{\cos(kx_1)}{k} \quad (k \text{ real}). \tag{74}$$

Upon combining (73) and (74) we are led to the following expression:

$$\int_{-\infty}^{\infty} dx_2 \int_{-\infty}^{\infty} dx_3 \frac{\exp\{ik\sqrt{(x_1^2+x_2^2+x_3^2)}\}}{\sqrt{(x_1^2+x_2^2+x_3^2)}} = 2\pi i \frac{\exp(ik\,|\,x_1\,|)}{k},$$

which is also valid for complex k, as can be demonstrated by using (67) instead of (68). The last formula shows that a plane wave propagating itself at both sides of, and in a direction normal to, an infinite plane may be considered as a sum of spherical wavelets originating from certain sources lying on the plane†.

Example 2. *Discontinuous factor for the n-dimensional sphere.* This factor is defined as a function that has unit value inside, and zero value outside, the n-sphere

$$x_1^2+x_2^2+\ldots+x_n^2 = a^2.$$

It is an extension of Dirichlet's well-known discontinuous factor which involves only one variable. The generalized factor may be represented by $U(a-r_n)$ if

$$r_n = \sqrt{(x_1^2+x_2^2+\ldots+x_n^2)}.$$

For this function a simultaneous-operational relation is readily deduced. Since $U(a-r_n)$ is radially symmetric, the Laplacian of this function reduces to

$$\left(\frac{\partial^2}{\partial x_1^2}+\ldots+\frac{\partial^2}{\partial x_n^2}\right) U(a-r_n) = \left\{\frac{d^2}{dr_n^2}+\frac{n-1}{r_n}\frac{d}{dr_n}\right\} U(a-r_n) = \left\{\frac{d^2}{dr_n^2}+\frac{n-1}{a}\frac{d}{dr_n}\right\} U(a-r_n).$$

Now, it is very important that the differential quotients in the right-hand member with respect to r_n can be transformed into those with respect to a, so as to produce

$$\left(\frac{\partial^2}{\partial x_1^2}+\ldots+\frac{\partial^2}{\partial x_n^2}\right) U(a-r_n) = \left\{\frac{\partial^2}{\partial a^2}-\frac{n-1}{a}\frac{\partial}{\partial a}\right\} U(a-r_n),$$

since transposition of the last equation yields an *ordinary* differential equation for the image $f_n(p_1,\ldots,p_n)$ of $U(a-r_n)$, viz.

$$\frac{d^2 f_n}{da^2}-\frac{n-1}{a}\frac{df_n}{da}-(p_1^2+\ldots+p_n^2)f_n = 0.$$

It may be noted that this differential equation is closely related to (XV, 49). From the condition that f_n shall be zero when $a = 0$, it follows that the solution required for our purpose takes on the form

$$\alpha_n(p_1,\ldots,p_n)\, a^{\frac{1}{2}n} I_{\frac{1}{2}n}\{a\sqrt{(p_1^2+\ldots+p_n^2)}\} \doteqdot U(a-r_n). \tag{75}$$

† See, for instance, Balth. van der Pol, *Handelingen 24ste Ned. Natuur- en Genees-kundig Congres*, Haarlem, 1933, p. 110.

In order to specify the function α_n we apply to (75) the operator $\lim_{a\to 0} 1/a^n$. Then the right-hand member becomes equal to the following impulse function:

$$\lim_{a\to 0} \frac{V_n(a)}{a^n} \delta(x_1)\,\delta(x_2)\ldots\delta(x_n) = \frac{\pi^{\frac{1}{2}n}}{\Pi(\frac{1}{2}n)} \delta(x_1)\,\delta(x_2)\ldots\delta(x_n),$$

in which $V_n(a)$ denotes the volume of the n-dimensional sphere of radius a according to (IV, 24). With the aid of this limit as $a \to 0$ we determine α_n, which is independent of a. After some calculations, the final result from (75) proves to be

$$U(a-r_n) \doteqdot (2\pi a)^{\frac{1}{2}n} \frac{p_1 p_2 \cdots p_n}{(p_1^2+p_2^2+\ldots+p_n^2)^{\frac{1}{4}n}} I_{\frac{1}{2}n}\{a\sqrt{(p_1^2+p_2^2+\ldots+p_n^2)}\},$$
$$-\infty < \mathrm{Re}\, p_{1,2,\ldots,n} < \infty. \quad (76)$$

As to the discontinuous factor itself, we obtain from the inversion integral of (76), when the paths of integration of all the variables p are taken along the respective imaginary axes ($p_1 = is_1$, etc.),

$$U(a-r_n) = \left(\frac{a}{2\pi}\right)^{\frac{1}{2}n} \int_{-\infty}^{\infty} ds_1 \int_{-\infty}^{\infty} ds_2 \ldots \int_{-\infty}^{\infty} ds_n$$
$$\times \frac{\exp\{i(s_1x_1+s_2x_2+\ldots+s_nx_n)\}}{(s_1^2+s_2^2+\ldots+s_n^2)^{\frac{1}{4}n}} J_{\frac{1}{2}n}\{a\sqrt{(s_1^2+s_2^2+\ldots+s_n^2)}\}.$$

Regarding the general character of (76), there is again a striking difference between spaces of an even and of an odd number of dimensions. Only in the last case does the image become an elementary function. For instance, when $n = 3$,

$$U\{a-\sqrt{(x_1^2+x_2^2+x_3^2)}\} \doteqdot 4\pi a^2 \frac{p_1p_2p_3}{p_1^2+p_2^2+p_3^2} \frac{d}{da}\left(\frac{\sinh\{a\sqrt{(p_1^2+p_2^2+p_3^2)}\}}{a\sqrt{(p_1^2+p_2^2+p_3^2)}}\right),$$
$$-\infty < \mathrm{Re}\, p_{1,2,3} < \infty. \quad (77)$$

We finally mention the following relation originating from (76) by differentiation with respect to a,

$$\delta(a-r_n) \doteqdot (2\pi a)^{\frac{1}{2}n} \frac{p_1 p_2 \cdots p_n}{(p_1^2+p_2^2+\ldots+p_n^2)^{\frac{1}{4}(n-2)}} I_{\frac{1}{2}n-1}\{a\sqrt{(p_1^2+p_2^2+\ldots+p_n^2)}\},$$
$$-\infty < \mathrm{Re}\, p_{1,2,\ldots,n} < \infty. \quad (78)$$

In particular, for $n = 3$,

$$\frac{\delta\{a-\sqrt{(x_1^2+x_2^2+x_3^2)}\}}{4\pi a^2} \doteqdot p_1p_2p_3 \frac{\sinh\{a\sqrt{(p_1^2+p_2^2+p_3^2)}\}}{a\sqrt{(p_1^2+p_2^2+p_3^2)}}, \quad -\infty < \mathrm{Re}\, p_{1,2,3} < \infty, \quad (79)$$

the original of which represents a function that is concentrated at the surface, $r_3 = a$, of the sphere; the volume integral of the function is equal to unity. With the aid of this operational relation it is possible to handle problems of potential theory for arbitrarily prescribed radially symmetric charge distributions. It is also remarkable that such radially symmetric problems, depending only on the variable r_n, can be studied by *simultaneous* operational calculus—by considering the functions involved in their dependence on n variables.

In concluding this section, we would emphasize that analogous discontinuous factors can be derived for ellipsoids, by simply applying the similarity rule to the result just obtained for spheres.

6. Simultaneous partial differential equations

In the quasi-stationary treatment of coaxial cables we saw how simultaneous operational calculus may be applied to the investigation of a set of two simultaneous linear differential equations (see example 2 of § 3).

Obviously the method is by no means restricted to a set of two such equations. An illustrative example is formed by Maxwell's equations of the electromagnetic field. They may be solved upon their reduction to a system of linear algebraic equations, in the case of arbitrary space distributions of charges (represented by the density $P(x_1, x_2, x_3, t)$ and current-density components: $I_1(x_1, x_2, x_3, t)$, $I_2(\ )$, $I_3(\ )$).

Let E_1, E_2, E_3 denote the Cartesian components of the electric field strength E, and let H_1, H_2, H_3 be those of the magnetic field strength H. When all quantities are measured in units of the Gaussian system, then

$$\operatorname{curl} E + \frac{1}{c}\frac{\partial H}{\partial t} = 0, \quad \operatorname{curl} H - \frac{1}{c}\frac{\partial E}{\partial t} = \frac{4\pi}{c} I,$$

or, written out in components,

$$\frac{\partial E_3}{\partial x_2} - \frac{\partial E_2}{\partial x_3} + \frac{1}{c}\frac{\partial H_1}{\partial t} = 0, \quad \frac{\partial H_3}{\partial x_2} - \frac{\partial H_2}{\partial x_3} - \frac{1}{c}\frac{\partial E_1}{\partial t} = \frac{4\pi}{c} I_1,$$

$$\frac{\partial E_1}{\partial x_3} - \frac{\partial E_3}{\partial x_1} + \frac{1}{c}\frac{\partial H_2}{\partial t} = 0, \quad \frac{\partial H_1}{\partial x_3} - \frac{\partial H_3}{\partial x_1} - \frac{1}{c}\frac{\partial E_2}{\partial t} = \frac{4\pi}{c} I_2,$$

$$\frac{\partial E_2}{\partial x_1} - \frac{\partial E_1}{\partial x_2} + \frac{1}{c}\frac{\partial H_3}{\partial t} = 0, \quad \frac{\partial H_2}{\partial x_1} - \frac{\partial H_1}{\partial x_2} - \frac{1}{c}\frac{\partial E_3}{\partial t} = \frac{4\pi}{c} I_3,$$

where c represents the velocity of light in free space.

Now, by putting

$$E_n(x_1, x_2, x_3, t) \overset{::}{\underset{::}{=}} e_n(p_1, p_2, p_3, q),$$

$$H_n(x_1, x_2, x_3, t) \overset{::}{\underset{::}{=}} h_n(p_1, p_2, p_3, q),$$

$$I_n(x_1, x_2, x_3, t) \overset{::}{\underset{::}{=}} i_n(p_1, p_2, p_3, q) \quad (n = 1, 2, 3),$$

and transposing the differential equations above, we obtain a set of six linear equations in six unknowns, namely,

$$\begin{aligned}
-p_3 e_2 + p_2 e_3 + \frac{q}{c} h_1 &= 0,\\
p_3 e_1 \qquad - p_1 e_3 \qquad + \frac{q}{c} h_2 &= 0,\\
-p_2 e_1 + p_1 e_2 \qquad\qquad + \frac{q}{c} h_3 &= 0,\\
-\frac{q}{c} e_1 \qquad\qquad - p_3 h_2 + p_2 h_3 &= \frac{4\pi}{c} i_1,\\
-\frac{q}{c} e_2 \qquad + p_3 h_1 \qquad - p_1 h_3 &= \frac{4\pi}{c} i_2,\\
-\frac{q}{c} e_3 \; - p_2 h_1 + p_1 h_2 \qquad &= \frac{4\pi}{c} i_3.
\end{aligned}$$

These equations are easy to solve, leading to

$$e_1 = 4\pi \frac{\left(p_1\rho+\frac{q}{c^2}i_1\right)}{\left(p_1^2+p_2^2+p_3^2-\frac{q^2}{c^2}\right)}, \quad \text{etc.} \tag{80}$$

$$h_1 = \frac{4\pi}{c} \frac{(p_3 i_2 - p_2 i_3)}{\left(p_1^2+p_2^2+p_3^2-\frac{q^2}{c^2}\right)}, \quad \text{etc.,} \tag{81}$$

in which ρ is an abbreviation for

$$\rho = -\frac{1}{q}(p_1 i_1 + p_2 i_2 + p_3 i_3). \tag{82}$$

The quantity ρ turns out to be the image of the charge density P; if (82) is multiplied through by q, it becomes the image of the familiar equation of continuity:

$$\frac{\partial P}{\partial t}+\frac{\partial I_1}{\partial x_1}+\frac{\partial I_2}{\partial x_2}+\frac{\partial I_3}{\partial x_3} = 0,$$

representing the conservation of charge. It may further be remarked that the images of the subsidiary equations $\operatorname{div} H = 0$, $\operatorname{div} E = 4\pi\rho$ are consistent with the system (80), (81) and (82).

The conventional formulae for the electromagnetic field, in terms of the retarded potentials, can be derived from the originals of (80) and (81). Concerning the magnetic field, two different expressions may be obtained by applying the rule of the composition product according to (10). First, (81) will be factorized in the following manner:

$$(a) \qquad h_1 = \frac{1}{p_1 p_2 p_3 q} \times \frac{-(4\pi/c)p_1 p_2 p_3 q}{p_1^2+p_2^2+p_3^2-\frac{q^2}{c^2}} \times (p_2 i_3 - p_3 i_2), \quad \text{etc.} \tag{83}$$

We now use the relation (57), taken for $n = 3$ and $a = 0$, with regard to the first factor of the composition product; thus

$$-4\pi \frac{p_1 p_2 p_3 q}{p_1^2+p_2^2+p_3^2-\frac{q^2}{c^2}} \doteqdot \frac{\delta\left\{t-\frac{1}{c}\sqrt{(x_1^2+x_2^2+x_3^2)}\right\}}{\sqrt{(x_1^2+x_2^2+x_3^2)}},$$

$$\operatorname{Re} q > c\,|\operatorname{Re}\sqrt{(p_1^2+p_2^2+p_3^2)}\,|, \tag{84}$$

where the domain of convergence is chosen so as to apply to the solution that vanishes rapidly enough as $t \to -\infty$. For, if the same does hold for the given currents, the Laplace integrals with respect to the variable t will converge at their lower limit of integration for all values of q, which implies that their domains of convergence with respect to q extend to infinity on the right. We may then infer from (83) that

$$H_1(x_1,x_2,x_3,t) = \frac{1}{c}\int_{-\infty}^{\infty} d\xi_1 \int_{-\infty}^{\infty} d\xi_2 \int_{-\infty}^{\infty} d\xi_3 \int_{-\infty}^{\infty} d\tau \times \left\{\frac{\partial I_3}{\partial \xi_2}(\xi_1,\xi_2,\xi_3,\tau) - \frac{\partial I_2}{\partial \xi_3}(\xi_1,\xi_2,\xi_3,\tau)\right\} \frac{\delta(t-\tau-R/c)}{R},$$

in which $R = \sqrt{\{(x_1-\xi_1)^2+(x_2-\xi_2)^2+(x_3-\xi_3)^2\}}$. Carrying out the integration with respect to τ we are led to

$$H_1(x_1,x_2,x_3,t) = \frac{1}{c}\int_{-\infty}^{\infty} d\xi_1 \int_{-\infty}^{\infty} d\xi_2 \int_{-\infty}^{\infty} d\xi_3 \frac{\dfrac{\partial I_3}{\partial \xi_2}\left(\xi_1,\xi_2,\xi_3,t-\dfrac{R}{c}\right) - \dfrac{\partial I_2}{\partial \xi_3}\left(\xi_1,\xi_2,\xi_3,t-\dfrac{R}{c}\right)}{R}.$$

This formula, together with the analogues for H_2 and H_3, may be written in conventional vector notation as follows:

$$H(x_1,x_2 x_3,t) = \frac{1}{c}\iiint_{-\infty}^{\infty} \frac{\operatorname{curl} I(\xi_1,\xi_2,\xi_3,t-R/c)}{R}\,dS, \tag{85}$$

where $dS = d\xi_1 d\xi_2 d\xi_3$ is the space element. The result (85) shows that time variations of the current distribution manifest themselves at a later moment (retardation effect), the disturbances travelling at a velocity equal to c.

Another useful factorization of h is the following:

$$(b) \quad h_1 = \frac{1}{cp_1p_2p_3q}\left\{-\frac{4\pi p_1p_2^2p_3q}{p_1^2+p_2^2+p_3^2-\dfrac{q^2}{c^2}} \times i_3 + \frac{4\pi p_1p_2p_3^2q}{p_1^2+p_2^2+p_3^2-\dfrac{q^2}{c^2}} \times i_2\right\}, \quad \text{etc.}$$

In this case the first factors of the two composition products correspond to operational relations obtained by differentiation of (84) with respect to x_2 and x_3. Thus we now obtain

$$H_1 = \frac{1}{c}\int_{-\infty}^{\infty} d\xi_1 \int_{-\infty}^{\infty} d\xi_2 \int_{-\infty}^{\infty} d\xi_3 \int_{-\infty}^{\infty} d\tau \left[I_3(\xi,\eta,\zeta,\tau)\frac{\partial}{\partial(x_2-\xi_2)}\left\{\frac{\delta(t-\tau-R/c)}{R}\right\} - I_2(\xi,\eta,\zeta,\tau)\frac{\partial}{\partial(x_3-\xi_3)}\left\{\frac{\delta(t-\tau-R/c)}{R}\right\}\right].$$

Owing to the special form of the functions to be differentiated, the derivatives with respect to $x_2-\xi_2$ may be replaced by those with respect to x_2, etc. Further, the latter differentiations may be performed outside the integral. When, moreover, the integration over τ is carried out explicitly, and the remaining components H_2 and H_3 are dealt with similarly, the result may be summarized in a single-vector formula, as follows:

$$H(x_1,x_2,x_3,t) = \frac{1}{c}\operatorname{curl}\iiint_{-\infty}^{\infty} \frac{I(\xi_1,\xi_2,\xi_3,t-R/c)}{R}\,dS. \tag{86}$$

It should be kept in mind that the operator 'curl' in (86) is understood to act upon the co-ordinates of the field point (x_1, x_2, x_3), whereas under the sign of integration in (85) it refers to the co-ordinates of the source (ξ_1, ξ_2, ξ_3). The integral given in (86) is the familiar expression for the vector potential A, which here was seen to emerge from a composition product. By well-known methods of vector analysis, (85) and (86) are easily shown to be equivalent, under proper boundary conditions, at infinity.

The equation (80) determining the electric field strength can be treated in an analogous manner. The final expression then becomes

$$E = -\operatorname{grad}\phi - \frac{1}{c}\frac{\partial A}{\partial t},$$

in which the scalar potential is given by

$$\phi(x,y,z,t) = \iiint_{-\infty}^{\infty} \frac{P(\xi,\eta,\zeta,t-R/c)}{R}\,dS.$$

7. Partial difference equations

A few examples may illustrate that for partial difference equations also the simultaneous operational calculus is a useful tool in rapidly obtaining the solution required. Let us consider for a while the difference analogue of the two-dimensional elliptic wave equation (59), that is,

$$(\Delta_x^2 + \Delta_y^2 + k^2)\,u(x,y) = -\varphi(x,y), \tag{87}$$

where, according to (XIII, 1), the symmetric difference quotients are defined by

$$\Delta_x u(x,y) = u(x+\tfrac{1}{2},y) - u(x-\tfrac{1}{2},y),$$
$$\Delta_y u(x,y) = u(x,y+\tfrac{1}{2}) - u(x,y-\tfrac{1}{2}).$$

When (87) is written out in full, it becomes

$$u(x+1,y) + u(x-1,y) + u(x,y+1) + u(x,y-1) + (k^2-4)\,u(x,y) = -\varphi(x,y),$$

thus expressing a definite relationship between the functional values of u at five (if $k^2 \neq 4$) neighbouring points. To construct a particular solution of (87) by operational methods, we put

$$u(x,y) \doteqdot f(p,q), \quad \varphi(x,y) \doteqdot \phi(p,q).$$

Further, according to (XIII, 2), the effect of the difference operators Δ_x and Δ_y is equivalent to multiplying the image by $2\sinh(\frac{1}{2}p)$ and $2\sinh(\frac{1}{2}q)$ respectively. Therefore, the image equation of (87) is given by

$$4\{\sinh^2(\tfrac{1}{2}p) + \sinh^2(\tfrac{1}{2}q) + \tfrac{1}{4}k^2\}\,f(p,q) = -\phi(p,q),$$

whence it follows that possible particular solutions of (87) are to be deduced from the operational relation

$$u(x,y) \doteqdot -\frac{\phi(p,q)}{4\{\sinh^2(\frac{1}{2}p) + \sinh^2(\frac{1}{2}q) + \frac{1}{4}k^2\}}. \tag{88}$$

The rule of the composition product enables us to obtain from (88) the original u in the case of an arbitrary right-hand member $-\varphi$, whenever the original corresponding to some *special* function ϕ is known. It is thus natural, by analogy to the Green functions of differential equations, to take that special function φ as simple as possible, by letting it vanish almost everywhere. Then restricting ourselves to the most usual case, that is, when the solution of the difference equation is required only at the integral points (with co-ordinates $x = m$, $y = n$; n, m integers), it is sufficient to consider a Green function, G, for which φ is zero at all integral points except the origin, $m = n = 0$, where it takes the value 1. In so doing we have the advantage that the values to be assumed by φ at the intermediate, non-integral, points are still open; it is profitable to choose them so as to make the image of G as simple as possible. For instance, two simple Green functions originate from the following two possible choices of φ:

(1) $$\varphi_1(x,y) = \frac{\sin(\pi x)}{\pi x}\frac{\sin(\pi y)}{\pi y} \doteqdot pqU(\pi^2+p^2)\,U(\pi^2+q^2) = \phi_1(p,q),$$
$$\operatorname{Re} p = \operatorname{Re} q = 0.$$

In this case the limits of integration in the inversion integral of (88) are finite, since the image is a cut-off function with respect to both p and q. When in this integral we make the substitutions $p = is_1$, $q = is_2$, the Green function in question is found to be (for convenience, the corresponding image is also given)

$$\begin{aligned} u_1(x,y) &= \frac{1}{16\pi^2}\int_{-\pi}^{\pi} ds_1 \int_{-\pi}^{\pi} ds_2 \frac{e^{i(s_1x+s_2y)}}{\sin^2(\frac{1}{2}s_1)+\sin^2(\frac{1}{2}s_2)-\frac{1}{4}k^2} \\ &= \frac{1}{4\pi^2}\int_0^{\pi} ds_1 \int_0^{\pi} ds_2 \frac{\cos(s_1x)\cos(s_2y)}{\sin^2(\frac{1}{2}s_1)+\sin^2(\frac{1}{2}s_2)-\frac{1}{4}k^2} \\ &\doteqdot -\frac{pqU(\pi^2+p^2)\,U(\pi^2+q^2)}{4\{\sinh^2(\frac{1}{2}p)+\sinh^2(\frac{1}{2}q)+\frac{1}{4}k^2\}}, \qquad \operatorname{Re} p = \operatorname{Re} q = 0. \end{aligned} \tag{89}$$

Concerning the dependence on k, the double integral converges for all complex or real values of k outside the real segment $-\sqrt{8} \leqslant k \leqslant \sqrt{8}$; the excluded segment of the real axis is a cut of the function.

(2) $$\varphi_2(x,y) = U(\tfrac{1}{4}-x^2)\,U(\tfrac{1}{4}-y^2) \doteqdot 4\sinh\tfrac{1}{2}p\,\sinh\tfrac{1}{2}q = \phi_2(p,q),$$
$$-\infty < \operatorname{Re} p,\ q < \infty.$$

From (88) and the corresponding inversion integral for $\operatorname{Re} p = \operatorname{Re} q = 0$ we now find the solution

$$\begin{aligned} u_2(x,y) &= \frac{1}{4\pi^2}\int_{-\infty}^{\infty} ds_1 \int_{-\infty}^{\infty} ds_2 \frac{\sin(\frac{1}{2}s_1)\sin(\frac{1}{2}s_2)\, e^{i(s_1x+s_2y)}}{s_1s_2\{\sin^2(\frac{1}{2}s_1)+\sin^2(\frac{1}{2}s_2)-\frac{1}{4}k^2\}} \\ &\doteqdot -\frac{\sinh(\frac{1}{2}p)\sinh(\frac{1}{2}q)}{\sinh^2(\frac{1}{2}p)+\sinh^2(\frac{1}{2}q)+\frac{1}{4}k^2}, \qquad \operatorname{Re} p = \operatorname{Re} q = 0. \end{aligned} \tag{90}$$

The interrelation of the two solutions u_1 and u_2 is derived with the aid of the series

$$4\sinh\tfrac{1}{2}p\,\sinh\tfrac{1}{2}q \sum_{m=-\infty}^{\infty}\sum_{n=-\infty}^{\infty} u_1(m,n)\,e^{-pm-qn}$$

$$= -\frac{\sinh(\tfrac{1}{2}p)\sinh(\tfrac{1}{2}q)}{\sinh^2(\tfrac{1}{2}p)+\sinh^2(\tfrac{1}{2}q)+\tfrac{1}{4}k^2} \doteqdot u_2(x,y) \qquad (\operatorname{Re}p = \operatorname{Re}q = 0),$$

which can be proved by means of (VI, 33). Next, in virtue of (XII, 1) extended to two variables, we have the simple formula

$$u_2(x,y) = u_1([x+\tfrac{1}{2}],[y+\tfrac{1}{2}]).$$

Consequently, u_2 is a step function of two variables that is identical with u_1 at the integral points, that is, $u_1(m,n) = u_2(m,n)$. Whereas u_1 is everywhere continuous, the function u_2 changes abruptly across a system of lines parallel to the co-ordinate axes. At the points outside these lines the value of u_2 is equal to that at the nearest lattice point. For integral points the function u_1 is most easily evaluated, since in (89) the integration with respect to s_2 can be carried out explicitly after substituting $w = e^{is_2}$ and applying Cauchy's theorem. Introducing the variable $u = \cos s_1$, the function u_1 is found as an elliptic integral depending on m, n and k:

$$u_1(m,n;k) = -\frac{1}{2\pi}\int_{-1}^{1} \frac{T_m(u)\,[\sqrt{\{(u+\tfrac{1}{2}k^2-2)^2-1\}}-u-\tfrac{1}{2}k^2+2]^{|n|}}{\sqrt{(1-u^2)}\,\sqrt{\{(u+\tfrac{1}{2}k^2-2)^2-1\}}}\,du, \qquad (91)$$

in which $T_m(u) = \cos(m \arccos u)$ is the mth polynomial of Tchebycheff, and the imaginary part of the square root in the numerator of the integrand is understood to have the same sign as $\operatorname{Im}k^2$. By means of the integral representation (91) it is easily verified that u_1 is logarithmically infinite at the five branch points $k = 0, \pm 2, \pm 2\sqrt{2}$.

Whereas u_1 is most useful in numerical calculations, the function u_2 allows to investigate how the solution of the difference equation at the integral points is related to the solution of the corresponding elliptic differential equation (60). To see this, the operational relation (62) for the Green function of the differential equation is rewritten as

$$f_0(p,q) = -\frac{pq}{p^2+q^2+k^2} \doteqdot \frac{i}{4}H_0^{(1)}\{k\sqrt{(x^2+y^2)}\}, \qquad |\operatorname{Re}\sqrt{(p^2+q^2)}| < \operatorname{Im}k,$$

whence it follows that (90) is equivalent to

$$u_2(x,y) \doteqdot f_0(2\sinh\tfrac{1}{2}p, 2\sinh\tfrac{1}{2}q), \qquad \operatorname{Re}p = \operatorname{Re}q = 0. \qquad (92)$$

Let us now expand the image into a Taylor series of two variables, thus

$$f_0\Big(p+\frac{p^3}{24}+\ldots,\ q+\frac{q^3}{24}+\ldots\Big) = f_0(p,q)+\frac{1}{24}\Big(p^3\frac{\partial f_0}{\partial p}+q^3\frac{\partial f_0}{\partial q}\Big)+\ldots.$$

Then transposing back leads to

$$u_2(x,y) = \frac{i}{4}H_0^{(1)}\{k\sqrt{(x^2+y^2)}\} - \frac{i}{96}\left\{\frac{\partial^2}{\partial x^2}\left(x\frac{\partial}{\partial x}\right) + \frac{\partial^2}{\partial y^2}\left(y\frac{\partial}{\partial y}\right)\right\} H_0^{(1)}\{k\sqrt{(x^2+y^2)}\}\ldots \qquad (\operatorname{Im} k > 0).$$

From this expansion it becomes evident that u_2 (and consequently also u_1 at the integral points) is approximated for complex values of k having a small modulus by the Hankel-function solution of the corresponding differential equation. The restriction laid upon k implies that the wave-length, $2\pi/k$, corresponding to the differential equation is large compared to 1, i.e. to the basic distance in the lattice formed by the integral points. Similarly, for *real* values of k with small modulus, the function u_2 tends to the original $-\frac{1}{4}Y_0\{|k|\sqrt{(x^2+y^2)}\}$ of (63); this holds even for the cut, $-2\sqrt{2} \leqslant k \leqslant 2\sqrt{2}$, provided u_2 is there replaced by the arithmetic mean of its values assumed immediately above and below the real axis of k.

It has to be remarked that, in transforming the double integral (89) into the simple integral (91), we have lost the symmetry with respect to m and n. In order to obtain a symmetric integral representation we may proceed as follows. Replace k in the double integral of (89) by $\sqrt{(p+4)}$, and multiply through by p (assume $x = m$, $y = n$). Then determine the original with respect to the operational variable p. Thereupon the integrations over s_1 and s_2 can be carried out in terms of Bessel functions. We so obtain

$$pu_1\{m,n;\sqrt{(p+4)}\} \fallingdotseq (-1)^{m+n+1} I_m(2t)\, I_n(2t)\, U(t), \qquad \operatorname{Re} p > 4,$$

whose definition integral is equivalent to

$$u_1(m,n;k) = \tfrac{1}{2}(-1)^{m+n+1}\int_0^\infty e^{-(\frac{1}{2}k^2-2)t} I_m(t)\, I_n(t)\, dt \qquad (\operatorname{Re} k^2 > 8).$$

It will be obvious from the considerations given above that the operational knowledge so far obtained is an extremely useful tool in deriving a great many properties of partial difference equations with constant coefficients.

Example 1. As an illustration of the theory of the difference equation (87) let us consider the homogeneous two-dimensional low-pass filter of fig. 93*a*. It consists of an infinite net of squares, the sides of which all contain a self-inductance L, while the vertices (m,n) are connected to a perfectly conducting plate at zero potential via capacitances C. Let $V(m,n;t)$ denote the potential at the lattice point (m,n) at the moment t, and assume

$$V(m,n;t) \fallingdotseq \phi(m,n;p).$$

Then, according to Ohm's law (VIII, 17), the image of the current flowing through the connexion between the vertices (m,n) and $(m+1,n)$ is given by

$$\frac{\phi(m+1,n;p)-\phi(m,n;p)}{Lp}.$$

Analogous expressions hold for the image currents flowing through the remaining three branches joining (m, n). By addition, the image of the total current entering from all sides of the lattice at (m, n) is found to be

$$\frac{(\Delta_m^2 + \Delta_n^2)\,\phi(m, n; p)}{Lp}.$$

This expression has to be equal to the image $Cp\,\phi(m, n; p)$ of the current flowing off through the capacitance C between the vertex (m, n) and the conducting plate. This yields the following partial difference equation:

$$(\Delta_m^2 + \Delta_n^2 - LCp^2)\,\phi(m, n; p) = 0. \tag{93}$$

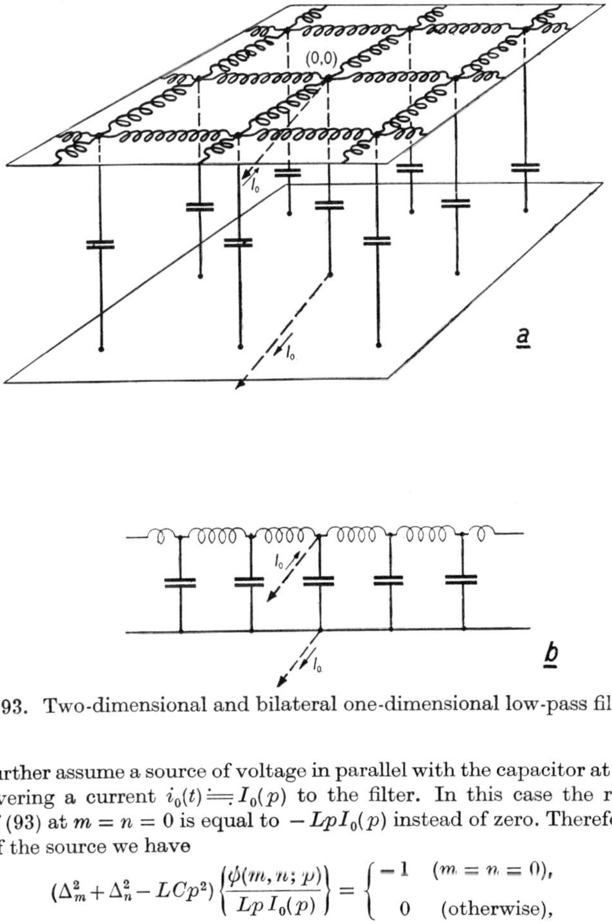

Fig. 93. Two-dimensional and bilateral one-dimensional low-pass filter.

Let us further assume a source of voltage in parallel with the capacitor at the origin $(0, 0)$, delivering a current $i_0(t) \doteqdot I_0(p)$ to the filter. In this case the right-hand member of (93) at $m = n = 0$ is equal to $-LpI_0(p)$ instead of zero. Therefore, in the presence of the source we have

$$(\Delta_m^2 + \Delta_n^2 - LCp^2)\left\{\frac{\phi(m, n; p)}{Lp\,I_0(p)}\right\} = \begin{cases} -1 & (m = n = 0), \\ 0 & (\text{otherwise}), \end{cases}$$

this being the type of difference equation solved above. Accordingly we at once have for the voltage function at the vertices of the lattice:

$$\frac{\phi(m, n; p)}{I_0(p)} = Lp\,u_1\left\{m, n; 2\sqrt{\left(-\frac{p^2}{\omega_0^2}\right)}\right\} \qquad \left(\omega_0 = \frac{2}{\sqrt{(LC)}}\right), \tag{94}$$

the function u_1 being given by (91). This function does satisfy the condition that the corresponding branch currents at infinity $(m, n \to \infty)$ vanish. Moreover, the left-hand side of (94) is just the transfer impedance between the source at $(0, 0)$ and the capacitance at (m, n). In particular, for the driving-point impedance (obtained for $m = n = 0$) we have, in the case of real frequencies $\omega = p/i$, the following elliptic integral:

$$Z(i\omega) = Li\omega\, u_1\left(0, 0;\, 2\frac{\omega}{\omega_0}\right) = \frac{L\omega}{2\pi i}\int_{-1}^{1} \frac{du}{\sqrt{(1-u^2)}\sqrt{\left\{\left(u+2\frac{\omega^2}{\omega_0^2}-2\right)^2-1\right\}}}$$

$$= \frac{1}{\pi i}\sqrt{\left(\frac{L}{C}\right)}\frac{\omega}{\omega_0}k\int_0^{\frac{1}{2}\pi} \frac{d\phi}{\sqrt{(1-k^2\sin^2\phi)}}, \tag{95}$$

in which the last integral is the complete elliptic integral of the first kind of modulus

$$k = \frac{1}{\dfrac{\omega^2}{\omega_0^2}-1}.$$

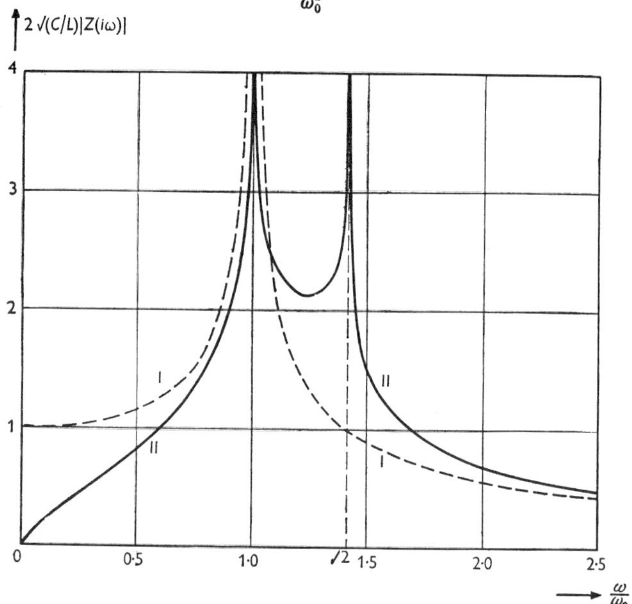

Fig. 94. The driving-point impedances versus frequency for the low-pass filters of fig. 93.

It is worth while comparing the driving-point impedance (95) with the function

$$\frac{1}{2}\sqrt{\left(\frac{L}{C}\right)}\frac{1}{\sqrt{\left(1-\dfrac{\omega^2}{\omega_0^2}\right)}}$$

which represents the driving-point impedance of the one-dimensional, two-sided, infinite filter of fig. 93*b*. Fig. 94 shows the absolute values of these two different impedances against frequency (they are marked with II and I respectively). Most striking is that the driving-point impedance of the two-dimensional filter possesses, besides the pole at $\omega = \omega_0$, a second pole at $\omega = \omega_0\sqrt{2}$. The transfer impedances, which are the more important from the technical point of view, can be dealt with similarly.

The two-dimensional filter is an interesting and very illustrative example of realizing the complete elliptic integral in physical problems. In addition, by replacing the self-inductances and capacitances by general impedances, we merely change the modulus of the elliptic integral (95). In particular if all the filter elements are resistances the elliptic integral is real for all real frequencies.

Example 2. In conclusion, consider the special case $C = 0$ of the preceding example. We obtain a two-dimensional lattice into which, at the origin $(0, 0)$, a current I_0 enters. This current spreads itself over the successive meshes of the lattice. Let us, moreover, assume the impedances Lp replaced by simple resistances R. Now, according to (94), the potentials inside the new lattice (see fig. 95) are expected to be determinable from

$$\phi(m,n;p) = RI_0(p)\,u_1(m,n;0),$$

whereas the parameter p is insignificant in this network containing resistances only. Unfortunately, however, the integrals (91) for u_1 do not converge when $k = 0$. To overcome this difficulty, replace $\phi(m,n;k)$ by $\phi^*(m,n;k) = \phi(m,n;k) - \phi(0,0;k)$, and take the limit as $k \to 0$. This process does not affect the currents, any of which is completely determined by the potential *difference* across its branch. At the same time the elliptic integrals reduce to elementary functions:

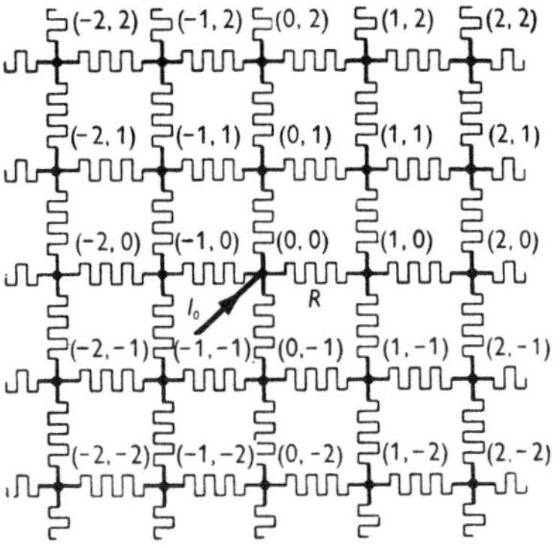

Fig. 95. Two-dimensional lattice consisting of resistances.

$$\phi^*(m,n;0) = \frac{RI_0}{2\pi}\int_{-1}^{1} \frac{1 - T_m(u)\,e^{-|n|\,\mathrm{arc\,cosh}(2-u)}}{(1-u)\sqrt{\{(1+u)(3-u)\}}}\,du. \tag{96}$$

The currents flowing across the lattice are obtained when the potential differences corresponding to (96) are divided by R. The function ϕ^*, which may be called a 'discrete potential', satisfies the following difference analogue of the two-dimensional potential equation:

$$(\Delta_m^2 + \Delta_n^2)\,\phi^*(m,n;0) = \begin{cases} -RI_0 & (m = n = 0), \\ 0 & (\text{otherwise}). \end{cases}$$

The numerical values of ϕ^*, which can be simply calculated from (96), are given in fig. 96 in the neighbourhood of the origin; RI_0 is taken equal to unity. Moreover, in fig. 97 we have plotted the currents, which are obtained from the differences of the function drawn in fig. 96, supposing that the input current i_0 at $m = n = 0$ has unit value.

This theory can be extended for the case in which the current is leaving the network at any point $m = M$, $n = N$ instead of at infinity. From an addition of the solutions (96) corresponding to a current $+J_0$ entering at $m = n = 0$ and to a current $-J_0$ entering at $m = M$, $n = N$ respectively, we derive the following expression for the resistance as measured between the points $(0, 0)$ and (m, n)

$$\frac{2}{J_0}\phi^*(m,n;0). \tag{97}$$

Thus, e.g. we find for the resistance between the neighbouring points $(0,0)$ and $(0,1)$ the value $\frac{1}{2}R$, for that between $(0,0)$ and the first diagonal point $(1,1)$ the value $\frac{2}{\pi}R$, and for that between two points separated by a 'Knights move' (i.e. between

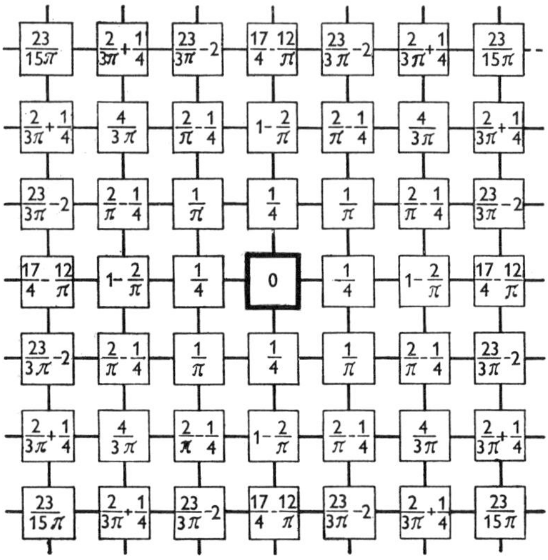

Fig. 96. The 'discrete potential' in the neighbourhood of a source.

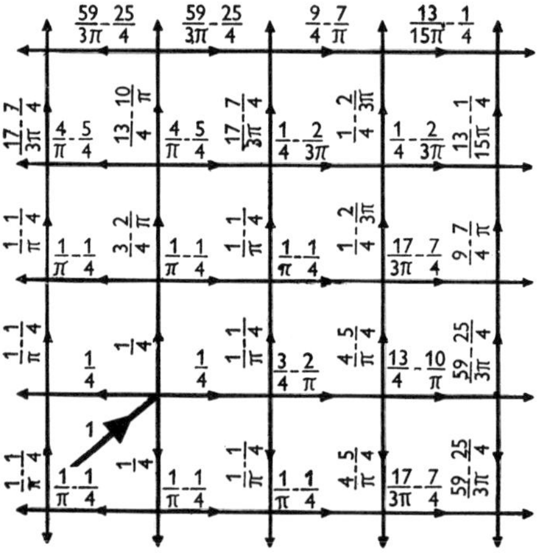

Fig. 97. The current distribution in a two-dimensional homogeneous lattice.

$(0,0)$ and $(1,2)$ or, in general, between (m,n) and $(m+1,n+2)$) the value $\left(\frac{4}{\pi}-\frac{1}{2}\right)R$. Thus (97) enables us to find the resistance between any pair of points of the square meshed network.

CHAPTER XVII

'GRAMMAR'

When the strip of convergence is given in parentheses, the relation concerned is at best valid inside the said strip; but the relation will actually hold there only so far as the corresponding definition integral does converge (that is, the relation holds in a part of the strip mentioned). For relations without a specified strip of convergence, the latter has to be investigated, given the original, for each individual case separately. The rules treated in the text have been provided with their original numbers.

1. *Rules of a general character*

$h(t) \fallingdotseq f(p)$	$\alpha < \mathrm{Re}\, p < \beta$	(II, 19)
$Ch(t) \fallingdotseq Cf(p)$	$\alpha < \mathrm{Re}\, p < \beta$	
$h(\lambda t) \fallingdotseq f\left(\frac{p}{\lambda}\right)$	$\lambda\alpha < \mathrm{Re}\, p < \lambda\beta \ (\lambda > 0)$	(IV, 1*a*)
$h(\lambda t) \fallingdotseq -f\left(\frac{p}{\lambda}\right)$	$\lambda\beta < \mathrm{Re}\, p < \lambda\alpha \ (\lambda < 0)$	(IV, 1*b*)
$\frac{\partial h(t,\nu)}{\partial \nu} \fallingdotseq \frac{\partial f(p,\nu)}{\partial \nu}$	$(\alpha < \mathrm{Re}\, p < \beta)$	(III, 24)
$\frac{dh(t)}{dt} \fallingdotseq pf(p)$	$(\alpha < \mathrm{Re}\, p < \beta)$	(IV, 31)
$\frac{d^n h(t)}{dt^n} \fallingdotseq p^n f(p)$	$(\alpha < \mathrm{Re}\, p < \beta)$	(IV, 33)
$\frac{i}{t} f\left(\frac{t}{i}\right) \fallingdotseq 2\pi p\, h\left(\frac{p}{i}\right)$	$\mathrm{Re}\, p = 0 \ (\alpha \leqslant 0 \leqslant \beta)$	(VI, 60)

2. *Transformation of the original*

$h(t+\lambda) \fallingdotseq e^{\lambda p} f(p)$	$\alpha < \operatorname{Re} p < \beta$	(IV, 4)
$e^{-\lambda t} h(t) \fallingdotseq \dfrac{p}{p+\lambda} f(p+\lambda)$	$\alpha - \operatorname{Re}\lambda < \operatorname{Re} p < \beta - \operatorname{Re}\lambda$	(IV, 12)
$\left(e^{t} \dfrac{d}{dt}\right)^{n} h(t) \fallingdotseq p(p-1)\ldots(p-n+1) f(p-n)$	$\alpha + n < \operatorname{Re} p < \beta + n$	(IV, 33*b*)
$t^{n} h(t) \fallingdotseq p\left(-\dfrac{d}{dp}\right)^{n} \left\{\dfrac{f(p)}{p}\right\}$	$\alpha < \operatorname{Re} p < \beta$	(IV, 38)
$\left(t \dfrac{d}{dt}\right)^{n} h(t) \fallingdotseq \left(-p \dfrac{d}{dp}\right)^{n} f(p)$	$\alpha < \operatorname{Re} p < \beta$	(IV, 39)
$h(-e^{-t}) \fallingdotseq \dfrac{\pi p}{\sin(\pi p)} c(-p)$	$\left(h(x) = \sum_{0}^{\infty} c_n x^n\right)$	(XI, 40)
$h(e^{-t} + i0) - h(e^{-t} - i0) \fallingdotseq 2\pi i p\, c(-p)$	(positive axis cross-cut of h)	(XI, 41)
$h([t]) \fallingdotseq (1 - e^{-p}) \sum_{n=-\infty}^{\infty} h(n) e^{-np}$	$-\log \rho < \operatorname{Re} p < \infty$	(XII, 1)
(ρ radius of convergence of power series for $h(x)$)		
$[h(t)]\, U\{h(t)\} \fallingdotseq \sum_{n=1}^{[h(\infty)]} e^{-p h^{-1}(n)}$	$(0 < \operatorname{Re} p < \infty)$	(XII, 5)
($h(t)$ increasing monotonically)		
$h(t - [t])\, U(t) \fallingdotseq \dfrac{f(p)}{1 - e^{-p}}$	$(0 < \operatorname{Re} p < \infty)$	(IV, 6)
($h(t)$ different from zero only for $0 < t < 1$)		

For one-sided originals $h(t)$ only:

$$\frac{h(t)}{t}\,U(t) \doteqdot p\int_p^\infty \frac{f(s)}{s}\,ds \qquad (0<\mathrm{Re}\,p<\infty) \qquad (\text{IV}, 42)$$

$$h\left(\frac{1}{t}\right)U(t) \doteqdot \int_0^\infty J_1\{2\sqrt{(ps)}\}\sqrt{\left(\frac{p}{s}\right)}\,f(s)\,ds \qquad (0<\mathrm{Re}\,p<\infty)\ (\alpha\leqslant 0) \qquad (\text{XI}, 14)$$

$$h(t^2)\,U(t) \doteqdot \frac{1}{\sqrt{\pi}}\int_0^\infty \mathrm{e}^{-s} f\left(\frac{p^2}{4s}\right)\frac{ds}{\sqrt{s}} \qquad (0<\mathrm{Re}\,p<\infty)\ (\alpha\leqslant 0) \qquad (\text{XI}, 18)$$

$$h(\mathrm{e}^t) \doteqdot \frac{1}{\Gamma(p)}\int_0^\infty s^{p-1} f(s)\,ds \qquad (\alpha\leqslant 0) \qquad (\text{XI}, 19)$$

$$h\{\lambda(\mathrm{e}^t-1)\}\,U(t) \doteqdot \frac{1}{\Gamma(p)}\int_0^\infty \mathrm{e}^{-s}s^{p-1} f\left(\frac{s}{\lambda}\right)ds \qquad (\alpha\leqslant 0) \qquad (\text{XI}, 17)$$

$$h(\lambda\sinh t)\,U(t) \doteqdot p\int_0^\infty J_p(\lambda s)\frac{f(s)}{s}\,ds \qquad (\alpha\leqslant 0)$$

$$U(t)\sum_{n=1}^\infty \frac{h(t/n)}{n} \doteqdot p\int_0^\infty \frac{h(s)}{\mathrm{e}^{ps}-1}\,ds \qquad (0<\mathrm{Re}\,p<\infty) \qquad (\text{XI}, 50)$$

For one-sided originals $h(t)$ having an *image* expansible in a Laurent series:

$$U(t)\,h(t+a) \doteqdot [[\mathrm{e}^{ap} f(p)]] \qquad (0<\mathrm{Re}\,p<\infty) \qquad (\text{IV}, 4a)$$

$$U(t)\frac{d^n}{dt^n}h(t) \doteqdot [[p^n f(p)]] \qquad (0<\mathrm{Re}\,p<\infty) \qquad (\text{IV}, 33a)$$

For originals $h(t)$ that are one-sided step-functions:

$$[t]^n h([t])\,U(t) \doteqdot (1-\mathrm{e}^{-p})\left(-\frac{d}{dp}\right)^n\left\{\frac{f(p)}{1-\mathrm{e}^{-p}}\right\}$$

3. *Transformation of the image*

$$\int_{-\infty}^{t} h(\tau)\,d\tau \doteq \frac{f(p)}{p} \qquad \max(\alpha,0)<\operatorname{Re}p<\beta \qquad (\text{IV},\ 34a)$$

$$\int_{+\infty}^{t} h(\tau)\,d\tau \doteq \frac{f(p)}{p} \qquad \alpha<\operatorname{Re}p<\min(\beta,0) \qquad (\text{IV},\ 34b)$$

$$\frac{1}{(n-1)!}\int_{-\infty}^{t} h(\tau)\,(t-\tau)^{n-1}\,d\tau \doteq \frac{f(p)}{p^n} \qquad (0<\operatorname{Re}p<\infty) \qquad (\text{IV},\ 35)$$

$$\int_{t}^{\infty\,\operatorname{sgn}t} \frac{h(\tau)}{\tau}\,d\tau \doteq \int_0^p \frac{f(s)}{s}\,ds \qquad (\text{IV},\ 43)$$

$$\frac{1}{2}\int_{-\infty}^{\infty} J_{[t]}(s)\,h(s)\,ds \doteq \frac{f(\sinh p)}{e^p+1}$$

For one-sided originals only:

$$U(t)\int_t^{\infty} \frac{h(\tau)}{\tau}\,d\tau \doteq \int_0^p \frac{f(s)}{s}\,ds \qquad (\alpha<0) \qquad (\text{IV},\ 40)$$

$$U(t)\int_0^{t} \frac{h(\tau)}{\tau}\,d\tau \doteq \int_p^{\infty} \frac{f(s)}{s}\,ds \qquad (0<\operatorname{Re}p<\infty) \qquad (\text{IV},\ 41a)$$

$$\int_{\infty}^{\max(0,t)} \frac{h(\tau)}{\tau}\,d\tau \doteq \int_p^{\infty} \frac{f(s)}{s}\,ds \qquad (-\infty<\operatorname{Re}p<0) \qquad (\text{IV},\ 41b)$$

$$U(t)\int_0^{\infty} \frac{\sin\{2\sqrt{(ts)}\}}{\sqrt{(\pi s)}}\,h(s)\,ds \doteq \sqrt{p}\,f\left(\frac{1}{p}\right) \qquad (0<\operatorname{Re}p<\infty)\ (\alpha\leqslant 0)$$

$$U(t)\int_0^{\infty} \left(\frac{t}{s}\right)^{\frac{1}{2}\nu} J_\nu\{2\sqrt{(ts)}\}\,h(s)\,ds \doteq \frac{f(1/p)}{p^{\nu-1}} \qquad (0<\operatorname{Re}p<\infty)\ (\alpha\leqslant 0;\ \operatorname{Re}\nu>-1) \qquad (\text{XI},\ 2)$$

$$U(t)\int_0^t J_0(2\sqrt{\{s(t-s)\}})\,h(s)\,ds \fallingdotseq \frac{f\left(p+\dfrac{1}{p}\right)}{p+\dfrac{1}{p}} \qquad (0<\operatorname{Re}p<\infty)\ (\alpha\leqslant 0) \qquad (\text{XI}, 6)$$

$$\frac{U(t)}{\sqrt{(\pi t)}}\int_0^\infty e^{-s^2/4t}\,h(s)\,ds \fallingdotseq f(\sqrt{p}) \qquad (0<\operatorname{Re}p<\infty)\ (\alpha\leqslant 0) \qquad (\text{XI}, 5)$$

$$\frac{U(t)}{2\sqrt{\pi}\,t^{3/2}}\int_0^\infty e^{-s^2/4t}\,h(s)s\,ds \fallingdotseq \sqrt{p}f(\sqrt{p}) \qquad (0<\operatorname{Re}p<\infty)\ (\alpha\leqslant 0)$$

$$U(t)\int_0^t J_0\{\sqrt{(t^2-s^2)}\}\,h(s)\,ds \fallingdotseq \frac{p}{p^2+1}f\{\sqrt{(p^2+1)}\} \qquad (0<\operatorname{Re}p<\infty)\ (\alpha\leqslant 0) \qquad (\text{XI}, 7)$$

4. *Triplets*

$$\left.\begin{aligned} h(t)\,U(t) &\fallingdotseq f(p) \\ \frac{U(t)}{\sqrt{(\pi t)}}\int_0^\infty e^{-s^2/4t}\,h(s)\,ds \equiv h_1(t)\,U(t) &\fallingdotseq f(\sqrt{p}) \\ \frac{h(\sqrt{t})}{\sqrt{t}}\,U(t) &\fallingdotseq \sqrt{(\pi p)}\,h_1\left(\frac{1}{4p}\right) \end{aligned}\right\} \qquad (0<\operatorname{Re}p<\infty) \qquad (\text{XI}, 8)$$

$$\left.\begin{aligned} h(t)\,U(t) &\fallingdotseq f(p) \\ h(t^2)\,U(t) &\fallingdotseq \frac{1}{\sqrt{\pi}}\int_0^\infty e^{-s}f\left(\frac{p^2}{4s}\right)\frac{ds}{\sqrt{s}} \equiv f_1(p) \\ \frac{f\left(\dfrac{1}{4t}\right)}{\sqrt{(\pi t)}} &\fallingdotseq \sqrt{p}f_1(\sqrt{p}) \end{aligned}\right\} \qquad (0<\operatorname{Re}p<\infty) \qquad (\text{XI}, 18)$$

$$\left.\begin{aligned} h(t)\,U(t) &\fallingdotseq f(p) \\ h(e^t) &\fallingdotseq \frac{1}{\Gamma(p)}\int_0^\infty s^{p-1}f(s)\,ds \equiv f_1(p) \\ f(e^{-t}) &\fallingdotseq \Pi(p)f_1(p) \end{aligned}\right\} \qquad (\text{XI}. 20)$$

5. *Rules depending on two or more given relations*

[Unless otherwise stated, the relations mentioned are valid in the common strip of convergence of the given relations $f_1 \risingdotseq h_1(t)$ $(\alpha_1 < \operatorname{Re} p < \beta_1)$, $f_2(p) \risingdotseq h_2(t)$ $(\alpha_2 < \operatorname{Re} p < \beta_2)$, etc., provided this common strip exists.]

$$h_1(t) + h_2(t) \fallingdotseq f_1(p) + f_2(p) \qquad \text{(III, 1)}$$

$$\int_{-\infty}^{\infty} h_1(\tau)\, h_2(t-\tau)\, d\tau \fallingdotseq \frac{1}{p} f_1(p)\, f_2(p) \qquad \text{(IV, 13)}$$

$$\int_{-\infty}^{\infty} d\tau_2 \int_{-\infty}^{\infty} d\tau_3 \ldots \int_{-\infty}^{\infty} d\tau_n\, h_1(t-\tau_2-\tau_3-\ldots-\tau_n)\, h_2(\tau_2)\, h_3(\tau_3) \ldots h_n(\tau_n) \fallingdotseq \frac{f_1(p)\, f_2(p) \ldots f_n(p)}{p^{n-1}} \qquad \text{(IV, 21)}$$

$$\underset{\tau_1+\tau_2+\ldots+\tau_n<t}{\int d\tau_1 \int d\tau_2 \ldots \int} d\tau_n\, h_1(\tau_1)\, h_2(\tau_2) \ldots h_n(\tau_n) \fallingdotseq \frac{f_1(p)\, f_2(p) \ldots f_n(p)}{p^n} \qquad \text{(IV, 23)}$$

$$\max(\alpha_1, \alpha_2, \ldots, \alpha_n, 0) < \operatorname{Re} p < \min(\beta_1, \beta_2, \ldots, \beta_n)$$

$$h_1(t)\, h_2(t) \fallingdotseq \frac{p}{2\pi i} \int_{c-i\infty}^{c+i\infty} \frac{f_1(s)}{s} \frac{f_2(p-s)}{p-s}\, ds \qquad \alpha_2 + c < \operatorname{Re} p < \beta_2 + c \quad (\alpha_1 < c < \beta_1) \qquad \text{(IV, 29)}$$

For one-sided originals $h_1(t)$ and $h_2(t)$ only:

$$U(t) \int_0^t h_1(\tau)\, h_2(t-\tau)\, d\tau \fallingdotseq \frac{1}{p} f_1(p)\, f_2(p) \qquad \text{(IV, 15)}$$

For originals $h_1(t)$ and $h_2(t)$ that are step-functions:

$$\operatorname{Lin} \sum_{n=-\infty}^{\infty} h_1(n)\, h_2([t]-n) \fallingdotseq \frac{1}{p} f_1(p)\, f_2(p) \qquad \text{(XII, 47)}$$

and in particular for one-sided step-functions:

$$U(t) \operatorname{Lin} \sum_{n=0}^{[t]} h_1(n)\, h_2([t]-n) \fallingdotseq \frac{1}{p} f_1(p)\, f_2(p) \qquad \text{(XII, 48)}$$

6. *Rules depending on a known series*

The series $\lambda(x) = \sum_{n=0}^{\infty} c_n x^n \, (|x| < \rho)$ yields:

$$U(t) \sum_{n=0}^{\infty} \frac{c_n}{n!} L_n(t) \risingdotseq \lambda\left(1 - \frac{1}{p}\right) \qquad \left(\frac{1}{\rho+1} < \operatorname{Re} p < \infty\right) \qquad (\text{XI}, 54)$$

$$\frac{U(t)}{\sqrt{(\pi t)}} \sum_{n=0}^{\infty} c_n \, e^{-n^2/4t} \risingdotseq \sqrt{p}\, \lambda(e^{-\sqrt{p}}) \qquad (0 < \operatorname{Re} p < \infty) \qquad (\text{XI}, 55)$$

$$U(t)\, t^{\frac{1}{2}\nu} \sum_{n=0}^{\infty} \frac{c_n}{n^{\frac{1}{2}\nu}} J_\nu\{2\sqrt{(nt)}\} \risingdotseq \frac{\lambda(e^{-1/p})}{p^\nu} \qquad (0 < \operatorname{Re} p < \infty)\ (\operatorname{Re} \nu > -1) \qquad (\text{XI}, 52)$$

$$\lambda(1 - e^{-t})\, U(t) \risingdotseq \sum_{n=0}^{\infty} \frac{n!\, c_n}{(p+1)(p+2)\ldots(p+n)} \qquad \max(\alpha, 0) < \operatorname{Re} p < \infty \qquad (\text{XI}, 58)$$

$$\left(\alpha = \text{abscissa of convergence of } \sum_{1}^{\infty} \frac{c_n}{n^p}\right)$$

$$\lambda(e^{-e^{-t}}) - \lambda(0) \risingdotseq \Pi(p) \sum_{n=1}^{\infty} \frac{c_n}{n^p} \qquad \max(\alpha, 0) < \operatorname{Re} p < \infty \qquad (\text{XI}, 56)$$

$$U(t) \sum_{n=0}^{\infty} c_n \frac{d}{dn}\left\{\frac{t^n}{\Pi(n)}\right\} \risingdotseq -\log p\, \lambda\left(\frac{1}{p}\right) \qquad \frac{1}{\rho} < \operatorname{Re} p < \infty \qquad (\text{XI}, 59)$$

The series $\chi(x, \alpha) = \sum_{n=-\infty}^{\infty} \phi_n(\alpha)\, x^n \, (a < |x| < b)$ yields:

$$\phi_{[t]}(\alpha) \risingdotseq (1 - e^{-p})\, \chi(e^{-p}, \alpha) \qquad -\log b < \operatorname{Re} p < -\log a \qquad (\text{XII}, 40)$$

The series $\lambda(x)\, e^{\alpha x} = \sum_{n=0}^{\infty} \phi_n(\alpha)\, x^n$ yields:

$$\phi_n(t)\, U(t) \risingdotseq \left[\left[\frac{\lambda(p)}{p^n}\right]\right] \qquad (0 < \operatorname{Re} p < \infty) \qquad (\text{XI}, 61)$$

7. *Equalities referring to a single relation*

$$\sum_{n=-\infty}^{\infty} h(n) = \sum_{n=-\infty}^{\infty} \frac{f(2\pi i n)}{2\pi i n} \qquad (\alpha \leqslant 0 \leqslant \beta) \qquad (\text{XI}, 62)$$

$$\int_{-a}^{a} h(s)\,ds = \frac{1}{\pi i}\int_{-\infty}^{\infty} f(is)\sin(as)\frac{ds}{s^2} \qquad (\alpha \leqslant 0 \leqslant \beta) \qquad (\text{XI}, 69)$$

$$\int_{-\infty}^{\infty} \frac{e^{-as}-e^{-bs}}{s} h(s)\,ds = \int_a^b \frac{f(s)}{s}\,ds \qquad ((a,b) \subset (\alpha,\beta)) \qquad (\text{XI}, 68)$$

For one-sided originals $h(t)$ only:

$$\int_0^{\infty} \frac{h(s)}{s^{\nu+1}}\,ds = \frac{1}{\Pi(\nu)}\int_0^{\infty} f(s)\,s^{\nu-1}\,ds \qquad (\alpha \leqslant 0;\ \operatorname{Re}\nu > -1) \qquad (\text{XI}, 70)$$

$$\tfrac{1}{2}\{h(t+0)+h(t-0)\} = \operatorname*{Lim}_{N\to\infty}\left\{\frac{p^{N+1}}{N!}\left(-\frac{d}{dp}\right)^N \frac{f(p)}{p}\right\}_{p=N/t} \qquad (t>0) \qquad (\text{VII}, 58)$$

$$\tfrac{1}{2}\{h(t+0)+h(t-0)\} = \operatorname*{Lim}_{p\to\infty}[[e^{pt}f(p)]] \qquad (\text{VII}, 61)$$

$$\int_{t-0}^{t+0} h(s)\,ds = \operatorname*{Lim}_{N\to\infty}\left(-\frac{e}{t}\right)^N \left\{\frac{d^N}{dp^N}\frac{f(p)}{p}\right\}_{p=N/t} \qquad (t>0) \qquad (\text{VII}, 62)$$

$$\operatorname*{Lim}_{t\to+0}\frac{h(t)}{t^\nu} \Longrightarrow \operatorname*{Lim}_{p\to+\infty}\frac{p^\nu f(p)}{\Pi(\nu)} \qquad (\nu > -1) \qquad (\text{VII}, 4)$$

$$\operatorname*{Lim}_{t\to+\infty}\frac{h(t)}{t^\nu} \Longrightarrow \operatorname*{Lim}_{p\to+0}\frac{p^\nu f(p)}{\Pi(\nu)} \qquad (\alpha \leqslant 0;\ \nu > -1) \qquad (\text{VII}, 5)$$

$$\operatorname*{Lim}_{t\to+0}\frac{\int_0^t h(s)\left(1-\frac{s}{t}\right)^n ds}{n!\,t^{\nu+1}} \Longrightarrow \operatorname*{Lim}_{p\to+\infty}\frac{p^\nu f(p)}{\Pi(\nu+n+1)} \qquad (\nu > -n-2) \qquad (\text{VII}, 8)$$

$$\operatorname*{Lim}_{t\to+\infty} \frac{\int_0^t h(s)\left(1-\frac{s}{t}\right)^n ds}{n!\,t^{\nu+1}} \Longrightarrow \operatorname*{Lim}_{p\to+0} \frac{p^\nu f(p)}{\Pi(\nu+n+1)} \qquad (\alpha \leqslant 0;\ \nu > -n-2) \qquad (\text{VII}, 9)$$

$$(C,n)\int_0^\infty h(s)\,ds \Longrightarrow \operatorname*{Lim}_{p\to+0} \frac{f(p)}{p} \qquad (\alpha \leqslant 0) \qquad (\text{VII}, 11)$$

$$\operatorname*{Lim}_{t\to+\infty} \frac{h(t)}{t^\nu} \Longleftarrow \operatorname*{Lim}_{p\to+0} \frac{p^\nu f(p)}{\Pi(\nu)} \qquad (\alpha \leqslant 0;\ \nu \geqslant -1;\ h(t) \text{ monotonic for } t>0) \qquad (\text{VII}, 15)$$

$$\operatorname*{Lim}_{t\to+\infty} \frac{\int_0^t h(s)\,ds}{t^{\nu+1}} \Longleftarrow \operatorname*{Lim}_{p\to+0} \frac{p^\nu f(p)}{\Pi(\nu+1)} \qquad (\alpha \leqslant 0;\ \nu \geqslant -1;\ Kt^\nu + h(t) \geqslant 0 \text{ for } t>0) \qquad (\text{VII}, 12)$$

8. *Equalities referring to two simultaneous relations*

$$\int_{-\infty}^\infty h_1(\lambda s)\frac{f_2(s)}{s}\,ds = \int_{-\infty}^\infty h_2(\lambda s)\frac{f_1(s)}{s}\,ds \qquad (\alpha_1 = \alpha_2 = -\infty;\ \beta_1 = \beta_2 = \infty;\ \lambda > 0) \qquad (\text{XI}, \S 8)$$

$$\int_{-\infty}^\infty h_1(s)\,h_2(-s)\,ds = \frac{1}{2\pi i}\int_{c-i\infty}^{c+i\infty} \frac{f_1(s)\,f_2(s)}{s^2}\,ds \qquad \max(\alpha_1,\alpha_2) < c < \min(\beta_1,\beta_2) \qquad (\text{XI}, 73)$$

For one-sided originals $h_1(t)$ and $h_2(t)$ only:

$$\int_0^\infty h_1(\lambda s)\frac{f_2(s)}{s}\,ds = \int_0^\infty h_2(\lambda s)\frac{f_1(s)}{s}\,ds \qquad (\alpha_1 \leqslant 0;\ \alpha_2 \leqslant 0;\ \lambda > 0) \qquad (\text{XI}, 71)$$

For originals $h_1(t)$ and $h_2(t)$ differing from zero only in the intervals $a_1 < t < b_1$ and $a_2 < t < b_2$ respectively:

$$\int_{a_1}^{b_1} h_1(s)\frac{f_2(s)}{s}\,ds = \int_{a_2}^{b_2} h_2(s)\frac{f_1(s)}{s}\,ds \qquad (\text{XI}, 72)$$

9. *Simultaneous operational calculus*

$$h(x,y) \overset{..}{\underset{..}{=}} f(p,q) \qquad \text{(XVI, 1)}$$

$$h\{\min(x,y)\} \overset{..}{\underset{..}{=}} f(p+q) \qquad (\mathrm{Re}\, p>0;\ \mathrm{Re}\, q>0) \qquad \text{(XVI, 9)}$$

$$\int_{-\infty}^{\min(x,y)} h(s)\, ds \overset{..}{\underset{..}{=}} \frac{f(p+q)}{p+q} \qquad (\mathrm{Re}\, p>0;\ \mathrm{Re}\, q>0)$$

$$h(x)\, U(y-x) \overset{..}{\underset{..}{=}} \frac{pf(p+q)}{p+q} \qquad (\mathrm{Re}\, q>0) \qquad \text{(XVI, 8)}$$

$$h(\alpha x+\beta y, \gamma x+\delta y) \overset{..}{\underset{..}{=}} \frac{pq}{|\Delta|}\left\{\frac{f(p',q')}{p'q'}\right\}_{p'=\frac{\delta p-\gamma q}{\Delta},\ q'=\frac{-\beta p+\alpha q}{\Delta}} \qquad (\Delta = \alpha\delta-\beta\gamma) \qquad \text{(XVI, 7}a\text{)}$$

$$\frac{1}{|\Delta|} h\left(\frac{\delta x-\gamma y}{\Delta},\ \frac{-\beta x+\alpha y}{\Delta}\right) \overset{..}{\underset{..}{=}} pq\left\{\frac{f(p',q')}{p'q'}\right\}_{p'=\alpha p+\beta q,\ q'=\gamma p+\delta q} \qquad \text{(XVI, 7}b\text{)}$$

$$\int_{-\infty}^{\infty} d\xi \int_{-\infty}^{\infty} d\eta\, h_1(\xi,\eta)\, h_2(x-\xi, y-\eta) \overset{..}{\underset{..}{=}} \frac{f_1(p,q)\, f_2(p,q)}{pq} \qquad \text{(XVI, 10).}$$

CHAPTER XVIII

'DICTIONARY'

Where the following relations have been treated in the text, the number of the corresponding formula is indicated. For the sake of simplicity, special values are often substituted for the parameters occurring; especially if the formula for the general value of the parameter can be found by mere application of one of the elementary rules. For the same reason we have omitted the relations that can easily be derived from other listed relations by using one of these rules. Moreover, many relations not dealt with in the text have been added. In general, the relations given are classified with respect to either their original or their image according to which of the two is the more intricate. This implies that the relations referring to some particular function will connect this function with a simpler one. Sometimes one and the same relation is contained in two groups (possible differences in parameter values being ignored). Restrictions with respect to the validity of the relations are omitted whenever they are obvious from the specification of the strip of convergence $\alpha < \operatorname{Re} p < \beta$, i.e. in so far as they are determinable from the condition $\alpha < \beta$.

1. *Polynomials*

$t^{\nu}\,U(t) \mathrel{\dot{=}} \dfrac{\Pi(\nu)}{p^{\nu}}$	$0 < \operatorname{Re} p < \infty$ $(\operatorname{Re} \nu > -1)$	(III, 3)
$L_n(t)\,U(t) \mathrel{\dot{=}} n!\left(1-\dfrac{1}{p}\right)^n$	$0 < \operatorname{Re} p < \infty$	(VI, 17)
$\dfrac{L_{[t]}(\alpha)}{[t]!} \mathrel{\dot{=}} \mathrm{e}^{\alpha/(1-\mathrm{e}^{p})}$	$0 < \operatorname{Re} p < \infty$	(XII, 44a)
$\mathrm{He}_n\!\left(\dfrac{t}{2}\right) U(t) \mathrel{\dot{=}} n!\left[\left[\dfrac{\mathrm{e}^{-p^2}}{p^n}\right]\right]$	$0 < \operatorname{Re} p < \infty$	(X, 8)

$e^{-t^2}\mathrm{He}_n(-t) \fallingdotseq \sqrt{\pi}\, p^{n+1} e^{\frac{1}{4}p^2}$	$-\infty < \mathrm{Re}\, p < \infty$	(V, 65)
$\mathrm{He}_{2n+1}(\sqrt{t})\, U(t) \fallingdotseq \dfrac{2^{2n+1}\Pi(n+\frac{1}{2})}{\sqrt{p}} \left(\dfrac{1}{p}-1\right)^n$	$0 < \mathrm{Re}\, p < \infty$	(X, 9)
$\dfrac{B_n(t)}{n!}\, U(t) \fallingdotseq \left[\left[\dfrac{1}{p^{n-1}(e^p-1)}\right]\right]$	$0 < \mathrm{Re}\, p < \infty$	(XI, 67)
$\cosh(n \operatorname{arc\,sinh} t)\, U(t) \fallingdotseq p O_n(p)$	$0 < \mathrm{Re}\, p < \infty$ (n even)	
(O_n = Neumann's polynomial)		
$\sinh(n \operatorname{arc\,sinh} t)\, U(t) \fallingdotseq p O_n(p)$	$0 < \mathrm{Re}\, p < \infty$ (n odd)	

2. *Elementary discontinuous functions*

$U(t) \fallingdotseq 1$	$0 < \mathrm{Re}\, p < \infty$	(II, 21)
$-U(-t) \fallingdotseq 1$	$-\infty < \mathrm{Re}\, p < 0$	(IV, 2)
$\operatorname{sgn} t \fallingdotseq 2$ for $p \neq 0$, 0 for $p = 0$	$\mathrm{Re}\, p = 0$	
$[t]\, U(t) \fallingdotseq \dfrac{1}{e^p - 1}$	$0 < \mathrm{Re}\, p < \infty$	(V, 8)
$\dfrac{1-e^{-\alpha[t]}}{e^\alpha - 1}\, U(t) \fallingdotseq \dfrac{1}{e^{p+\alpha}-1}$	$\max(0, -\mathrm{Re}\, \alpha) < \mathrm{Re}\, p < \infty$	
$\lambda^{[t]}\, U(t) \fallingdotseq \dfrac{1-e^{-p}}{1-\lambda e^{-p}}$	$\log \lambda < \mathrm{Re}\, p < \infty$	(XII, 3)
$\dfrac{U(t-1)}{[t]} \fallingdotseq -(1-e^{-p}) \log(1-e^{-p})$	$0 < \mathrm{Re}\, p < \infty$	

$\sum_{n=1}^{[t]} \frac{1}{n} \doteqdot -\log(1-\mathrm{e}^{-p})$	$0<\mathrm{Re}\,p<\infty$	
$\mathfrak{S}\mathrm{in}\, t\, U(t) \doteqdot \tanh\left(\frac{\pi p}{2}\right)$	$0<\mathrm{Re}\,p<\infty$	(IV, 7)
$\mathrm{Sa}(t)\, U(t) \doteqdot \frac{1}{\mathrm{e}^{p}-1} - \frac{1}{p} + \frac{1}{2}$	$0<\mathrm{Re}\,p<\infty$	(XII, 7)
$\frac{1}{t} \doteqdot \pi \mid p \mid$	$\mathrm{Re}\,p = 0$	(VI, 53)
$\frac{\sin t}{t} \doteqdot \pi p\, U(p^2+1)$	$\mathrm{Re}\,p = 0$	(VI, 52)
$\delta(t) \doteqdot p$	$-\infty<\mathrm{Re}\,p<\infty$	(V, 25)
$\delta^{(n)}(t) \doteqdot p^{n+1}$	$-\infty<\mathrm{Re}\,p<\infty$	(V, 62)
$\delta^{(n)}(1-\mathrm{e}^{-t}) \doteqdot p(p-1)(p-2)\ldots(p-n)$	$-\infty<\mathrm{Re}\,p<\infty$	(V, 68)
$1 \doteqdot 2\pi p\, \delta(ip)$	$\mathrm{Re}\,p = 0$	(VI, 50)

3. *Logarithmic functions*

$\log t\, U(t) \doteqdot -\log p - C$	$0<\mathrm{Re}\,p<\infty$	(III, 27)
$(\log t + C)\, U(t) \doteqdot -\log p$	$0<\mathrm{Re}\,p<\infty$	(III, 26)
$\frac{1-\mathrm{e}^{-t}}{t} U(t) \doteqdot p \log\left(1+\frac{1}{p}\right)$	$0<\mathrm{Re}\,p<\infty$	(VII, 30)
$\log(\mathrm{e}^{t}+1) \doteqdot \frac{\pi}{\sin(\pi p)}$	$0<\mathrm{Re}\,p<1$	

4. *Exponential functions*

$e^{-\alpha t}\,U(t) \doteq \dfrac{p}{p+\alpha}$	$-\operatorname{Re}\alpha < \operatorname{Re}p < \infty$	(III, 8)
$\tfrac{1}{2}e^{-\alpha\lvert t\rvert} \doteq \dfrac{\alpha p}{\alpha^2-p^2}$	$-\operatorname{Re}\alpha < \operatorname{Re}p < \operatorname{Re}\alpha$	(III, 11)
$\cosh(\alpha t)\,U(t) \doteq \dfrac{p^2}{p^2-\alpha^2}$	$\lvert\operatorname{Re}\alpha\rvert < \operatorname{Re}p < \infty$	(III, 9)
$\sinh(\alpha t)\,U(t) \doteq \dfrac{\alpha p}{p^2-\alpha^2}$	$\lvert\operatorname{Re}\alpha\rvert < \operatorname{Re}p < \infty$	(III, 10)
$(1-e^{-t})^n\,U(t) \doteq \dfrac{n!}{(p+1)(p+2)\dots(p+n)}$	$0 < \operatorname{Re}p < \infty$	(IV, 26)
$e^{-t^2} \doteq \sqrt{\pi}\,p\,e^{\frac{1}{4}p^2}$	$-\infty < \operatorname{Re}p < \infty$	(II, 24)
$\dfrac{\epsilon}{\pi(t^2+\epsilon^2)} \doteq p\,e^{-\epsilon\lvert p\rvert}$	$\operatorname{Re}p = 0\ (\epsilon > 0)$	(VI, 51)
$\dfrac{e^{-1/4t}}{\sqrt{(\pi t)}}\,U(t) \doteq \sqrt{p}\,e^{-\sqrt{p}}$	$0 < \operatorname{Re}p < \infty$	(XI, 4)
$\dfrac{e^{-1/4t}}{2\sqrt{\pi}\,t^{\frac{3}{2}}}\,U(t) \doteq p\,e^{-\sqrt{p}}$	$0 < \operatorname{Re}p < \infty$	
$\dfrac{\sin(2\sqrt{t})}{\sqrt{\pi}}\,U(t) \doteq \dfrac{e^{-1/p}}{\sqrt{p}}$	$0 < \operatorname{Re}p < \infty$	(VII, 27)
$\dfrac{\cos(2\sqrt{t})}{\sqrt{(\pi t)}}\,U(t) \doteq \sqrt{p}\,e^{-1/p}$	$0 < \operatorname{Re}p < \infty$	

5. *Trigonometric functions*

$\sin t\, U(t) \fallingdotseq \dfrac{p}{p^2+1}$	$0<\mathrm{Re}\, p<\infty$	(III, 22)
$\cos t\, U(t) \fallingdotseq \dfrac{p^2}{p^2+1}$	$0<\mathrm{Re}\, p<\infty$	(III, 21)
$\sin t \fallingdotseq 2\pi\delta(p^2+1)$	$\mathrm{Re}\, p = 0$	(VI, 55)
$\cos t \fallingdotseq 2\pi p \quad \delta(p^2+1)$	$\mathrm{Re}\, p = 0$	(VI, 56)
$\sin(\lvert t \rvert) \fallingdotseq \dfrac{2p}{p^2+1}$	$\mathrm{Re}\, p = 0$	(XVI, 72)
$\sin(\theta[t])\, U(t) \fallingdotseq \dfrac{\sin\theta(1-\mathrm{e}^{-p})}{2(\cosh p - \cos\theta)}$	$0<\mathrm{Re}\, p<\infty$	(XII, 42*a*)
$\sin^{2n} t\, U(t) \fallingdotseq \dfrac{(2n)!}{(p^2+2^2)(p^2+4^2)\dots(p^2+4n^2)}$	$0<\mathrm{Re}\, p<\infty$	(III, 23)
$\sin^{2n+1} t\, U(t) \fallingdotseq \dfrac{(2n+1)!\,p}{(p^2+1^2)(p^2+3^2)\dots\{p^2-(2n+1)^2\}}$	$0<\mathrm{Re}\, p<\infty$	
$(2\sin\tfrac{1}{2}t)^{2n}\, U(t) \fallingdotseq \dfrac{(2n)!}{(p^2+1^2)(p^2+2^2)\dots(p^2+n^2)}$	$0<\mathrm{Re}\, p<\infty$	(IV, 3)
$\dfrac{1}{\mathrm{e}^t+1} \fallingdotseq -\dfrac{\pi p}{\sin(\pi p)}$	$-1<\mathrm{Re}\, p<0$	(III, 12)
$\dfrac{1}{\mathrm{e}^t-1} \fallingdotseq \pi p \cot(\pi p)$	$-1<\mathrm{Re}\, p<0$	

$\dfrac{t}{e^t - 1} \risingdotseq p\left\{\dfrac{\pi}{\sin(\pi p)}\right\}^2$	$-1 < \mathrm{Re}\, p < 0$	
$\lvert \sin t \rvert\, U(t) \risingdotseq \dfrac{p}{p^2+1} \coth\left(\dfrac{\pi p}{2}\right)$	$0 < \mathrm{Re}\, p < \infty$	(VI, 29)
$\tfrac{1}{2} \coth \tfrac{1}{2}t \risingdotseq \pi p \cot(\pi p)$	$\mathrm{Re}\, p = 0$	
$\dfrac{\operatorname{sgn} t}{e^{\lvert t \rvert} - 1} \risingdotseq \pi p \cot(\pi p) - 1$	$-1 < \mathrm{Re}\, p < 1$	(VII, 53)
$\log \lvert \coth \tfrac{1}{2}t \rvert \risingdotseq \pi \tan\left(\dfrac{\pi p}{2}\right)$	$-1 < \mathrm{Re}\, p < 1$	

6. *Error functions*

$\operatorname{erfc}(-t) \risingdotseq 2 e^{\frac{1}{4}p^2}$	$0 < \mathrm{Re}\, p < \infty$	(V, 6)
$e^{-\frac{1}{4}t^2} U(t) \risingdotseq \sqrt{\pi}\, p\, e^{p^2} \operatorname{erfc} p$	$-\infty < \mathrm{Re}\, p < \infty$	(II, 25)
$\operatorname{erf}(\sqrt{t})\, U(t) \risingdotseq \dfrac{1}{\sqrt{(p+1)}}$	$0 < \mathrm{Re}\, p < \infty$	(VII, 37)
$e^t \operatorname{erfc}(\sqrt{t})\, U(t) \risingdotseq \dfrac{\sqrt{p}}{\sqrt{p}+1}$	$0 < \mathrm{Re}\, p < \infty$	
$\operatorname{erfc}\left(\dfrac{1}{2\sqrt{t}}\right) U(t) \risingdotseq e^{-\sqrt{p}}$	$0 < \mathrm{Re}\, p < \infty$	(VII, 40)
$e^{-e^{-t}} \operatorname{erfc}(e^{-\frac{1}{2}t}) \risingdotseq \dfrac{\Pi(p)}{\cos(\pi p)}$	$0 < \mathrm{Re}\, p < \tfrac{1}{2}$	
$U(t) \dfrac{2}{\sqrt{\pi}} \displaystyle\int_0^{\sqrt{t}} \sin(s^2)\, ds \risingdotseq \dfrac{\{\sqrt{(p^2+1)} - p\}^{\frac{1}{2}}}{\sqrt{\{2(p^2+1)\}}}$	$0 < \mathrm{Re}\, p < \infty$	
$U(t) \dfrac{2}{\sqrt{\pi}} \displaystyle\int_0^{\sqrt{t}} \cos(s^2)\, ds \risingdotseq \dfrac{\{\sqrt{(p^2+1)} - p\}^{-\frac{1}{2}}}{\sqrt{\{2(p^2+1)\}}}$.	$0 < \mathrm{Re}\, p < \infty$	

7. *Exponential and trigonometric integrals*

$-\mathrm{Ei}(-t)\,U(t) \doteqdot \log(p+1)$	$-1<\mathrm{Re}\,p<\infty$	(IV, 46)
$\dfrac{U(t-1)}{t} \doteqdot -p\,\mathrm{Ei}(-p)$	$0<\mathrm{Re}\,p<\infty$	(III, 34)
$\log t\,U(t-1) \doteqdot -\mathrm{Ei}(-p)$	$0<\mathrm{Re}\,p<\infty$	
$\overline{\mathrm{Ei}}(t)\,U(t) \doteqdot -\log(p-1)$	$1<\mathrm{Re}\,p<\infty$	(VI, 4)
$\mathrm{Ei}(-\mathrm{e}^{-t}) \doteqdot -\Gamma(p)$	$0<\mathrm{Re}\,p<\infty$	
$\mathrm{e}^{\mathrm{e}^{-t}}\,\mathrm{Ei}(-\mathrm{e}^{-t}) \doteqdot -\pi\dfrac{\Pi(p)}{\sin(\pi p)}$	$0<\mathrm{Re}\,p<1$	
$\mathrm{Si}(t)\,U(t) \doteqdot \operatorname{arc\,cot} p$	$0<\mathrm{Re}\,p<\infty$	(IV, 45)
$\mathrm{Ci}(t)\,U(t) \doteqdot -\log\{\sqrt{(p^2+1)}\}$	$0<\mathrm{Re}\,p<\infty$	(IV, 44)

8. *Ordinary Bessel functions*

(positive integral orders are indicated by n and general complex orders by ν)

$J_\nu(t)\,U(t) \doteqdot \dfrac{p}{\sqrt{(p^2+1)}}\{\sqrt{(p^2+1)}-p\}^\nu$	$0<\mathrm{Re}\,p<\infty$ $(\mathrm{Re}\,\nu>-1)$	(X, 12)
$I_\nu(t)\,U(t) \doteqdot \dfrac{p}{\sqrt{(p^2-1)}}\{p-\sqrt{(p^2-1)}\}^\nu$	$1<\mathrm{Re}\,p<\infty$ $(\mathrm{Re}\,\nu>-1)$	
$\dfrac{J_\nu(t)}{t}\,U(t) \doteqdot \dfrac{p}{\nu}\{\sqrt{(p^2+1)}-p\}^\nu$	$0<\mathrm{Re}\,p<\infty$ $(\mathrm{Re}\,\nu>0)$	(X, 16)

$t^{\nu} J_{\nu}(t)\, U(t) \doteq \dfrac{2^{\nu}\Pi(\nu-\frac{1}{2})}{\sqrt{\pi}} \dfrac{p}{(p^2+1)^{\nu+\frac{1}{2}}}$	$0<\mathrm{Re}\,p<\infty$ $(\mathrm{Re}\,\nu>-\frac{1}{2})$	(x, 18)
$t^{\nu} I_{\nu}(t)\, U(t) \doteq \dfrac{2^{\nu}\Pi(\nu-\frac{1}{2})}{\sqrt{\pi}} \dfrac{p}{(p^2-1)^{\nu+\frac{1}{2}}}$	$1<\mathrm{Re}\,p<\infty$ $(\mathrm{Re}\,\nu>-\frac{1}{2})$	(x, 25)
$t^{\frac{1}{2}\nu} J_{\nu}(2\sqrt{t})\, U(t) \doteq \dfrac{e^{-1/p}}{p^{\nu}}$	$0<\mathrm{Re}\,p<\infty$ $(\mathrm{Re}\,\nu>-1)$	(x, 20)
$t^{\frac{1}{2}\nu} I_{\nu}(2\sqrt{t})\, U(t) \doteq \dfrac{e^{1/p}}{p^{\nu}}$	$0<\mathrm{Re}\,p<\infty$ $(\mathrm{Re}\,\nu>-1)$	(x, 26)
$t^{\frac{1}{2}\nu}\, \mathrm{ber}_{\nu}(2\sqrt{t})\, U(t) \doteq \dfrac{\cos\left(\frac{1}{p}+\frac{3}{4}\pi\nu\right)}{p^{\nu}}$	$0<\mathrm{Re}\,p<\infty$ $(\mathrm{Re}\,\nu>-1)$	(x, 34)
$t^{\frac{1}{2}\nu}\, \mathrm{bei}_{\nu}(2\sqrt{t})\, U(t) \doteq \dfrac{\sin\left(\frac{1}{p}+\frac{3}{4}\pi\nu\right)}{p^{\nu}}$	$0<\mathrm{Re}\,p<\infty$ $(\mathrm{Re}\,\nu>-1)$	(x, 34)
$\dfrac{J_n(2\sqrt{t})}{t^{\frac{1}{2}n}}\, U(t) \doteq [[(-p)^n e^{-1/p}]]$	$0<\mathrm{Re}\,p<\infty$	(x, 22)
$t^{\frac{1}{2}(n-m)} J_m(2\sqrt{t})\, U(t) \doteq (-1)^m \dfrac{(n-m)!}{n!} \dfrac{e^{-1/p}}{p^n} L_n^{(m)}\left(\dfrac{1}{p}\right)$	$0<\mathrm{Re}\,p<\infty$	
(n and m integral and positive)		
$\dfrac{(1-t^2)^{\nu-\frac{1}{2}}}{\sqrt{\pi}\, 2^{\nu}\Pi(\nu-\frac{1}{2})}\, U(1-t^2) \doteq \dfrac{I_{\nu}(p)}{p^{\nu-1}}$	$-\infty<\mathrm{Re}\,p<\infty$ $(\mathrm{Re}\,\nu>-\frac{1}{2})$	(x, 28)
$J_{[t]}(\alpha) \doteq (1-e^{-p})\, e^{-\alpha \sinh p}$	$-\infty<\mathrm{Re}\,p<\infty$	(xii, 41)
$J_{\nu}(2e^{\frac{1}{2}t}) \doteq p\, \dfrac{\Gamma(\frac{1}{2}\nu-p)}{\Gamma(\frac{1}{2}\nu+p+1)}$	$-\frac{3}{4}<\mathrm{Re}\,p<\frac{1}{2}\mathrm{Re}\,\nu$	(xi, 23)
$\dfrac{J_{\nu}\{\sqrt{(a^2+2a\,e^{-t})}\}}{\left(1+\frac{2}{a}e^{-t}\right)^{\frac{1}{2}\nu}} \doteq \Pi(p)\, J_{\nu-p}(a)$	$0<\mathrm{Re}\,p<\frac{1}{2}\mathrm{Re}\,\nu+\frac{3}{4}$	

$(1-e^{-t})^{\frac{1}{2}\mu} J_\mu\{z\sqrt{(1-e^{-t})}\}\, U(t) \doteqdot \left(\frac{2}{z}\right)^p \Pi(p)\, J_{\mu+p}(z)$	$0<\operatorname{Re}p<\infty\ (\operatorname{Re}\mu>-1)$	
$e^{-\frac{1}{2}e^t} I_\nu(\tfrac{1}{2}e^t) \doteqdot \frac{p}{\sqrt{\pi}} \frac{\Gamma(p+\frac{1}{2})\,\Gamma(\nu-p)}{\Gamma(\nu+p+1)}$	$-\tfrac{1}{2}<\operatorname{Re}p<\operatorname{Re}\nu$	
$J_0\{\sqrt{(t^2-a^2)}\}\, U(t-a) \doteqdot \frac{p}{\sqrt{(p^2+1)}} e^{-a\sqrt{(p^2+1)}}$	$0<\operatorname{Re}p<\infty$	(XV, 13)
$I_0\{\sqrt{(t^2-a^2)}\}\, U(t-a) \doteqdot \frac{p}{\sqrt{(p^2-1)}} e^{-a\sqrt{(p^2-1)}}$	$1<\operatorname{Re}p<\infty$	(XV, 15)
$\left(\frac{t-a}{t+a}\right)^{\frac{1}{2}\nu} J_\nu\{\sqrt{(t^2-a^2)}\}\, U(t-a) \doteqdot \frac{p}{\sqrt{(p^2+1)}} \{\sqrt{(p^2+1)}-p\}^\nu e^{-a\sqrt{(p^2+1)}}$	$0<\operatorname{Re}p<\infty\ (\operatorname{Re}\nu>-1)$	(X, 48)
$\left(\frac{t-a}{t+a}\right)^{\frac{1}{2}\nu} I_\nu\{\sqrt{(t^2-a^2)}\}\, U(t-a) \doteqdot \frac{p}{\sqrt{(p^2-1)}} \{p-\sqrt{(p^2-1)}\}^\nu e^{-a\sqrt{(p^2-1)}}$	$1<\operatorname{Re}p<\infty\ (\operatorname{Re}\nu>-1)$	
$J_0\{\sqrt{(a^2-t^2)}\}\, U(a^2-t^2) \doteqdot \frac{2p}{\sqrt{(p^2-1)}} \sinh\{\lvert a\rvert \sqrt{(p^2-1)}\}$	$-\infty<\operatorname{Re}p<\infty$	
$I_0\{\sqrt{(a^2-t^2)}\}\, U(a^2-t^2) \doteqdot \frac{2p}{\sqrt{(p^2+1)}} \sinh\{\lvert a\rvert \sqrt{(p^2+1)}\}$	$-\infty<\operatorname{Re}p<\infty$	
$\frac{\cosh\{a\sqrt{(1-t^2)}\}}{\sqrt{(1-t^2)}} U(1-t^2) \doteqdot \pi p I_0\{\sqrt{(p^2+a^2)}\}$	$-\infty<\operatorname{Re}p<\infty$	(XVI, 54)
$\frac{\sin\{\lvert a\rvert\sqrt{(t^2-1)}\}}{\sqrt{(t^2-1)}} \doteqdot \pi p I_0\{\sqrt{(p^2+a^2)}\}\, U(p^2+a^2)$	$\operatorname{Re}p=0$	
$J_\nu\{2\sqrt{(at)}\}\, J_\nu\{2\sqrt{(bt)}\}\, U(t) \doteqdot e^{-(a+b)/p} I_\nu\left(\frac{2\sqrt{(ab)}}{p}\right)$	$0<\operatorname{Re}p<\infty\ (\operatorname{Re}\nu>-1)$	(XI, 9)
$\{J_0(\tfrac{1}{2}t)\}^2\, U(t) \doteqdot \frac{2p}{\pi} \frac{K\left(\frac{1}{\sqrt{(p^2+1)}}\right)}{\sqrt{(p^2+1)}}$ (K complete elliptic integral)	$0<\operatorname{Re}p<\infty$	

$$J_\mu(2\,e^{\frac{1}{2}t})\,J_\nu(2\,e^{\frac{1}{2}t}) \doteqdot \frac{p\Gamma(2p+1)\,\Gamma\left(\frac{\mu+\nu}{2}-p\right)}{\Gamma\left(\frac{\mu+\nu}{2}+1+p\right)\Gamma\left(\frac{\mu-\nu}{2}+1+p\right)\Gamma\left(\frac{\nu-\mu}{2}+1+p\right)} \qquad -\tfrac{1}{2}<\operatorname{Re}p<\tfrac{1}{2}\operatorname{Re}(\mu+\nu)$$

$$I_{2\alpha}(2z\cos\tfrac{1}{2}t)\,U(\pi^2-t^2) \doteqdot 2\pi p\,I_{\alpha+ip}(z)\,I_{\alpha-ip}(z) \qquad -\infty<\operatorname{Re}p<\infty \ (\operatorname{Re}\alpha>-\tfrac{1}{2})$$

$$U(t)\int_t^\infty \frac{J_0(s)}{s}\,ds \doteqdot \log\{\sqrt{(p^2+1)}+p\} \qquad 0<\operatorname{Re}p<\infty \qquad (x, 54)$$

$$U(t)\int_t^\infty \frac{J_\nu(s)}{s}\,ds \doteqdot \frac{1-\{\sqrt{(p^2+1)}-p\}^\nu}{\nu} \qquad 0<\operatorname{Re}p<\infty \ (\operatorname{Re}\nu>0)$$

9. *Bessel functions of the second and third kinds*

$$\overset{(1)}{H}{}^{(2)}_0(t)\,U(t) \doteqdot \frac{p}{\sqrt{(p^2+1)}}\left(1\mp\frac{2i}{\pi}\operatorname{arc\,sinh}p\right) \qquad 0<\operatorname{Re}p<\infty \qquad (x, 15)$$

$$\frac{\sin(\nu\pi)}{2}\,e^{\pm(\nu+1)\frac{1}{2}\pi i}\,\overset{(1)}{H}{}^{(2)}_\nu(t)\,U(t) \doteqdot \frac{p}{\sqrt{(p^2+1)}}\sinh(\nu\operatorname{arc\,sinh}p\pm\tfrac{1}{2}\nu\pi i) \qquad 0<\operatorname{Re}p<\infty \ (-1<\operatorname{Re}\nu<1)$$

$$Y_0(t)\,U(t) \doteqdot \frac{2}{\pi}\frac{p}{\sqrt{(p^2+1)}}\log\{\sqrt{(p^2+1)}-p\} \qquad 0<\operatorname{Re}p<\infty$$

$$\mathbf{Y}_0(2\sqrt{t})\,U(t) \doteqdot e^{-1/p}\,\overline{\operatorname{Ei}}\left(\frac{1}{p}\right) \qquad 0<\operatorname{Re}p<\infty \qquad (x, 50)$$

$$t^{\frac{1}{2}n}\mathbf{Y}_n(2\sqrt{t})\,U(t) \doteqdot \frac{1}{p^n}\left\{e^{-1/p}\,\overline{\operatorname{Ei}}\left(\frac{1}{p}\right)-\sum_{k=1}^{n}(k-1)!\,p^k\right\} \qquad 0<\operatorname{Re}p<\infty \qquad (x, 23)$$

$\dfrac{(t^2 \mp 2it)^{\nu-\frac{1}{2}}}{\sqrt{\pi}\, 2^{\nu-1}\Pi(\nu-\frac{1}{2})} U(t) \doteqdot \dfrac{e^{\mp i(p-\frac{1}{2}\pi)} H_\nu^{(1)\atop(2)}(p)}{p^{\nu-1}}$	$0 < \operatorname{Re} p < \infty$ $(\operatorname{Re}\nu > -\frac{1}{2})$	
$\dfrac{\sqrt{\pi}}{2^\nu \Pi(\nu-\frac{1}{2})} (t^2-1)^{\nu-\frac{1}{2}} U(t-1) \doteqdot \dfrac{K_\nu(p)}{p^{\nu-1}}$	$0 < \operatorname{Re} p < \infty$ $(\operatorname{Re}\nu > -\frac{1}{2})$	(x, 30)
$\sinh(\nu \operatorname{arc\,cosh} t)\, U(t-1) \doteqdot \nu K_\nu(p)$	$0 < \operatorname{Re} p < \infty$	
$\lvert \frac{1}{2}t \rvert^\nu K_\nu(\lvert t \rvert) \doteqdot \sqrt{\pi}\, \Pi(\nu-\frac{1}{2}) \dfrac{p}{(1-p^2)^{\nu+\frac{1}{2}}}$	$-1 < \operatorname{Re} p < 1$ $(\operatorname{Re}\nu > -\frac{1}{2})$	
$(2t)^{\nu-1} e^{-1/4t}\, U(t) \doteqdot p^{1-\frac{1}{2}\nu} K_\nu(\sqrt{p})$	$0 < \operatorname{Re} p < \infty$	(III, 31)
$e^{-\alpha \cosh t} \doteqdot 2p K_p(\alpha)$	$-\infty < \operatorname{Re} p < \infty$ $(\operatorname{Re}\alpha > 0)$	(III, 30)
$H_\nu^{(1)\atop(2)}(2 e^{-\frac{1}{2}t}) \doteqdot \mp \dfrac{i}{\pi} p\, e^{\pm i\pi(p-\frac{1}{2}\nu)} \Gamma(p+\frac{1}{2}\nu)\, \Gamma(p-\frac{1}{2}\nu)$	$\frac{1}{2}\lvert \operatorname{Re}\nu \rvert < \operatorname{Re} p < \frac{3}{4}$	(XI, 24)
$K_\nu(2 e^{-\frac{1}{2}t}) \doteqdot \frac{1}{2} p \Gamma(p+\frac{1}{2}\nu)\, \Gamma(p-\frac{1}{2}\nu)$	$\frac{1}{2}\lvert \operatorname{Re}\nu \rvert < \operatorname{Re} p < \infty$	(XI, 25)
$Y_\nu(2 e^{-\frac{1}{2}t}) \doteqdot -\dfrac{p}{\pi} \cos\{\pi(p-\frac{1}{2}\nu)\}\, \Gamma(p+\frac{1}{2}\nu)\, \Gamma(p-\frac{1}{2}\nu)$	$\frac{1}{2}\lvert \operatorname{Re}\nu \rvert < \operatorname{Re} p < \frac{3}{4}$	
$\dfrac{\sqrt{\pi}}{\cos(\nu\pi)} e^{\frac{1}{2}e^{-t}} K_\nu(\frac{1}{2}e^{-t}) \doteqdot p\Gamma(p+\nu)\, \Gamma(p-\nu)\, \Gamma(\frac{1}{2}-p)$	$\lvert \operatorname{Re}\nu \rvert < \operatorname{Re} p < \frac{1}{2}$	(XI, 46)
$\dfrac{K_\nu\{\sqrt{(a^2+2a e^{-t})}\}}{\left(1+\dfrac{2}{a}e^{-t}\right)^{\frac{1}{2}\nu}} \doteqdot \Pi(p)\, K_{\nu-p}(a)$	$0 < \operatorname{Re} p < \infty$	
$K_{2\nu}(2z \cosh \frac{1}{2}t) \doteqdot 2p K_{\nu+p}(z)\, K_{\nu-p}(z)$	$-\infty < \operatorname{Re} p < \infty$ $(\lvert \arg z \rvert < \frac{1}{2}\pi)$	
$-\dfrac{2}{\pi} K_0(2z \sinh \frac{1}{2}t)\, U(t) \doteqdot p\left\{J_p(z) \dfrac{\partial}{\partial p} Y_p(z) - Y_p(z) \dfrac{\partial}{\partial p} J_p(z)\right\}$	$-\infty < \operatorname{Re} p < \infty$ $(\operatorname{Re} z > 0)$	
$\dfrac{2}{\pi^2} K_0(2z \lvert \sinh \frac{1}{2}t \rvert) \doteqdot p\{J_p^2(z) + Y_p^2(z)\}$	$-\infty < \operatorname{Re} p < \infty$ $(\operatorname{Re} z > 0)$	

$\dfrac{e^{-(a+b)/t}}{t} K_\nu\left(\dfrac{2\sqrt{(ab)}}{t}\right) U(t) \doteqdot 2pK_\nu\{2\sqrt{(ap)}\}\, K_\nu\{2\sqrt{(bp)}\}$	$0 < \operatorname{Re} p < \infty$	
$\dfrac{e^{-(a+b)/t}}{t} I_\nu\left(\dfrac{2\sqrt{(ab)}}{t}\right) U(t) \doteqdot 2pI_\nu\{2\sqrt{(ap)}\}\, K_\nu\{2\sqrt{(bp)}\}$	$0 < \operatorname{Re} p < \infty$ $(\operatorname{Re}\nu > -1;\ a < b)$	
$K_0\{\sqrt{(a^2+b^2+2ab\cosh t)}\} \doteqdot 2pK_p(a)\, K_p(b)$	$-\infty < \operatorname{Re} p < \infty$ $(\operatorname{Re} a > 0;\ \operatorname{Re} b > 0)$	(IV, 19)
$e^{it^3} \doteqdot \dfrac{2}{3i}(ip)^{\frac{3}{2}} K_{\frac{1}{3}}\left\{2\left(\dfrac{ip}{3}\right)^{\frac{3}{2}}\right\}$	$\operatorname{Re} p = 0$	(XIII, 31)

10. *Spherical harmonics*†

(positive integral orders are indicated by n and m, and general complex orders by ν and μ)

$P_\nu(t)\, U(t-1) \doteqdot \sqrt{\left(\dfrac{2p}{\pi}\right)} K_{\nu+\frac{1}{2}}(p)$	$0 < \operatorname{Re} p < \infty$	(x, 57)
$\dfrac{P_\nu^\mu(t)}{(t^2-1)^{\frac{1}{2}\mu}} U(t-1) \doteqdot \sqrt{\left(\dfrac{2}{\pi}\right)} p^{\mu+\frac{1}{2}} K_{\nu+\frac{1}{2}}(p)$	$0 < \operatorname{Re} p < \infty$ $(\operatorname{Re}\mu < 1)$	(x, 84)
$P_n(t)\, U(t) \doteqdot \left[\left[p^{n+1}\left(-\dfrac{1}{p}\dfrac{d}{dp}\right)^n \left(\dfrac{e^{-p}}{p}\right)\right]\right]$	$0 < \operatorname{Re} p < \infty$	(x, 59)
$P_n(t)\, U(1-t^2) \doteqdot (-1)^n \sqrt{(2\pi p)}\, I_{n+\frac{1}{2}}(p)$	$-\infty < \operatorname{Re} p < \infty$	(x, 60)
$\dfrac{\sqrt{2}}{\sqrt{\pi}\,\Gamma(\nu-\mu+1)\,\Gamma(-\nu-\mu)} \dfrac{K_{\nu+\frac{1}{2}}(t)}{t^{\mu+\frac{1}{2}}} U(t) \doteqdot p(p^2-1)^{\frac{1}{2}\mu} P_\nu^\mu(p)$	$-1 < \operatorname{Re} p < \infty$ $(\operatorname{Re}\mu - 1 < \operatorname{Re}\nu < -\operatorname{Re}\mu)$	
$\dfrac{t^n L_n(t)}{(n!)^2} U(t) \doteqdot \dfrac{P_n\left(1-\dfrac{2}{p}\right)}{p^n}$	$0 < \operatorname{Re} p < \infty$	(x, 83)

† See also § 21.

$\sqrt{\left(\frac{\pi}{2t}\right)}\, I_{\nu+\frac{1}{2}}(t)\, U(t) \doteqdot p Q_\nu(p)$	$1 < \operatorname{Re} p < \infty$ $(\operatorname{Re}\nu > -1)$	(x, 64)
$Q_n(t) \doteqdot \pi \mid p \mid p^n \left(-\frac{1}{p}\frac{d}{dp}\right)^n \left(\frac{\sinh p}{p}\right)$	$\operatorname{Re} p = 0$	(x, 65)
$P_{2n}(e^t)\, U(t) \doteqdot \frac{(p+1)(p+3)\dots(p+2n-1)}{(p-2)(p-4)\dots(p-2n)}$	$2n < \operatorname{Re} p < \infty$	(xi, 28)
$P_{2n+1}(e^t)\, U(t) \doteqdot \frac{p(p+2)(p+4)\dots(p+2n)}{(p-1)(p-3)(p-5)\dots(p-2n-1)}$	$2n+1 < \operatorname{Re} p < \infty$	(xi, 28)
$P_\nu(e^t)\, U(t) \doteqdot \frac{2^{p-1}}{\sqrt{\pi}} \frac{\Gamma\left(\frac{p-\nu}{2}\right)\Gamma\left(\frac{p+\nu+1}{2}\right)}{\Gamma(p)}$	$\lvert \operatorname{Re}\nu + \frac{1}{2} \rvert - \frac{1}{2} < \operatorname{Re} p < \infty$	(xi, 26)
$\frac{P_\nu^\mu(e^t)}{(e^{2t}-1)^{\frac{1}{2}\mu}}\, U(t) \doteqdot \frac{2^{p+\mu-1}}{\sqrt{\pi}} \frac{\Gamma\left(\frac{p+\mu-\nu}{2}\right)\Gamma\left(\frac{p+\mu+\nu+1}{2}\right)}{\Gamma(p)}$	$\lvert \operatorname{Re}\nu + \frac{1}{2} \rvert - \operatorname{Re}\mu - \frac{1}{2} < \operatorname{Re} p < \infty$ $(\operatorname{Re}\mu < 1)$	(xi, 26)
$P_{2n}(2e^{-t}-1)\, U(t) \doteqdot \frac{(p-1)(p-2)\dots(p-n)}{(p+1)(p+2)\dots(p+n)}$	$0 < \operatorname{Re} p < \infty$	(xi, 36)
$P_\nu(2e^{-t}-1)\, U(t) \doteqdot \frac{\Gamma(p)\,\Gamma(p+1)}{\Gamma(p+\nu+1)\,\Gamma(p-\nu)}$	$0 < \operatorname{Re} p < \infty$	(xi, 35)
$\frac{(1-e^t)^{\frac{1}{2}\mu}}{(1-e^{-t})^\mu} P_\nu^\mu(2e^{-t}-1)\, U(t) \doteqdot \frac{\Gamma(p+1)\,\Gamma(p-\mu)}{\Gamma(p+\nu-\mu+1)\,\Gamma(p-\nu-\mu)}$	$0 < \operatorname{Re} p < \infty$ $(\operatorname{Re}\mu < 1)$	
$P_n(\cos t)\, U(t) \doteqdot \frac{(p^2+1^2)(p^2+3^2)\dots\{p^2+(n-1)^2\}}{(p^2+2^2)(p^2+4^2)\dots(p^2+n^2)}$	$0 < \operatorname{Re} p < \infty$ (n even)	(xiii, 37 a)
$P_n(\cos t)\, U(t) \doteqdot \frac{(p^2+0^2)(p^2+2^2)\dots\{p^2+(n-1)^2\}}{(p^2+1^2)(p^2+3^2)\dots(p^2+n^2)}$	$0 < \operatorname{Re} p < \infty$ (n odd)	(xiii, 37 b)

$\dfrac{(n-m)!}{(n+m)!}\dfrac{2^m m!}{(2m)!}\sin^m t\, P_n^m(\cos t)\, U(t) \doteqdot \dfrac{(p^2+1^2)(p^2+3^2)\ldots\{p^2+(n-m-1)^2\}}{(p^2+2^2)(p^2+4^2)\ldots\{p^2+(n+m)^2\}}$	$0<\operatorname{Re}p<\infty$ ($n+m$ even)	
$\dfrac{(n-m)!}{(n+m)!}\dfrac{2^m m!}{(2m)!}\sin^m t\, P_n^m(\cos t)\, U(t) \doteqdot \dfrac{(p^2+0^2)(p^2+2^2)\ldots\{p^2+(n-m-1)^2\}}{(p^2+1^2)(p^2+3^2)\ldots\{p^2+(n+m)^2\}}$	$0<\operatorname{Re}p<\infty$ ($n+m$ odd)	
$\dfrac{e^{-\frac{1}{2}t}}{\sqrt{2}\,(\cosh t+\cosh\psi)^{\frac{1}{2}}} \doteqdot -\dfrac{\pi p}{\sin(\pi p)}\, P_p(\cosh\psi)$	$-1<\operatorname{Re}p<0$	
$\dfrac{e^{-\frac{1}{2}t}\sinh^\mu\psi}{\sqrt{(2\pi)}\,(\cosh t+\cosh\psi)^{\mu+\frac{1}{2}}} \doteqdot p\,\dfrac{\Gamma(\mu+p+1)\,\Gamma(\mu-p)}{\Gamma(\mu+\frac{1}{2})}\, P_p^{-\mu}(\cosh\psi)$	$-1-\operatorname{Re}\mu<\operatorname{Re}p<\operatorname{Re}\mu$	
$\dfrac{U(\psi^2-t^2)}{(1-e^{t+\psi})^{\frac{1}{2}}(1-e^{t-\psi})^{\frac{1}{2}}} \doteqdot \pi i p\, P_p(\cosh\psi)$	$-\infty<\operatorname{Re}p<\infty$	
$\dfrac{e^{-\frac{1}{2}t}\,U(t-\psi)}{\sqrt{2}\,(\cosh t-\cosh\psi)^{\frac{1}{2}}} \doteqdot p\,Q_p(\cosh\psi)$	$-1<\operatorname{Re}p<\infty$	
$\dfrac{(e^{\pi i}\sinh\psi)^\mu}{\Pi(-\mu-\frac{1}{2})}\,\dfrac{e^{-\frac{1}{2}t}\,U(t-\psi)}{(\cosh t-\cosh\psi)^{\mu+\frac{1}{2}}} \doteqdot \sqrt{\left(\dfrac{2}{\pi}\right)}\, Q_p^\mu(\cosh\psi)$	$-1-\operatorname{Re}\mu<\operatorname{Re}p<\infty$ ($\operatorname{Re}\mu<\frac{1}{2}$)	
$J_\nu(at)\,J_\nu(bt)\,U(t) \doteqdot \dfrac{p}{\pi\sqrt{(ab)}}\, Q_{\nu-\frac{1}{2}}\left(\dfrac{p^2+a^2+b^2}{2ab}\right)$	$0<\operatorname{Re}p<\infty$ ($\operatorname{Re}\nu>-\frac{1}{2}$)	(XI, 10)
$P_{[t]}(\cos\theta)\,U(t) \doteqdot \dfrac{\sqrt{2}\sinh\frac{1}{2}p}{(\cosh p-\cos\theta)^{\frac{1}{2}}}$	$0<\operatorname{Re}p<\infty$	(XII, 43*a*)
$P_{[t]}^m(\cos\theta)\,U(t) \doteqdot \sqrt{\left(\dfrac{2}{\pi}\right)}\,\Pi(m-\tfrac{1}{2})\sin^m\theta\,\dfrac{\sinh\frac{1}{2}p}{(\cosh p-\cos\theta)^{m+\frac{1}{2}}}$	$0<\operatorname{Re}p<\infty$	
$\dfrac{t^n}{(n-m)!}\,J_m(t\rho)\,U(t) \doteqdot p\,\dfrac{P_n^m\left(\dfrac{p}{\sqrt{(p^2+\rho^2)}}\right)}{(p^2+\rho^2)^{\frac{1}{2}(n+1)}}$	$0<\operatorname{Re}p<\infty$	(XV, 21)

$\dfrac{t^\nu J_\mu(t)}{\Gamma(\mu+\nu+1)}\,U(t) \fallingdotseq \dfrac{p}{(p^2+1)^{\frac12(\nu+1)}}\,P_\nu^{-\mu}\left(\dfrac{p}{\sqrt{(p^2+1)}}\right)$	$0<\operatorname{Re}p<\infty$ $(\operatorname{Re}(\mu+\nu)>-1)$	
$\dfrac{t^\nu I_\mu(t)}{\Gamma(\mu+\nu+1)}\,U(t) \fallingdotseq \dfrac{p}{(p^2-1)^{\frac12(\nu+1)}}\,P_\nu^{-\mu}\left(\dfrac{p}{\sqrt{(p^2-1)}}\right)$	$1<\operatorname{Re}p<\infty$ $(\operatorname{Re}(\mu+\nu)>-1)$	
$(t^2+\rho^2)^{\frac12 n}\,P_n^m\left(\dfrac{t}{\sqrt{(t^2+\rho^2)}}\right)U(t) \fallingdotseq (n+m)!\left[\left[\dfrac{J_m(p\rho)}{p^n}\right]\right]$	$0<\operatorname{Re}p<\infty$	(xv, 23)

11. *Whittaker functions*

$t^{m-\frac12}M_{k,m}(t)\,U(t) \fallingdotseq \Pi(2m)\,p\,\dfrac{(p-\frac12)^{k-m-\frac12}}{(p+\frac12)^{k+m+\frac12}}$	$\frac12<\operatorname{Re}p<\infty$ $(m>-\frac12)$	(x, 81)
$\dfrac{(\frac12+t)^{k+m-\frac12}}{(\frac12-t)^{k-m+\frac12}}\,U(\tfrac14-t^2) \fallingdotseq \dfrac{\Pi(k+m-\frac12)\,\Pi(-k+m-\frac12)}{\Pi(2m)}\,\dfrac{M_{k,m}(p)}{p^{m-\frac12}}$	$-\infty<\operatorname{Re}p<\infty$ $(-m-\frac12<k<m+\frac12)$	
$\dfrac{(t+\frac12)^{k+m-\frac12}}{(t-\frac12)^{k-m+\frac12}}\,U(t-\tfrac12) \fallingdotseq \Pi(m-k-\tfrac12)\,\dfrac{W_{k,m}(p)}{p^{m-\frac12}}$	$0<\operatorname{Re}p<\infty$ $(k-m<\frac12)$	
$\dfrac{\Pi(2m)}{\Pi(m+k-\frac12)}\,t^{k-\frac12}J_{2m}(2\sqrt{t})\,U(t) \fallingdotseq \dfrac{e^{-1/2p}}{p^{k-1}}\,M_{k,m}\left(\dfrac{1}{p}\right)$	$0<\operatorname{Re}p<\infty$ $(m+k>-\frac12)$	
$\dfrac{\Pi(2m)}{\Pi(m-k-\frac12)}\,\dfrac{I_{2m}(2\sqrt{t})}{t^{k+\frac12}}\,U(t) \fallingdotseq p^{k+1}e^{1/2p}\,M_{k,m}\left(\dfrac{1}{p}\right)$	$0<\operatorname{Re}p<\infty$ $(m-k>-\frac12)$	(x, 80)
$e^{-\frac12 e^{-t}}M_{k,m}(e^{-t}) \fallingdotseq \dfrac{\Gamma(2m+1)}{\Gamma(k+m+\frac12)}\,\dfrac{p\Gamma(m+\frac12+p)\,\Gamma(k-p)}{\Gamma(m+\frac12-p)}$	$-m-\frac12<\operatorname{Re}p<k$	
$e^{\frac12 e^{-t}}W_{k,m}(e^{-t}) \fallingdotseq \dfrac{p\Gamma(p+m+\frac12)\,\Gamma(\frac12-m+p)\,\Gamma(-k-p)}{\Gamma(m-k+\frac12)\,\Gamma(-m-k+\frac12)}$	$\lvert m\rvert-\frac12<\operatorname{Re}p<-k$	

12. *Ordinary hypergeometric functions*

$$\frac{t^{\gamma-1}}{\Gamma(\gamma)} F(\alpha,\beta;\gamma;-t)\,U(t) \risingdotseq e^{\frac{1}{2}p}\, p^{\frac{1}{2}(\alpha+\beta+1)-\gamma}\, W_{\frac{1}{2}(1-\alpha-\beta),\,\frac{1}{2}(\alpha-\beta)}(p) \qquad 0<\operatorname{Re}p<\infty\ (\operatorname{Re}\gamma>0)$$

$$\frac{e^{\frac{1}{2}t}\, t^{\alpha-\frac{1}{2}\gamma-1}}{\Gamma(\alpha)} M_{\frac{1}{2}\gamma-\beta,\,\frac{1}{2}(\gamma-1)}(t)\,U(t) \risingdotseq p^{1-\alpha} F\left(\alpha,\beta;\gamma;\frac{1}{p}\right) \qquad 1<\operatorname{Re}p<\infty\ (\operatorname{Re}\alpha>0) \qquad (\text{x}, 82)$$

$$e^{\frac{1}{2}t}\, t^{\frac{1}{2}(\alpha+\beta-3)} W_{\frac{1}{2}(\alpha+\beta+1)-\gamma,\,\frac{1}{2}(\alpha-\beta)}(t)\,U(t) \risingdotseq \frac{\Gamma(\alpha)\,\Gamma(\beta)}{\Gamma(\gamma)}\, pF(\alpha,\beta;\gamma;1-p) \qquad 0<\operatorname{Re}p<\infty\ (\operatorname{Re}\alpha>0;\ \operatorname{Re}\beta>0)$$

$$\frac{\Gamma(\alpha)\,\Gamma(\beta)}{\Gamma(\gamma)} F(\alpha,\beta;\gamma;-e^{-t}) \risingdotseq \frac{\Gamma(p+1)\,\Gamma(\alpha-p)\,\Gamma(\beta-p)}{\Gamma(\gamma-p)} \qquad 0<\operatorname{Re}p<\min(\operatorname{Re}\alpha,\operatorname{Re}\beta) \qquad (\text{xi}, 47)$$

$$\frac{\Gamma(\alpha)\,\Gamma(\beta)}{\Gamma(\gamma)} F(\alpha,\beta;\gamma;e^{-t}\pm i0) \risingdotseq e^{\pm\pi i p}\frac{\Gamma(p+1)\,\Gamma(\alpha-p)\,\Gamma(\beta-p)}{\Gamma(\gamma-p)} \qquad 0<\operatorname{Re}p<\min(\operatorname{Re}\alpha,\operatorname{Re}\beta)$$

$$\frac{\Gamma(\alpha)\,\Gamma(\beta)}{\Gamma(\gamma)} F(\alpha,\beta;\gamma;1-e^{-t}) \risingdotseq \frac{\Gamma(p+1)\,\Gamma(\alpha-p)\,\Gamma(\beta-p)\,\Gamma(p+\gamma-\alpha-\beta)}{\Gamma(\gamma-\alpha)\,\Gamma(\gamma-\beta)} \qquad \max\{0,\operatorname{Re}(\alpha+\beta-\gamma)\}<\operatorname{Re}p<\min(\operatorname{Re}\alpha,\operatorname{Re}\beta) \qquad (\text{xi}, 48)$$

$$\frac{(1-e^{-t})^{\gamma-1}}{\Gamma(\gamma)} F(\alpha,\beta;\gamma;1-e^{-t})\,U(t) \risingdotseq \frac{\Gamma(p+1)\,\Gamma(p-\alpha-\beta+\gamma)}{\Gamma(p-\alpha+\gamma)\,\Gamma(p-\beta+\gamma)} \qquad \max\{0,\operatorname{Re}(\alpha+\beta-\gamma)\}<\operatorname{Re}p<\infty\ (\operatorname{Re}\gamma>0) \qquad (\text{xi}, 34)$$

$$2\left(\frac{b}{a}\right)^{\mu}\Gamma(\mu+1)\,J_\mu(2a\,e^{-t})\,J_\nu(2b\,e^{-t}) \risingdotseq \frac{p}{b^p}\,\frac{\Gamma\left(\frac{p+\mu+\nu}{2}\right)}{\Gamma\left(\frac{\nu-\mu-p}{2}+1\right)}\,F\left(\frac{p+\mu+\nu}{2},\frac{p+\mu-\nu}{2};\mu+1;\frac{a^2}{b^2}\right) \qquad -\operatorname{Re}(\mu+\nu)<\operatorname{Re}p<2\ (0<a<b)$$

$$4\left(\frac{a}{b}\right)^{\nu}\Gamma(\nu+1)\,K_\mu(2a\,e^{-t})\,J_\nu(2b\,e^{-t})$$
$$\risingdotseq \frac{p}{a^p}\,\Gamma\left(\frac{p+\nu+\mu}{2}\right)\Gamma\left(\frac{p+\nu-\mu}{2}\right)F\left(\frac{p+\nu+\mu}{2},\frac{p+\nu-\mu}{2};\nu+1;-\frac{b^2}{a^2}\right) \qquad |\operatorname{Re}\mu|-\operatorname{Re}\nu<\operatorname{Re}p<\infty\ (\operatorname{Re}a>|\operatorname{Im}b|)$$

13. *Generalized hypergeometric functions*

$$t^{\nu-1}{}_rF_s(\alpha;\gamma;\pm t)\,U(t) \doteq \frac{\Gamma(\nu)}{p^{\nu-1}}\,{}_{r+1}F_s(\alpha,\nu;\gamma;\pm 1/p) \qquad \left.\begin{array}{l} 0<\mathrm{Re}\,p<\infty\ (r<s) \\ 1<\mathrm{Re}\,p<\infty\ (r=s)\end{array}\right\}(\mathrm{Re}\,\nu>0) \qquad (\text{x}, 79)$$

$$t^{\nu-1}{}_rF_{s+1}(\alpha;\gamma,\nu;\pm t)\,U(t) \doteq \frac{\Gamma(\nu)}{p^{\nu-1}}\,{}_rF_s(\alpha;\gamma;\pm 1/p) \qquad \left.\begin{array}{l} 0<\mathrm{Re}\,p<\infty\,(r\leqslant s) \\ 1<\mathrm{Re}\,p<\infty\,(r=s+1)\end{array}\right\}(\mathrm{Re}\,\nu>0) \qquad (\text{x}, 78)$$

$$\frac{\Gamma(\alpha_1)\ldots\Gamma(\alpha_r)}{\Gamma(\gamma_1)\ldots\Gamma(\gamma_s)}\,{}_rF_s(\alpha;\gamma;-e^{-t}) \doteq \frac{\Gamma(p+1)\,\Gamma(\alpha_1-p)\ldots\Gamma(\alpha_r-p)}{\Gamma(\gamma_1-p)\ldots\Gamma(\gamma_s-p)} \qquad 0<\mathrm{Re}\,p<\min(\mathrm{Re}\,\alpha_1,\ldots,\mathrm{Re}\,\alpha_r) \qquad (\text{xi}, 42)$$

$$\frac{\Gamma(\alpha_1)\ldots\Gamma(\alpha_r)}{\Gamma(\gamma_1)\ldots\Gamma(\gamma_s)}\,{}_rF_s(\alpha;\gamma;-\lambda-e^{-t}) \doteq \frac{\Gamma(p+1)\,\Gamma(\alpha_1-p)\ldots\Gamma(\alpha_r-p)}{\Gamma(\gamma_1-p)\ldots\Gamma(\gamma_s-p)}\,{}_rF_s(\alpha-p;\gamma-p;-\lambda) \qquad 0<\mathrm{Re}\,p<\min(\mathrm{Re}\,\alpha_1,\ldots,\mathrm{Re}\,\alpha_r) \qquad (\text{xi}, 44)$$

14. *Gamma function*

$$e^{-e^{-t}} \doteq \Pi(p) \qquad 0<\mathrm{Re}\,p<\infty \qquad (\text{iii}, 16)$$

$$\cos(e^{-t}) \doteq \cos\left(\frac{\pi}{2}p\right)\Pi(p) \qquad 0<\mathrm{Re}\,p<1 \qquad (\text{vi}, 23)$$

$$\sin(e^{-t}) \doteq \sin\left(\frac{\pi}{2}p\right)\Pi(p) \qquad -1<\mathrm{Re}\,p<1 \qquad (\text{vi}, 24)$$

$$\cos(2\pi e^{-t}) \doteq \frac{\cos\left(\frac{\pi}{2}p\right)}{(2\pi)^p}\Pi(p) = \frac{p}{2}\frac{\zeta(1-p)}{\zeta(p)} \qquad 0<\mathrm{Re}\,p<1$$

$$\frac{1}{\pi}\sin(2\pi e^{t}) \doteq \frac{2^{p-1}\pi^p}{\cos\left(\frac{\pi}{2}p\right)\Gamma(p)} = \frac{\zeta(p)}{\zeta(1-p)} \qquad -1<\mathrm{Re}\,p<1$$

$\cos(e^{-t}+\alpha) \doteqdot \cos\left(\frac{\pi}{2}p+\alpha\right)\Pi(p)$	$0<\operatorname{Re}p<1$	
$e^{-e^{-t}}U(t) \doteqdot pP(p,1)$	$0<\operatorname{Re}p<\infty$	(III, 19a)
$e^{-e^{-t}}U(-t) \doteqdot pQ(p,1)$	$-\infty<\operatorname{Re}p<\infty$	(III, 19b)
$\frac{\lambda^{[t]}}{[t]!} \doteqdot (1-e^{-p})\,e^{\lambda e^{-p}}$	$-\infty<\operatorname{Re}p<\infty$	(XII, 4)

15. *Relations involving several gamma functions*

$\frac{(1-e^{-t})^{a-1}}{\Gamma(a)}U(t) \doteqdot \frac{\Gamma(p+1)}{\Gamma(p+a)}$	$0<\operatorname{Re}p<\infty$ $(\operatorname{Re}a>0)$	(XI, 30)
$\frac{e^{-at}(1-e^{-t})^{b-a-1}}{\Gamma(b-a)}U(t) \doteqdot p\frac{\Gamma(p+a)}{\Gamma(p+b)}$	$-a<\operatorname{Re}p<\infty$ $(b>a)$	
$e^{\frac{1}{2}(a-b+1)t}J_{a+b-1}(2e^{\frac{1}{2}t}) \doteqdot p\frac{\Gamma(a-p)}{\Gamma(b+p)}$	$\frac{1}{2}\operatorname{Re}(a-b)-\frac{1}{4}<\operatorname{Re}p<\operatorname{Re}a$	(XI, 23)
$\frac{\Gamma(a+b)\,e^{-bt}}{(1+e^{-t})^{a+b}} \doteqdot p\Gamma(a-p)\,\Gamma(b+p)$	$-\operatorname{Re}b<\operatorname{Re}p<\operatorname{Re}a$	(XI, 22)
$2K_{2a}(2e^{-\frac{1}{2}t}) \doteqdot p\Gamma(p+a)\,\Gamma(p-a)$	$\lvert\operatorname{Re}a\rvert<\operatorname{Re}p<\infty$	(XI, 25)
$e^{-\frac{1}{2}at}\,Y_a(2e^{-\frac{1}{2}t}) \doteqdot -\frac{p}{\pi}\cos(\pi p)\,\Gamma(p)\,\Gamma(p+a)$	$\max(0,-\operatorname{Re}a)<\operatorname{Re}p<\frac{3}{4}-\frac{1}{2}\operatorname{Re}a$	
$\sqrt{\pi}\,e^{\frac{1}{2}t-\frac{1}{2}e^t}\,I_{a-\frac{1}{2}}(\frac{1}{2}e^t) \doteqdot \frac{\Gamma(p+1)\,\Gamma(a-p)}{\Gamma(a+p)}$	$0<\operatorname{Re}p<\operatorname{Re}a$	
$2\sqrt{\pi}\,\frac{P^{a+b-\frac{1}{2}}_{a-b-\frac{1}{2}}(\frac{1}{2}e^t)}{(e^t-4)^{\frac{1}{2}(a+b)-\frac{1}{4}}}U(t-\log 4) \doteqdot \frac{\Gamma(p+a)\,\Gamma(p+b)}{\Gamma(2p)}$	$-\min(\operatorname{Re}a,\operatorname{Re}b)<\operatorname{Re}p<\infty$ $(\operatorname{Re}(a+b)<\frac{3}{2})$	(XI, 26)

$$e^{\frac{1}{2}(c-a)t-\frac{1}{2}e^{-t}} M_{b+\frac{1}{2}(a-c),\frac{1}{2}(a+c-1)}(e^{-t}) \doteqdot \frac{\Gamma(a+c)}{\Gamma(a+b)} p \frac{\Gamma(p+a)\,\Gamma(b-p)}{\Gamma(c-p)} \qquad -\operatorname{Re} a < \operatorname{Re} p < \operatorname{Re} b$$

$$\frac{\sqrt{\pi}}{\cos(\pi a)} e^{\frac{1}{2}e^{-t}} K_a(\tfrac{1}{2}e^{-t}) \doteqdot p\Gamma(p+a)\,\Gamma(p-a)\,\Gamma(\tfrac{1}{2}-p) \qquad |\operatorname{Re} a| < \operatorname{Re} p < \tfrac{1}{2} \qquad (\text{XI}, 46)$$

$$e^{\frac{1}{2}(e^{-t}-at)} W_{-\frac{1}{2}a-b,\frac{1}{2}(1-a)}(e^{-t}) \doteqdot \frac{p\Gamma(p+1)\,\Gamma(p+a)\,\Gamma(b-p)}{\Gamma(a+b)\,\Gamma(b+1)} \qquad -\min(\operatorname{Re} a, 1) < \operatorname{Re} p < \operatorname{Re} b$$

$$\frac{(1-e^{t})^{\frac{1}{2}a}}{(e^{t}-1)^{a}} P_b^a(2e^{-t}-1)\,U(t) \doteqdot \frac{\Gamma(p+1)\,\Gamma(p+a)}{\Gamma(p-b)\,\Gamma(p+b+1)} \qquad \max(0, -\operatorname{Re} a) < \operatorname{Re} p < \infty \ (\operatorname{Re} a < 1)$$

$$\frac{(1-e^{-t})^{b+c-a-1}}{\Gamma(b+c-a)} F(b-a, c-a; b+c-a; 1-e^{-t})\,U(t) \doteqdot \frac{\Gamma(p+1)\,\Gamma(p+a)}{\Gamma(p+b)\,\Gamma(p+c)} \qquad \max(0, -\operatorname{Re} a) < \operatorname{Re} p < \infty \ (\operatorname{Re}(b+c-a) > 0) \qquad (\text{XI}, 34)$$

$$\frac{\Gamma(a)\,\Gamma(b)}{\Gamma(c)} F(a, b; c; -e^{-t}) \doteqdot \frac{\Gamma(p+1)\,\Gamma(a-p)\,\Gamma(b-p)}{\Gamma(c-p)} \qquad 0 < \operatorname{Re} p < \min(\operatorname{Re} a, \operatorname{Re} b) \qquad (\text{XI}, 47)$$

$$\frac{\Gamma(a)\,\Gamma(b)\,\Gamma(a+c)\,\Gamma(b+c)}{\Gamma(a+b+c)} F(a, b; a+b+c; 1-e^{-t}) \doteqdot \Gamma(p+1)\,\Gamma(a-p)\,\Gamma(b-p)\,\Gamma(p+c) \qquad \max(0, -\operatorname{Re} c) < \operatorname{Re} p < \min(\operatorname{Re} a, \operatorname{Re} b) \qquad (\text{XI}, 48)$$

$$\frac{\Gamma(a_1)\ldots\Gamma(a_r)}{\Gamma(c_1)\ldots\Gamma(c_s)}\, {}_rF_s(a; c; -e^{-t}) \doteqdot \frac{\Gamma(p+1)\,\Gamma(a_1-p)\ldots\Gamma(a_r-p)}{\Gamma(c_1-p)\ldots\Gamma(c_s-p)} \qquad 0 < \operatorname{Re} p < \min(\operatorname{Re} a_1, \ldots, \operatorname{Re} a_r) \qquad (\text{XI}, 42)$$

$$\{r \geqslant s;\ r = s-1 \ \text{if} \ \operatorname{Re}(\gamma_1 + \ldots + \gamma_s - \alpha_1 - \ldots - \alpha_r) > 1\}$$

$$\frac{\Pi([t]-\nu)}{\Pi(-\nu)\,[t]!} \doteqdot (1-e^{-p})^{\nu} \qquad 0 < \operatorname{Re} p < \infty$$

$$\frac{1}{[t]!\,\Pi(\nu+[t])} \doteqdot e^{\frac{1}{2}\nu p}(1-e^{-p})\,I_{\nu}(2e^{-\frac{1}{2}p}) \qquad 0 < \operatorname{Re} p < \infty$$

$$\frac{\Gamma(2a-1)}{\Gamma(a+t)\,\Gamma(a-t)} \doteqdot p(2\cosh\tfrac{1}{2}p)^{2a-2}\,U(\pi^2+p^2) \qquad \operatorname{Re} p = 0 \ (\operatorname{Re} a > \tfrac{1}{2})$$

16. *Logarithmic derivative of the gamma function*

$-\log(1-e^{-t})\,U(t) \risingdotseq \psi(p)+C$	$-1<\operatorname{Re}p<\infty$	(VII, 50)
$\{\log t-\psi(\nu)\}\dfrac{t^{\nu}}{\Pi(\nu)}\,U(t) \risingdotseq -\dfrac{\log p}{p^{\nu}}$	$0<\operatorname{Re}p<\infty$ $(\operatorname{Re}\nu>-1)$	(III, 25)
$\dfrac{t}{e^{t}-1}\,U(t) \risingdotseq p\psi'(p)$	$-1<\operatorname{Re}p<\infty$	(VII, 56)
$\left(\dfrac{1}{e^{t}-1}-\dfrac{1}{t}\right)U(t) \risingdotseq p\{\log p-\psi(p)\}$	$0<\operatorname{Re}p<\infty$	
$\left(\dfrac{1}{e^{t}-1}-\dfrac{1}{t}+\dfrac{1}{2}\right)U(t) \risingdotseq -p\mu'(p)$	$0<\operatorname{Re}p<\infty$	(XII, 16)
$\dfrac{1}{t}\left(\dfrac{1}{e^{t}-1}-\dfrac{1}{t}+\dfrac{1}{2}\right)U(t) \risingdotseq p\mu(p)$	$0<\operatorname{Re}p<\infty$	(VII, 31)
$\dfrac{U(t)}{e^{t}+1} \risingdotseq \dfrac{p}{2}\left\{\psi\left(\dfrac{p}{2}\right)-\psi\left(\dfrac{p-1}{2}\right)\right\}$	$-1<\operatorname{Re}p<\infty$	(III, 15)
$\dfrac{e^{-\nu\lvert t\rvert}}{1-e^{t}} \risingdotseq p\{\psi(\nu+p)-\psi(\nu-1-p)\}$	$-1-\operatorname{Re}\nu<\operatorname{Re}p<\operatorname{Re}\nu$	(VI, 2)
$\dfrac{e^{-at}-e^{-bt}}{1-e^{t}}\,U(t) \risingdotseq p\{\psi(p+a)-\psi(p+b)\}$	$-1-\min(a,b)<\operatorname{Re}p<\infty$	(XI, 29)
$\{\psi([t])+C\}\,U(t-1) \risingdotseq -\log(1-e^{-p})$	$0<\operatorname{Re}p<\infty$	

17. *Riemann's ζ-function*

$[e^t] \fallingdotseq \zeta(p)$	$1 < \operatorname{Re} p < \infty$	(VI, 11)
$[e^t] - e^t \fallingdotseq \zeta(p)$	$0 < \operatorname{Re} p < 1$	(VI, 42)
$[e^t] - e^t + \frac{1}{2} = \operatorname{Sa}(e^t) \fallingdotseq \zeta(p)$	$-1 < \operatorname{Re} p < 0$	(VI, 43)
$[e^t] - e^t\, U(t) \fallingdotseq \zeta(p) - \dfrac{p}{p-1}$	$0 < \operatorname{Re} p < \infty$	(VI, 45)
$\dfrac{t^{\nu-1}}{1-e^{-t}}\, U(t) \fallingdotseq \Gamma(\nu)\, p\, \zeta(\nu, p)$	$0 < \operatorname{Re} p < \infty$ $(\operatorname{Re} \nu > 1)$	(III, 38)
$\dfrac{1}{e^{e^{-t}} - 1} \fallingdotseq \Pi(p)\, \zeta(p)$	$1 < \operatorname{Re} p < \infty$	(III, 40)
$\dfrac{1}{e^{e^{-t}} - 1} - e^t \fallingdotseq \Pi(p)\, \zeta(p)$	$0 < \operatorname{Re} p < 1$	(VI, 47)
$\dfrac{1}{e^{e^{-t}} - 1} - e^t + \frac{1}{2} \fallingdotseq \Pi(p)\, \zeta(p)$	$-1 < \operatorname{Re} p < 0$	(VI, 47)
$\dfrac{1}{e^{e^{-t}} - 1} - \sum\limits_{n=0}^{2N} \dfrac{B_n}{n!} e^{-(n-1)t} \fallingdotseq \Pi(p)\, \zeta(p)$	$-2N - 1 < \operatorname{Re} p < -2N + 1$ $(N \geqslant 1)$	(VI, 47)
$\log\left\{\dfrac{\sinh(\pi e^t)}{\pi e^t}\right\} \fallingdotseq \dfrac{\pi}{\sin(\frac{1}{2}\pi p)} \zeta(p)$	$1 < \operatorname{Re} p < 2$	
$\log\{\Pi(e^t)\} + C e^t \fallingdotseq -\dfrac{\pi}{\sin(\pi p)} \zeta(p)$	$1 < \operatorname{Re} p < 2$	
$\psi([e^t]) + C \fallingdotseq \zeta(p+1)$	$0 < \operatorname{Re} p < \infty$	

$U(t)\sum_{n=1}^{[e^t]}\frac{1}{n^\nu} \doteqdot \zeta(p+\nu)$	$\max(0, 1-\mathrm{Re}\,\nu) < \mathrm{Re}\,p < \infty$	
$U(t)\sum_{n=1}^{[e^t]} n^\nu \doteqdot \zeta(p-\nu)$	$\max(0, 1+\mathrm{Re}\,\nu) < \mathrm{Re}\,p < \infty$	
$\sum_{n=1}^{[e^t]}\frac{\log n}{n} \doteqdot -\zeta'(p+1)$	$0 < \mathrm{Re}\,p < \infty$	
$\log([e^t]!) \doteqdot -\zeta'(p)$	$1 < \mathrm{Re}\,p < \infty$	
$\psi_T(e^t) \doteqdot -\frac{\zeta'(p)}{\zeta(p)}$	$1 < \mathrm{Re}\,p < \infty$	(XII, 30)
$\sum_{n=1}^{[e^t]}\mu(n) \doteqdot \frac{1}{\zeta(p)}$	$1 < \mathrm{Re}\,p < \infty$	(XII, 26*a*)
$\sum_{n=1}^{[e^t]}\mu^2(n) \doteqdot \frac{\zeta(p)}{\zeta(2p)}$	$1 < \mathrm{Re}\,p < \infty$	(XII, 27*a*)
$\pi(e^t) \doteqdot \sum_{n=1}^{\infty}\frac{\mu(n)}{n}\log\zeta(np)$	$1 < \mathrm{Re}\,p < \infty$	(XII, 37)
$\sum_{n=1}^{\infty}\frac{\pi(e^{t/n})}{n} \doteqdot \log\zeta(p)$	$1 < \mathrm{Re}\,p < \infty$	
$\frac{U(t)}{\Gamma(\nu)}\int_0^t \frac{e^{-ax}x^{\nu-1}}{1-e^{-x}}\,dx \doteqdot \zeta(\nu, p+a)$	$-\mathrm{Re}\,a < \mathrm{Re}\,p < \infty$ $(\mathrm{Re}\,\nu > 1)$	(IV, 37)
$\int_0^{e^{-t}} \frac{e^{-ax}x^{\nu-1}}{1-e^{-x}}\,dx \doteqdot -\Gamma(p+\nu)\,\zeta(p+\nu, a)$	$1-\mathrm{Re}\,\nu < \mathrm{Re}\,p < 0$ $(\mathrm{Re}\,a > 0)$	
$\int_{e^{-t}}^{\infty} \frac{e^{-ax}x^{\nu-1}}{1-e^{-x}}\,dx \doteqdot \Gamma(p+\nu)\,\zeta(p+\nu, a)$	$\max(0, 1-\mathrm{Re}\,\nu) < \mathrm{Re}\,p < \infty$ $(\mathrm{Re}\,a > 0)$	

18. *Theta functions*

$\theta_0(0,t)\,U(t) \doteqdot \dfrac{\sqrt{(\pi p)}}{\sinh\{\sqrt{(\pi p)}\}}$	$0<\mathrm{Re}\,p<\infty$	
$\theta_1'(0,t)\,U(t) \doteqdot -\dfrac{2\pi p}{\cosh\{\sqrt{(\pi p)}\}}$	$0<\mathrm{Re}\,p<\infty$	
$\theta_2(0,t)\,U(t) \doteqdot \sqrt{(\pi p)}\tanh\{\sqrt{(\pi p)}\}$	$0<\mathrm{Re}\,p<\infty$	
$\theta_3(0,t)\,U(t) \doteqdot \sqrt{(\pi p)}\coth\{\sqrt{(\pi p)}\}$	$0<\mathrm{Re}\,p<\infty$	(VII, 51)
$\theta_0(v,t)\,U(t) \doteqdot \sqrt{(\pi p)}\dfrac{\cosh\{2v\sqrt{(\pi p)}\}}{\sinh\{\sqrt{(\pi p)}\}}$	$0<\mathrm{Re}\,p<\infty\ (-\tfrac{1}{2}\leqslant v\leqslant\tfrac{1}{2})$	
$\theta_1(v,t)\,U(t) \doteqdot -\sqrt{(\pi p)}\dfrac{\sinh\{2v\sqrt{(\pi p)}\}}{\cosh\{\sqrt{(\pi p)}\}}$	$0<\mathrm{Re}\,p<\infty\ (-\tfrac{1}{2}\leqslant v\leqslant\tfrac{1}{2})$	
$\theta_2(v,t)\,U(t) \doteqdot \sqrt{(\pi p)}\dfrac{\sinh\{(1-2v)\sqrt{(\pi p)}\}}{\cosh\{\sqrt{(\pi p)}\}}$	$0<\mathrm{Re}\,p<\infty\ (0\leqslant v\leqslant 1)$	
$\theta_3(v,t)\,U(t) \doteqdot \sqrt{(\pi p)}\dfrac{\cosh\{(1-2v)\sqrt{(\pi p)}\}}{\sinh\{\sqrt{(\pi p)}\}}$	$0<\mathrm{Re}\,p<\infty\ (0\leqslant v\leqslant 1)$	(XI, 13)
$\left[\sqrt{\dfrac{t}{\pi}}\right]U(t) \doteqdot \tfrac{1}{2}\{\theta_3(0,p)-1\}$	$0<\mathrm{Re}\,p<\infty$	(VI, 39)
$-\tfrac{1}{2}\{\theta_3(0,\mathrm{e}^{-2t})-1\} \doteqdot \dfrac{\Pi(\frac{1}{2}p)}{\pi^{\frac{1}{2}p}}\zeta(p)$	$1<\mathrm{Re}\,p<\infty$	(VI, 48)
$\left(\dfrac{d^2}{dt^2}-\dfrac{1}{4}\right)\{\mathrm{e}^{\frac{1}{2}t}\theta_3(0,\mathrm{e}^{2t})\} \doteqdot 2p\,\xi(p+\tfrac{1}{2})$	$-\infty<\mathrm{Re}\,p<\infty$	(VI, 49)
$A_2(t) \doteqdot \{\theta_3(0,p/\pi)\}^2$	$0<\mathrm{Re}\,p<\infty$	(XII, 19)
$A_n(t) \doteqdot \{\theta_3(0,p/\pi)\}^n$	$0<\mathrm{Re}\,p<\infty$	(XII, 19)

19. *Elementary simultaneous relations*†

$$-\tfrac{1}{2}U(x_2-|x_1|) \stackrel{\cdot\cdot}{\underset{\cdot\cdot}{=}} \frac{p_1p_2}{p_1^2-p_2^2} \qquad \operatorname{Re}p_2>|\operatorname{Re}p_1| \qquad (\text{XVI}, 45b)$$

$$-\tfrac{1}{2}I_0(\sqrt{\{a(x_2^2-x_1^2)\}})\,U(x_2-|x_1|) \stackrel{\cdot\cdot}{\underset{\cdot\cdot}{=}} \frac{p_1p_2}{p_1^2-p_2^2+a} \qquad \operatorname{Re}p_2>|\operatorname{Re}\sqrt{(p_1^2+a)}| \qquad (\text{XVI}, 53)$$

$$-\tfrac{1}{2}J_0(\sqrt{\{a(x_2^2-x_1^2)\}})\,U(x_2-|x_1|) \stackrel{\cdot\cdot}{\underset{\cdot\cdot}{=}} \frac{p_1p_2}{p_1^2-p_2^2-a} \qquad \operatorname{Re}p_2>|\operatorname{Re}\sqrt{(p_1^2-a)}| \qquad (\text{XVI}, 45c)$$

$$\frac{x_1}{x_2}J_{x_1}(x_2)\,U(x_1)\,U(x_2) \stackrel{\cdot\cdot}{\underset{\cdot\cdot}{=}} \frac{p_1p_2}{p_1+\operatorname{arc\,sinh}p_2} \qquad \begin{cases}\operatorname{Re}(p_1+\operatorname{arc\,sinh}p_2)>0\\ \operatorname{Re}p_2>0\end{cases} \qquad (\text{XVI}, 4)$$

$$\theta_3(\tfrac{1}{2}x_1, \pi x_2)\,U(x_1)\,U(x_2) \stackrel{\cdot\cdot}{\underset{\cdot\cdot}{=}} \frac{p_1^2\sqrt{p_2}\coth(\sqrt{p_2})-p_1p_2\coth p_1}{p_1^2-p_2} \qquad \operatorname{Re}p_{1,2}>0$$

$$\phi_1=\arccos\left(\frac{x_2+x_3-x_1}{2\sqrt{(x_2x_3)}}\right) \stackrel{\cdots}{\underset{\cdots}{=}} \frac{\pi}{p_1\left(\dfrac{1}{p_1}+\dfrac{1}{p_2}+\dfrac{1}{p_3}\right)} \qquad \operatorname{Re}p_{1,2,3}>0 \qquad (\text{XVI}, 21)$$

$$\frac{U(\Delta)}{\Delta(2\sqrt{x_1}, 2\sqrt{x_2}, 2\sqrt{x_3})} \stackrel{\cdots}{\underset{\cdots}{=}} \frac{\pi}{\dfrac{1}{p_1}+\dfrac{1}{p_2}+\dfrac{1}{p_3}} \qquad \operatorname{Re}p_{1,2,3}>0 \qquad (\text{XVI}, 19)$$

$$\frac{\{\Delta(2\sqrt{x_1}, 2\sqrt{x_2}, 2\sqrt{x_3})\}^{2\nu-1}}{\Pi(2\nu)}U(\Delta) \stackrel{\cdots}{\underset{\cdots}{=}} \frac{\pi}{(p_1p_2p_3)^\nu\left(\dfrac{1}{p_1}+\dfrac{1}{p_2}+\dfrac{1}{p_3}\right)^{\nu+1}} \qquad \operatorname{Re}p_{1,2,3}>0\ (\operatorname{Re}\nu>-\tfrac{1}{2}) \qquad (\text{XVI}, 18)$$

† In this and the following sections the operational variables $p_1, p_2, \ldots, p_n, q$ correspond to $x_1, x_2, \ldots, x_n, t$ respectively.

20. *Two-dimensional wave and potential functions*

$$[\rho = \sqrt{(x_1^2+x_2^2)}]$$

$$-\frac{U(t-\rho)}{2\pi\sqrt{(t^2-\rho^2)}} \overset{\cdots}{\underset{\cdots}{=}} \frac{p_1 p_2 q}{p_1^2+p_2^2-q^2}$$ $\mathrm{Re}\, q > |\mathrm{Re}\sqrt{(p_1^2+p_2^2)}|$ (XVI, 56)

$$-\frac{\cosh(\sqrt{\{a(t^2-\rho^2)\}})}{2\pi\sqrt{(t^2-\rho^2)}}\, U(t-\rho) \overset{\cdots}{\underset{\cdots}{=}} \frac{p_1 p_2 q}{p_1^2+p_2^2-q^2+a}$$ $\mathrm{Re}\, q > |\mathrm{Re}\sqrt{(p_1^2+p_2^2+a)}|$ (XVI, 55)

$$-\tfrac{1}{4} i H_0^{(1)}(k\rho) \overset{\cdot\cdot}{\underset{\cdot\cdot}{=}} \frac{p_1 p_2}{p_1^2+p_2^2+k^2}$$ $|\mathrm{Re}\sqrt{(p_1^2+p_2^2)}| < \mathrm{Im}\, k$ (XVI, 62)

$$\tfrac{1}{4} Y_0(|k|\rho) \overset{\cdot\cdot}{\underset{\cdot\cdot}{=}} \frac{p_1 p_2}{p_1^2+p_2^2+k^2}$$ $\mathrm{Re}\sqrt{(p_1^2+p_2^2)} = 0$ (k real) (XVI, 63)

$$J_0(k\rho) \overset{\cdot\cdot}{\underset{\cdot\cdot}{=}} 4\pi p_1 p_2 \delta(p_1^2+p_2^2+k^2)$$ $\mathrm{Re}\sqrt{(p_1^2+p_2^2)} = 0$ (k real) (XVI, 64)

$$J_0(k\rho)\, U(x_1)\, U(x_2) \overset{\cdot\cdot}{\underset{\cdot\cdot}{=}} \frac{p_1 p_2}{(p_1^2+p_2^2+k^2)}\left(\frac{p_1}{\sqrt{(p_2^2+k^2)}}+\frac{p_2}{\sqrt{(p_1^2+k^2)}}\right)$$ $\mathrm{Re}\, p_{1,2} > 0$

$$\log\rho\, U(x_1)\, U(x_2) \overset{\cdot\cdot}{\underset{\cdot\cdot}{=}} \frac{\frac{1}{2}\pi p_1 p_2 - p_1^2 \log p_2 - p_2^2 \log p_1}{p_1^2+p_2^2} - C$$ $\mathrm{Re}\, p_{1,2} > 0$ (XVI, 65)

21. *Three-dimensional potential and wave functions*

Polar coordinates: $r = \sqrt{(x_1^2+x_2^2+x_3^2)}$, $\theta = \arccos(x_3/r)$, $\phi = \arctan(x_2/x_1)$

$$-\frac{1}{4\pi r} \overset{\cdots}{\underset{\cdots}{=}} \frac{p_1 p_2 p_3}{p_1^2+p_2^2+p_3^2}$$ $\mathrm{Re}\sqrt{(p_1^2+p_2^2+p_3^2)} = 0$ (XVI, 70)

$$\frac{P_n(\cos\theta)}{r^{n+1}} \overset{\cdots}{\underset{\cdots}{=}} \frac{4\pi}{n!}\frac{p_1 p_2 (-p_3)^{n+1}}{p_1^2+p_2^2+p_3^2}$$ $\mathrm{Re}\sqrt{(p_1^2+p_2^2+p_3^2)} = 0$

$\dfrac{Q_n(\cos\theta)}{r^{n+1}} \overset{\cdots}{\underset{\cdots}{=}} \dfrac{2\pi^2}{in!}\dfrac{p_1p_2(-p_3)^{n+1}}{p_1^2+p_2^2+p_3^2}\operatorname{sgn}\left(\dfrac{p_3}{i}\right)$	$\begin{cases}\operatorname{Re}\sqrt{(p_1^2+p_2^2+p_3^2)}=0\\ \operatorname{Re}p_3=0\end{cases}$	
$\dfrac{P_n(\cos\theta)\pm\dfrac{2i}{\pi}Q_n(\cos\theta)}{r^{n+1}} \overset{\cdots}{\underset{\cdots}{=}} \dfrac{8\pi}{n!}\dfrac{p_1p_2(-p_3)^{n+1}}{p_1^2+p_2^2+p_3^2}U\left(\pm\dfrac{p_3}{i}\right)$	$\begin{cases}\operatorname{Re}\sqrt{(p_1^2+p_2^2+p_3^2)}=0\\ \operatorname{Re}p_3=0\end{cases}$	
$\dfrac{(n-m)!}{4\pi}\dfrac{P_n^m(\cos\theta)\,\mathrm{e}^{im\phi}}{r^{n+1}} \overset{\cdots}{\underset{\cdots}{=}} (p_1+ip_2)^m\dfrac{p_1p_2(-p_3)^{n-m+1}}{p_1^2+p_2^2+p_3^2}$	$\operatorname{Re}\sqrt{(p_1^2+p_2^2+p_3^2)}=0$	
$-\dfrac{\mathrm{e}^{ikr}}{4\pi r} \overset{\cdots}{\underset{\cdots}{=}} \dfrac{p_1p_2p_3}{p_1^2+p_2^2+p_3^2+k^2}$	$\lvert\operatorname{Re}\sqrt{(p_1^2+p_2^2+p_3^2)}\rvert<\operatorname{Im}k$	(XVI, 67)
$-\dfrac{\cos(kr)}{4\pi r} \overset{\cdots}{\underset{\cdots}{=}} \dfrac{p_1p_2p_3}{p_1^2+p_2^2+p_3^2+k^2}$	$\lvert\operatorname{Re}\sqrt{(p_1^2+p_2^2+p_3^2)}\rvert=0$ (k real)	(XVI, 68)
$\dfrac{\sin(kr)}{4\pi^2 r} \overset{\cdots}{\underset{\cdots}{=}} p_1p_2p_3\,\delta(p_1^2+p_2^2+p_3^2+k^2)$	$\lvert\operatorname{Re}\sqrt{(p_1^2+p_2^2+p_3^2)}\rvert=0$ (k real)	(XVI, 69)
$\dfrac{i^{n-1}}{4}\sqrt{\left(\dfrac{k}{2\pi r}\right)}H_{n+\frac12}^{(1)}(kr)\,P_n^m(\cos\theta)\,\mathrm{e}^{im\phi} \overset{\cdots}{\underset{\cdots}{=}} (p_1+ip_2)^m\dfrac{p_1p_2p_3}{p_1^2+p_2^2+p_3^2+k^2}\dfrac{d^m}{dp_3^m}P_n\left(\dfrac{p_3}{ik}\right)$	$\lvert\operatorname{Re}\sqrt{(p_1^2+p_2^2+p_3^2)}\rvert<\operatorname{Im}k$	
$-\dfrac{\delta(t-r)}{4\pi r} \overset{::}{\underset{::}{=}} \dfrac{p_1p_2p_3q}{p_1^2+p_2^2+p_3^2-q^2}$	$\lvert\operatorname{Re}\sqrt{(p_1^2+p_2^2+p_3^2)}\rvert<\operatorname{Re}q$	(XVI, 84)
$\dfrac{\delta(a-r)}{4\pi a^2} \overset{\cdots}{\underset{\cdots}{=}} p_1p_2p_3\dfrac{\sinh\{a\sqrt{(p_1^2+p_2^2+p_3^2)}\}}{a\sqrt{(p_1^2+p_2^2+p_3^2)}}$	$-\infty<\operatorname{Re}p_{1,2,3}<\infty$	(XVI, 79)

22. *n-dimensional potential and wave functions*

$$[r = \sqrt{(x_1^2 + x_2^2 + \ldots + x_n^2)}]$$

$$-\frac{\Gamma(\frac{1}{2}n-1)}{4\pi^{\frac{1}{2}n}}\frac{1}{r^{n-2}} \fallingdotseq \frac{p_1 p_2 \cdots p_n}{p_1^2+p_2^2+\ldots+p_n^2} \qquad \operatorname{Re}\sqrt{(p_1^2+p_2^2+\ldots+p_n^2)} = 0 \quad (n \geqslant 3) \qquad (\text{XVI}, 71)$$

$$\frac{U(a-r)}{(2\pi a)^{\frac{1}{2}n}} \fallingdotseq \frac{p_1 p_2 \cdots p_n}{(p_1^2+p_2^2+\ldots+p_n^2)^{\frac{1}{2}n}} I_{\frac{1}{2}n}\{a\sqrt{(p_1^2+p_2^2+\ldots+p_n^2)}\} \qquad -\infty < \operatorname{Re} p_{1,2,\ldots,n} < \infty \qquad (\text{XVI}, 76)$$

$$\frac{\delta(a-r)}{(2\pi a)^{\frac{1}{2}n}} \fallingdotseq \frac{p_1 p_2 \cdots p_n}{(p_1^2+p_2^2+\ldots+p_n^2)^{\frac{1}{4}(n-2)}} I_{\frac{1}{2}n-1}\{a\sqrt{(p_1^2+p_2^2+\ldots+p_n^2)}\} \qquad -\infty < \operatorname{Re} p_{1,2,\ldots,n} < \infty \qquad (\text{XVI}, 78)$$

For n odd:

$$-\frac{1}{2}\left(-\frac{1}{2\pi r}\frac{d}{dr}\right)^{\frac{1}{2}(n-1)} U(t-r) \fallingdotseq \frac{p_1 p_2 \cdots p_n q}{p_1^2+p_2^2+\ldots+p_n^2-q^2} \qquad |\operatorname{Re}\sqrt{(p_1^2+p_2^2+\ldots+p_n^2)}| < \operatorname{Re} q \qquad (\text{XVI}, 57)$$

$$-\frac{1}{2}\left(-\frac{1}{2\pi r}\frac{d}{dr}\right)^{\frac{1}{2}(n-1)} \left[I_0(\sqrt{\{a(t^2-r^2)\}})\, U(t-r)\right] \fallingdotseq \frac{p_1 p_2 \cdots p_n q}{p_1^2+p_2^2+\ldots+p_n^2-q^2+a} \qquad |\operatorname{Re}\sqrt{(p_1^2+\ldots+p_n^2+a)}| < \operatorname{Re} q \qquad (\text{XVI}, 57)$$

For n even:

$$-\frac{U(t-r)}{2\pi}\left(\frac{1}{2\pi r}\frac{d}{dt}\right)^{\frac{1}{2}n-1} \left[\frac{\cosh\{(\frac{1}{2}n-1)\operatorname{arc\,cosh}(t/r)\}}{\sqrt{(t^2-r^2)}}\right] \fallingdotseq \frac{p_1 p_2 \cdots p_n q}{p_1^2+p_2^2+\ldots+p_n^2-q^2} \qquad |\operatorname{Re}\sqrt{(p_1^2+p_2^2+\ldots+p_n^2)}| < \operatorname{Re} q \qquad (\text{XVI}, 57)$$

$$-\frac{U(t-r)}{2\pi} e^{t\sqrt{a}}\left(\frac{1}{2\pi r}\frac{d}{dt}\right)^{\frac{1}{2}n-1} \left[\frac{e^{-t\sqrt{a}}\cosh\{(\frac{1}{2}n-1)\operatorname{arc\,cosh}(t/r)+\sqrt{[a(t^2-r^2)]}\}}{\sqrt{(t^2-r^2)}}\right]$$

$$\fallingdotseq \frac{p_1 p_2 \cdots p_n q}{p_1^2+p_2^2+\ldots+p_n^2-q^2+a} \qquad |\operatorname{Re}\sqrt{(p_1^2+\ldots+p_n^2+a)}| < \operatorname{Re} q \qquad (\text{XVI}, 57)$$

LIST OF AUTHORS QUOTED

(THE NUMBERS REFER TO THE PAGES)

GENERAL INDEX

The numbers refer to the pages. If an item is the main subject of several pages the beginning page has been indicated. For subjects occurring frequently only the most important references are given. Principal references are in black type.

CHELSEA

SCIENTIFIC

BOOKS

ANALYSIS

(See also: Differential Equations, General, Group Theory)

BANACH, Stefan: Theorie des Operations Lineaires. (French) xii + 250 pp. 5⅜ x 8. ISBN -0110-1, ∞G

BERBERIAN, Sterling K.: Introduction to Hilbert Space, 2nd ed. vi + 206 pp. 5⅜ x 8. CIP. ISBN -0287-6, ∞G

BOCHNER, Salomon: Fouriersche Integrale. (Germ.) vi + 229 pp. 5½ x 8½. ISBN -0042-3, G

BOHR, Harald: Almost Periodic Functions. 120 pp. 6 x 9. (lithotyped). ISBN -0027-X, ∞G

BOLZA, Oskar: Lectures on the Calculus of Variations, 3rd ed. 280 pp. 5⅜ x 8. ISBN -0145-4, ∞G

———Vorlesungen ueber Variationsrechnung. (Germ.) ix + 715 pp. 5⅜ x 8. ISBN -0160-8, ∞G

CARATHEODORY, Constantin: Calculus of Variations and Partial Differential Equations, 2nd (revised) ed. xvi + 401 pp. 6 x 9. ISBN -0318-X, ∞G

CORDUNEANU, Constantin: Principles of Differential and Integral Equations, 2nd ed., x + 205 pp. 6 x 9. ISBN -0295-7, T.O.P.

EDWARDS, Joseph: The Integral Calculus, 2 vols. 1,922 pp. 5 x 8. Vol. I, ISBN -0102-0, ∞G Vol. II, ISBN -0105-5, ∞G

EVANS, Griffith C.: Logarithmic Potential and Other Monographs. Logarithmic Potential, 2nd ed.; & BLISS, G. A. Fundamental Existence Theorems, & KASNER, E. Differential-Geometric Aspects of Dynamics. 3 vols in 1. 399 pp. 5⅜ x 8. ISBN-0305-8, ∞G

FORD, Lester R.: Automorphic Functions. 343 pp. 5⅜ x 8. ISBN -0085-7, ∞G

FORD, Walter B.: Studies in Divergent Series and Summability; Asymptotic Developments of Functions. 371 pp. 6 x 9. ISBN -0143-8, ∞G

GAMELIN, Theodore W.: Uniform Algebras, 2nd ed. xiii + 263 pp. 6 x 9. ISBN -0311-2, ∞G

GELFAND, Israel M., RAIKOV, D. A. & SHILOV, G. E.: Commutative Normed Rings. 306 pp. 6 x 9. ISBN -0170-5, ∞G

GOFFMAN, Caspar & PEDRICK, George: A First Course in Functional Analysis, 2nd. ed. xiv + 284 pp. 6 x 9. ISBN -0319-8, ∞G

GUNTHER, N.: Integrales de Stieltjes (French) iv + 494 pp. 5⅜ x 8. ISBN -0063-6, G

HALMOS, Paul R.: Introduction to Hilbert Space. 120 pp. 6 x 9. ISBN -0082-2, ∞G

———Lectures on Ergodic Theory, viii + 101 pp. 5⅜ x 8. ISBN -0142-X, ∞G

ANALYSIS

(See also: Differential Equations, General, Group Theory)

HOBSON, Ernest W.: Spherical and Ellipsoidal Harmonics. xi + 500 pp. 5⅜ x 8. ISBN -0104-7, ∞G

KRAZER, Adolf: Lehrbuch der Thetafunktionen. (Germ.) xxiv + 509 pp. 5⅜ x 8. ISBN -0244-2, ∞G

LANDAU, Edmund: Differential and Integral Calculus, 372 pp. 6 x 9. ISBN -0078-4, ∞G

———Foundations of Analysis, xiv + 136 pp. 6 x 9. ISBN -0079-2, ∞G

———Grundlagen der Analysis, with a complete German-English Vocabulary, 4th ed. 173 pp. 5½ x 8½. ISBN -0141-1

LEBESGUE, Henri: Lecons sur L'Integration et la Recherche des Fonctions Primitives, 3rd ed. (French) xii + 340 pp. 5⅜ x 8. CIP. ISBN -0267-1 ∞G

LORENTZ, George G.. Bernstein Polynomials, 2nd ed., x + 134 pp. 6 x 9. ISBN -0323-6, ∞G

MINKOWSKI, Hermann: Gesammelte Abhandlungen, 2 v. in 1. (Germ.) 871 pp. 6 x 9. ISBN -0208-6, ∞G

NIELSEN, Niels: Die Gammafunktion, 2 v. in 1. (Germ.) 448 pp. 6 x 9. ISBN -0188-8, ∞G $19.50

RAINVILLE, Earl D.: Special Functions. xxi + 365 pp. 6 x 9. ISBN -0258-2, ∞G

ROGOSINSKI, Werner: Fourier Series, 2nd ed. 183 pp. 4½ x 6½. CIP. ISBN -0067-9, ∞

SCHOUTEN, Jan. A. & W. VAN DER KULK: Pfaff's Problem and its Generalisations. xvi + 542 pp. 5⅜ x 8. ISBN -0221-3, ∞G

SEIFERT, Herbert & THRELFALL, W.: Variationsrechnung im Grossen. (Germ.) iv + 116 pp. 6 x 9. ISBN -0049-0, ∞G

STRUTT, M. J. O.: Lame, Mathieu Funktionen. (Germ.) viii + 116 pp. 6 x 9. ISBN -0203-5, ∞G

TRICOMI, F. C.: Integral Equations, viii + 238 pp. 6 x 9. ∞G prob.

YOUNG, Laurence C.: Lectures on the Calculus of Variations and Optimal Control Theory. 2nd ed. xi + 311 pp. 6½ x 9¾. ISBN -0304-X, ∞G

TOPOLOGY, MEASURE AND INTEGRATION

ALEXANDROFF (Alexandrov), Paul S. & HOPF, H.: Topologie (Germ.) xii + 696 pp. 6 x 9. ISBN -0197-7, ∞G

BERBERIAN, Sterling K.: Measure and Integration. xx + 312 pp. 6 x 9. ISBN -0241-8, ∞G

BONNESEN, Tommy & FENCHEL, W.: Theorie der Konvexen Koerper. (Germ.) 171 pp. 5½ x 8½. ISBN -0054-7, ∞G

CARATHEODORY, Constantin: Algebraic Theory of Measure and Integration. 378 pp. 6 x 9. ISBN -0161-6, ∞G

———Vorlesungen ueber Reelle Funktionen, 3rd ed. (Germ.) 728 pp. 5⅜ x 8. ISBN -0038-5, ∞G $17.95

HAUSDORFF, Felix: Set Theory, 2nd ed. 352 pp. 6 x 9. ISBN -0119-5, ∞G

———Grundzuege der Megenlehre. (Germ.) 484 pp. 5⅜ x 8 ISBN -0061-X, ∞G

HAWKINS, Thomas: Lebesgue's Theory of Integration: Its Origins and Development, 3rd ed. xv + 227 pp. 6 x 9. ISBN -0282-5, ∞G

KAPLANSKY Irving: Set Theory and Metric Spaces, 2nd ed. xiii + 140 pp. 6 x 9. ISBN -0298-1, ∞G

LEBESGUE, Henri: Lecons sur l'Integration et la Recherche des Fonctions Primitives, 3rd. ed. (French) xii + 340 pp. 5⅜ x 8. CIP. ISBN -0267-1, ∞G

LEFSCHETZ, Solomon: Topology, 2nd ed. 410 pp. 5⅜ x 8. ISBN -0116-0, ∞G

———Selected Papers. 639 pp. 6 x 9. ISBN -0234-5, ∞G

MENGER, Karl: Kurventheorie, 2nd ed. 376 pp. 6½ x 9¼. ISBN -0172-1, ∞G

REIDEMEISTER, Kurt: Kombinatorische Topologie. (Germ.) 221 pp. 5⅜ x 8, ISBN -0076-8

SEIFERT, Herbert & THRELFALL, W.: Lehrbuch der Topologie. (Germ.) xvii + 353 pp. 5⅜ x 8. ISBN -0031-8, ∞G

WANG, Hao: Logic, Computers and Sets. x + 651 pp. 6 x 9. ISBN -0245-0, ∞G

YOUNG, Grace Chisholm and William Henry YOUNG: The Theorv of Sets of Points, 2nd ed. xvi + 326 pp. 6 x 9. ISBN -0259-0, ∞G

HISTORY

BOCHENSKI, Innocenty M.: History of Formal Logic, 2nd ed. xx + 567 pp. 5⅜ x 8. ISBN -0238-8, ∞G

CAJORI, Florian: History of Mathematics. 4th ed. xi + 524 pp. 6 x 9. CIP. ISBN -1303-7

DICKSON, Leonard Eugene: History of the Theory of Numbers, 3 vols. xii + 468 pp.; xxv + 803 pp.; V + 313 pp. 5⅜ x 8. ISBN -0086-5. ∞G

GOW, James: Short History of Greek Mathematics. xii + 325 pp. 5⅜ x 8. ISBN -0218-3, ∞G

HAWKINS, Thomas: Lebesgue's Theory of Integration: Its Origins and Development, 3rd ed. xv + 227 pp. 6 x 9. ISBN -0282-5, ∞G

KLEIN, Felix: Entwicklung der Mathematik im 19. Jahrhundert, 2 vols. in 1. (Germ.) 619 pp. 5¼ x 8. ISBN -0074-1, ∞G

MIKAMI, Yashio: The Development of Mathematics in China and Japan. 2nd ed. x + 383 pp. 5⅜ x 8. ISBN -0149-7, ∞G

SCOTT, Joseph Frederick: The Mathematical Work of John Wallis. 2nd (unaltered) ed. xii + 240 pp. 6 x 9. ISBN -0314-7, ∞G

STIRLING, James: The Differential Method; or, A Treatise Concerning Summation and Interpolation of Infinite Series [with an Introduction by Charles Tweedie], in prep.

TODHUNTER, Isaac: History of the Calculus of Variations in the 19th Century. xii + 532 pp. 5⅜ x 8. ISBN -0164-0, ∞G

———History of the Mathematical Theory of Probability. xvi + 624 pp. 5⅜ x 8. ISBN -0057-1, ∞G

WOODHOUSE, Robert: History of the Calculus of Variations in the 18th Century. ix + x + 154 pp. 5 x 8¼. ISBN -0177-2, ∞G

CONTINUED FRACTIONS

CHRYSTAL, George: Textbook of Algebra, 7th ed., 2 vols. xlviii + 1,210 pp. 5⅜ x 8. ISBN -0084-9, ∞G

WALL, Hubert S.: Analytic Theory of Continued Fractions. xiv + 433 pp. 5⅜ x 8. ISBN -0207-8, ∞G

PROBABILITY AND STATISTICS

BERTRAND, Joseph: Calcul des Probabilites. 3rd ed. (French) 57 + 322 pp. 5⅜ x 8. ISBN -0262-0, ∞G

CONDORCET, M. J.: Essai sur L'Application de L'Analyse aux Probabilites. (French) 191 + 304 pp. 6 x 9. ISBN -0252-3, ∞G

DE MOIVRE, Abraham: The Doctrine of Chances, 3rd ed. xii + 368 pp. 6 x 9. ISBN -0200-0, ∞G

———A Treatise on the Annuities of Lives. included in: DE MOIVRE: The Doctrine of Chances, 3rd ed.

GNEDENKO, Boris V.: The Theory of Probability and the Elements of Statistics (with answers to the exercises), 5th ed., 527 pp. 6 x 9. ISBN -2320-1, ∞

GRENANDER, Ulf & ROSENBLATT, Murray: Statistical Analysis of Stationary Time Series, 2nd (corr.) ed. 308 pp. 6 x 9. ISBN -0320-1, ∞G

GRENANDER, Ulf & SZEGO, Gabor: Toeplitz Forms and Their Applications, 2nd (corr.) ed. x + 245 pp. 5⅜ x 8. ISBN -0321-X, ∞G

KOLMOGOROV, Andrei N.: Foundations of the Theory of Probability. viii + 84 pp. 6 x 9. ISBN -0023-7, ∞G

MONTMORT, R. de: Essay sur les Jeux de Hazard, (French) 458 pp. + plates. 6½ x 10. ISBN -0307-4, ∞G

TODHUNTER, Isaac: History of the Mathematical Theory of Probability. xvi + 624 pp. 5⅜ x 8. ISBN -0057-1, ∞G

VENN, John: The Logic of Chance, 4th ed. xxix + 508 pp. 5⅜ x 8. ISBN -0173-X, ∞G

WALD, Abraham: Statistical Decision Functions. ix + 179 pp. 5⅜ x 8. ISBN -0243-4, ∞G

BIOGRAPHY

INFELD, Leopold: Quest: An Autobiography, 2nd ed., viii + 361 pp. 6 x 9. ISBN -0309-0, ∞G

ORE, Oystein: Niels Henrik Abel, Mathematician Extraordinary. 285 pp. + plates. 6 x 9. CIP. ISBN -0274-4, ∞G

SCOTT, Joseph Frederick: The Mathematical Work of John Wallis. 2nd (unaltered) ed. xii + 240 pp. 6 x 9. ISBN -0314-7, ∞G

THOMPSON, Silvanus P: The Life of Lord Kelvin, 2 vols., 2nd ed. 1,329 pp. + 16 plates. 5⅜ x 8. CIP. ISBN -0292-2, ∞G

ALGEBRA

(See also: Group Theory, Linear Algebra)

CHRYSTAL, George: Textbook of Algebra, 7th ed., 2 vols. xlviii + 1,210 pp. 5⅜ x 8. ISBN -0084-9, ∞G

DICKSON, Leonard Eugene: Introduction to the Theory of Algebraic Equations (See: Sierpinski, Congruence of Sets. (General)).

ELLIOTT, Edwin B.: Algebra of Quantics. xvi + 416 pp. 5⅜ x 8. ISBN -0184-5, ∞G

GAAL, Lisl: Classical Galois Theory with Examples, 3rd ed. 255 pp. 6 x 9¾. CIP. ISBN -0268-X, ∞G

GORDAN, P.A.: Invariantentheorie, 2 v. in 1. (Germ.) 583 pp. 5⅜ x 8. CIP., in prep.

GRACE, John H. & YOUNG, Alfred: The Algebra of Invariants, vii + 384 pp. 5⅜ x 8. ISBN -0180-2, ∞G

HAMILTON, William R.: Elements of Quaternions, 3rd ed., 2 vols. 1,185 pp. 6 x 9. ISBN 0-0219-1, 2 vol. set ∞G

HENSEL, Kurt & LANDSBERG, G.: Algebraische Funktionen. (Germ.) xvi + 707 pp. 6 x 9. ISBN -0179-9, ∞G

KOWALEWSKI, Gerhard: Determinantentheorie, 3rd ed. (Germ.) vii + 320 pp. 5⅜ x 8. ISBN -0039-3, G

LEVI, Howard: Elements of Algebra, 4th ed. 189 pp. 5⅜ x 8. ISBN -0103-9, ∞

McCARTHY, Paul J.: Algebraic Extensions of Fields. ix + 166 pp. 6 x 9. CIP. ISBN -1284-7, ∞G

McCOY, Neal H.: The Theory of Rings x + 161 pp. 6 x 9. CIP. ISBN -0266-3, ∞G

MacMAHON, Percy A.: Combinatory Analysis, 2 vols. in 1. 660 pp. 5⅜ x 8. ISBN -0137-3, ∞G

———Introduction to Combinatory Analysis, Included in: KLEIN, Felix, Famous Problems. (General).

NETTO, Eugen: The Theory of Substitutions. 304 pp. 5⅜ x 8. ISBN -0165-9 ∞G

PICARD, Emile & SIMART, G.: Fonctions Algebriques. 2 v. in 1. (French) 784 pp. 5⅜ x 8. ISBN -0248-5, ∞G

SALMON, George: Modern Higher Algebra, 5th ed. 376 pp. 5⅜ x 8. ISBN -0150-0

WEBER. Heinrich: Lehrbuch der Algebra, 3 vols. (Germ.) 3rd ed. 2,345 pp. 5⅜ x 8. ISBN -0144-6, ∞G

YOUNG, Alfred. See GRACE, John H.

DIFFERENTIAL EQUATIONS

BATEMAN, Harry: Differential Equations. xi + 306 pp. 5⅜ x 8 ISBN -0190-X, ∞G

BLISS, Gilbert A.: Fundamental Existence Theorems. See: EVANS, G. C.

CARATHEODORY, Constantin: Calculus of Variations and Partial Differential Equations, 2nd (revised) ed. xvi + 401 pp. 6 x 9. ISBN -0318-X, ∞G

EVANS, Griffith C.: Logarithmic Potential, 2nd ed.; & BLISS, G. A. Fundaamental Existence Theorems, & KASNER, E. Differential-Geometric Aspects of Dynamics. 3 vols in 1. 399 pp. 5⅜ x 8. ISBN -0305-8, ∞G

GARABEDIAN, Paul R.: Partial Differential Equations, xii + 672 pp. 6 x 9. ISBN -0325-2, ∞G

KAMKE, Erich: Differentialgleichungen: Loesungsmethoden und Loesungen. Vol. II. Partial D. E. (Germ.) xv + 243 pp. 6 x 9. CIP. ISBN -0277-9, ∞G

LAPPO-DANILEVSKII, J. A.: Systemes des Equations Differentielles, 3 vols. in 1 (French) 689 pp. 5⅜ x 8. ISBN -0094-6, ∞'G

GROUP THEORY

(Including: Continuous Groups)

CAMPBELL, John R.: Introductory Treatise on Lie's Theory. xx + 416 pp. 5⅜ x 8. ISBN -0183-7, T.O.P.

GORENSTEIN, Daniel: Finite Groups, 2nd (corr.) ed. xvii + 517 pp. 6 x 9. CIP., ISBN -0301-5, ∞G

HALL, Marshall: The Theory of Groups. xiii + 434 pp. 5⅜ x 8. CIP. ISBN -0288-4, ∞G

KOWALEWSKI, Gerhard: Kontinuierliche Gruppen. (Germ.) viii + 396 pp. 5⅜ x 8. ISBN -0070-9, G

KUROSH, Alexander G.: The Theory of Groups, Vol. I, 271 pp. 6 x 9. ISBN -0107-1, ∞G . Vol. II, 308 pp. 6 x 9. ISBN -109-8, ∞G

LIE, Sophus: Transformationsgruppen, 2nd ed., 3 vols. (Germ.) 2,090 pp. 6 x 9. ISBN -0232-9, ∞G

———Differentialgleichungen. (Germ.) xiv + 568 pp. 6 x 9. ISBN -0206-X, ∞G

———Continuierliche Gruppen (Germ.) xii + 810 pp. 6 x 9. ISBN -0199-3, ∞G

———Geometrie der Beruehrungstransformationen. (Germ.) 705 pp. 6 x 9. CIP. ISBN -0291-4, ∞G

ZASSENHAUS, Hans J.: The Theory of Groups, 2nd ed. viii + 265 pp. 6 x 9. ISBN -0053-9, ∞G

FUNCTION THEORY

(See also: Analysis)

APPELL, Paul and Edouard GOURSAT: Fonctions Algebriques et leurs Integrales, 3rd ed. Vol. I (French) xxv + 526 pp. 5⅜ x 8. CIP. ISBN -0285-X, ∞G Vol. II (French) xii + 521 pp. 5⅜ x 8. ISBN -0299-X, ∞G

CARATHEODORY, Constantin: Theory of Functions, 2nd ed. Vol. I, 310 pp. ISBN -0097-0, ∞G 'ol. II, 220 pp. 6 x 9. ISBN -0106-3, ∞G

FORD, Lester R.: Automorphic Functions. 343 pp. 5⅜ x 8. ISBN -0085-7, ∞G

HENSEL, Kurt & LANDSBERG, G.: Algebraische Funktionen. (Germ.) xvi + 707 pp. 6 x 9. ISBN -0179-9, ∞G

HILLE, Einar: Analytic Function Theory, 2nd ed. Vol. I, 319 pp. 6 x 9. CIP. ISBN -0269-8, ∞G

HOBSON, Ernest W.: Spherical and Ellipsoidal Harmonics. xi + 500 pp. 5⅜ x 8. ISBN -0104-7, ∞G

KRAZER, Adolf: Lehrbuch der Thetafunktionen. (Germ.) xxiv + 509 pp. 5⅜ x 8. ISBN -0244-2, ∞G

MARKUSHEVICH, Aleksei I.: Theory of Functions of a Complex Variable, 2nd ed., 3 vols. in 1. xxiv + 1,128 pp. CIP. ISBN -0296-5, ∞G

MONTEL, Paul: Lecons sur les Familles Normales de Fonctions Analytiques. (French) xiii + 301 pp. 5⅜ x 8. CIP. ISBN -0271-X, ∞G

NEVANLINNA, Rolf: Le Theoreme de Picard-Borel. (French) x + 171 pp. 5⅜ x 8. CIP. ISBN -0272-8, ∞G

———and PAATERO, Viekko: Introduction to Complex Analysis, 2nd ed. ix + 350 pp. 6 x 9. ISBN -0310-4, ∞G

NIELSEN, Niels: Die Gammafunktion, 2 v. in 1. (Germ.) 448 pp. 6 x 9. ISBN -0188-8, ∞G

OSGOOD, William F.: Funktionentheorie. 5th ed. Vol. I (Germ.) 818 pp. 5 x 8. ISBN -0193-4, ∞G 2nd ed. Vol. II (Germ.) 685 pp. 5 x 8. ISBN -0182-9, ∞G

PICARD, Emile & SIMART, G.: Fonctions Algebriques. 2 v. in 1. (French) 784 pp. 5⅜ x 8. ISBN -0248-5, ∞G

SCHWARZ, Hermann Amandus: Gesammelte Mathematische Abhandlungen, 2 vols. in 1. (Germ.) 726 pp. 6 x 9. ISBN -0260-4, ∞G

SPRINGER, George: Introduction to Riemann Surfaces, 2nd ed. viii + 307 pp. 6 x 9. ISBN -313-9, ∞G

TANNERY, Jules and Jules MOLK: Fonctions Elliptiques, 4 vols. in 2 (French) 2nd ed., 1,146 pp. 5⅜ x 8. ISBN -0257-4, ∞G

TSUJI, MASATSUGU: Potential Theory in Modern Function Theory, 2nd ed. x + 590 pp. 5⅜ x 8. CIP. ISBN -0281-7, ∞G

WEBER, Heinrich: Lehrbuch der Algebra, 3 vols. (Germ.) 3rd ed. 2,345 pp. 5⅜ x 8. ISBN -0144-6, ∞G (Vol. III: Elliptische Funktionen.)

NUMBER THEORY

BACHMANN, Paul: Niedere Zahlentheorie, 2 v. in 1. (Germ.) 902 pp. 6 x 9. ISBN -0217-5, ∞G

CHEBYSHEV, Pafnuty L.: Theorie der Congruenzen, 2nd ed. (Germ.) xvii + 366 pp. 5 x 8 ISBN -0254-X, ∞G

DICKSON, Leonard Eugene: Studies in the Theory of Numbers, viii + 230 pp. 5⅜ x 8. ISBN -0151-9, ∞G

———History of the Theory of Numbers, 3 vols. xii + 468 pp.; xxv + 803 pp.; V + 313 pp. 5⅜ x 8. ISBN -0086-5, ∞G

DIRICHLET, P.G. Lejeune & DEDEKIND, Richard: Zahlentheorie, 4th ed. (Germ.) xv + 657 pp. 5⅜ x 8. ISBN -0213-2, ∞G

GAUSS, Carl Friedrich: Untersuchungen ueber Hoehere Arithmetik (including: Disquisitiones Arithmeticae), 2nd ed. (Germ.) xv + 695 pp. 6 x 9. ISBN -0191-8, ∞G

HARDY, Godfrey H.: Ramanujan. 3rd ed., viii + 236 pp. 6 x 9. ISBN -0136-5, ∞G

HILBERT, David: Gesammelte Abhandlungen, Vol. I (Number Theory), 4th ed. xi + 539 pp. 6 x 9. ISBN -1195-6, ∞G

LANDAU, Edmund: Elementary Number Theory. 256 pp. 6 x 9. ISBN -0125-X, in prep.

———Elementare Zahlentheorie. (Germ.) vii + 180 + iv pp. 5½ x 8½. ISBN -0026-1, G

———Foundations of Analysis, xiv + 136 pp. 6 x 9. ISBN -0079-2, ∞G

———Grundlagen der Analysis, with a complete German-English Vocabulary, 4th ed. 173 pp 5½ x 8½. ISBN -0141-1

———Vorlesungen ueber Zahlentheorie, 2½ vols, in 1. (Germ.) 876 pp. 5¼ x 8¼. ISBN -0032-6, ∞G

———Handbuch der Verteilung der Primzahlen, 2 vols. in 1, 2nd ed. (Germ.) 1,028 pp. 5⅜ x 8. ISBN -0096-2, ∞G

———Algebraische Zahlen. (Germ.) vii + 147 pp. 5⅜ x 8. ISBN -0062-8, G

MATHEWS, George B.: Theory of Numbers, 335 pp. 5⅜ x 8. ISBN -0156-X, ∞G

MINKOWSKI, Hermann: Diophantische Approximationen. (Germ.) viii + 235 pp. 5⅜ x 8. ISBN -0118-7, ∞G

MORDELL, L. J.: Fermat's Last Theorem. Included in: KLEIN, F. Famous Problems.

NAGELL, Trygve: Introduction to Number Theory. 309 pp. 5⅜ x 8. ISBN -0163-2, ∞G

SHANKS, Daniel: Solved and Unsolved Problems in Number Theory, 3rd ed., xiv + 304 pp. 6 x 9. ISBN -1297-9, ∞G

APPLIED MATHEMATICS AND PHYSICS

(See also: Biography, Collected Papers)

CORSON, Edward M.: Introduction to Tensors, Spinors, and Relativistic Wave Equations. 2nd (unaltered) ed. 222 pp. 6½ x 9¾. ISBN -0315-5, ∞G

CRABTREE, Harold: Spinning Tops and Gyroscopic Motion, 2nd ed. 203 pp. 6 x 9. ISBN -0204-3, ∞G

EDDINGTON, Arthur S.: The Mathematical Theory of Relativity, 3rd (unaltered) ed. ix + 270 pp. 6 x 9. CIP. ISBN -0278-8, ∞G

EVANS, Griffith C.: Logarithmic Potential, 2nd ed.; & BLISS, G. A. Fundamental Existence Theorems, & KASNER, E. Differential-Geometric Aspects of Dynamics. 3 vols in 1. 399 pp. 5⅜ x 8. ISBN -0305-8, ∞G

GANTMACHER, Felix R.: Lectures on Analytical Mechanics, 2nd ed. in prep.

GREENHILL, George: Gyroscopic Theory. vi + 277 pp. + fold-out plates. 6½ x 10¾. ISBN -0205-1, ∞G

GRENANDER, Ulf & SZEGO, Gabor: Toeplitz Forms and Their Applications, 2nd (corr.) ed. x + 245 pp. 5⅜ x 8. ISBN -0321-X, ∞G

HEAVISIDE, Oliver: Electrical Papers, 2 vols. 1,183 pp. 5⅜ x 8. ISBN -0235-3, ∞G

———Electromagnetic Theory, including an account of Heaviside's Unpublished Notes, 3rd ed. 3 vols. 1,791 pp. 5¼ x 8. ISBN -0237-X, ∞G

JACOBI, Carl G. J.: Gesammelte Werke, 2nd ed. (Germ., French) Vols. 1-7, 4,032 pp. 6½ x 9¼. ISBN -0226-4; Vol. 8, (Germ.) 308 pp. 6½ x 9¼. ISBN -0227-2, ∞G

———Vorlesungen ueber Dynamik (vol. 8 of Gesammelte Werke), (Germ.) 308 pp. 6½ x 9¼. ISBN -0227-2, ∞G

KASNER, Edward: Differential-Geometric Aspects of Dynamics. See EVANS, G.C.

KLEIN, Felix: The Mathematical Theory of the Top. Included in SIERPINSKI, Waclaw: Congruence of Sets. (General).

LAPLACE, Pierre S.: Celestial Mechanics, Vols. 1-4. xxiv + 136 + 746 pp.; xvii + 990 pp.; xxx + 910 + 117 pp. of tables; xxvi + 1,018 pp. ISBN -0194-2, ∞G Vol. 5 (in French), ix + 508 pp. 6 x 9. ISBN -0214-0, ∞G

SCHROEDINGER, Erwin: Collected Papers on Wave Mechanics, including Four Lectures on Wave Mechanics, 3rd (augmented) ed. xii + 212 pp. 6 x 9. ISBN -1302-9, ∞G

LINEAR ALGEBRA AND MATRIX THEORY

FORDER, Henry G.: The Calculus of Extension, xvi + 490 pp. 5⅜ x 8. ISBN -0135-7, ∞G

GANTMACHER, Felix R.: The Theory of Matrices. Vol. I, x + 374 pp. 6 x 9. ISBN -0131-4, ∞G Vol. II, x + 277 pp. 6 x 9. ISBN -0133-0, ∞G

GRASSMANN, Hermann G.: Die Ausdehnungslehre von 1844/1878, 4th ed. (Germ.) xii + 435 pp. 6 x 9. ISBN -0222-1, ∞G

KAPLANSKY, Irving: Linear Algebra and Geometry, 2nd ed. 151 pp. 6 x 9. CIP. ISBN -0279-5, ∞G

NOMIZU, Katsumi: Fundamentals of Linear Algebra. 2nd ed. (with answers). xii + 361 pp. 6 x 9. CIP. ISBN -0276-0, ∞G

PEDOE, Daniel: A Geometric Introduction to Linear Algebra, 2nd ed. xi + 224 pp. 6 x 9. CIP. ISBN -0286-8, ∞G

LOGIC

BOCHENSKI, Innocenty M.: History of Formal Logic, 2nd ed. xx + 567 pp. 5⅜ x 8. ISBN -0238-8, ∞G

HALMOS, Paul R.: Algebraic Logic, 271 pp. 6 x 9. ISBN -0154-3, ∞G

HILBERT, David & ACKERMANN, Wilhelm: Principles of Mathematical Logic. xii + 172 pp. 6 x 9. ISBN -0069-5, ∞G

ROSSER, John Barkley: Logic for Mathematicians, 2nd ed., xv + 574 pp. 6 x 9. ISBN -0294-9, ∞G

SCHROEDER, Ernst: Algebra der Logik, 5 vols. in 3. (Germ.) 2nd ed. 2,192 pp. 6 x 9. ISBN -0171-3, ∞G

VENN, John: The Principles of Inductive Logic, 2nd ed. xx + 604 pp. 5⅜ x 8. CIP. ISBN -0265-5, ∞G

———Symbolic Logic, 2nd ed. xxxviii + 540 pp. 5⅜ x 8. ISBN -0251-5, ∞G

WANG, Hao: Logic, Computers and Sets. x + 651 pp. 6 x 9. ISBN -0245-0, ∞G

GEOMETRY

BAKER, H. F.: Introduction to Plane Geometry, with many Examples, 2nd. ed. 382 pp. 5⅜ x 8. ISBN -0247-7, ∞G

BIRKHOFF, George D. & BEATLEY R.: Basic Geometry, 3rd ed. 294 pp. 5⅜ x 8. ISBN -0120-9, ∞ Teacher's Manual, Answer Book,

BLUMENTHAL, Leonard M.: Theory and Applications of Distance Geometry, 2nd ed. 359 pp. 5⅜ x 8. ISBN -0242-6, ∞G

BONNESEN, Tommy & FENCHEL, W.: Theorie der Konvexen Koerper. (Germ.) 171 pp. 5½ x 8½. ISBN -0054-7, ∞G

COOLIDGE, Julian L.: A Treatise on the Circle and the Sphere, 2nd (corr.) ed. 602 pp. 5⅜ x 8. ISBN -0236-1, ∞G

COURT, Nathan: Modern Pure Solid Geometry, xiv + 353 pp. 5½ x 8¼. ISBN -0147-0, ∞G

DARBOUX, Gaston: Théorie Générale des Surfaces, 3rd ed. 4 vols. (French) 2,283 pp. 5⅜ x 8. ISBN -0216-7, Four vol. set, ∞G

GOMES TEIXEIRA, Francisco: Traité des Courbes Spéciales Planes et Gauches, 2nd ed. 3 vols. (French) 1,337 pp. 6½ x 10. ISBN -0255-8, ∞G

HILBERT, David: & COHN-VOSSEN. S.: Geometry and the Imagination. 358 pp. 6 x 9. ISBN -0087-3, ∞G

JESSOP, Charles H.: Treatise on the Line Complex. xv + 364 pp. 5⅜ x 8. ISBN -0223-X, ∞G

GEOMETRY

KLEIN, Felix: Hoehere Geometrie, 3rd ed. (Germ.) vi + 405 pp. 5⅜ x 8. ISBN -0065-2, G

———Nicht-Euklidische Geometrie. (Germ.) xii + 326 pp. 5⅜ x 8. ISBN -0129-2, ∞G

———et al. Famous Problems and Other Monographs, 4 vols. in 1. *Contains*: KLEIN, Famous Problems of Elementary Geometry; SHEPPARD, From Determinant to Tensor; MACMAHON, Intro. to Combinatory Analysis, MORDELL, L. J., Three Lectures on Fermat's Last Theorem. 350 pp. 5⅜ x 8. ISBN -108-X, ∞

SACCHERI, Girolamo: Euclides Vindicatus, 2nd ed. (translated by G. B. Halsted and with added notes by P. Staeckel and F. Engel), xxx + 250 pp. 5⅜ x 8. ISBN -0289-2, ∞G

SALMON, George: Analytic Geometry of Three Dimensions. 7th ed. Vol. I. xxiv + 470 pp. 5⅜ x 8. ISBN -0122-5, O.P.; 5th ed. Vol. II. xvi + 334 pp. 5⅜ x 8. ISBN -0196-9, ∞G

———Conic Sections, 6th ed. 415 pp. 5⅜ x 8. ISBN -0099-7, ∞G cl., ISBN -0098-9, pa.

SCOTT, Charlotte A.: Projective Methods in Plane Analytic Geometry. 3rd ed. xii + 290 pp. 5⅜ x 8. ISBN -0146-2, ∞G

SNYDER, Virgil, et al: Selected Topics in Algebraic Geometry, 2 vols. in 1. 490 pp. 6 x 9. ISBN -0189-6, ∞G

STERNBERG, Shlomo: Lectures on Differential Geometry. 2nd (corrected and supplemented) ed. xiii + 421 pp. 6 x 9. ISBN -0316-3, ∞G

COLLECTED PAPERS

BOLTZMANN, Ludwig: Wissenschaftliche Abhandlungen, 3 vols. (Germ.) 1,976 pp. 5⅜ x 8. ISBN -0215-9, 3 vol set, ∞G

BOLZA, Oskar: Festschrift Schwarz. (Germ.) x + 451 pp. 6 x 9. ISBN -0275-2, ∞G

CHEBYSHEV. Pafnuty L.: Collected Works (Oeuvres), 2 vols. (French) 1,480 pp. 5½ x 8¼. ISBN -0157-8, ∞G

CLIFFORD, William K.: Mathematical Papers. 70 + 658 pp. 5⅜ x 8. ISBN -0210-8, ∞G

DICKSON, Leonard Eugene: Collected Mathematical Papers, 6 vols. 4.100 pp. 6½ x 10. CIP. ISBN -0273-6 (vols. 1-5) and -0306-6 (vol. 6), ∞G

DIRICHLET, P.G. Lejeune: Werke, 2 vols. in 1. (Germ.) 1,086 pp. 6½ x 9¼. ISBN -0225-6, ∞G

EISENSTEIN, Gotthold: Mathematische Werke, 2nd ed. 2 vols. (Germ.) xiii + 929 pp. 6½ x 10. CIP. ISBN -1280-6, in prep.

GREEN, George: Mathematical Papers. xii + 336 pp. 5⅜ x 8. ISBN -0229-9, ∞G

HEAVISIDE, Oliver: See "Applied Mathematics."

HESSE, Otto: Gesammelte Werke. (Germ.) 739 pp. 6½ x 9. ISBN -0261-2, ∞G

HILBERT, David: Gesammelte Abhandlungen, Vol. I (Number Theory), 4th ed. (Germ.) xi + 538 pp. 6 x 9. ISBN -1195-6, ∞G

JACOBI, Carl G. J.: Gesammelte Werke, 2nd ed. (Germ., French) Vols. 1-7, 4,032 pp. 6½ x 9¼. ISBN -0226-4; Vol. 8, (Germ.) 308 pp. 6½ x 9¼. ISBN -0227-2, ∞G

KRONECKER, Leopold: Werke, 6 vols. in 5 (Germ.) 2,530 pp. 6½ x 8½. ISBN-0224-8, ∞G

LAGUERRE, Edmond: Oeuvres (Collected Mathematical Works), 2 vols. (French) 1,202 pp. 5⅜ x 8. ISBN -0263-9, ∞G

LEFSCHETZ, Solomon: Selected Papers. 639 pp. 6 x 9. ISBN -0234-5, $32.50

LEJEUNE-DIRICHLET. See DIRICHLET, P. J. Lejeune

MINKOWSKI, Hermann: Gesammelte Abhandlungen, 2 v. in 1. (Germ.) 871 pp. 6 x 9. ISBN -0208-6, ∞G